IAL

AL

ENGINEERING

THERMAL ENVIRONMENTAL ENGINEERING

Second Edition

JAMES L. THRELKELD

Professor of Mechanical Engineering
University of Minnesota

PRENTICE-HALL, INC.

Englewood Cliffs, New Jersey

Current printing (last digit):
10 9 8 7 6 5 4

13–914721–7

Library of Congress Catalog Card Number: 77–105450

Printed in the United States of America

PRENTICE-HALL INTERNATIONAL, INC., London
PRENTICE-HALL OF AUSTRALIA, PTY. LTD., Sydney
PRENTICE-HALL OF CANADA, LTD., Toronto
PRENTICE-HALL OF INDIA PRIVATE LTD., New Delhi
PRENTICE-HALL OF JAPAN, INC., Tokyo

to

MY MOTHER

AND FATHER

PREFACE

This book is intended as a text for undergraduate or graduate students in mechanical engineering and as a basic reference for practicing engineers. Principally, the book covers refrigeration, psychrometrics, solar radiation, and heat transmission in buildings. A few selected application topics are included in the last two chapters. Theory and analysis are emphasized throughout.

It is intended that this book may serve as a text for a course of about two semesters; for shorter courses, certain chapters may be deleted without loss of continuity. The reader should have a background in elementary thermodynamics and heat transfer.

Certain features of this book should be mentioned. Where appropriate, an attempt has been made to exploit advantages of the enthalpy-concentration diagram in solution of problems. The h-x diagram is a key factor in Chapter 5, *Absorption Refrigeration*, and in the treatment of air separation in Chapter 7. The h-W diagram for moist air (psychrometric chart) is extensively used in Chapters 9–12, and in Chapter 16.

Chapters 10–12 involve heat transfer and/or water vapor transfer processes with moist air. Beginning with Chapter 10 on psychrometry, a common approach to such problems is presented.

Considerable emphasis is also given to solar radiation problems in this book. Chapter 13 is a basic treatment of solar radiation; effects of solar radiation upon structures are discussed in Chapter 14. Solar energy collectors are covered in Chapter 17.

Substantial changes appear in this second edition. Improvements have been made throughout, but particular strength has been added to the treatment of thermoelectric cooling in Chapter 6, air separation in Chapter 7, properties of moist air in Chapter 8, psychrometry and humidity measurement in Chapter 10, and the estimating of thermal loads in Chapter 16. Important additions have been made to the problems of Chapters 7, 8, 10, and 13 through 16. A new *h-x* diagram has been included for nitrogen-oxygen mixtures. The three ASHRAE sea-level pressure psychrometric charts have been used in place of the multiple-pressure charts of the first edition.

Many persons have aided me in the preparation of this revision. I am particularly indebted to my wife, who assisted me in many ways. I wish to thank the American Society of Heating, Refrigerating and Air Conditioning Engineers, and other publishers and authors who, as indicated in the book, permitted me to include many charts and other data.

J. L. THRELKELD

CONTENTS

Part II *REFRIGERATION*

3. *MECHANICAL VAPOR COMPRESSION REFRIGERATION CYCLES*   *41*

4. *REFRIGERANTS*   *77*

5. *ABSORPTION REFRIGERATION*   *85*

6. MISCELLANEOUS REFRIGERATION PROCESSES 116

7. ULTRA-LOW-TEMPERATURE REFRIGERATION: CRYOGENICS 136

Part III *PSYCHROMETRICS*

11. DIRECT CONTACT TRANSFER PROCESSES BETWEEN MOIST AIR AND WATER   **215**

12. HEATING AND COOLING OF MOIST AIR BY EXTENDED SURFACE COILS   **235**

Part IV **SOLAR RADIATION**

Part V *APPLICATIONS*

CONTENTS

Part I

INTRODUCTION

1

ELEMENTARY
THERMODYNAMIC CONSIDERATIONS

1.1. INTRODUCTION

This book is concerned with the control of the thermal environment within enclosed spaces and therefore involves transfers of heat and work. It is assumed throughout that the reader has completed a course in mechanical engineering thermodynamics of about one year's length. The purpose of this chapter is to review a few thermodynamic principles pertinent to air conditioning and refrigeration.

1.2. THERMODYNAMIC PROPERTIES

A thermodynamic property is any observable, measurable, or calculable characteristic of a substance which depends only upon the state of the substance. In an air conditioning system, moist air is always a working substance, but associated systems may involve refrigerants, steam, water, and other fluids. In order to study an air conditioning or refrigeration system, it is necessary that the state of the working substance always be defined in terms of its thermodynamic properties.

1.3. SYSTEM OF UNITS

The system of units used throughout this book is the *pound-second-foot* system. The unit of mass is the *pound mass* which is determined by reference to a standard quantity of material. The unit of force is the *pound force* which is the force required to accelerate the pound mass at the standard acceleration of gravity of 32.17 ft per (sec)². Mass, force, and acceleration are related by *Newton's second law of motion*

$$F = kma \tag{1.1}$$

where F = force in pounds (lb_f), m = mass in pounds (lb_m), a = acceleration of mass m (ft per sec²), k = proportionality constant (lb_f-sec² per lb_m-ft).

The weight of a body in pounds, which is the force of gravity upon the body at its location, is numerically equal to its mass in pounds at the earth's surface if k in Eq. (1.1) is set equal to 1/32.17. Thus,

$$F = \frac{ma}{32.17} \tag{1.2}$$

and the weight W of a body at any location where the acceleration of gravity is g, is

$$W = \frac{mg}{32.17} \tag{1.3}$$

Even at high altitudes g differs from 32.17 by only a fraction of one per cent. Thus, for practical purposes, weight in pounds and mass in pounds are numerically equal.

1.4. SPECIFIC VOLUME, DENSITY, SPECIFIC WEIGHT,
AND SPECIFIC GRAVITY

The *specific volume* of a substance is its volume per unit mass and is expressed in cu ft per lb. The *density* of a substance is its mass per unit volume expressed in lb per cu ft and is the reciprocal of specific volume. The *specific weight* of a substance is its weight per unit volume measured in lb per cu ft.

The terms density and specific weight are often confused. Density involves mass whereas specific weight involves force. In the pound-second-foot system of units as discussed in Sec. 1.3, density and specific weight are numerically equal for practical purposes, provided g in Eq. (1.3) is equal or almost equal to 32.17. The distinction between density and specific weight should always be recognized.

The specific volume and density of a vapor or gas are affected by both pressure and temperature, and values must be found from tables of properties, or often, in the case of gases, calculated. The density of liquids is, to some extent, affected by pressure, but is usually taken as a function of temperature only.

The *specific gravity* of a liquid is defined as the ratio of the weight of the given liquid to the weight of an equal volume of water at some standard temperature, usually 39.2F. For practical purposes, the specific gravity of a liquid is its specific weight in lb per cu ft divided by 62.4.

1.5. PRESSURE

The *pressure* of a substance is the force it exerts per unit area of its boundaries. In engineering, pressures may be designated as absolute, gauge, or vacuum. *Absolute pressure* is a thermodynamic property. If the absolute pressure is higher than atmospheric or barometric pressure, the *gauge pressure* is the difference between the absolute pressure and atmospheric pressure. If the absolute pressure is less than atmospheric pressure, the *vacuum pressure* is the difference between atmospheric pressure and the absolute pressure.

A variety of units are used for expressing pressure, those commonly used being lb per sq ft, lb per sq in., inches of mercury, and feet or inches of water. Since the specific weight of liquids varies with temperature, pressure in terms of liquid height implies a standard liquid specific weight. Taking the specific weight of water as 62.4 lb per cu ft and the specific gravity of mercury as 13.6, we have the following conversion relations:

$$1 \text{ psi} = 144 \text{ psf} = 2.04 \text{ in. Hg} = 27.7 \text{ in. water}$$
$$1 \text{ in. Hg} = 0.491 \text{ psi} = 13.6 \text{ in. water}$$
$$1 \text{ in. water} = 0.0361 \text{ psi} = 0.0735 \text{ in. Hg}$$
$$1 \text{ standard atmosphere} = 14.696 \text{ psia} = 29.92 \text{ in. Hg}$$

If fluids other than mercury or water are used for measuring pressure heads, or if mercury or water is used at temperatures other than standard temperatures, a useful relation is

$$z_1 w_1 = z_2 w_2 \tag{1.4}$$

where z_1 and z_2 are heights of the two fluids for respective specific weights w_1 and w_2. Consistent units for z and w must be used.

1.6. TEMPERATURE

The temperature of a substance may be expressed in either relative or absolute units. On the Fahrenheit scale, the temperature of melting ice is arbitrarily called 32 degrees, and the temperature of boiling water at 14.696 psia is called 212 degrees. On the Centigrade scale, similar markings are 0 degrees for melting ice and 100 degrees for boiling water. Conversions between scales are given by

$$t_F = \frac{9}{5} t_C + 32 \tag{1.5}$$

$$t_C = \frac{5}{9} (t_F - 32) \tag{1.6}$$

By means of the second law of thermodynamics, we can prove that there is a "lowest conceivable temperature." This temperature is *absolute zero* and any temperature reckoned above it is an *absolute temperature*. The absolute Fahrenheit scale (commonly called the Rankine scale) and the absolute Centigrade scale (or Kelvin scale) are used for expressing absolute temperatures. Absolute zero occurs at -459.69 degrees on the ordinary Fahrenheit scale and at -273.16 degrees on the ordinary Centigrade scale. Absolute temperature on the two scales may be found by

$$T_F = t_F + 459.69 \tag{1.7}$$

$$T_C = t_C + 273.16 \tag{1.8}$$

In most calculations, the constants used are 460 and 273.

1.7. HEAT AND POWER UNITS

Heat is energy transferred from one body to another because of a temperature difference. The unit used to quantitatively express heat is the *British thermal unit* (Btu). In ordinary use, one Btu is defined as the quantity of heat required to raise the temperature of one pound of water through one degree Fahrenheit.

The *specific heat* of a substance is the heat required to raise the temperature of unit mass of the substance one degree. In engineering, units used are Btu per lb per degree F. The specific heat of most substances varies with temperature, but over limited temperature ranges, average values may be used.

In the case of gases, two specific heats are of importance, specific heat at constant pressure, c_p, and at constant volume, c_v. In air conditioning work, values of c_p are usually needed.

Calculation of heat quantities involving specific heats is given by

$$Q = mc(t_2 - t_1) \tag{1.9}$$

where Q = heat transfer in Btu, m = mass in lb, c = specific heat in Btu per (lb)(F), $(t_2 - t_1)$ = temperature change in F.

In air conditioning work, two types of heat are often considered. *Sensible heat* is heat added to a substance resulting in a temperature rise; *latent heat* is heat associated with a change of phase at constant temperature. The heat necessary to melt a solid without change of temperature is called the *latent heat of fusion*, while the heat required to vaporize a liquid without temperature change is called the *latent heat of vaporization*.

Work is energy transferred between a system and its surroundings when either of them exerts a force that acts on the other through a distance. *Power* is defined as the time rate at which work is performed. The basic unit for expressing work is the ft-lb. Power may be expressed in terms of ft-lb per min, horsepower, or in electrical units such as watts or kilowatts. The following conversion units are useful:

1 hp = 33,000 ft-lb per min = 2545 Btu per hr = 746 watts
1 kw = 3413 Btu per hr = 1.34 hp

A *ton of refrigeration* is the removal of heat at a rate of 200 Btu per min. This unit is customarily used for expressing the capacity of refrigeration systems.

1.8. FIRST AND SECOND LAWS OF THERMODYNAMICS

The science of engineering thermodynamics is based upon two empirical principles called the *first and second laws of thermodynamics.* Neither of these principles can be proved, but since no exceptions to them have ever been observed, they are accepted as correct.

The *first law of thermodynamics* states that if any system undergoes a process during which work or heat is added to or removed from it, none of the energy added is destroyed within the system and none of the energy removed is created within the system. According to the first law, heat and work are interconvertible.

It is a matter of experience that heat will not flow spontaneously out of one system into another at higher temperature. In order to transfer heat to a higher temperature level, a refrigerating machine is needed wherein energy from an external source is supplied. The above statements embody the principles of the *second law of thermodynamics* which, according to Clausius, states that it is impossible for a self-acting machine, unaided by any external agency, to transfer heat from one body to another at higher temperature.

1.9. INTERNAL ENERGY, ENTHALPY, AND ENTROPY

In many thermodynamic processes, the amount of energy added to a system in the form of work or heat is not equal to the heat and/or work removed from it. The first law of thermodynamics states that it is impossible to create or destroy energy, and thus it is possible to store energy in a fluid. As a result of the first law, the concept of the property of *internal energy* was obtained. The internal energy of a substance includes all types of energy stored within its molecules in both kinetic and potential form.

The *enthalpy* of a substance is a composite energy term defined by the following equation:

$$h = u + \frac{Pv}{J} \tag{1.10}$$

where h = enthalpy, Btu per lb, u = internal energy, Btu per lb, P = pressure, lb per sq ft, v = specific volume, cu ft per lb, J = 778 ft-lb per Btu. The importance of enthalpy is due to its presence in all steady flow problems.

As a consequence of the second law of thermodynamics, the concept of a property called *entropy* is established. Entropy s is defined by the following equation:

$$ds = \frac{dQ}{T} \quad \text{(reversible process)}$$

where ds = differential change in entropy, Btu per (lb)(R), dQ = differential heat reversibly added, Btu per lb, and T = absolute temperature, R. For a finite amount of heat added,

$$s_2 - s_1 = \int_1^2 \frac{dQ}{T} \tag{1.11}$$

Entropy, like enthalpy, is a mathematical property and is not evident by direct measurement. In engineering, entropy is useful in the solution of problems involving isothermal or adiabatic reversible processes. In more advanced phases of thermodynamics, entropy is used as a criterion of equilibrium.

1.10. THE PERFECT GAS

An equation of state expresses the relationship between pressure, specific volume, and temperature of a substance. It must be determined at least in part by experiment. Such equations are available for many fluids. In the case of a perfect gas, an important special equation of state is:

$$Pv = RT \tag{1.12}$$

where P = absolute pressure, lb per sq ft, v = specific volume, cu ft per lb, T = absolute temperature, R, and R = gas constant, ft-lb per (lb)(R). The gas constant R varies with different gases and is approximately

$$R = \frac{1545}{M} \tag{1.13}$$

where M is the molecular weight of the gas.

Equation (1.12) is satisfactory for real gases at relatively high temperatures and at low pressures. In air conditioning work, Eq. (1.12) is a valuable tool since moist air behaves very much like a perfect gas.

Several useful relationships involving specific heats may be derived for perfect gases. For any process, we may prove that the change of internal energy is a sole function of temperature, given by

$$u_2 - u_1 = c_v(t_2 - t_1) \tag{1.14}$$

Likewise, the change of enthalpy of a perfect gas is, for any process,

$$h_2 - h_1 = c_p(t_2 - t_1) \tag{1.15}$$

A useful relationship between c_p and c_v for a perfect gas is

$$c_p - c_v = \frac{R}{J} \tag{1.16}$$

Numerous other relations for perfect gases may be derived from fundamental thermodynamic concepts. For complete discussion and derivation, reference should be made to any standard text on engineering thermodynamics.

1.11. *MIXTURES OF PERFECT GASES*

The air conditioning engineer is continually involved with gaseous mixtures. In this section some basic concepts regarding mixtures of perfect gases will be discussed.

Let us first consider a given volume of a mixture of two perfect gases x and y. Each gas occupies the entire volume V, and each gas is at the same temperature T. Since we are dealing with perfect gases, there are no interactions between them, and each separately follows Eq. (1.12). The following relations will then apply:

$$m = m_x + m_y$$
$$V = V_x = V_y$$
$$T = T_x = T_y$$
$$P = P_x + P_y$$

Since each gas is assumed to behave as if the other gas were not present, by Eq. (1.12) we have

$$P_x V = m_x R_x T \qquad (1.17)$$
$$P_y V = m_y R_y T \qquad (1.18)$$

and for the mixture of the two gases,

$$PV = mRT \qquad (1.19)$$

Thus, from Eqs. (1.17)–(1.19), we have

$$R = \frac{m_x R_x + m_y R_y}{m} \qquad (1.20)$$

The gas constant of the perfect-gas mixture is thus the weighted mean of the gas constants of the components.

When gases are mixed adiabatically without work being done, the first law of thermodynamics requires that the enthalpy of the system remain constant. Thus, we may write

$$h = \frac{m_x h_x + m_y h_y}{m} \qquad (1.21)$$

If Eq. (1.21) is written in differential form,

$$m\,dh = m_x\,dh_x + m_y\,dh_y$$

and since, for perfect gases

$$dh = c_p\,dT$$

we have the following relation for the specific heat c_p of the mixture of gases:

$$c_p = \frac{m_x c_{p,x} + m_y c_{p,y}}{m} \qquad (1.22)$$

1.12. DRY AIR

A clean atmosphere is composed of dry air and water (principally in vapor form). On a volume basis, dry air contains 78.08 per cent nitrogen, 20.95 per cent oxygen, and traces of approximately 15 other gases. The composition of dry air is nearly uniform over the earth's surface and is scarcely changed by altitude until a height of 100 miles or more is reached.

The molecular weight of dry air is 28.966 and its gas law constant R is 53.34 ft-lb per (lb)(R). Precise measurements have shown that Eq. (1.12) is only approximately correct for dry air. At sufficiently low pressures and high temperatures, the correlation is excellent. For other conditions, the departure from perfect gas behavior may be considerable.

A reliable equation of state for real gases may be derived from fundamental considerations of statistical mechanics. This equation, called a *virial equation of state*, is given by the power series

$$\frac{Pv}{RT} = 1 + A_2(T)P + A_3(T)P^2 + \cdots = Z \tag{1.23}$$

The sum of the series Z is called the *compressibility factor*. The coefficients $A_2(T)$, $A_3(T)$, etc., are called *virial coefficients* and represent the departure from the perfect gas law caused by reactions between molecules. These coefficients are functions of temperature.

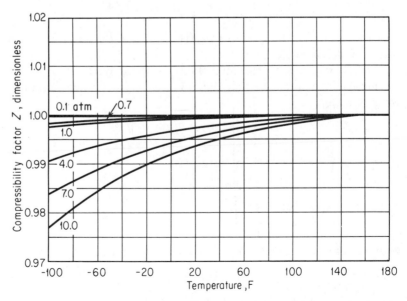

Figure 1.1 Compressibility factor for dry air.

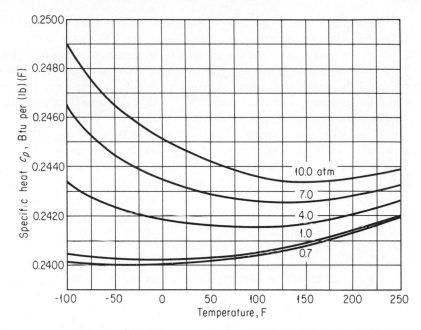

Figure 1.2 Specific heat c_p for dry air.

The National Bureau of Standards publication *Tables of Thermal Properties of Gases** shows accurately determined thermodynamic properties for dry air and several other gases. It was the source of information for Figs. 1.1 and 1.2. Figure 1.1 shows the compressibility factor for dry air at various pressures and temperatures. At ordinary atmospheric pressures, Z is essentially unity in the range of -100 F to 180 F. Thus, for this range, Eq. (1.12) is an excellent equation of state for dry air. Figure 1.2 shows the specific heat c_p for dry air as a function of pressure and temperature. At ordinary atmospheric pressures, c_p varies from about 0.240 to 0.242 Btu per (lb)(F) in the range of -100 F to 250 F.

1.13. PROPERTIES OF WATER AND STEAM

A thorough understanding of the properties of ice, water, and water vapor (steam) is essential for air conditioning engineers. Water vapor is a constituent of the atmosphere, and a knowledge of its characteristics is required for almost all air conditioning calculations. Steam and hot water are common media for heating buildings, while chilled water is a popular means of cooling atmospheric air in summer. The air conditioning engineer may be involved, to a lesser extent, with solid water or ice.

**Tables of Thermal Properties of Gases*, U.S. Department of Commerce, National Bureau of Standards Circular 564 (Washington, D.C.: Government Printing office, 1955).

Thermodynamic properties of water and steam must be found from tables or charts based on experimental data. Appendix table A.1* shows thermodynamic properties of water and steam for the saturation temperature range of -160 F to 212 F. These data were abstracted from the *ASHRAE Handbook of Fundamentals*† and were compiled by Goff and Gratch.‡ Appendix table A.2 shows properties of water and steam for saturation temperatures of 212 F to 500 F. These data were abstracted from *Thermodynamic Properties of Steam* by Keenan and Keyes.§

If water (or any other pure fluid) is heated at constant pressure, the temperature at which it boils is called its *saturation temperature*. Tables A.1 and A.2 show, for various saturation temperatures, the pressure, specific volume of saturated liquid v_f and of saturated vapor v_g, enthalpy of saturated liquid h_f and of saturated vapor h_g, and entropy of saturated liquid s_f and of saturated vapor s_g. All properties with the subscript fg are values equal to the difference between properties with subscripts g and f. In the case of enthalpy, h_{fg} is also called the *latent heat of vaporization*. Enthalpy and entropy values in Tables A.1 and A.2 are reckoned above an arbitrary zero for saturated liquid at 32 F.

Figure 1.3 shows properties of water and steam on temperature-entropy coordinates and was adapted from a diagram in Keenan and Keyes' *Thermodynamic Properties of Steam*. The saturated liquid line is a locus of points representing liquid water at the boiling temperature for various pressures. The saturated vapor line is a locus of points showing steam at the condensation temperature for various pressures. Any state point located between the two saturation lines, such as 1 or 2, is a mixture of saturated water and saturated steam with each component at the same temperature. Any state point to the right of the saturated vapor line, such as 3, is superheated since its temperature is higher than the saturation temperature corresponding to its pressure. The diagram reaches a peak at the *critical point* of 705.4 F and 3206.2 psia. The critical point is the maximum saturation condition at which evaporation and condensation may occur.

The specific volume, enthalpy, and entropy of a mixture state may be found if the *quality* x, in pounds of saturated vapor per pound of mixture, is known. For each pound of mixture there are x pounds of vapor and $(1 - x)$ pounds of liquid. The following equations are then evident:

$$v = (1 - x)v_f + xv_g$$
$$h = (1 - x)h_f + xh_g$$
$$s = (1 - x)s_f + xs_g$$

*For all tables with the letter A preceding the number, such as Table A.1, refer to Appendix tables.

†*ASHRAE Handbook of Fundamentals* (New York: American Society of Heating, Refrigerating and Air Conditioning Engineers, 1967) pp. 365–368.

‡J. A. Goff and S. Gratch, "Low Pressure Properties of Water in the Range -160 to 212 F," *Trans. ASHAE*, Vol. 52, 1946, p. 95.

§J. H. Keenan and F. G. Keyes, *Thermodynamic Properties of Steam* (New York: John Wiley & Sons, Inc., 1936).

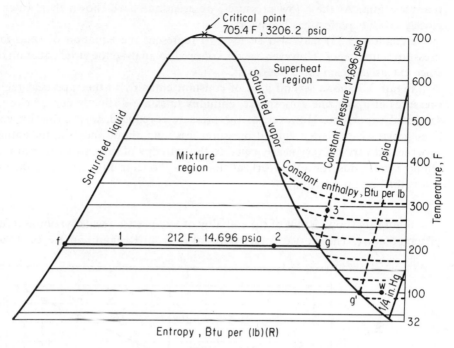

Figure 1.3 Temperature-entropy diagram for water.

The above equations may be changed into the following forms:

$$v = v_g - (1 - x)v_{fg} \tag{1.24}$$

$$h = h_g - (1 - x)h_{fg} \tag{1.25}$$

$$s = s_g - (1 - x)s_{fg} \tag{1.26}$$

or alternatively:

$$v = v_f + xv_{fg} \tag{1.27}$$

$$h = h_f + xh_{fg} \tag{1.28}$$

$$s = s_f + xs_{fg} \tag{1.29}$$

Although the two sets of equations are identical, Eqs. (1.24)–(1.26) are preferred for qualities x close to unity when slide rule calculations are made.

In Eqs. (1.24) and (1.27), since v_f is very small compared to v_g, it is usually satisfactory to use the approximation

$$v = xv_g \tag{1.30}$$

1.14. LOW-PRESSURE WATER VAPOR

The properties of low-pressure water vapor are of particular significance in air conditioning since water vapor existing in the atmosphere typically exerts a pressure less than one psia. At these low pressures, observations have shown that water vapor closely exhibits perfect-gas behavior.

Equation (1.23) may also be used to represent the equation of state for low-pressure water vapor. However, we do not need to analyze the virial equation of state since extensive steam table data exist.

Figure 1.3 shows several lines of constant enthalpy in the superheat region. At pressures of about one atmosphere, enthalpy lines are distinctly curved close to the saturated vapor line. However, as the pressure is reduced, we see that the enthalpy lines progressively flatten and, at pressures less than about one psia, the enthalpy of superheated vapor is very nearly equal to the enthalpy of saturated vapor at the same temperature. Thus, for any superheat state w as shown in Fig. 1.3 where the pressure is less than about one psia,

$$h_w = h_{g'} \tag{1.31}$$

Figure 1.4 shows a plot of the enthalpy of low-pressure water vapor as a function of temperature. The plotted values were taken from Keenan and Keyes' *Thermody-*

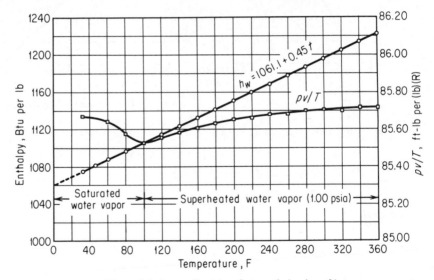

Figure 1.4 Approximate perfect-gas behavior of low-pressure water vapor.

namic Properties of Steam. We see that a straight line fits the plotted points quite well. The equation of this line is approximately

$$h_w = 1061 + 0.45t \tag{1.32}$$

If low-pressure water vapor behaved as a perfect gas, its gas-law constant by Eq. (1.13) would be

$$R_w = \frac{1545}{18.016} = 85.76 \text{ ft-lb/(lb)(R)}$$

Figure 1.4 shows a plot of Pv/T for water vapor using Keenan and Keyes values. Saturated vapor values are shown up to 100 F, and at higher temperatures superheat values at one psia are plotted. In the temperature range of 32 F to 200 F, excellent correlation is obtained at a value of Pv/T equal to 85.6. Thus, we have further proof that at pressures of less than about one psia, water vapor exhibits approximate perfect-gas behavior. Although use of Eq. (1.12) with water vapor is not accurate enough for many thermodynamic applications, nevertheless it is a valid and valuable method for many problems in air conditioning.

1.15. THERMODYNAMIC PROPERTIES OF REFRIGERANTS

Broadly, we may define a refrigerant as any substance which absorbs heat from another substance or from a space. However, we will restrict our discussion here to *vapor refrigerants*. There are many available vapor refrigerants, two of the most common being *dichlorodifluoromethane* (*Refrigerant* 12) and *ammonia*.

Table A.3 of the Appendix shows thermodynamic properties for saturated-liquid and saturated-vapor states for ammonia, and Table A.4 shows similar properties for Refrigerant 12. Figure E-1 (in the envelope) is a *Mollier* diagram (*P-h* coordinates) for ammonia. Figure E-2 is a Mollier diagram for Refrigerant 12.

All vapor refrigerants behave similarly in a qualitative sense. Nomenclature and equations given in Section 1.13 for water and steam apply to any vapor refrigerant. Figure 1.3, although restricted quantitatively to water, represents qualitatively and schematically a *T-s* diagram for any vapor refrigerant.

The most common diagram of thermodynamic properties used in refrigeration-cycle calculations is the Mollier or *P-h* diagram. However, the *T-s* diagram is highly useful, particularly as a schematic tool. From the definition of entropy, it follows that the area under a reversible-process line on *T-s* coordinates represents the heat in Btu per lb added or withdrawn during the process. Figure 1.5 shows a schematic case where a superheated vapor (State 1) is condensed to a saturated liquid (State 2).

By the general property relation

$$dh = T\,ds + \frac{v\,dP}{J}$$

it follows that the change of enthalpy Δh for a reversible process is equal to the area under the process line on *T-s* coordinates if the *pressure is constant*. Thus, assuming constant pressure for the process 1–2 in Fig. 1.5, the area under the process line is also equal to $(h_1 - h_2)$.

A frequent problem in refrigeration cycle calculations is that of evaluating the properties of a *subcooled* liquid. In Fig. 1.5, suppose that the saturated liquid at State (2) is further cooled at constant pressure to some subcooled condition (State 3). We now consider how the thermodynamic properties at State (3) differ from those of a saturated liquid at the same temperature (State 3'). If we assume that the liquid is

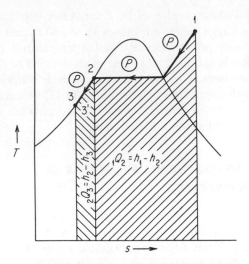

Figure 1.5 Schematic heat removal processes on *T-s* coordinates.

incompressible and that the internal energy is a function of temperature only, from the general property relation it follows that since

$$T \, ds = du + \frac{P \, dv}{J}$$

$s_3 = s_{3'}$. Therefore, States (3) and (3') are coincident on the *T-s* diagram. The enthalpy h_3 of the subcooled state may be calculated by

$$h_3 = h_{3'} + \frac{(P_3 - P_{3'})v_{3'}}{J} \tag{1.33}$$

However, for subcooled refrigerant liquids, the term $(P_3 - P_{3'})v_{3'}/J$ is usually so small compared to $h_{3'}$ that it may be neglected. Thus, for most cases, the enthalpy of a subcooled liquid may be assumed equal to that of the saturated liquid at the *same temperature*.

1.16. THE STEADY-FLOW ENERGY EQUATION

Most of the thermodynamic processes considered in this book are of a flow nature. Usually these processes are approximately steady with respect to time. The *steady-flow energy equation* is a powerful tool in the solution of many problems throughout this text and, because of its importance, we will review it here.

Figure 1.6 shows a schematic thermodynamic system to illustrate the steady-flow process. Fluid at a constant rate of *m* lb per sec crosses the boundaries of the system at Section 1 in a uniform state P_1, t_1, v_1, etc. The fluid leaves the system at Section 2 at the same rate and in a uniform state. Also crossing the boundaries of the system

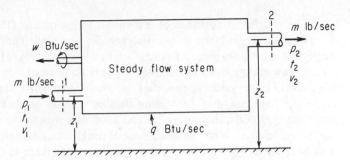

Figure 1.6 Schematic thermodynamic system for steady flow.

between Sections 1 and 2 may be heat at a constant rate of q Btu per sec (assumed inward) and shaft work at a constant rate of w Btu per sec (assumed outward). The thermodynamic state of the fluid at any location within the system is constant with respect to time.

For each pound of fluid that crosses the boundaries of the system, certain energy quantities are convected by the fluid. These energy quantities in Btu per lb are internal energy u, flow work Pv/J, potential energy due to elevation z/J, and kinetic energy due to velocity of flow $V^2/2gJ$.

By the first law of thermodynamics, the sum of all energy quantities entering the system must be equal to the sum of all energy quantities leaving the system since, for steady operation, no energy may be stored in the system. Thus:

$$mu_1 + m\frac{P_1v_1}{J} + m\frac{V_1^2}{2gJ} + m\frac{z_1}{J} + q = mu_2 + m\frac{P_2v_2}{J} + m\frac{V_2^2}{2gJ} + \frac{mz_2}{J} + w$$

Since by definition, enthalpy $h = u + Pv/J$, we have

$$mh_1 + m\frac{V_1^2}{2gJ} + \frac{mz_1}{J} + q = mh_2 + \frac{mV_2^2}{2gJ} + \frac{mz_2}{J} + w \qquad (1.34)$$

Equation (1.34) is the general steady-flow energy equation on a unit time basis. The use of this system of units is particularly advantageous if multiple streams of fluid enter and/or leave the system. Thus, if in Fig. 1.6 one fluid stream m_1 entered the system and two streams m_2 and m_3 left the system, we would have

$$m_1\left(h_1 + \frac{V_1^2}{2gJ} + \frac{z_1}{J}\right) + q = m_2\left(h_2 + \frac{V_2^2}{2gJ} + \frac{z_2}{J}\right) + m_3\left(h_3 + \frac{V_3^2}{2gJ} + \frac{z_3}{J}\right) + w \quad (1.35)$$

$$m_1 = m_2 + m_3 \qquad (1.36)$$

The use of units of energy per unit mass is often more convenient if only a single stream enters and leaves the system. If each term of Eq. (1.34) is divided by m, we have, in Btu per lb,

$$h_1 + \frac{V_1^2}{2gJ} + \frac{z_1}{J} + Q = h_2 + \frac{V_2^2}{2gJ} + \frac{z_2}{J} + W \qquad (1.37)$$

where Q and W are respectively heat added and work removed in Btu per lb of fluid.

The relative importance of individual terms in Eq. (1.34) or (1.37) depends on the steady-flow problem to be analyzed. In most problems some terms are usually negligible or non-existent. Because of the wide variety of problems for which the steady-flow energy equation applies, there is no general rule which defines the negligible terms. Experience shows that in many problems the potential energy term $(z_2 - z_1)/J$ is negligible. In heating devices such as steam boilers, warm-air furnaces and duct-type coils, the potential- and kinetic-energy terms are usually negligible and the work term is zero. In fluid flow through nozzles and venturi meters the work and heat terms are zero and the potential-energy term is negligible but the kinetic-energy term is significant. In many locations throughout this book the steady-flow energy equation will be used and its application to special problems will be discussed.

PROBLEMS

1.1. Convert a barometric pressure of 28.75 in. Hg at 32 F to (a) psia, (b) inches of water, and (c) feet of liquid whose specific gravity is 0.8.

1.2. Moist air at 70 F and 14.32 psia total pressure contains 0.012 lb of water vapor per lb of dry air. Assuming perfect-gas behavior, determine (a) the partial pressure of the water vapor in psi, (b) the density of the mixture, (c) gas-law constant for the mixture, and (d) the specific heat of the mixture per lb of dry air.

1.3. Determine the per cent error if Eq. (1.12) is used in calculating the density of dry air at a temperature of -20 F and at a pressure of 10 atmospheres.

1.4. Calculate the compressibility factor for saturated water vapor at 100 F.

1.5. Steam enters a radiator at 16 psia and 0.97 quality. The steam flows through the radiator, is condensed, and leaves as liquid water at 200 F. If the heating capacity of the radiator is 5000 Btu per hr, at what rate in lb per hr must the steam be supplied?

SYMBOLS USED IN CHAPTER 1

A Virial coefficient in Eq. (1.23).

a Acceleration, ft per sec^2.

c Specific heat, Btu per (lb) (F); c_p at constant pressure; c_v at constant volume.

F Force, lb.

g Acceleration of gravity, ft per sec^2.

h Specific enthalpy, Btu per lb; h_f for saturated liquid; h_g for saturated vapor; $h_{fg} = h_g - h_f$; h_w for low-pressure water vapor.

J Mechanical equivalent of heat, 778 ft-lb per Btu.

k F/ma, (lb force) (sec^{-2}) per (lb mass) (ft).

M Molecular weight.

m Mass, lb or lb per sec.

P Pressure, lb per sq ft.

Q Quantity of heat, Btu; dQ in Eqs. (1.10)–(1.11) refers to Btu per lb.

q Rate of heat transfer, Btu per sec.

R Pv/T for perfect gas, ft-lb per (lb) (R).

s Specific entropy, Btu per (lb)(R); s_f for saturated liquid; s_g for saturated vapor; $s_{fg} = s_g - s_f$.

T Absolute temperature, R.

t Temperature, F.

u Specific internal energy, Btu per lb.

V Volume, cu ft.

V Velocity, ft per sec.

v Specific volume, cu ft per lb; v_f for saturated liquid; v_g for saturated vapor; $v_{fg} = v_g - v_f$.

W Weight, lb.

w Specific weight, lb per cu ft.

w Rate of work transfer (power), Btu per sec.

x Quality, lb of saturated vapor per lb of mixture.

Z Pv/RT for gas, dimensionless.

z Height or elevation above arbitrary datum, ft.

2

ELEMENTARY
HEAT TRANSFER CONSIDERATIONS

2.1. INTRODUCTION

An understanding of the basic principles and practical applications of heat transfer is of vital importance to an engineer dealing with thermal environmental problems. The object of this chapter is to review some of the elementary fundamentals of heat transfer; in later chapters several special heat-transfer situations will be studied more thoroughly.

The three general methods of heat transfer are *conduction*, *convection*, and *radiation*. They will be dealt with separately in this chapter. Practical problems are seldom limited to a single method of heat transfer. Usually two, and often all three, methods are involved in an *overall* heat-transfer problem.

2.2. CONDUCTION

Thermal conduction is the transfer of heat from one part of a body to another part at lower temperature with negligible movement of the particles of the body, heat being transferred from molecule to molecule. Although conduction may occur in liquids and gases, most practical conduction problems involve solid materials.

Solutions of thermal conduction problems are based upon the empirical expression

$$q = -kA\frac{dt}{dx} \tag{2.1}$$

called *Fourier's law*. The proportionality factor k is called the *thermal conductivity*. In ordinary engineering units k is usually expressed in Btu per (hr) (sq ft) (F) per ft. A special case is that of building materials where a thickness of one inch is used.

Figure 2.1 shows a volume element with which we will derive an equation for

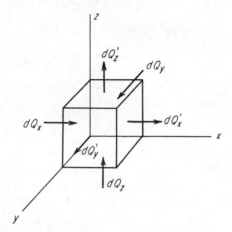

Figure 2.1 Three-dimensional heat conduction in a volume element.

heat conduction. The rate of heat flow into the element at the left-hand side (yz plane) is

$$dQ_x = -k\,dy\,dz\left(\frac{\partial t}{\partial x}\right)$$

The rate of heat flow out of the element at the right-hand side is

$$dQ'_x = dQ_x + \left(\frac{\partial\,dQ_x}{\partial x}\right)dx$$

Assuming k is a constant, we have

$$dQ_x - dQ'_x = -\left(\frac{\partial\,dQ_x}{\partial x}\right)dx = k\,dx\,dy\,dz\left(\frac{\partial^2 t}{\partial x^2}\right)$$

Likewise

$$dQ_y - dQ'_y = k\,dx\,dy\,dz\left(\frac{\partial^2 t}{\partial y^2}\right)$$

$$dQ_z - dQ'_z = k\,dx\,dy\,dz\left(\frac{\partial^2 t}{\partial z^2}\right)$$

The rate at which heat is stored in the element is

$$dQ_s = c\rho \, dx \, dy \, dz \left(\frac{\partial t}{\partial \theta}\right)$$

where c denotes specific heat, ρ density, and θ time.

With no generation of heat within the element, an energy balance gives

$$dQ_s = (dQ_x - dQ'_x) + (dQ_y - dQ'_y) + (dQ_z - dQ'_z)$$

or

$$\frac{\partial t}{\partial \theta} = \alpha\left[\left(\frac{\partial^2 t}{\partial x^2}\right) + \left(\frac{\partial^2 t}{\partial y^2}\right) + \left(\frac{\partial^2 t}{\partial z^2}\right)\right] \tag{2.2}$$

where α, the thermal diffusivity, is given by

$$\alpha = \frac{k}{\rho c} \tag{2.3}$$

Equation (2.2) applies for unsteady-state heat conduction in three dimensions. For the *steady-state*

$$\frac{\partial^2 t}{\partial x^2} + \frac{\partial^2 t}{\partial y^2} + \frac{\partial^2 t}{\partial z^2} = 0 \tag{2.4}$$

Many steady-state heat-conduction problems may be satisfactorily solved by assuming heat transfer in one dimension only. For a flat wall of thickness x whose face dimensions are large compared to its thickness, integration of Eq. (2.1) with the assumption of constant k, gives

$$q = \frac{kA}{x}(t_1 - t_2) \tag{2.5}$$

For a long cylinder, we obtain

$$q = \frac{2\pi k(t_1 - t_2)}{\ln r_o/r_i} = \frac{2\pi k(t_1 - t_2)}{\ln D_o/D_i} \tag{2.6}$$

where r_o/r_i is the ratio of the outside radius of the pipe to the inside radius, and D_o/D_i is the diameter ratio.

The thermal conductivity k for a material must be found by experiment. Its numerical value is somewhat dependent upon temperature. Table 2.1 shows approximate values of thermal conductivity for several substances.

TABLE 2.1

APPROXIMATE THERMAL CONDUCTIVITY VALUES

Material	k Btu/(hr)(sq ft)(F/ft)
Mineral wool	0.023
Wood	0.13
Glass	0.4
Lead	20
Steel	26
Aluminum	120
Copper	220

2.3. THERMAL RADIATION

Thermal radiation involves the transfer of heat from one body to another at lower temperature by electromagnetic waves passing through a separating medium. Thermal radiation waves have properties similar to other types of electromagnetic waves, differing only in wavelength. Most thermal radiation problems involve infrared rays. In the case of solar radiation, a significant amount of energy transfer occurs in the visible range of wavelengths. Although the various types of radiation waves have similar properties, effects produced by them vary a great deal.

Thermal radiation travels through a vacuum with the speed of light (186,000 miles per sec), travels in straight lines through a homogeneous medium, is converted to heat when it strikes any body which can absorb it, is reflected according to the same rules as light, and is refracted according to the same rules as light.

Figure 2.2 shows the changes which may occur when a ray of thermal radiation strikes a surface. Part of the incident radiation I may be reflected (I_r), part may be

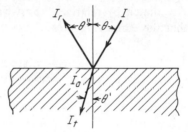

Figure 2.2 Reflection, absorption, and transmission of incident thermal radiation.

absorbed (I_a), and part may be transmitted (I_t). We define the *reflectivity* by $\rho = I_r/I$, the *absorptivity* by $\alpha = I_a/I$, and the *transmissivity* by $\tau = I_t/I$. Thus

$$\rho + \alpha + \tau = 1 \tag{2.7}$$

Equation (2.7) applies to diathermanous materials such as glass and some plastics. For *opaque materials*,

$$\rho + \alpha = 1 \tag{2.8}$$

We may also use Fig. 2.2 for further discussion of reflection and refraction of radiation rays. The case shown is for *specular* reflection. The *reflection angle* θ'' equals the *incidence angle* θ. Furthermore, the incident beam, the reflected beam, and the normal to the surface lie in the same plane. If the surface is rough, *diffuse* reflection may result where the reflected radiation is distributed in all directions.

Radiation rays are refracted (bent) when passing through a substance whose density is different from that of the substance through which the radiation is coming. Refraction causes the transmitted ray to be bent towards the perpendicular to the surface of higher density. The *refraction angle* θ' is determined by the *index of refraction*

$$n = \frac{\sin \theta}{\sin \theta'} \tag{2.9}$$

An important concept in thermal radiation is that of the *black body*. A black body or black surface absorbs all incident radiation ($\alpha = 1$). The term black should not be confused with the color of the same name, since a white-colored surface may have an absorptivity for infrared radiation as high as a black-colored surface. There is no perfect black body for thermal radiation. An approximation to one is a hollow enclosure with a small opening.

The *emissive power E* of a unit surface is the amount of heat radiated by the surface per unit time. The *emissivity ϵ* of a surface is defined by

$$\epsilon = \frac{E}{E_b} \qquad (2.10)$$

where E_b is the emissive power of a black body at the same temperature.

Kirchhoff's law states that the emissivity ϵ of a surface is equal to the absorptivity α of the surface. Kirchhoff's law is valid in ordinary thermal radiation problems where the incident radiation upon a surface has wavelengths similar to those being emitted by the surface. The law is not valid for a surface emitting long-wave radiation while being irradiated by relatively short-wave radiation. Furthermore, Kirchhoff's law shows that the emissive power of any body must be less than that of a black

TABLE 2.2*
EMISSIVITY AND ABSORPTIVITY VALUES FOR VARIOUS SURFACES

Surface	Emissivity or Absorptivity		Absorptivity for Solar Radiation
	50–100 F	1000 F	
A small hole in a large box, sphere, furnace, or enclosure	0.97 to 0.99	0.97 to 0.99	0.97 to 0.99
Black non-metallic surfaces such as asphalt, carbon, slate, paint, paper	0.90 to 0.98	0.90 to 0.98	0.85 to 0.98
Red brick and tile, concrete and stone, rusty steel and iron, dark paints (red, brown, green, etc.)	0.85 to 0.95	0.75 to 0.90	0.65 to 0.80
Yellow and buff brick and stone, firebrick, fire clay	0.85 to 0.95	0.70 to 0.85	0.50 to 0.70
White or light-cream brick, tile, paint or paper, plaster, whitewash	0.85 to 0.95	0.60 to 0.75	0.30 to 0.50
Window glass	0.90 to 0.95	. . . .	. . . .
Bright aluminum paint; gilt or bronze paint	0.40 to 0.60	. . . .	0.30 to 0.50
Dull brass, copper, or aluminum; galvanized steel; polished iron	0.20 to 0.30	0.30 to 0.50	0.40 to 0.65
Polished brass, copper, monel metal	0.02 to 0.05	0.05 to 0.15	0.30 to 0.50
Highly polished aluminum, tin plate, nickel, chromium	0.02 to 0.04	0.05 to 0.10	0.10 to 0.40

*Abstracted by permission from *ASHRAE Handbook of Fundamentals* (New York: American Society of Heating, Refrigerating and Air Conditioning Engineers, 1967) p. 38.

body at the same temperature. Therefore, the emissivity ϵ for any real surface must be less than unity. Table 2.2 shows emissivity and absorptivity values for various surfaces.

Three important laws for the emission of black-body radiation exist. The *Stefan-Boltzmann law* states that for a black body

$$E_b = \sigma T^4 \tag{2.11}$$

where E_b is the emissive power, Btu per (hr) (sq ft), T is absolute temperature, R, and $\sigma = (0.1713)(10^{-8})$ Btu per (hr) (sq ft) (R^4).

Thermal radiation emitted by any body includes rays of various wavelengths. The *Planck law* states that for a black body

$$E_{b\lambda} = \frac{c_1 \lambda^{-5}}{e^{c_2/\lambda T} - 1} \tag{2.12}$$

where $E_{b\lambda}$ is the monochromatic emissive power of a black body, Btu per (hr) (sq ft) (μ), $c_1 = (1.19)(10^8)$ Btu μ^4 per (hr) (sq ft), $c_2 = 25{,}900\ \mu$R, λ = wavelength, μ (microns), and T is absolute temperature, R.

Wien's displacement law, which may be derived from Eq. (2.12), states that the product of the wavelength at which maximum monochromatic emissive power is attained and the absolute temperature is a constant, or

$$\lambda_{max} T = c_3 \tag{2.13}$$

where $c_3 = 5216 \mu$R.

Figure 2.3 schematically illustrates the three black-body radiation laws given by Eqs. (2.11)–(2.13). Points for each curve may be calculated by using Eq. (2.12). As the temperature is increased, the maximum monochromatic emissive power is shifted to a shorter wavelength. This is required by Eq. (2.13). The area under any curve is the total emissive power which is equal to the value calculated by Eq. (2.11).

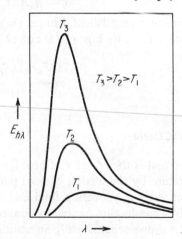

Figure 2.3 Schematic variation of monochromatic emissive power for a black body.

When two bodies within visual range of each other are separated by a medium which does not absorb radiation, an energy exchange occurs as a result of reciprocal processes of emission and absorption. If the two bodies are at the same temperature, the net heat exchange is zero. If the bodies are at different temperatures, there is a net transfer of heat from the warmer body to the colder body.

The net rate of transfer of heat q from one surface at T_1 to another surface at T_2 may be calculated by

$$q = \sigma A_1 F_A F_E (T_1^4 - T_2^4) \tag{2.14}$$

where q is the rate of heat transfer, Btu per hr, $\sigma = (0.1713)(10^{-8})$ Btu per (hr) (sq ft) (R^4), A_1 is the surface area, sq ft, of one surface, F_A is a dimensionless factor which accounts for the shapes and relative orientations of the two surfaces, F_E is a dimensionless factor which accounts for the emission and absorption characteristics of the two surfaces, and T_1 and T_2 are the respective absolute temperatures of the two surfaces, R.

For the purpose of calculation, Eq. (2.14) may be more conveniently written as

$$q = 0.1713 A_1 F_A F_E \left[\left(\frac{T_1}{100} \right)^4 - \left(\frac{T_2}{100} \right)^4 \right] \tag{2.15}$$

For the case of a small body 1 completely enclosed by a much larger body 2, Eq. (2.15) reduces to

$$q = 0.1713 A_1 \epsilon_1 \left[\left(\frac{T_1}{100} \right)^4 - \left(\frac{T_2}{100} \right)^4 \right] \tag{2.16}$$

Other solutions to Eq. (2.15) may be found in various references on heat transfer.

In many problems it is convenient to express the rate of heat transfer by radiation by

$$q = h_R A_1 (t_1 - t_2) \tag{2.17}$$

where h_R is a radiation coefficient, Btu per (hr) (sq ft) (F) and t_1 and t_2 are temperatures expressed in degrees F. By Eqs. (2.15) and (2.17), we have

$$h_R = \frac{0.1713 F_A F_E \left[\left(\frac{T_1}{100} \right)^4 - \left(\frac{T_2}{100} \right)^4 \right]}{(t_1 - t_2)} \tag{2.18}$$

2.4. CONVECTION

Thermal convection is the transfer of heat from one part of a fluid to another part at lower temperature by the mixing of fluid particles. Practical problems in convection deal with heat transfer between a fluid and a solid surface. The actual heat transfer process involves conduction as well as convection.

Figure 2.4 shows schematically an assumed set of conditions for convection heat transfer. A boundary layer of fluid in which both temperature and velocity may vary exists next to the surface. Resistance to heat transfer is influenced by the thickness and other characteristics of the boundary layer.

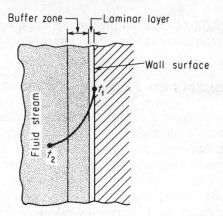

Figure 2.4 Schematic temperature gradient for convective heat transfer.

Convection heat-transfer calculations are made by the relation

$$q = h_c A(t_1 - t_2) \tag{2.19}$$

where q is the rate of heat transfer, Btu per hr; h_c is the convection coefficient, Btu (hr) (sq ft) (F); A is the solid surface area, sq ft; and t_1 and t_2 are respectively the wall-surface temperature and the bulk-fluid temperature, F.

Two types of convection heat transfer may exist: *forced* convection occurs when the fluid is forced or pumped past the wall surface; *free* or *natural* convection occurs when fluid motion is caused by density differences within the fluid. The *ASHRAE Handbook of Fundamentals** has presented an extensive summary of equations for h_c for both forced convection and for free convection. For longitudinal forced-convection flow in a circular cylinder, where the cylinder length $L > 4.4D$ and $DG/\mu > 2200$,

$$\frac{h_c D}{k} = 0.0225\left(\frac{DG}{\mu}\right)^{0.8}\left(\frac{c_p \mu}{k}\right)^{0.33} \tag{2.20}$$

where D is the pipe inside diameter, k is thermal conductivity, G is mass velocity, μ is viscosity, and c_p is specific heat at constant pressure of the fluid. Consistent units are needed so that the three groupings may be dimensionless. For liquid water (32 F–400 F), Eq. (2.20) may be simplified to

$$h_c = 13.5 t^{0.54} \frac{V^{0.8}}{D^{0.2}} \tag{2.21}$$

where t is the arithmetic mean of the surface temperature and water temperature, F; V is water velocity, ft per sec; and D is inside diameter of the pipe, ft. For forced-convection flow normal to a single cylinder, and where $1000 < DG/\mu < 50,000$,

$$\frac{h_c D}{k} = 0.26\left(\frac{DG}{\mu}\right)^{0.6}\left(\frac{c_p \mu}{k}\right)^{0.3} \tag{2.22}$$

**ASHRAE Handbook of Fundamentals*, 1967, pp. 42–46.

In Eqs. (2.20)–(2.22), fluid properties should be evaluated at the arithmetic mean of the fluid temperature and the pipe temperature.

2.5. CONDENSING VAPORS

Heat transfer from a condensing vapor to a cooler surface may be considered as a special case of convection heat transfer. In thermal environmental work, problems involving condensing vapors usually occur in refrigerant condensers and in steam coils used for heating air. In shell-and-tube condensers the vapor condenses on the outside of tubes, with cooling water passing on the inside of the tubes. In condensers which are air cooled, the refrigerant vapor condenses inside the tubes while air circulates over the outside surfaces. Heat transfer calculations with condensing vapors are made by an equation similar to Eq. (2.19) where t_1 is the temperature of the condensing vapor and t_2 is the surface temperature of the tube.

Figure 2.5 shows schematically a horizontal tube on which vapor is condensed. A continuous liquid film is assumed. Liquid flows downward from the tube by gravity,

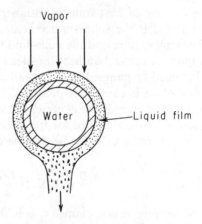

Figure 2.5 Schematic illustration of film-wise condensation.

and the liquid film increases in thickness in the downward direction. Primary resistance to heat transfer occurs in the liquid film. Since this film should be as thin as possible, rapid rate of removal of condensate is required.

In some laboratory experiments, *dropwise* condensation has been achieved through the use of special non-wetting agents. Such conditions result in condensing coefficients many times greater than those for *filmwise* condensation. In practical cases, filmwise condensation occurs almost always.

The *ASHRAE Handbook of Fundamentals** contains a summary of equations for

---

ASHRAE Handbook of Fundamentals, 1967, p. 54.

h_c for filmwise condensation. For condensation outside horizontal tubes,

$$h_c = 0.79F_1\left(\frac{h_{fg}}{ND_o\,\Delta t}\right)^{1/4} \tag{2.23}$$

where h_{fg} is the latent heat of vaporization, Btu per lb; N is the number of rows of tubes in a vertical plane; D_o is the outside diameter of the tubes, ft; Δt is the temperature difference between vapor and tube surface, F; and F_1 is given by

$$F_1 = \left(\frac{k^3\rho^2 g}{\mu}\right)^{1/4} \tag{2.24}$$

where k is the liquid thermal conductivity, Btu per (hr) (sq ft) (F per ft); ρ is liquid density, lb per cu ft; μ is liquid viscosity, lb per ft hr; and g is the acceleration of gravity, ft per hr². Table 2.3 shows values of F_1 in Eqs. (2.23) and (2.24) for three refrigerants.

TABLE 2.3*

FACTOR F_1 FOR VARIOUS REFRIGERANTS

Refrigerant	Film† Temperature t_f F	F_1
Ammonia	75	409
	100	408
	125	408
Refrigerant 12 . .	75	133
	100	122
	125	112
Refrigerant 22 . .	75	153
	100	144
	125	132

*Abstracted by permission from *ASHRAE Handbook of Fundamentals*, 1967, p. 55.

†$t_f = t_v - 0.75\,\Delta t$ where t_v is the saturated vapor temperature, F, and Δt is the temperature difference between vapor and tube surface, F.

In refrigerant or steam condensers, the entering vapor may be superheated. However, usual practice is to base the coefficient h_c upon saturated vapor properties. Magnitudes of condensing coefficients h_c for most condensing vapors are relatively large compared to forced- or free-convection coefficients for gases. In some overall heat transfer problems, sufficiently accurate final answers may be obtained by assuming approximate condensing coefficients. For steam condensing inside the tubes of an air-heating coil, a value of $h_c = 1200$ Btu per (hr) (sq ft) (F) is often used. For refrigerant vapors condensing upon the outside surfaces of tubes in shell-and-tube condensers, h_c is of the order of 200 to 400 Btu per (hr) (sq ft) (F) for halocarbons such as Refrigerant 12, whereas for ammonia it is of the order of 1000 Btu per (hr) (sq ft) (F).

2.6. BOILING LIQUIDS

Heat transfer from a surface to a boiling liquid may be considered as a special case of convection heat transfer. An important problem in thermal environmental work is the boiling of liquid refrigerants in evaporators.

Two important classes of boiling are *pool boiling* and *forced-convection boiling*. Pool boiling occurs when a relatively stagnant liquid is heated by a submerged surface. Liquid movement occurs only because of natural convection and by the stirring action of bubbles. Forced convection boiling occurs when the fluid is forced through a tube or passage with considerable velocity.

Most experimental work has been conducted on pool boiling. Such studies have shown that the boiling action is largely dependent on the temperature difference Δt between the submerged surface and the boiling liquid. Figure 2.6 shows schematically

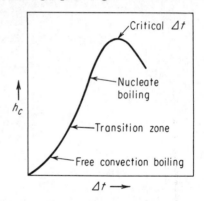

Figure 2.6 Schematic variation of the boiling heat transfer coefficient h_c with temperature difference for pool boiling.

how the heat transfer coefficient h_c may vary with Δt for pool boiling. When Δt is small there is little bubble formation and evaporation occurs primarily at the liquid surface. Heat transfer is essentially a problem in free convection. With increase of Δt, *nucleate boiling* occurs with a rapid increase of h_c. At a critical Δt, h_c decreases because the liquid droplets become covered by a vapor film. With forced-convection boiling, bubble formation and separation of vapor from liquid are greatly influenced by vapor velocity. The heat transfer coefficient depends upon the fraction of vapor present and upon the parameters for forced convection heat transfer.

Unfortunately, available information on boiling liquids is limited. A summary of information on boiling heat transfer, particularly for refrigerants, may be found in the *ASHRAE Handbook of Fundamentals.**

**ASHRAE Handbook of Fundamentals*, 1967, pp. 46–53.

2.7. OVERALL TRANSFER OF HEAT

Most practical problems in heat transfer involve two or more of the transfer methods. Such problems must be analyzed from a total or overall aspect. Two important determinations are the *overall heat transfer coefficient U* and the *true mean temperature difference* Δt_m.

Figure 2.7 shows schematically two fluids separated by a pipe. We will consider a short section dL where the various temperatures have constant values. We will also

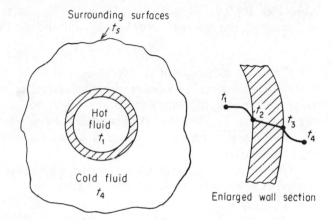

Figure 2.7 Schematic illustration of overall heat transfer.

assume the steady state where the rate of heat transfer through each part of the system is the same. We may write the following equations for the rate of heat transfer:

$$dq = h_i\, dA_i(t_1 - t_2) = \frac{2\pi k(t_2 - t_3)\, dL}{\ln(D_o/D_i)} = h_o\, dA_o(t_3 - t_4) \qquad (2.25)$$

The coefficient h_i may be determined from an appropriate equation for the heat-transfer case involved, such as for forced convection or for a condensing vapor. If the cold fluid shown in Fig. 2.7 is a gas such as air, h_o must be a *combined* coefficient, including radiation, as well as convection, effects. For these conditions, we may write

$$dq = h_{c,o}\, dA_o(t_3 - t_4) + h_{R,o}\, dA_o(t_3 - t_s) = h_o\, dA_o(t_3 - t_4)$$

Thus

$$h_o = h_{c,o} + h_{R,o}\left(\frac{t_3 - t_s}{t_3 - t_4}\right) \qquad (2.26)$$

With $t_s = t_4$, which is often approximately true,

$$h_o = h_{c,o} + h_{R,o} \qquad (2.27)$$

The equalities of Eq. (2.25) are inconvenient, since the pipe surface temperatures t_2 and t_3 are involved. By definition of U_o, we write

$$dq = U_o\, dA_o(t_1 - t_4) \qquad (2.28)$$

where U_o is the *overall heat transfer coefficient* based upon the *outside* surface area. By Eqs. (2.25) and (2.28) we may show that

$$\frac{1}{U_o} = \frac{dA_o}{h_i \, dA_i} + \frac{dA_o \ln (D_o/D_i)}{2\pi k \, dL} + \frac{1}{h_o}$$

Noting that $dA_i = \pi D_i \, dL$ and $dA_o = \pi D_o \, dL$, we have

$$\frac{1}{U_o} = \frac{D_o}{D_i h_i} + \frac{D_o \ln (D_o/D_i)}{2k} + \frac{1}{h_o} \tag{2.29}$$

The quantity $1/U_o$ may be called the *total thermal resistance* R_t. Likewise, the quantities $D_o/D_i h_i$, $[D_o \ln (D_o/D_i)]/2k$, and $1/h_o$ are respectively the inside surface resistance R_i, the *tube wall resistance* R_w, and the *outside surface resistance* R_o. Thus

$$1/U_o = R_t = R_i + R_w + R_o = \sum R \tag{2.30}$$

The resistance concept to overall heat transfer is important. Through simple addition of individual resistances, the total resistance may be found. In a multiple-layer pipe or wall, the total wall resistance may be found by adding the similar terms for each layer. Furthermore, in approximate calculations, the numerical significance of each resistance term may be studied, and the importance of each separate coefficient may be analyzed.

We will now summarize a few expressions for the overall heat transfer coefficient. For pipes or cylinders, by Eq. (2.29) we have

$$U_o = \frac{1}{\dfrac{D_o}{D_i h_i} + \dfrac{D_o \ln (D_o/D_i)}{2k} + \dfrac{1}{h_o}} \tag{2.31}$$

For metal pipes where D_i/D_o is *greater than about* 0.6, Eq. (2.31) may be satisfactorily written as

$$U_o = \frac{1}{\dfrac{D_o}{D_i h_i} + \dfrac{x}{k} + \dfrac{1}{h_o}} \tag{2.32}$$

where x is the thickness of the pipe $(D_o - D_i)/2$. For a flat wall, Eq. (2.31) reduces to

$$U = \frac{1}{\dfrac{1}{h_i} + \dfrac{x}{k} + \dfrac{1}{h_o}} \tag{2.33}$$

Equations (2.31) and (2.32) make no allowance for fouling, since we assumed that the tube of Fig. 2.7 was perfectly clean. This condition would be true for new equipment only. After use, heat transfer surfaces may accumulate deposits of scale, oil, or other foreign matter. Such deposits are usually thin but provide resistance to heat transfer and should be accounted for in the equipment design. Conventional practice is to express the resistance of the deposit in terms of a *fouling coefficient* or *deposit factor* h_d. Values of h_d are determined experimentally.

Assuming deposits on both sides of the tube, Eqs. (2.31) and (2.32) would be modified to read

$$U_o = \frac{1}{\dfrac{D_o}{D_i h_i} + \dfrac{D_o}{D_i h_{d,\,i}} + \dfrac{D_o \ln (D_o/D_i)}{2k} + \dfrac{1}{h_{d,\,o}} + \dfrac{1}{h_o}} \tag{2.34}$$

and

$$U_o = \frac{1}{\dfrac{D_o}{D_i h_i} + \dfrac{D_o}{D_i h_{d,\,i}} + \dfrac{x}{k} + \dfrac{1}{h_{d,\,o}} + \dfrac{1}{h_o}} \tag{2.35}$$

Table 2.4 shows a summary of deposit coefficients recommended for ordinary heat exchanger design problems. The values given are thermal resistances $(1/h_d)$.

TABLE 2.4*

RECOMMENDED DEPOSIT COEFFICIENTS $(1/h_d)$

Application	Non-Ferrous Tubes	Ferrous Tubes
Brine coolers (brine side)		
Inhibited salt brines	0.0005	0.001
Non-inhibited salt brines	0.001	0.002
Solvent brines (methylene chloride, halocarbon		
refrigerants).	0.0	0.0
Water coolers (water side)		
Recirculated water, closed system	0.0005	0.001
Recirculated water, open system	0.001	0.002
Condensers (water side)		
City or well water (typical)	0.001	0.002
River water (typical)	0.003	0.005
Brackish water	0.002	0.002
Condensers (refrigerant side)		
Ammonia	. . . .	0.0003
Halocarbon refrigerants (miscible with oil) . .	0.0	0.0002
Evaporators (refrigerant side)		
Ammonia	. . . .	0.001
Halocarbon refrigerants (miscible with oil) . .	0.0	0.0002

*Abstracted by permission from *Air Conditioning, Refrigerating Data Book Design Volume* (American Society of Refrigerating Engineers, 1957) pp. 20-14, 20-16.

So far, we have discussed overall heat transfer for only a local section in which temperatures were assumed to be constant. In a finite heat exchanger, fluid temperatures may vary, of course. Figure 2.8 shows a schematic counterflow heat exchanger with which we will derive an expression for the logarithmic mean temperature difference. We will assume that: (1) the overall heat transfer coefficient is constant, (2)

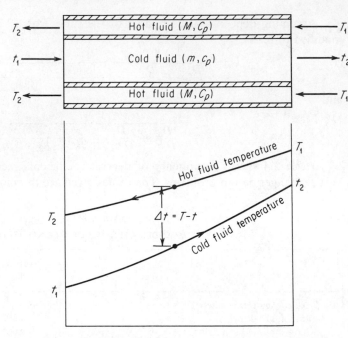

Figure 2.8 Schematic temperature changes in a counter-flow heat exchanger.

the mass rate of flow of each fluid is constant, (3) the specific heat of each fluid is constant, (4) each fluid undergoes no phase change, and (5) there is negligible heat loss from the outside of the heat exchanger. For any location, we may write

$$U_o \, dA_o(T - t) = mc_p \, dt$$

and

$$T = T_1 - \frac{mc_p}{MC_P}(t_2 - t)$$

Thus, we obtain

$$\frac{dt}{K_1 - K_2 t} = \frac{U_o \, dA_o}{mc_p} \tag{2.36}$$

where

$$K_1 = T_1 - \frac{mc_p}{MC_P}t_2$$

$$K_2 = 1 - \frac{mc_p}{MC_P}$$

Integration of Eq. (2.36) gives

$$\ln\left(\frac{K_1 - K_2 t_1}{K_1 - K_2 t_2}\right) = K_2\frac{U_o A_o}{mc_p}$$

Substitution back for K_1 and K_2 and noting that

$$\frac{mc_p}{MC_P} = \frac{T_1 - T_2}{t_2 - t_1}$$

and

$$q = mc_p(t_2 - t_1)$$

gives

$$\ln\left(\frac{T_2 - t_1}{T_1 - t_2}\right) = [(T_2 - t_1) - (T_1 - t_2)]\frac{U_o A_o}{q} \tag{2.37}$$

By definition of Δt_m

$$q = U_o A_o \, \Delta t_m \tag{2.38}$$

Thus, by Eqs. (2.37) and (2.38)

$$\Delta t_m = \frac{(T_2 - t_1) - (T_1 - t_2)}{\ln\left(\dfrac{T_2 - t_1}{T_1 - t_2}\right)} = \frac{\Delta t_A - \Delta t_B}{\ln\left(\dfrac{\Delta t_A}{\Delta t_B}\right)} \tag{2.39}$$

Equation (2.39) shows that, for pure counterflow, the true mean temperature difference is the *logarithmic mean temperature difference*. Equation (2.39) also applies to pure parallel flow and to any heat exchanger arrangement where one fluid temperature remains constant if we take Δt_A as the larger end-temperature difference and Δt_B as the smaller end-temperature difference. For other cases, such as cross-flow and where both fluid temperatures change, Eq. (2.39) must be modified in order to give the true mean temperature difference between the fluids. Kern* and Bowman *et al.*† have given solutions for a number of arrangements.

The design of a practical heat exchanger involves solution of Eq. (2.38). The heat transfer capacity q is usually known. Other factors such as one or more of the fluid temperatures may also be known. In general, it is desirable to have a large magnitude for U_o since the required area A_o may then be made smaller. High fluid velocities may give high values of U_o, but the drops in pressure of the fluids may be excessive, causing unduly large pumping power. The optimum situation is a design where the sum of the fixed charges and the operating costs is a minimum.

SYMBOLS USED IN CHAPTER 2

A	Surface area, sq ft; A_i for inside surface of pipe; A_o for outside surface of pipe.
C_P	Specific heat at constant pressure, Btu per (lb) (F), for hot fluid in Sec. 2.7.
c	Specific heat, Btu per (lb) (F); c_p for constant pressure.
c_1	$(1.19)(10^8)(Btu)(\mu^4)$ per (hr)(sq ft).

*Donald Q. Kern, *Process Heat Transfer* (New York: McGraw-Hill Book Company, 1950).

†R. A. Bowman, A. C. Mueller, and W. M. Nagle, "Mean Temperature Difference in Design," *ASME Trans.*, Vol. 62, 1940, pp. 283–294.

c_2	25,900 μR.
c_3	5216 μR.
D	Diameter of pipe, ft; D_o for outside diameter; D_i for inside diameter.
E	Emissive power, Btu per (hr) (sq ft); E_b for black body; $E_{b\lambda}$ for black body and monochromatic radiation.
F_A	Shape factor for radiation exchange, dimensionless.
F_E	Emission and absorption factor for radiation exchange, dimensionless.
G	Mass velocity of fluid, lb per (hr) (sq ft).
g	Acceleration of gravity, ft per hr^2.
h_c	Convection heat transfer coefficient, Btu per (hr) (sq ft) (F); $h_{c,o}$ for outside surface.
h_d	Deposit coefficient, Btu per (hr) (sq ft) (F); $h_{d,i}$ for inside surface; $h_{d,o}$ for outside surface.
h_{fg}	Latent heat of vaporization, Btu per lb.
h_i	Combined heat transfer coefficient for inside surface, Btu per (hr) (sq ft) (F).
h_o	Combined heat transfer coefficient for outside surface, Btu per (hr) (sq ft) (F).
h_R	Radiation heat transfer coefficient, Btu per (hr) (sq ft) (F); $h_{R,o}$ for outside surface.
I	Incident radiation, Btu per (hr) (sq ft); $I_R = \rho I$; $I_a = \alpha I$; $I_t = \tau I$.
K_1	$T_1 - (mc_p/MC_P)t_2$, dimensionless.
K_2	$1 - mc_p/MC_P$, dimensionless.
k	Thermal conductivity, Btu per (hr) (sq ft) (F per ft).
L	Length, ft.
M	Mass rate of flow of hot fluid, lb per hr.
m	Mass rate of flow of cold fluid, lb per hr.
N	Number of rows of tubes in a vertical plane.
n	Index of refraction, dimensionless.
q	Rate of heat transfer, Btu per hr.
R	Thermal resistance, (hr) (sq ft) (F) per Btu; R_i for inside surface; R_w for wall; R_o for outside surface; R_t for total.
r	Radius of pipe, ft; r_i for inside radius; r_o for outside radius.
T	Absolute temperature, R.
T	Temperature of hot fluid in Sec. 2.7, F (not absolute temperature).
t	Temperature, F.
Δt	Temperature difference, F.
Δt_A	Larger end-temperature difference between fluids, F.
Δt_B	Smaller end-temperature difference between fluids, F.
Δt_m	Mean temperature difference between fluids, F.
U	Overall heat transfer coefficient, Btu per (hr) (sq ft) (F); U_o based on outside surface area.
V	Velocity, ft per sec.
x	Thickness, ft.
α	$k/\rho c$, sq ft per hr.
α	Absorptivity, dimensionless.
ϵ	Emissivity, dimensionless.

θ	Time, hr.
θ	Angle of incidence.
θ'	Angle of refraction.
θ''	Angle of specular reflection.
λ	Wavelength, microns (μ).
μ	Viscosity, lb per (ft) (hr).
ρ	Density, lb per cu ft.
ρ	Reflectivity, dimensionless.
σ	(0.1713) (10^{-8}) Btu per (hr) (sq ft) (R^4)
τ	Transmissivity, dimensionless.

Part II

REFRIGERATION

3

MECHANICAL VAPOR COMPRESSION REFRIGERATION CYCLES

3.1. INTRODUCTION

One of the most important aspects of thermal environmental engineering is refrigeration. *Refrigeration* is the withdrawal of heat, producing in a substance or within a space a temperature lower than that of the natural surroundings. Thus any methods available for lowering temperature, in the range from ambient temperature to absolute zero, involve refrigerating processes.

Refrigeration may be produced by several means including (1) thermoelectric means, (2) vapor compression systems, (3) gas compression systems involving expansion of the compressed gas to produce work, and (4) gas compression systems involving throttling or unrestrained expansion of the compressed gas. All four methods will be studied in this text. Method (2) is used in the majority of refrigeration systems, with compression accomplished mechanically or through absorption methods. Method (3) has commercial application in air refrigeration systems used for cooling aircraft spaces. Methods (3) and (4) have both found application in the liquefaction of various gases.

Mechanical compression systems using vapor refrigerants are predominant among refrigerating methods. Cooling is accomplished by evaporation of a liquid refrigerant under reduced pressure and temperature. The saturation temperature of the vapor is then elevated by mechanical compression, allowing the vapor to be condensed by heat rejection to ordinary cooling water or atmospheric air. The relatively high-pressure liquid is then expanded to the heat exchanger where evaporation occurs. The expansion process is usually accomplished by throttling through a valve. A mechanical refrigeration system is a closed thermodynamic cycle.

Several arbitrary terms are used in expressing the performance of a refrigeration cycle. Except in small systems, the common unit of capacity is the *ton of refrigeration*. One ton of refrigeration is defined as the useful withdrawal of heat at a rate of 200 Btu per min. The word *ton* is an anachronism and is a carry-over from the days when ice was the principal means of refrigeration. By taking the heat of fusion of water as 144 Btu per lb, the heat absorbed by one ton (2000 lb) of ice in melting during a 24-hour period would be 288,000 Btu. This rate of heat absorption is equivalent to an average of 200 Btu per min.

The *coefficient of performance* expresses the effectiveness of a refrigeration system. It is a dimensionless ratio defined by the expression

$$\text{C.O.P.} = \frac{\text{Useful refrigerating effect}}{\text{Net energy supplied from external sources}} \qquad (3.1)$$

For a mechanical-compression system, work must be supplied by an external source. Thus

$$\text{C.O.P.} = \frac{Q}{W} \qquad (3.2)$$

It is common to express the performance of a mechanical compression system in terms of the horsepower required per ton of refrigeration produced. By Eq. (3.2),

$$W = \frac{Q}{\text{C.O.P.}}$$

If we let Q equal one ton of refrigeration

$$\text{Hp/ton} = \frac{12,000}{(2545)(\text{C.O.P.})} = \frac{4.72}{\text{C.O.P.}} \qquad (3.3)$$

The *refrigerating efficiency* η_R expresses the approach of the cycle or system, to that of an ideal reversible refrigerating cycle. By definition,

$$\eta_R = \frac{\text{C.O.P.}}{(\text{C.O.P.})_{\text{Rev}}} \qquad (3.4)$$

The *Carnot cycle* serves as the ideal reversible cycle.

This chapter deals principally with theoretical cycles. To understand the function-ing of an actual refrigerating system, we must first become thoroughly familiar with its theoretical cycle.

3.2. *THE CARNOT REFRIGERATION CYCLE*

The Carnot cycle, being completely reversible, is a model of perfection for a refrigera-tion cycle. Two important concepts involving reversible cycles are: (1) no refrigerating cycle may have a coefficient of performance higher than that for a reversible cycle which would be operated between the same temperatures of source and receiver, and (2) all reversible cycles when operated between the same temperatures would have the same coefficient of performance. Proof of both statements may be found in almost any book on elementary engineering thermodynamics.

Figure 3.1 shows the processes of the Carnot cycle on temperature-entropy coordinates. Heat is withdrawn at the constant temperature T_2 from the region to be

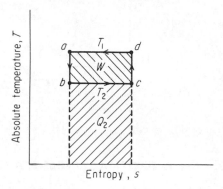

Figure 3.1 Processes of the Carnot refrigeration cycle.

refrigerated. Heat is rejected at the constant temperature T_1. The cycle is completed by an isentropic expansion and an isentropic compression. The energy transfers are given by

$$Q_1 = T_1(s_d - s_a)$$
$$Q_2 = T_2(s_d - s_a)$$
$$W = Q_1 - Q_2$$

Thus, by Eq. (3.2),

$$\text{C.O.P.} = \frac{T_2}{T_1 - T_2} \tag{3.5}$$

For vapor refrigerants it is desirable to consider the Carnot cycle as shown by Fig. 3.2. By applying the steady-flow energy equation on a one-pound basis, we obtain

$$_1W_b = h_1 - h_b$$
$$_3W_c = h_c - h_3$$
$$_cW_d = T_c(s_c - s_d) - (h_c - h_d)$$
$$_bQ_3 = h_3 - h_b = \text{Area } bef3b$$

The net work for the cycle is

$$W = {}_3W_c + {}_cW_d - {}_1W_b = \text{Area } b3cd1b$$

and

$$\text{C.O.P.} = \frac{_bQ_3}{W} = \frac{T_2}{T_1 - T_2}$$

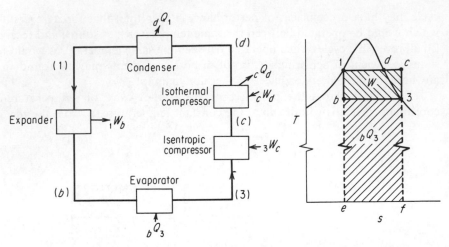

Figure 3.2 Carnot vapor compression cycle.

3.3. THE THEORETICAL SINGLE-STAGE CYCLE

It would appear desirable to design a system to approach the ideal model shown in Fig. 3.2. Practical considerations cause the expander (engine or turbine) to be replaced by a simple throttling valve and the use of one compressor instead of two. Figure

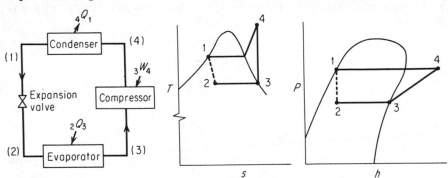

Figure 3.3 The theoretical single-stage vapor compression cycle.

3.3 shows the theoretical single-stage cycle which is used as a model for actual systems. By applying the steady-flow energy equation on a one-pound basis, we obtain

$$_2Q_3 = h_3 - h_2$$
$$_3W_4 = h_4 - h_3$$
$$_4Q_1 = h_4 - h_1$$
$$h_1 = h_2$$

The coefficient of performance is

$$\text{C.O.P.} = \frac{h_3 - h_2}{h_4 - h_3} \qquad (3.6)$$

The theoretical piston displacement P.D. (100 per cent volumetric efficiency) in cu ft per (min) (ton) is given by

$$\text{P.D.} = \frac{200 v_3}{h_3 - h_2} \qquad (3.7)$$

Example 3.1. An ammonia refrigerating plant following the theoretical single-stage cycle operates with a condensing temperature of 90 F and an evaporating temperature of 0 F. The system produces 15 tons of refrigeration. Determine (a) the coefficient of performance, (b) refrigerating efficiency, (c) horsepower per ton of refrigeration, (d) rate of refrigerant flow in lb per min, (e) theoretical horsepower input to compressor, and (f) theoretical piston displacement of compressor in cu ft per min.

SOLUTION: Figure 3.4 shows a schematic *P-h* diagram for the problem with numerical property data. Saturated liquid and saturated vapor properties were

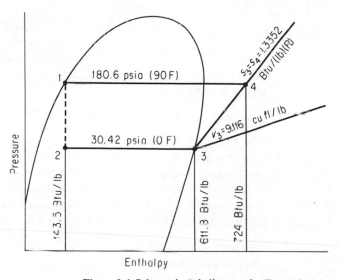

Figure 3.4 Schematic *P-h* diagram for Example 3.1.

read from Table A.3. The enthalpy $h_4 = 724$ Btu/lb was read from Fig. E-1* (Mollier chart for ammonia) at the intersection of $s_3 = 1.3352$ Btu/(lb) (R) and $P_4 = 180.6$ psia. The property data are tabulated below.

*All figures with the letter E preceding the number, such as Fig. E-1, refer to figures in the envelope.

$$h_1 = h_2 = 143.5 \text{ Btu/lb}, \qquad h_3 = 611.8 \text{ Btu/lb}, \qquad v_3 = 9.116 \text{ cu ft/lb},$$
$$s_3 = 1.3352 \text{ Btu/(lb) (R)}, \qquad h_4 = 724 \text{ Btu/lb}.$$

(a) By Eq. (3.6)

$$\text{C.O.P.} = \frac{611.8 - 143.5}{724 - 611.8} = 4.17$$

(b) By Eq. (3.4)

$$\eta_R = \frac{\text{C.O.P.} (T_1 - T_2)}{T_2} = \frac{(4.17)(90)}{459.6} = 0.82 \text{ or 82 per cent}$$

(c) By Eq. (3.3)

$$\text{Hp/ton} = \frac{4.72}{4.17} = 1.13$$

(d) The rate of refrigerant flow is obtained from an energy balance on the evaporator. Thus

$$m(h_3 - h_2) = (\text{tons})(200)$$

and

$$m = \frac{(15)(200)}{h_3 - h_2} = \frac{3000}{468.3} = 6.41 \text{ lb/min}$$

(e) The theoretical horsepower may be found by

$$\text{Hp} = \frac{m(h_4 - h_3)}{42.4} = \frac{(6.41)(112.2)}{42.4} = 17.0$$

or

$$\text{Hp} = (\text{Hp/ton})(\text{tons}) = (1.13)(15) = 17.0$$

(f) The theoretical piston displacement (100 per cent volumetric efficiency) may be found by

$$\text{P.D.} = mv_3 = (6.41)(9.116) = 58.5 \text{ cu ft/min}$$

Frequently, the liquid refrigerant entering the expansion valve may be *subcooled*, whereby its temperature is less than the saturation temperature corresponding to its pressure. Subcooling may occur within the condenser or within the liquid line by heat exchange with the ambient surroundings. The vapor may be superheated a few degrees in the evaporator, and may be further superheated in the compressor suction line. However, *the only useful refrigerating effect is the enthalpy change within the evaporator.*

Example 3.2. Liquid refrigerant leaves the condenser of an ammonia vapor compression system under a pressure of 153.0 psia and at a temperature of 70 F. Evaporating pressure is 23.74 psia and the vapor leaves the evaporator at 5 F. Otherwise, the system follows the theoretical single-stage cycle. Find (a) the enthalpy of the liquid entering the expansion valve, (b) the refrigerating effect in Btu per lb, and (c) the coefficient of performance.

SOLUTION: Figure 3.5 shows schematic *T-s* and *P-h* diagrams for the problem. If the liquid is assumed to be incompressible, the entropy of the subcooled liquid is equal to the entropy of saturated liquid at the same temperature.

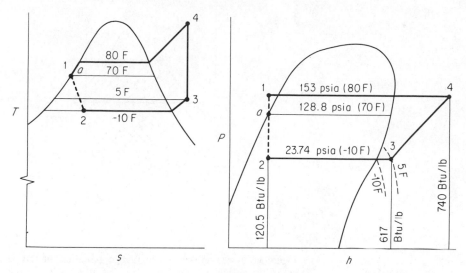

Figure 3.5 Schematic T-s and P-h diagrams for Example 3.2.

(a) Assuming the liquid to be incompressible, we may show that

$$h_1 = h_a + \frac{(P_1 - P_a)v_a}{J} - 120.5 + \frac{(153.0 - 128.8)(144)(0.0263)}{778} = 120.6 \text{ Btu/lb}$$

The above calculation shows that h_1 differs from h_a by a negligible amount. It is not necessary to correct for this small difference in ordinary refrigeration cycle problems. The enthalpy of a subcooled liquid refrigerant may be taken as equal to the enthalpy of the saturated liquid at the same *temperature*. Thus, use

$$h_1 = 120.5 \text{ Btu/lb}$$

(b) The evaporating temperature is -10 F and the vapor leaving the evaporator is at 5 F. Thus, the vapor is superheated 15 F. State (3) may be located on the Mollier chart by the intersection of $P_3 = 23.74$ psia and $t_3 = 5$ F. Thus $h_3 = 617$ Btu/lb and

$$_2Q_3 = h_3 - h_2 = 617 - 120.5 = 496.5 \text{ Btu/lb}$$

(c) State (4) for the vapor leaving the compressor may also be located through use of the Mollier chart. A sufficiently accurate procedure is as follows. Observe the position of State (3) with regard to the two lines of constant entropy which include it. With a pencil trace a path upward between the two lines of constant entropy. Locate State (4) on the line representing P_4 and at the same proportional position with respect to entropy. Thus, at $P_4 = 153$ psia, $h_4 = 740$ Btu/lb. By Eq. (3.6),

$$\text{C.O.P.} = \frac{617 - 120.5}{740 - 617} = 4.04$$

The saturation temperatures of the single-stage cycle have a strong influence on the magnitude of the coefficient of performance. This influence may be readily ap-

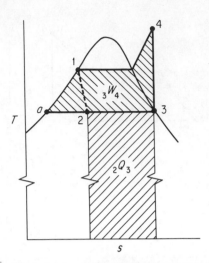

Figure 3.6 Areas on the T-s diagram representing the refrigerating effect, Btu per lb, and the work supplied, Btu per lb, for the theoretical single-stage cycle.

preciated by an area analysis on the T-s diagram. In Fig. 3.6, the area representing $_2Q_3$ follows directly from the definition of entropy. However, the work area is not so easily established. As discussed in Sec. 1.15, the area under a reversible constant-pressure process line on T-s coordinates is equal to the change in enthalpy for the process. We now make use of the excellent approximation that the process line on T-s coordinates for the constant pressure subcooling of a liquid coincides with the saturated liquid line. We may write

$$_3W_4 = h_4 - h_3 = (h_4 - h_a) - (h_3 - h_a)$$

By subtracting the area under the line $\overline{a\,3}$ from the area under the line $\overline{a\,14}$, we establish the area shown for $_3W_4$ in Fig. 3.6.

Since the C.O.P. $= {_2Q_3}/{_3W_4}$, we may observe how changes in evaporating temperature and condensing temperature affect the C.O.P. We find that a decrease in evaporating temperature significantly increases $_3W_4$ and slightly decreases $_2Q_3$. An increase in condensing pressure produces the same results but with lesser effect on $_3W_4$. For maximum coefficient of performance, the cycle should operate at the lowest possible condensing temperature and at the maximum possible evaporating temperature.

3.4. COMPARISON OF THE THEORETICAL SINGLE-STAGE CYCLE WITH THE CARNOT CYCLE

An interesting and informative analysis consists of comparing the processes of the single-stage cycle with those of the Carnot cycle on T-s coordinates. Figure 3.7 shows the single-stage cycle superimposed on the Carnot cycle. We assume that States (1) and (3) are common to both cycles. On an area basis, the single-stage cycle deviates from the Carnot cycle in three ways. Area A_1 represents additional work required for the single-stage cycle because of the superheat horn. Area A_2 represents additional work required for the single-stage cycle because no work is recovered in the expansion process. Area A_3 represents a loss of cooling effect caused by throttling as compared to the isentropic expansion of the Carnot cycle. On a Btu/lb basis

$$A_1 = (h_4 - h_d) - T_1(s_3 - s_d) \tag{3.8}$$

$$A_2 = h_1 - h_b \tag{3.9}$$

$$A_3 = h_2 - h_b = h_1 - h_b \tag{3.10}$$

The areas A_2 and A_3 are equal. Thus throttling causes an identical dual loss in Btu per lb. Since

$$_3W_4 = W_c + A_1 + A_2$$

and

$$_2Q_3 = Q_c - A_3$$

we may show that for the single-stage cycle

$$\eta_R = \frac{1 - (A_3/Q_c)}{1 + \dfrac{A_1 + A_2}{W_c}} \tag{3.11}$$

Example 3.3. A Refrigerant 12 theoretical single-stage cycle operates with a condensing temperature of 90 F and an evaporating temperature of 0 F. As shown in Fig. 3.7 we will assume a Carnot cycle operating between the same temperatures. Determine (a) the Carnot-cycle work of compression in Btu per lb, (b) the Carnot cycle refrigerating effect in Btu per lb, (c) the excess work of compression in Btu per lb for the single-stage cycle caused by the superheat horn, (d) the excess work of compression in Btu per lb for the single-stage cycle caused by throttling, (e) the loss in refrigerating effect in Btu per lb for the single-stage cycle caused by throttling, and (f) the refrigerating efficiency.

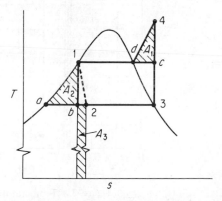

Figure 3.7 Area deviations on the T-s diagram for the theoretical single-stage cycle, with respect to the Carnot cycle.

SOLUTION: Following the nomenclature of Fig. 3.7, we have

(a) $W_c = (T_1 - T_2)(s_3 - s_1) = (90)(0.16888 - 0.05900) = 9.89$ Btu/lb

(b) $Q_c = T_2(s_3 - s_1) = (459.6)(0.10988) = 50.50$ Btu/lb

(c) By Eq. (3.8)

$$A_1 = (89.14 - 86.17) - (549.6)(0.16888 - 0.16353) = 0.03 \text{ Btu/lb}$$

(d) To evaluate A_2, we must first calculate h_b. We have

$$s_1 = s_b = s_a + x_b(s_3 - s_a)$$

$$x_b = \frac{s_1 - s_a}{s_3 - s_a} = \frac{0.05900 - 0.01932}{0.16888 - 0.01932} = 0.2653$$

$$h_b = h_a + x_b(h_3 - h_a) = 8.52 + (0.2653)(77.27 - 8.52) = 26.76 \text{ Btu/lb}$$

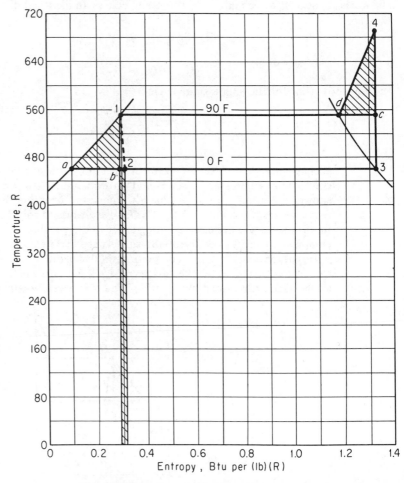

Figure 3.8 Temperature-entropy diagram for ammonia, showing area deviations of theoretical single-stage cycle with respect to the Carnot cycle.

By Eq. (3.9),

$$A_2 = 28.71 - 26.76 = 1.95 \text{ Btu/lb}$$

(e) Since $A_3 = A_2$, we have $A_3 = 1.95$ Btu/lb.

(f) The refrigerating efficiency may be calculated in two ways. Thus

$$\eta_R = \frac{(h_3 - h_2)(T_1 - T_2)}{(h_4 - h_3)T_2} = \frac{(48.56)(90)}{(11.87)(459.6)} = 0.80$$

or by Eq. (3.11),

$$\eta_R = \frac{1 - (1.95/50.50)}{1 + \dfrac{0.03 + 1.95}{9.89}} = 0.80$$

Thus for Refrigerant 12, the theoretical single-stage cycle has an extremely small superheat horn. The throttling process accounts for almost the entire deviation from the Carnot cycle.

The three deviations shown in Fig. 3.7 are dependent upon the shapes of the saturation lines on the T-s diagram. The areas A_2 and A_3 would disappear if the satu-

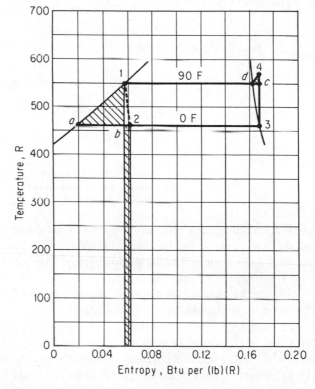

Figure 3.9 Temperature-entropy diagram for Refrigerant 12, showing area deviations of theoretical single-stage cycle with respect to the Carnot cycle.

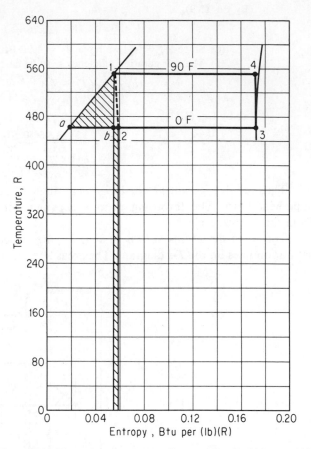

Figure 3.10 Temperature-entropy diagram for Refrigerant 113, showing area deviations of theoretical single-stage cycle with respect to the Carnot cycle.

rated liquid line were vertical. However, this would require the liquid to have a specific heat of zero. The superheat horn (A_1) would disappear if the saturated vapor line were vertical or if it had a positive slope. The shapes of the saturation lines vary widely among the various refrigerants.

Figure 3.8 shows that ammonia has symmetrically shaped saturation lines with all three area deviations being of significant size. Figure 3.9 shows that Refrigerant 12 has an extremely small superheat horn but relatively large deviation areas caused by throttling. Refrigerant 11 is closely similar in these respects. Refrigerant 22 is also similar but with a relatively greater superheat horn. Figure 3.10 shows that Refrigerant 113 has no superheat horn. Refrigerant 114 is a similar case.

Figure 3.11 shows the combined effects of the deviations from the Carnot cycle. Refrigerant 11 exhibits the highest refrigerating efficiency, while Refrigerant 22

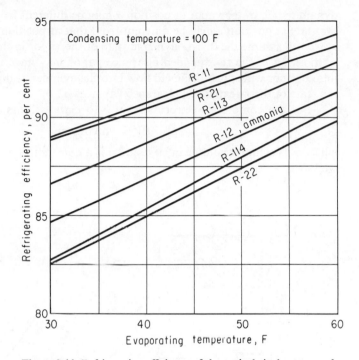

Figure 3.11 Refrigerating efficiency of theoretical single-stage cycle for various refrigerants.

shows the lowest refrigerating efficiency. However, the variation among the several refrigerants is not large.

3.5. A MORE PRACTICAL SINGLE-STAGE CYCLE

The thermodynamic cycle for an actual single-stage system may depart significantly from the theoretical cycle of Fig. 3.3. The principal departure occurs in the compressor. In the condenser, evaporator, and piping, departures from the theoretical cycle may be kept small by designing the components for small refrigerant pressure losses and by insulating components to reduce heat exchange with ambient surroundings. In this section we will be especially concerned with thermodynamic processes occurring within a reciprocating compressor.

Figure 3.12 is not a thermodynamic-state diagram but only shows the variation of pressure with position of the piston. The diagram is somewhat idealized with respect to compressor valve operation. Because of valve pressure drop, the cylinder pressure during intake $P_a = P_b$ may be less than the suction-line pressure P_3, and the cylinder pressure during discharge $P_c = P_d$ may be greater than the discharge-line pressure P_4. Because of heat exchange between the vapor and the cylinder walls, the thermodynamic state of the vapor at position a may be different from that at position b, and the state at position c may be different from that at position d. At positions b, c, and d, we will assume the thermodynamic state of the vapor in the cylinder to be uniform. At position a, the re-expanded clearance vapor may have a different state than the intake vapor which just begins to flow into the cylinder at this position. Let us denote the state of the re-expanded clearance vapor as a'.

For the compression process from b to c we will assume a polytropic process

$$PV^n = \text{a constant}$$

where V is cylinder volume in cu ft and n is a constant. Since the process is non-flow, we have

$$P_b v_b^n = P_c v_c^n$$

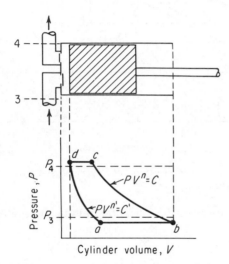

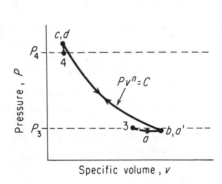

Figure 3.12 Schematic indicator diagram for a reciprocating compressor.

Figure 3.13 Pressure-specific volume diagram for compressor of Fig. 3.12.

where v is specific volume in cu ft per lb. Let us further assume that the temperature change from c to d is negligible. With this approximation, $v_c = v_d$. For the re-expansion process

$$P_d v_d^{n'} = P_a v_{a'}^{n'}$$

If we assume that $n = n'$, we have the result that $v_{a'} = v_b$. The above analysis with its several approximations establishes a P-v diagram as shown in Fig. 3.13. The compression and re-expansion curves are coincident. Thus, we may conclude that the net work required to compress the clearance vapor in ft-lb per lb is zero.

At the end of the intake stroke (position b), the total mass of vapor in the cylinder is V_b/v_b. The mass of clearance vapor is $V_a/v_{a'}$. Thus

$$\text{Mass of intake vapor per cycle} = \frac{V_b}{v_b} - \frac{V_a}{v_{a'}} = \frac{V_b - V_a}{v_b} \tag{3.12}$$

The compressor *volumetric efficiency* η_v may be defined as the mass of vapor actually pumped by the compressor divided by the mass of vapor which the compressor could pump if it handled a volume of vapor equal to its piston displacement and if no thermodynamic state changes occurred during the intake stroke. Thus, for the compressor of Fig. 3.12 and for one cycle of events in the compressor

$$\eta_v = \frac{(V_b - V_a)v_3}{(V_b - V_d)v_b} \tag{3.13}$$

But

$$V_b - V_a = (V_b - V_d) - (V_a - V_d)$$

$$V_a = V_d\left(\frac{P_d}{P_a}\right)^{1/n} = V_d\left(\frac{P_c}{P_b}\right)^{1/n}$$

Let us define the clearance factor C as

$$C = \frac{V_d}{V_b - V_d}$$

Thus

$$\frac{V_b - V_a}{V_b - V_d} = 1 + C - C\left(\frac{P_c}{P_b}\right)^{1/n}$$

and

$$\eta_v = \left[1 + C - C\left(\frac{P_c}{P_b}\right)^{1/n}\right]\frac{v_3}{v_b} \tag{3.14}$$

Equation (3.14) accounts for the three principal factors which affect compressor volumetric efficiency—re-expansion of clearance vapor, pressure drop in suction and discharge valves, and heating of the vapor on the intake stroke. If $C = 0$, the term in brackets is unity. When $C > 0$, valve pressure drops also contribute to decreasing the term in brackets. The quantity v_3/v_b may be less than unity because of pressure drop in the suction valve and because of cylinder-wall heating effects.

It also follows from the basic definition of volumetric effciency that

$$\eta_v = \frac{mv_3}{\text{P.D.}} \tag{3.15}$$

where m is the lb per min pumped by the compressor and P.D. is the compressor piston displacement in cu ft per min. By Eqs. (3.14) and (3.15),

$$m = \left[1 + C - C\left(\frac{P_c}{P_b}\right)^{1/n}\right]\frac{\text{P.D.}}{v_b} \tag{3.16}$$

Table 3.1 shows values of the isentropic exponent γ for several refrigerants. This exponent may be used as an approximation to the actual exponent n where more accurate data are not known.

TABLE 3.1
ISENTROPIC EXPONENT ($\gamma = c_p/c_v$) FOR VARIOUS REFRIGERANT VAPORS*

Refrigerant	11	12	22	113	114	Ammonia
Vapor temperature, F	86	50	86	160	110	70
$\gamma = c_p/c_v$	1.11	1.13	1.16	1.08	1.09	1.31

*14.696 psia.

Example 3.4. Saturated ammonia vapor at 10 F enters the compressor of a single-stage system. Discharge pressure is 200 psia. Pressure drop in the compressor suction valves is 2 psi. Pressure drop in the discharge valves is 4 psi. Assume that the vapor is superheated 20 F in the cylinder during the intake stroke. Compressor clearance is 5 per cent. Determine (a) the volumetric efficiency, and (b) the compressor pumping capacity in lb per min if the piston displacement is 50 cu ft per min.

SOLUTION: From the given data and Table A.3, we have

$$C = 0.05, \qquad P_c = 200 + 4 = 204 \text{ psia},$$

$$P_b = 38.5 - 2 = 36.5 \text{ psia}, \qquad v_3 = 7.304 \text{ cu ft/lb}$$

By Fig. E-1, at $t_b = 30$ F and $P_b = 36.5$ psia, we obtain

$$v_b = 8.2 \text{ cu ft/lb}$$

By Table 3.1, assume that $n = \gamma = 1.31$.

(a) By Eq. (3.14),

$$\eta_v = \left[1 + 0.05 - (0.05)\left(\frac{204}{36.5}\right)^{1/1.31}\right]\frac{7.304}{8.2} = 0.77$$

(b) By Eq. (3.15),

$$m = \frac{\eta_v(\text{P.D.})}{v_3} = \frac{(0.77)(50)}{7.304} = 5.27 \text{ lb/min}$$

We may derive an expression for the compressor work required, subject to the same approximations and limitations as used in the analysis for volumetric efficiency. Since the work in ft-lb per cycle is represented by the enclosed area of the diagram of Fig. 3.12, we may write

$$W \text{ ft-lb/cycle} = \int_b^c V \, dP - \int_a^d V \, dP$$

Using the relation $PV^n = a$ constant,

$$W \text{ ft-lb/cycle} = \frac{n}{n-1} P_b(V_b - V_a)\left[\left(\frac{P_c}{P_b}\right)^{(n-1)/n} - 1\right] \qquad (3.17)$$

By Eqs. (3.12) and (3.17),

$$W = \frac{n}{n-1} P_b v_b\left[\left(\frac{P_c}{P_b}\right)^{(n-1)/n} - 1\right] \qquad (3.18)$$

with W in ft-lb per lb.

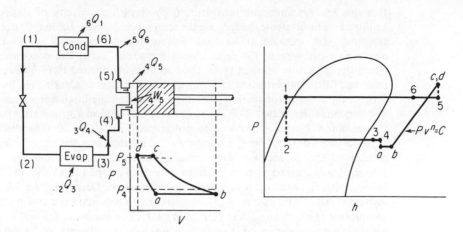

Figure 3.14 Schematic diagrams describing practical single-stage cycle.

Figure 3.14 shows schematic diagrams defining a single-stage cycle which deviates from the theoretical cycle in several respects. Pressure drop in compressor valves and heating of the vapor on the intake stroke are assumed. The compression process is assumed to be polytropic. Pressure drops in the condenser, evaporator, and piping are neglected. Heat addition in the compressor suction line and heat rejection in the compressor discharge line are assumed. The tons capacity and the horsepower requirement are important performance data. The equations which follow are associated with thermodynamic calculations for capacity and horsepower. State designations are given by Fig. 3.14.

$$\text{Tons} = \frac{m(h_3 - h_2)}{200} \tag{3.19}$$

$$m = \frac{\eta_v(\text{P.D.})}{v_4} \tag{3.20}$$

$$\text{Horsepower} = \frac{m\,_4W_s}{42.4\eta_m} \tag{3.21}$$

$$_4W_s = \frac{nP_b v_b}{(n-1)(778)}\left[\left(\frac{P_c}{P_b}\right)^{(n-1)/n} - 1\right] \tag{3.22}$$

$$\text{Hp/ton} = 0.00606\frac{nP_b v_b}{(n-1)\eta_m(h_3 - h_2)}\left[\left(\frac{P_c}{P_b}\right)^{(n-1)/n} - 1\right] \tag{3.23}$$

where in Eqs. (3.19)–(3.23), m = mass rate of refrigerant flow, lb per min; h = specific enthalpy, Btu per lb; η_v = compressor volumetric efficiency, dimensionless; P.D. = compressor piston displacement, cu ft per min; P = pressure, psfa; $_4W_s$ = compressor work, Btu per lb; η_m = compressor mechanical efficiency, dimensionless; n = polytropic exponent for compression path, dimensionless.

Example 3.5. An ammonia refrigeration plant with a capacity of 100 tons is equipped with a two-cylinder, single-acting compressor having 3 per cent clearance and operated at 560 rpm. Evaporating temperature is -20 F and condensing pressure is 194 psia. Liquid ammonia leaves the condenser and enters the expansion valve at 86 F. The vapor is superheated 10 F in the evaporator and further superheated 25 F more in the compressor suction line. Pressure drop in the compressor suction valves is 4 psi and in the discharge valves 6 psi. The vapor is superheated 20 F in the cylinder on the intake stroke after passing the suction valves. Compression is polytropic with $n = 1.20$. The discharge vapor is cooled 100 F in the compressor discharge line before entering the condenser. Ignore pressure drop in condenser, evaporator, and refrigerant piping. Ignore heat transfer in the liquid line to the expansion valve and in the line between the expansion valve and the evaporator. Determine the following items: (a) volumetric efficiency of compressor, (b) required bore and stroke for compressor if a stroke-to-diameter ratio of 1.25 is assumed, (c) brake horsepower input to compressor, assuming a mechanical efficiency of 75 per cent, (d) heat rejected to compressor jacket water in Btu per min, and (e) quantity of condensing water required in gallons per minute if water temperature rise is 12 F.

SOLUTION: The first step is to establish the various state points on the Mollier diagram. States (1), (2), (3), (4), (a), and (b) may be located from the given data. Thus

$$P_1 = 194 \text{ psia}, \quad t_1 = 86 \text{ F}, \quad h_1 = h_2 = 138.9 \text{ Btu/lb},$$

$$P_2 = P_3 = P_4 = 18.3 \text{ psia}, \quad t_3 = -10 \text{ F}, \quad h_3 = 610.5 \text{ Btu/lb},$$

$$t_4 = 15 \text{ F}, \quad h_4 = 625 \text{ Btu/lb}, \quad v_4 = 16 \text{ cu ft/lb},$$

$$P_a = P_4 - \Delta P_s = 18.3 - 4 = 14.3 \text{ psia}, \quad h_a = h_4 = 625 \text{ Btu/lb},$$

$$t_a = 14 \text{ F}, \quad P_b = P_a = 14.3 \text{ psia}, \quad t_b = t_a + 20 = 34 \text{ F},$$

and, $v_b = 21.5$ cu ft/lb. State (c) can be located on the Mollier diagram by the following method:

$$P_c = P_5 + \Delta P_d = 194 + 6 = 200 \text{ psia}$$

$$P_b v_b^n = P_c v_c^n$$

$$v_c = \frac{v_b}{\left(\dfrac{P_c}{P_b}\right)^{1/n}} = \frac{21.5}{\left(\dfrac{200}{14.3}\right)^{0.833}} = 2.39 \text{ cu ft/lb}$$

Thus

$$P_c = 200 \text{ psia}, \quad v_c = 2.39 \text{ cu ft/lb}, \quad h_c = 780 \text{ Btu/lb},$$

$$h_5 = h_c = 780 \text{ Btu/lb}, \quad P_5 = 194 \text{ psia}, \quad t_5 = 325 \text{ F},$$

$$P_6 = P_5 = 194 \text{ psia}, \quad t_6 = t_5 - 100 = 225 \text{ F}, \quad h_6 = 720 \text{ Btu/lb}.$$

(a) By Eq. (3.14) and using v_4 instead of v_3

$$\eta_v = \left[1.03 - 0.03\left(\frac{200}{14.3}\right)^{0.833}\right]\frac{16}{21.5} = 0.566$$

(b) By Eq. (3.19)

$$m = \frac{(100)(200)}{610.5 - 138.9} = 42.41 \text{ lb/min}$$

By Eq. (3.20)

$$\text{P.D.} = \frac{(42.41)(16)}{0.566} = 1199 \text{ cu ft/min}$$

and

$$\frac{2\pi D^2 (1.25)(D)(\text{RPM})}{(4)(144)(12)} = \text{P.D.}$$

Thus

$$D = \sqrt[3]{\frac{(1199)(1728)}{\pi(1.25)(280)}} = 12.3 \text{ in.}$$

$$L = (1.25)(12.3) = 15.4 \text{ in.}$$

(c) By Eq. (3.22)

$$_4W_s = \frac{(1.20)(14.3)(144)(21.5)}{(0.2)(778)}\left[\left(\frac{200}{14.3}\right)^{0.1667} - 1\right] = 188.8 \text{ Btu/lb}$$

By Eq. (3.21)

$$\text{Hp} = \frac{(42.41)(188.8)}{(42.4)(0.75)} = 252$$

(d) Per lb of refrigerant, we have by the steady-flow equation

$$_4Q_s = {}_4W_s - (h_5 - h_4) = 188.8 - (780 - 625) = 33.8 \text{ Btu/lb}$$

Thus

$$_4q_s = m\,_4Q_s = (42.41)(33.8) = 1433 \text{ Btu/min}$$

(e) $_6q_1 = m(h_6 - h_1) = (42.41)(720 - 138.9) = 24,640$ Btu/min

and

$$\text{GPM} = \frac{_6q_1}{(1)(12)(8.35)} = 246$$

Figure 3.15 shows variation of tons capacity, horsepower, and horsepower per ton with saturation-suction temperature and condensing temperature for a Refrigerant 12 condensing unit. Figure 3.15 was constructed from actual performance ratings obtained from a manufacturer. It shows that capacity decreases sharply with decrease of evaporating temperature. The horsepower required decreases and the horsepower per ton increases with decrease of evaporating temperature.

The trends of the curves in Fig. 3.15 may be readily explained by equations of this section. The decrease in capacity with decrease of saturation-suction temperature may be explained by examining Eqs. (3.14), (3.20), and (3.19). For this case assume a constant condensing pressure. With decrease in saturation-suction temperature, η_v will decrease due to the increase in the ratio P_c/P_b, while the specific volume at compressor intake, v_4, will increase. Both of these effects will cause a reduction in the mass rate of flow m. The refrigerating effect $(h_3 - h_2)$ will decrease with decrease in saturation-suction temperature. Thus, the reduction in tons capacity is due to three causes: (a) decrease of volumetric efficiency, (b) increase in specific volume, and (c) decrease

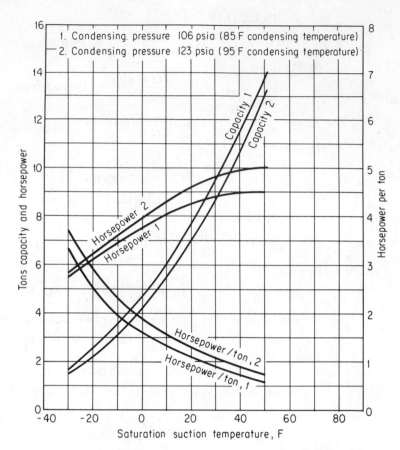

Figure 3.15 Capacity and horsepower characteristics for a typical Refrigerant-12 condensing unit equipped with a constant displacement compressor.

of refrigerating effect. Of these three factors, the increase in specific volume is by far the most important.

The trend of the horsepower curve may be explained with aid of Eqs. (3.21) and (3.22). With decrease of saturation-suction temperature, $_4W_5$ will increase due to increase in the pressure ratio P_c/P_b and increase in specific volume v_b. The cylinder suction pressure P_b will decrease as a compensating factor, but the net result will be an increase in $_4W_5$. However, the mass rate of flow m decreases at a much greater rate than the increase of $_4W_5$. The net result is a decrease in brake horsepower with decrease in saturation-suction temperature. The trend of the horsepower-per-ton curves may be justified by a similar analysis using Eq. (3.23).

The effect of an increase in condensing pressure can be readily explained by the equations. At a given saturation-suction temperature the capacity will decrease with increase of condensing pressure due to decrease of volumetric efficiency and decrease of refrigerating effect. The horsepower increases due to the increase of the work $_4W_5$.

3.6. PERFORMANCE OF SINGLE-STAGE CYCLE AT LOW EVAPORATING TEMPERATURES

The single-stage vapor compression cycle is an efficient refrigerating method at relatively high evaporating temperatures. As the evaporating temperature is reduced, the coefficient of performance decreases, horsepower per ton increases, and the

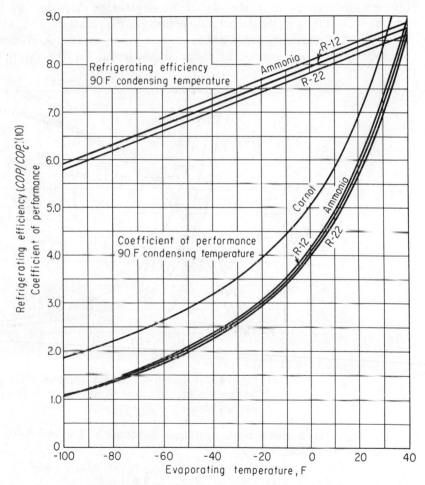

Figure 3.16 Variation of coefficient of performance and refrigerating efficiency for theoretical single-stage cycle.

required displacement in cu ft per (min) (ton) increases. There are three principal disadvantages of single-stage operation at low evaporating temperatures. These are (1) low refrigerating efficiency, (2) low compressor volumetric efficiency (reciprocating compressors), and (3) high compressor discharge temperature.

In Section 3.4, we discussed the three area deviations on the T-s diagram for the theoretical single-stage cycle as compared to the Carnot cycle. In Eq. (3.11), the ratios A_3/Q_c, A_1/W_c, and A_2/W_c all increase with decrease in evaporating temperature, thus causing the refrigerating efficiency η_R to decrease.

Figure 3.16 shows the variation of coefficient of performance and refrigerating efficiency for the theoretical single-stage cycle for Refrigerants 12 and 22, and for ammonia. A relatively low C.O.P. is inevitable at the lower evaporating temperatures. However, the decrease in η_R shows that the single-stage cycle deviates increasingly from the Carnot cycle as the evaporating temperature is reduced.

Figure 3.17 shows the rapid decrease of clearance volumetric efficiency and the correspondingly sharp increase of piston displacement for the theoretical single-

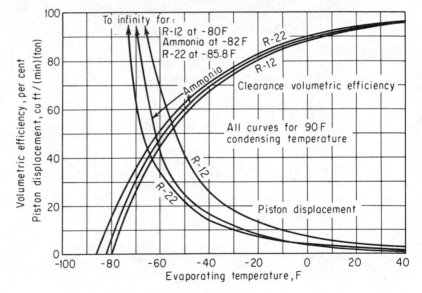

Figure 3.17 Variation of clearance volumetric efficiency and piston displacement for theoretical single-stage cycle with a compressor having a clearance of four per cent.

stage cycle as the evaporating temperature is reduced. The clearance volumetric efficiency η_{cv} was calculated by

$$\eta_{cv} = 1 + C - C\left(\frac{P_4}{P_3}\right)^{1/\gamma}$$

where P_4 is the condensing pressure and P_3 is the evaporating pressure. The total volumetric efficiency η_v as given by Eq. (3.14) would be somewhat smaller. Decrease of compressor volumetric efficiency imposes a practical limitation on the single-stage cycle and makes staging of compressors necessary in low temperature systems, irrespective of other considerations.

Figure 3.18 shows that at relatively low evaporating temperatures, the compressor

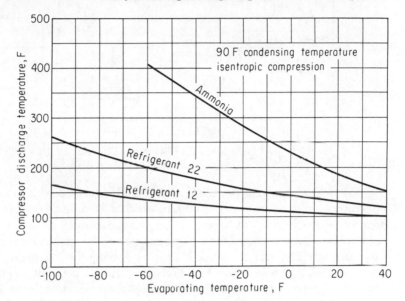

Figure 3.18 Variation of compressor discharge temperature for theoretical single-stage cycle.

discharge temperature in an ammonia system becomes extremely high. Practically, in Refrigerant 22 systems as well as in ammonia systems, staging of compressors and vapor intercooling becomes necessary at low evaporating temperatures in order to prevent overheating of the compressor.

3.7. THEORETICAL MULTI-STAGE, VAPOR-COMPRESSION CYCLES

By using multiple compressors and with intercooling of liquid and vapor refrigerant, the disadvantages of the single-stage cycle may be largely overcome. At low evaporating temperatures, staging of compressors is necessary because of reduced volumetric efficiency. With refrigerants such as ammonia, compressor staging and vapor intercooling are required to prevent excessive discharge temperature. Fortunately, the same arrangements allow a significant improvement in coefficient of performance also. In this section we will deal principally with multi-stage arrangements where ammonia and Refrigerant 12 are used.

REFRIGERATION

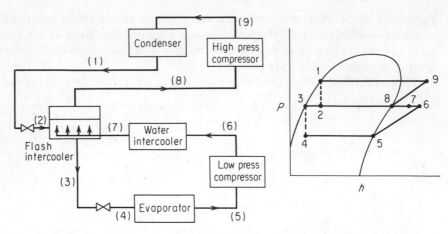

Figure 3.19 Schematic diagrams for cycle with two-stage compression, water intercooler, and flash intercooler.

Figure 3.19 shows a two-stage cycle suitable for use with ammonia. The flash intercooler is a pressure vessel or tank in which a fixed liquid level is maintained by a float valve which serves as an expansion valve. Saturated liquid at the intermediate pressure is withdrawn from the flash intercooler and is expanded through an expansion valve to the evaporator pressure. The vapor leaving the low-pressure compressor is first intercooled by water. This heat exchanger may or may not be feasible depending upon the temperature and availability of cooling water. After it leaves the water intercooler, the vapor is further cooled by direct contact with the relatively cold liquid in the flash intercooler.

The flash intercooler of Fig. 3.19 is of particular interest. The flash vapor formed by the throttling process in the float valve is removed to the suction line of the high pressure compressor. The vapor at State (7) is admitted to the bottom of the tank through a slotted pipe or through orifices, and bubbles up through the liquid. The vapor is de-superheated by evaporation of liquid. The extent of de-superheating depends upon the direct-contact heat-transfer rate. Complete de-superheating is assumed in Fig. 3.19, although this is impossible practically. The vapor entering the high-pressure compressor thus includes three components: (a) the flash vapor formed in the float valve, (b) the refrigerant circulated through the evaporator, and (c) the vapor formed by evaporation of liquid in the flash intercooler.

Since equipment arrangement in a multi-stage system may vary with the purpose of the system and with the refrigerant used, no development of special formulae will be given here. Solution of a typical problem involving the system of Fig. 3.19 is shown by the following example:

Example 3.6. Determine (a) the coefficient of performance, (b) the maximum cycle temperature, and (c) the total piston displacement in cu ft per (min) (ton) for the theoretical single-stage cycle and the two-stage cycle of Fig. 3.19. For each cycle assume an evaporating temperature of -50 F, a condensing temperature of 100 F, a compressor clearance of 5 per cent, and ammonia as the refrigerant. For the two-stage cycle, assume that the intermediate saturation temperature is 12 F and that the vapor leaves the water intercooler at 100 F.

SOLUTION: We will first consider the single-stage cycle. Figure 3.20(a) shows a schematic P-h diagram. By Table A.3 and Fig. E-1, we obtain

$$h_1 = h_2 = 155.2 \text{ Btu/lb}, \quad h_3 = 593.2 \text{ Btu/lb}, \quad v_3 = 33.08 \text{ cu ft/lb},$$

$$s_3 = 1.4487 \text{ Btu/(lb)(R)}, \quad h_4 = 827.5 \text{ Btu/lb}, \quad t_4 = 405 \text{ F}.$$

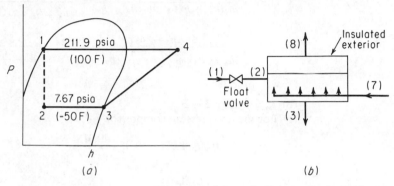

Figure 3.20 Schematic diagrams for Example 3.6.

(a) By Eq. (3.6)

$$\text{C.O.P.} = \frac{593.2 - 155.2}{827.5 - 593.2} = 1.87$$

(b) $t_{\max} = t_4 = 405$ F

(c)

$$\eta_{cv} = 1.05 - (0.05)\left(\frac{211.9}{7.67}\right)^{1/1.31} = 0.42$$

and

$$\text{P.D.} = \frac{(200)v_3}{(h_3 - h_2)\eta_{cv}} = \frac{(200)(33.08)}{(438.0)(0.42)} = 36.0 \text{ cu ft/(min)(ton)}$$

For the two-stage cycle of Fig. 3.19, we obtain the following property data:

$$h_1 = h_2 = 155.2 \text{ Btu/lb}, \quad h_3 = h_4 = 56.0 \text{ Btu/lb}, \quad h_5 = 593.2 \text{ Btu/lb},$$

$$v_5 = 33.08 \text{ cu ft/lb}, \quad s_5 = 1.4487 \text{ Btu/(lb)(R)}, \quad h_6 = 689 \text{ Btu/lb},$$

$$t_6 = 144 \text{ F}, \quad h_7 = 665 \text{ Btu/lb}, \quad h_8 = 615.5 \text{ Btu/lb}, \quad v_8 = 6.996 \text{ cu ft/lb},$$

$$s_8 = 1.3118 \text{ Btu/(lb)(R)}, \quad h_9 = 721.5 \text{ Btu/lb}, \quad t_9 = 230 \text{ F}.$$

(a) The coefficient of performance must be dimensionless. Since the mass flows through the two loops of the system are different, we may *not* express the C.O.P. as a ratio of enthalpy differences alone. For the low-pressure loop

$$m_3 = \frac{200}{h_5 - h_4} = \frac{200}{537.2} = 0.372 \text{ lb/(min)(ton)}$$

Figure 3.20(b) shows the flash intercooler as an isolated thermodynamic system. The fundamental steady-flow energy and material-balance equations are

$$m_2 h_2 + m_7 h_7 = m_3 h_3 + m_8 h_8$$

$$m_2 + m_7 = m_3 + m_8$$

However, for this cycle, $m_2 = m_8$ and $m_3 = m_7$. Thus

$$m_8(h_8 - h_2) = m_3(h_7 - h_3)$$

By Eq. (3.1),

$$\text{C.O.P.} = \frac{m_3(h_5 - h_4)}{m_3(h_6 - h_5) + m_8(h_9 - h_8)} = \frac{(h_5 - h_4)}{(h_6 - h_5) + \dfrac{m_8}{m_3}(h_9 - h_8)}$$

$$= \frac{(h_5 - h_4)}{(h_6 - h_5) + \dfrac{(h_7 - h_3)}{(h_8 - h_2)}(h_9 - h_8)} = \frac{537.2}{95.8 + \dfrac{(609)}{(460.3)}(106)}$$

$$= 2.28$$

(b) $t_{max} = t_9 = 230 \text{ F}$

(c) For the low-pressure compressor

$$\eta_{cv} = 1.05 - (0.05)\left(\frac{40.31}{7.67}\right)^{1/1.31} = 0.87$$

and

$$\text{P.D.} = \frac{m_3 v_5}{\eta_{cv}} = \frac{(0.372)(33.08)}{0.87} = 14.1 \text{ cu ft/(min)(ton)}$$

For the high-pressure compressor

$$\eta_{cv} = 1.05 - (0.05)\left(\frac{211.9}{40.31}\right)^{1/1.31} = 0.87$$

$$m_8 = m_3 \frac{(h_7 - h_3)}{(h_8 - h_2)} = (0.372)\frac{(609)}{(460.3)} = 0.493 \text{ lb/(min)(ton)}$$

and

$$\text{P.D.} = \frac{m_8 v_8}{\eta_{cv}} = \frac{(0.493)(6.996)}{0.87} = 3.96 \text{ cu ft/(min)(ton)}$$

The results of the example are summarized in Table 3.2.

TABLE 3.2

CALCULATED RESULTS FOR EXAMPLE 3.6

Cycle	C.O.P.	t_{max} F	η_{cv}		P.D. cu ft/(min)(ton)		
			L.P.	H.P.	L.P.	H.P.	Total
Single-stage	1.87	405	0.42	—	—	—	36.00
Two-stage	2.28	230	0.87	0.87	14.2	3.96	18.16

In Example 3.6, the reasons for increase of volumetric efficiency and decrease of maximum temperature with the two-stage cycle are obvious. We will further examine the reasons for the substantial increase in the coefficient of performance. The method of analysis, described in Sec. 3.4 and shown in Fig. 3.7, which was used for the single-stage cycle, is not straightforward for the two-stage cycle. Complications arise with areas on the T-s diagram because of different masses in the system.

For the two-stage cycle, we can easily show that the throttling loss (work recoverable by an isentropic expansion) is considerably less than it is for the single-stage cycle. This is apparent from practical reasoning as well, since the flash vapor formed in the throttling process from State (1) to State (2) is compressed in the high-pressure compressor only. Plank* has shown that the intercooling of vapor by evaporation of liquid in the flash intercooler results in an increase of C.O.P. except for conditions near the critical point.

A further matter of thermodynamic importance with the cycle of Fig. 3.19 is

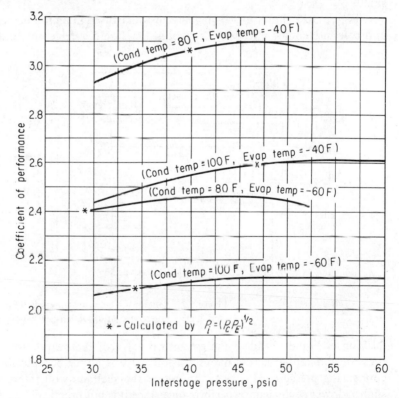

Figure 3.21 Optimum interstage pressure for cycle of Fig. 3.19.

*R. Plank, "Ueber den Ideal-Prozess von Kältemaschinen bei Verbund-Kompression," *Zeitschrift für die gesamte Kälte-Industrie*, Vol. 35, February 1928, pp. 17–24.

determination of the optimum interstage pressure. It is not possible to determine mathematically an optimum pressure for a vapor refrigerant, as would be the case for a perfect gas. For two-stage compression of a perfect gas with complete intercooling, we may show that, for minimum work, the interstage pressure P_i is given by

$$P_i = (P_A P_B)^{0.5} \tag{3.24}$$

where P_A is the suction pressure of the low-pressure compressor and P_B is the discharge pressure of the high-pressure compressor. Figure 3.21 shows that for maximum C.O.P. the optimum interstage pressure is slightly higher than that given by Eq. (3.24). However, Eq. (3.24) provides a simple and approximately correct procedure.

The flash intercooler of Fig. 3.19 has practical disadvantages. The liquid refrigerant in the tank is at the interstage pressure and is saturated. If the evaporator is above the flash intercooler, or if heat is absorbed in the liquid line to the expansion valve, some liquid will evaporate ahead of the expansion valve. In addition, the operation of the expansion valve may be sluggish because the pressure differential is too small.

Figure 3.22 shows a two-stage cycle widely used in ammonia systems. The shell-and-coil intercooler effectively subcools the liquid refrigerant and eliminates the

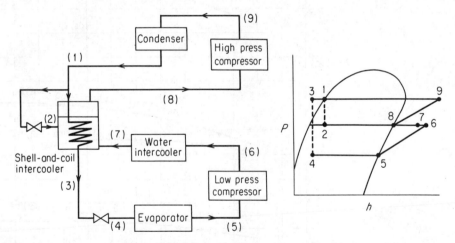

Figure 3.22 Schematic diagrams for cycle with two-stage compression, water intercooler, and flooded-type, shell-and-coil intercooler.

flashing of liquid ahead of the expansion valve of the evaporator. In addition, the pressure differential across the expansion valve is larger because the liquid is at the condensing pressure. However, the use of the shell-and-coil intercooler results in a slightly lower coefficient of performance since it is not practically possible to intercool the liquid as much as in the flash intercooler.

Figure 3.23 shows a two-stage cycle suitable for use with Refrigerant 12. The vapor discharged by the low-pressure compressor is not intercooled, except by mixing

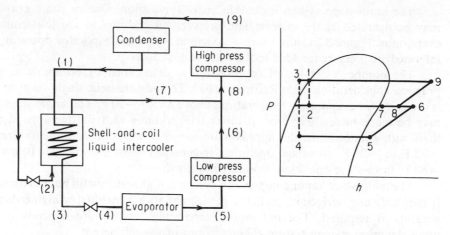

Figure 3.23 Schematic diagrams for cycle with two-stage compression and dry-type, shell-and-coil liquid intercooler.

with refrigerant from the intercooler. The intercooler is typically not of the flooded type. Instead of a float valve as shown in Figs. 3.19 and 3.22, a thermostatic expansion valve is used. The feeler bulb of the valve is usually placed in the suction line of the high-pressure compressor after the mixing of the two streams.

The two-stage systems of Figs. 3.19 and 3.22 have been frequently used with ammonia for industrial and commercial applications. In the food industry, ice cream holding rooms and sharp-freezer rooms are usually refrigerated in this way. Refrigerants 12 and 22 have been widely applied in systems similar to Fig. 3.23 for low-temperature test rooms and environmental chambers.

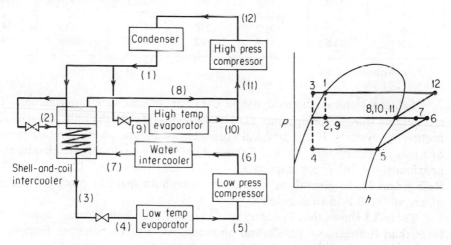

Figure 3.24 Schematic diagrams for two-stage cycle with both low-pressure and intermediate-pressure evaporators.

The multi-stage system is flexible in its application. One or more evaporators may be operated at the intermediate pressure, in addition to the low-temperature evaporator. Figure 3.24 shows such a system in which an evaporator operating at the intermediate pressure is added to the cycle of Fig. 3.22.

The number of stages of compression is determined by economic as well as practical considerations. For Refrigerants 12, 22, and ammonia, single-stage operation is indicated for evaporating temperatures above about -20 F. The single-stage system may be more successfully used at lower temperatures with Refrigerants 12 and 22 than with ammonia. In the approximate range of evaporating temperatures from -75 F to -25 F, two-stage operation is generally most economical. Below about -80 F, three-stage operation is usually indicated.

The multi-stage vapor compression system is highly successful but has limitations. It uses only one refrigerant, so that a refrigerant with a reasonably high critical temperature is required. The minimum evaporating temperature depends primarily upon the pressure-temperature characteristics of the refrigerant.

TABLE 3.3

PRESSURE AND TEMPERATURE DATA FOR VARIOUS REFRIGERANTS

Refrigerant	Saturation Pressure at 100 F psia	Saturation Temperature at 14.696 psia F	Saturation Temperature at 1.0 psia F*	Freezing Temperature F	Critical Temperature F	Critical Pressure psia
Ammonia	211.9	-28.0	-105	-107.9	271.4	1657.0
12	131.9	-21.6	-108	-252.0	232.7	582.0
13	—	-114.5	-184	-296.0	83.9	561.3
14	—	-198.2	-250	-312.0	-49.9	542.4
22	212.6	-41.4	-122	-256.0	204.8	716.0

*Approximate values.

Table 3.3 shows thermodynamic data for several vapor refrigerants. Of the refrigerants shown, Refrigerants 12, 22, and ammonia are the only ones suitable for multi-stage systems. If, for practical purposes, we assume an evaporating pressure of one psia as a minimum, Table 3.3 shows minimum evaporating temperatures of approximately -105 F for ammonia, -108 F for Refrigerant 12, and -122 F for Refrigerant 22. In general, we may obtain minimum space or product temperatures of about -100 F in such systems.

Table 3.3 shows that Refrigerants 13 and 14 evaporate at much lower temperatures than Refrigerants 12, 22, and ammonia for a given pressure. However, their low critical temperatures preclude their use in multi-stage systems. They are suitable for cascade systems, which are discussed in the next section.

3.8. THEORETICAL CASCADE VAPOR COMPRESSION CYCLES

We may produce space temperatures lower than about -100 F by mechanical compression cycles through cascade or binary arrangements. Figure 3.25 shows a simple scheme using two single-stage cycles. The two cycles are independent systems. The

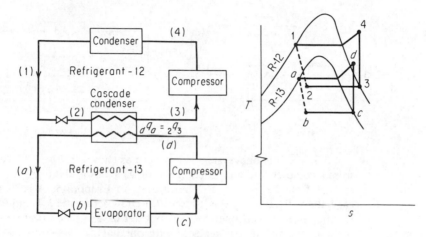

Figure 3.25 Schematic diagrams for cascade cycle composed of two single-stage cycles.

cascade condenser is a heat exchanger where the Refrigerant 13 vapor is condensed and the Refrigerant 12 liquid is evaporated. We may, of course, multi-stage either or both sides of the system. The analysis of a cascade system follows directly from the analyses for single-stage and/or multi-stage systems.

The cascade system does not compete with multi-stage systems. It only provides a vapor compression method for obtaining lower temperatures. The main thermodynamic disadvantage of a cascade system is the temperature overlap in the cascade condenser.

Cascade arrangements are applicable to various industrial processes including the liquefaction of petroleum vapors, the liquefaction of atmospheric gases, and the manufacture of dry ice.

PROBLEMS

3.1. Solve Example 3.1 for Refrigerant 12.

3.2. In a theoretical single-stage ammonia vapor-compression system, liquid leaves the condenser at 250 psia and 104 F. Evaporator pressure is 18.30

psia. Vapor leaves the evaporator at 0 F. The system produces 25 tons of refrigeration. Determine (a) the coefficient of performance, and (b) the piston displacement in cu ft per min, assuming a volumetric efficiency of 100 per cent.

3.3. A Refrigerant 12 system is arranged as shown in Fig. 3.26. Compression is isentropic. Assume frictionless flow. The known data are: $t_1 = 100$ F, $t_2 = 80$ F, and $t_4 = -10$ F. Find the system horsepower per ton.

3.4. Solve Example 3.3 for ammonia.

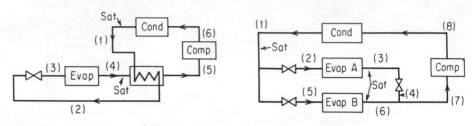

Figure 3.26   Figure 3.27

3.5. An ammonia system is arranged as shown in Fig. 3.27. Assume isentropic compression and frictionless flow. Known data are $P_1 = 160$ psia, $t_3 = 20$ F, and $t_6 = -30$ F. The capacity of Evaporator A is 5 tons and the capacity of Evaporator B is 10 tons. (a) Draw schematic P-h and T-s diagrams for the cycle. (b) Calculate the per cent decrease in theoretical power required if the above system was replaced with one having a separate compressor for each evaporator. Assume the same evaporating and condensing pressures and tons capacity.

3.6. A Refrigerant 12 plant operating at 0 F evaporating temperature and 70 F condensing temperature, uses a compressor having 4 per cent clearance. Under these conditions the plant produces 20 tons of refrigeration. (a) Calculate the capacity of the plant if the condensing temperature were to be increased to 100 F. Assume that other conditions are the same and the compressor is operated at the same RPM. Assume theoretical single-stage cycle but include effect of clearance. (b) Assume above system is operated at 100 F condensing temperature. What evaporating temperature would be required for refrigerating capacity to be zero?

3.7. An ammonia refrigeration plant utilizes a water-jacketed compressor. Saturated vapor at 10 F enters the compressor, and the vapor leaves the compressor at 180.6 psia and 170 F. Flow through the condenser and evaporator is at constant pressure. The liquid leaving the condenser is saturated. The system produces 25 tons of refrigeration and the power input (power delivered to refrigerant passing through compressor) is 23.2 horsepower. Find the lb per min of cooling water which must flow through the compressor jacket, if the water temperature rise is 10 F.

3.8. An industrial plant has available a 4-cylinder, 3-in. bore by 4-in. stroke, 800 RPM, single-acting compressor for use with Refrigerant 12. Proposed operating conditions for the compressor are 100 F condensing temperature and 40 F evaporating temperature. It is estimated that the refrigerant will enter the expansion valve as a saturated liquid, that vapor will leave the evaporator at a temperature of 45 F, and that vapor will enter the compressor at a temperature of 55 F. Assume a compressor volumetric efficiency of 70 per cent. Assume frictionless flow. Calculate the refrigerating capacity in tons for a system equipped with this compressor.

3.9. The following are the conditions for a single-stage ammonia refrigeration plant: condensing temperature 90 F with no liquid subcooling, evaporating temperature 0 F with saturated vapor at evaporator outlet, polytropic compression with $n = 1.24$, compressor clearance = 5 per cent. Assume no pressure drop in piping and compressor valves and no temperature changes either in piping or on intake stroke of compressor. (a) Calculate the coefficient of performance and compare with coefficient of performance calculated for isentropic compression. (b) Assume that the original system is operated with one additional change. The vapor picks up heat in the suction line of the compressor so that the temperature of the vapor entering the compressor is 60 F. Calculate the per cent decrease in the coefficient of performance. (c) Assume that the original system is operated with the only change being a 4 psi drop in the suction valves and 6 psi drop in the discharge valves of the compressor. Calculate the per cent decrease in the coefficient of performance.

3.10. Work Parts (b) and (c) of Problem 3.9 but calculate per cent decrease in tons capacity. Assume a constant displacement compressor.

3.11. Assume that you are the chief engineer of a creamery. You are in need of an ammonia compressor for an addition to your plant. You have been informed that another creamery operated by your company has a surplus compressor which will be made available to you if it is of adequate capacity for your installation. It is estimated that the proposed installation would operate under the following conditions: capacity = 18.5 tons, condensing pressure = 195 psia, ammonia liquid leaves condenser at saturated conditions, ammonia liquid enters expansion valve at 76 F, evaporating temperature = 10 F, vapor leaves evaporator at saturated conditions, vapor is superheated 20 F in compressor suction line. The data supplied to you about the surplus compressor are only the following: It is a 4-cylinder, vertical, reciprocating, single acting, single-stage, water-jacketed compressor, with a maximum RPM of 600. Cylinder diameter is 4 in. and the stroke is 5 in. Based on your past experience you make the following supplemental assumptions to allow further calculations: clearance = 4 per cent, pressure drop in suction valves = 4 psi, vapor is superheated 15 F in cylinder on the intake stroke after passing the suction valves, polytropic compression with $n = 1.27$, pressure drop in discharge valves = 6 psi, compressor mechanical efficiency = 80 per cent. (a) Determine whether the surplus compressor may be used and, if so, what RPM it should operate at. (b) Estimate the horsepower required to drive the compressor.

3.12. An ammonia vapor-compression system is arranged as shown in Fig. 3.28. Assume isentropic compression and frictionless flow. Given data are

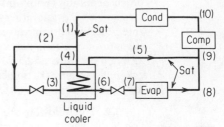

Figure 3.28

$t_1 = 90$ F, $t_6 = -10$ F, and $t_8 = -20$ F. (a) Compare the coefficient of performance for the cycle of Fig. 3.28 with that obtained if the liquid cooler were omitted. (b) What conclusion can be drawn from theoretical reasoning on the use of liquid intercoolers in single-stage systems? (c) Using practical reasoning, discuss the conditions under which the arrangement of Fig. 3.28 might be advantageous.

3.13. A two-stage ammonia system is arranged as shown in Fig. 3.22 except that the water intercooler is omitted. Condensing temperature is 90 F, intermediate saturation temperature is 15 F, and evaporating temperature is −40 F. Liquid leaves the intercooler at a temperature of 25 F. Saturated liquid leaves the condenser. Saturated vapor leaves the evaporator. Saturated vapor leaves the intercooler. It is known that the volumetric efficiency for each compressor is the same. The piston displacement of the second-stage (high-pressure) compressor is 100 cu ft per min. Calculate the piston displacement of the first-stage compressor.

3.14. A Refrigerant 12 system is arranged as shown in Fig. 3.23. Condensing pressure is 120 psia, intermediate pressure is 30 psia, and evaporating pressure is 7 psia. The following temperatures are known: $t_1 = 80$ F, $t_3 = 20$ F, $t_5 = -40$ F, and $t_7 = 20$ F. Assume isentropic compression and frictionless flow. Calculate the coefficient of performance.

3.15. An ammonia system is arranged as shown in Fig. 3.24. Condensing temperature is 100 F, intermediate saturation temperature is 12 F, low-side evaporating temperature is −50 F, $t_3 = 20$ F, and $t_7 = 100$ F. Saturation states occur as shown in the *P-h* diagram of Fig. 3.24. Assume isentropic compression and frictionless flow. Tons capacity of the high-temperature evaporator is four times that of the low-temperature evaporator. Assuming equal volumetric efficiencies, calculate the ratio of the piston displacement of the low-pressure compressor to that of the high-pressure compressor.

3.16. A Refrigerant 12 system is arranged as shown in Fig. 3.29. Assume isentropic compression and frictionless flow. The following data are given: condensing temperature = 90 F; Evaporator A has a capacity of 5 tons and an evaporating temperature of −20 F; Evaporator B has a capacity of 10 tons and an evaporating temperature of −70 F; vapor leaves each evaporator in dry and saturated condition; $t_1 = 84$ F, $t_6 = -5$ F, $t_{11} = 20$ F, $t_{16} = -5$ F, and $t_{17} = -55$ F; each compressor has 5 per cent clearance.

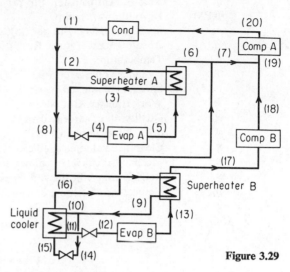

Figure 3.29

Draw schematic *P-h* and *T-s* diagrams for the cycle and determine (a) the C.O.P. for the *system*, (b) the piston displacement, cu ft per min, for each compressor, and (c) the theoretical horsepower input for each compressor.

SYMBOLS USED IN CHAPTER 3

A_1	Additional work compared to Carnot cycle, Btu per lb, required by theoretical single-stage cycle because of superheat horn on *T-s* diagram.
A_2	Additional work compared to Carnot cycle, Btu per lb, required by theoretical single-stage cycle because of throttling process.
A_3	Loss of refrigerating effect, Btu per lb, compared to Carnot cycle because of throttling process in theoretical single-stage cycle.
C	Compressor clearance fraction $V_d/(V_b - V_d)$, dimensionless.
C.O.P.	Coefficient of performance, dimensionless.
c_p	Specific heat at constant pressure, Btu per (lb)(F).
c_v	Specific heat at constant volume, Btu per (lb)(F).
D	Cylinder diameter, in.
GPM	Gallons per min.
Hp	Horsepower.
h	Specific enthalpy, Btu per lb.
J	Mechanical equivalent of heat, 778 ft-lb per Btu.
m	Mass rate of flow, lb per min.
n	Polytropic exponent, dimensionless.
P	Pressure, psia or psfa.
P.D.	Piston displacement, cu ft per min or cu ft per (min)(ton).

Q	Heat transfer, Btu per lb.
q	Rate of heat transfer, Btu per min.
RPM	Revolutions per min.
s	Specific entropy, Btu per (lb)(R).
T	Absolute temperature, R.
t	Temperature, F.
V	Cylinder volume, cu ft.
v	Specific volume, cu ft per lb.
W	Work transfer, Btu per lb.
x	Quality, lb of saturated vapor per lb of mixture.
γ	c_p/c_v, dimensionless.
η_{cv}	Clearance volumetric efficiency, dimensionless.
η_m	Mechanical efficiency, dimensionless.
η_R	Refrigerating efficiency, dimensionless.
η_v	Volumetric efficiency, dimensionless.

4

REFRIGERANTS

4.1. DESIGNATION OF REFRIGERANTS

In this chapter we will deal with the working mediums used in compression refrigeration systems. Such a medium is called a *refrigerant*. Most of our attention will be given to vapor refrigerants.

The design of a vapor compression refrigeration system is greatly influenced by the properties of the refrigerant employed. The suitability of a refrigerant for a certain application is determined by its physical, thermodynamic, and chemical properties and by various practical factors.

Many substances may be used as refrigerants including halocarbon compounds, hydrocarbon compounds, inorganic compounds, and others. The American Society of Heating, Refrigerating and Air Conditioning Engineers* has adopted a specific designation system for refrigerants. Table 4.1 shows this system for a few common compounds of commercial importance in refrigeration. The numbers assigned to the hydrocarbons and halocarbons have special meaning. The first digit on the right is the number of fluorine (F) atoms in the refrigerant. The second digit from the right is one more

*ASHRAE Handbook of Fundamentals (New York: American Society of Heating, Refrigerating and Air Conditioning Engineers, 1967) p. 213.

TABLE 4.1*

ASHRAE DESIGNATIONS FOR REFRIGERANTS

Number	Type of Compound and Chemical Name	Chemical Formula	Molecular Weight
	Halocarbons		
11	Trichloromonofluoromethane	$C Cl_3 F$	137.4
12	Dichlorodifluoromethane	$C Cl_2 F_2$	120.9
13	Monochlorotrifluoromethane	$C Cl F_3$	104.5
14	Carbontetrafluoride	$C F_4$	88.0
21	Dichloromonofluoromethane	$CH Cl_2 F$	102.9
22	Monochlorodifluoromethane	$CH Cl F_2$	86.5
113	Trichlorotrifluoroethane	$C Cl_2 F C Cl F_2$	187.4
114	Dichlorotetrafluoroethane	$C Cl F_2 C Cl F_2$	170.9
	Hydrocarbons		
50	Methane	CH_4	16.0
170	Ethane	$CH_3 CH_3$	30.0
290	Propane	$CH_3 CH_2 CH_3$	44.0
	Inorganic compounds		
717	Ammonia	NH_3	17.0
718	Water	H_2O	18.0
729	Air	...	29.0
744	Carbon dioxide	CO_2	44.0
764	Sulfur dioxide	SO_2	64.0

*Abstracted by permission from *ASHRAE Handbook of Fundamentals*, 1967, p. 214.

than the number of hydrogen (H) atoms present. The third digit from the right is one less than the number of carbon (C) atoms, but when this digit is zero it is omitted. Thus, ethane ($CH_3 CH_3$) is Refrigerant 170 while dichlorodifluoromethane ($C Cl_2 F_2$) is Refrigerant 12.

The inorganic refrigerants are designated by adding 700 to the molecular weight of the compound. This identification was adopted arbitrarily. Thus, ammonia is Refrigerant 717 while water is Refrigerant 718.

Table 4.2 shows various thermodynamic data for several common refrigerants. Refrigerants 113 and 114 have high boiling points and relatively high critical temperatures. Refrigerants 12 and 22 and ammonia have intermediate boiling points and moderately high critical temperatures, while Refrigerants 13 and 14 have low boiling points and low critical temperatures.

4.2. DESIRABLE PROPERTIES FOR AN IDEAL REFRIGERANT

Before proceeding further in discussing definite compounds, we will examine several properties which a desirable refrigerant should possess. The ideal refrigerant should have, at least, the following characteristics:

TABLE 4.2*
COMPARATIVE REFRIGERANT PRESSURE AND TEMPERATURE DATA

Refrigerant	Boiling Temperature at 14.696 psia F	Freezing Temperature F	Critical Temperature F	Critical Pressure psia
Refrigerant 11	74.9	−168.0	388.4	639.5
Refrigerant 12	−21.6	−252.0	233.6	596.9
Refrigerant 13	−114.6	−294.0	83.9	561.0
Refrigerant 14	−198.4	−299.0	−49.9	542.0
Refrigerant 21	48.0	−211.0	353.3	750.0
Refrigerant 22	−41.4	−256.0	204.8	721.9
Refrigerant 113 . . .	117.6	−31.0	417.4	495.0
Refrigerant 114 . . .	38.4	−137.0	294.3	474.0
Ammonia	−28.0	−107.9	271.4	1657.0
Water	212.0	32.0	706.1	3226.0
Air	−317.8	—	−221.0	547.0
Carbon Dioxide . . .	−109.3	−69.9	87.8	1071.1
Sulfur Dioxide . . .	14.0	−103.9	314.8	1141.5

*Abstracted by permission from *ASHRAE Handbook of Fundamentals*, 1967, p. 214.

1. *Positive evaporating pressures.* Positive evaporating or low-side pressures prevent possible leakage of atmospheric air into the system during operation.
2. *Moderately low condensing pressures.* This feature permits the use of lightweight equipment and piping on the high-pressure side of the system.
3. *Relatively high critical temperature.* The critical temperature of the refrigerant should be much higher than normal operating condensing temperatures in order to prevent unduly large power requirements.
4. *Low freezing temperature.* The freezing temperature should be sufficiently low so that solidification of the refrigerant cannot occur during normal operation.
5. *Low cost of refrigerant.* This feature is obvious, although its significance depends upon the size of the system. Cost of the refrigerant may be rather immaterial where the entire charge is a matter of a few ounces, as in a household refrigerator.
6. *High latent heat of vaporization.* A high latent heat of vaporization means a high refrigerating effect per pound of refrigerant circulated. This feature is generally desirable, although in small capacity systems too low rates of refrigerant flow may result in control difficulties.
7. *Inertness and stability.* The refrigerant should be inert to reactions with materials of the system. It should be non-corrosive in the presence of water. It should be entirely stable in its chemical make-up through the entire range of operating conditions.
8. *High dielectric strength of vapor.* This characteristic is important in her-

metically-sealed compressor units where the refrigerant vapor may come into contact with the motor windings.

9. *High heat transfer characteristics.* This is a broad statement involving such properties as density, specific heat, thermal conductivity, and viscosity. High heat transfer coefficients reduce the surface area required in heat exchangers.

10. *Satisfactory oil solubility.* This characteristic is separately discussed in Section 4.6.

11. *Low water solubility.* This factor is separately discussed in Section 4.5.

12. *Non-toxicity.* The refrigerant should be non-poisonous to human beings and to foodstuffs. Besides the relative toxicity of the refrigerant vapor, important factors are (a) the concentration or the percentage of vapor in the air, and (b) the duration of exposure.

13. *Non-irritability.* This characteristic involves human beings exposed to the refrigerant vapor. The vapor should not irritate the eyes, nose, lungs, or skin.

14. *Non-flammability.* The vapor should not burn or support combustion in any concentration with atmospheric air.

15. *Easy leakage detection.* Leaks in refrigerant lines and equipment should be detectable by a simple and positive method.

Unfortunately, there is no single compound available which is absolutely satisfactory for all refrigeration systems. We find that some refrigerants are well suited for particular applications but may be wholly unsuitable for others.

4.3. THE INORGANIC REFRIGERANTS

Until about 1930, refrigerants in use were almost exclusively inorganic compounds. Ammonia, carbon dioxide, and sulfur dioxide were commonly used in vapor compression systems. Of these three compounds, only ammonia is of commercial importance today. Various halocarbon refrigerants have largely taken over the applications where carbon dioxide and sulfur dioxide were once used.

(a) *Ammonia.* Although it is one of the oldest refrigerants, ammonia is still widely used, particularly in the larger industrial applications. Ammonia is an excellent refrigerant from a thermodynamic standpoint. Its boiling point of -28 F at 14.696 psia allows positive evaporating pressures in a majority of refrigeration needs. Its critical temperature of 271 F is relatively high, while its freezing temperature of -108 F is sufficiently low. One of its outstanding characteristics is its high latent heat of vaporization. This feature accounts for the relatively small compressor displacement required for ammonia systems, and for the modest size of pipe lines. Ammonia has excellent heat transfer characteristics; it is relatively cheap, too.

In the presence of water, ammonia strongly attacks copper and cuprous alloys. Since some water may exist in any refrigeration system, cuprous metals are never used with ammonia. Ferrous materials are used for equipment and piping. Ammonia is non-miscible or insoluble with mineral lubricating oils. However, it is soluble in all proportions with water.

The primary disadvantages of ammonia are certain rather hazardous characteristics it has. It is a strong irritant and is intolerable even in small concentrations. It is mildly toxic. In concentrations in air of between 16 to 25 per cent by volume, it will burn feebly.

(b) *Carbon dioxide.* The principal refrigeration use of CO_2 today is as dry ice. Carbon dioxide is non-toxic, non-irritating, and non-flammable. The triple point of CO_2 is -69.9 F, and at atmospheric pressure dry ice sublimes at -109 F. As a refrigerant in a vapor compression system, carbon dioxide has distinct disadvantages. Its critical temperature of 87.8 F is below normal condensing temperatures attainable in most systems. Compared to most other refrigerants, operating pressures are extremely high and power costs relatively large.

(c) *Sulfur dioxide.* In former years, SO_2 was widely used in household and small commercial systems. Sulfur dioxide is non-flammable, but it is a strong irritant and is mildly toxic. In the presence of water, sulfurous acid is formed, which strongly attacks metals, particularly ferrous materials.

(d) *Water.* The principal refrigeration use of water is as ice. The high freezing temperature of water limits its use in vapor compression systems. Water is used as the refrigerant vapor in some absorption systems and in systems with steam-jet compressors.

(e) *Air.* Dry air is used as a gaseous refrigerant in some compression systems, notably in aircraft air conditioning. It also serves refrigeration purposes as liquid air and is of commercial importance as a material source in the production of liquid oxygen and nitrogen.

4.4. *THE HALOCARBON REFRIGERANTS*

Although the American Society of Heating, Refrigerating and Air Conditioning Engineers identifies some 42 halocarbon compounds as refrigerants, only Refrigerants 11, 12, 13, 14, 21, 22, 113, and 114 will be considered here. These compounds are all synthetically produced and were developed as the *Freon** family of refrigerants. In 1930, the first of the family, Freon 12, was developed.† Introduction of the Freon

---

*Freon is a registered trademark of E. I. DuPont de Nemours & Co. Most of the compounds are now available from other manufacturers also under various trade names such as Genetron, Isotron, etc.

†Thomas Midgley, Jr. and Albert L. Henne, "Organic Fluorides as Refrigerants," *Industrial and Engineering Chemistry*, Vol. 22, 1930, p. 542.

refrigerants provided a tremendous impetus to the refrigeration industry. The outstanding characteristics of the family of compounds are their non-hazardous features. All of them are non-toxic, non-irritating, and non-flammable.

(*a*) *Refrigerant 11.* Trichloromonofluoromethane ($C Cl_3 F$) is a clear, colorless liquid with a relatively high boiling point of 74.9 F at 14.696 psia. It has a freezing temperature of -168 F and a critical temperature of 388 F. Refrigerant 11 is used mainly in relatively high temperature systems employing centrifugal compressors.

(*b*) *Refrigerant 12.* Dichlorodifluoromethane ($C Cl_2 F_2$) was the first Freon to be developed and is still one of the most widely used. It is a clear and colorless liquid with a boiling point of -21.6 F at 14.696 psia, thus allowing positive evaporating pressures for a wide range of applications. It has a relatively high critical temperature of 233.6 F and a relatively low freezing temperature of -252 F.

Refrigerant 12 has a low latent heat of vaporization. It is a highly inert and stable compound. Its vapor has a relatively high dielectric strength. Because of its high vapor density, compressor piston displacement is moderately low. Refrigerant 12 is completely miscible with mineral lubricating oil but is essentially insoluble with water.

(*c*) *Refrigerant 13.* Monochlorotrifluoromethane ($C Cl F_3$) has a boiling temperature of -114.6 F at 14.696 psia and has a critical temperature of 83.9 F. Refrigerant 13 is important as a safe, non-hazardous refrigerant for the low-temperature side of cascade systems.

(*d*) *Refrigerant 14.* Carbontetrafluoride ($C F_4$) has a boiling temperature of -198.4 F at 14.696 psia and a critical temperature of -49.9 F. It serves as an ultra-low-temperature refrigerant for use in cascade systems.

(*e*) *Refrigerant 21.* Dichloromonofluoromethane ($CH Cl_2 F$) has a boiling temperature of 48.0 F at 14.696 psia. It has found its principal use in centrifugal-compressor systems for relatively high-temperature refrigeration needs.

(*f*) *Refrigerant 22.* Monochlorodifluoromethane ($CH Cl F_2$) boils at -41.4 F at 14.696 psia. It has been widely applied to the lower-temperature refrigeration applications. Refrigerant 22 has a greater latent heat of vaporization than Refrigerant 12 and its vapor is also more dense. It is semi-miscible with oil, being highly miscible at room temperature but relatively immiscible at temperatures well below 0 F.

(*g*) *Refrigerant 113.* Trichlorotrifluoroethane ($C Cl_2 F C Cl F_2$) boils at the relatively high temperature of 117.6 F at 14.696 psia. It has been used to some extent as a refrigerant in centrifugal-compressor systems.

(*h*) *Refrigerant 114.* Dichlorotetrafluoroethane ($C Cl F_2 C Cl F_2$) boils at 38.4 F at 14.696 psia. It has been principally used as a refrigerant with fractional-horse-power, rotary-vane type compressors.

Of the eight halocarbon refrigerants discussed above, Refrigerants 11, 12, and 22 are the most important commercially. Refrigerant 11 has been extensively used in large-capacity, centrifugal-compressor type, water-chilling machines used for air conditioning large buildings.

Refrigerant 12 is probably the most widely used refrigerant today. Important applications include household refrigerators, small commercial systems for stores, restaurants, etc., and smaller-capacity air conditioning systems. Refrigerant 22 is also extensively used. Applications include household freezers and various low-temperature, multi-stage compressor systems. In addition, Refrigerant 22 is widely used in self-contained, small air conditioning equipment.

4.5. REACTION OF REFRIGERANTS WITH MOISTURE

A refrigeration system should be charged with only the pure compound intended as the refrigerant. However, it is extremely difficult to prevent traces of moisture from existing in systems. Various refrigerants react differently in the presence of water.

The two principal effects resulting from moisture in a refrigeration system are corrosion and freeze-up of expansion devices. Almost all refrigerants will form corrosive acids or bases in the presence of water; these corrosive compounds may be highly destructive to valves, seals, and other metallic parts.

When water comes in contact with a refrigerant, it may go into solution with the refrigerant or it may remain as free water. Water is soluble in ammonia in all proportions, so that free water does not occur in ammonia systems. Water is also highly soluble in carbon dioxide and sulfur dioxide. Figure 4.1 shows the solubility of water

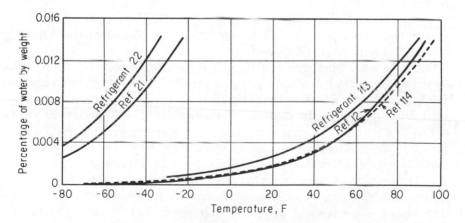

Figure 4.1 Solubility of water in several refrigerants. (Reproduced by permission from *Air Conditioning, Refrigerating Data Book*, Design Volume, 1957-58, p. 10-02.)

in several of the halocarbon refrigerants. Any water in excess of the soluble amount would occur as free water. Water is relatively insoluble in all of the compounds shown in Fig. 4.1, but particularly so in Refrigerants 12, 113, and 114. Freeze-up of expansion valves and formation of ice in evaporators result from presence of free water only. Refrigerant 12 systems are particularly susceptible to this difficulty.

An important practical problem is to keep a refrigeration system dry. Any system should be thoroughly dehydrated before being charged with a refrigerant. The refrigerant and the compressor oil used should be moisture-free. Every precaution should be observed to prevent possible entry of moisture, both during erection and during operation of a system. As a further protection, many systems, particularly those using Refrigerant 12, employ permanently installed dryers in the system. The dryer element typically uses a desiccant such as silica gel or activated alumina.

4.6. REACTION OF REFRIGERANTS WITH OIL

In systems using reciprocating compressors, the refrigerant comes into direct contact with lubricating oil. Depending upon the mutual solubility characteristics of the refrigerant and of the oil, some oil may go into solution with the refrigerant vapor and some refrigerant may be dissolved into the oil. Some oil may also be picked up physically by the vapor without going into solution.

Reaction of refrigerants with oils is important in several ways. Oil carried from the compressor into other parts of the system may reduce heat transfer in the condenser and evaporator. The pressure-temperature characteristic of the refrigerant-oil solution may differ from that of the pure refrigerant. Also, lubrication of the compressor may be affected, since viscosity of the oil changes by dilution with the refrigerant.

Some refrigerants are fully miscible with oil. The solution formed is homogeneous, and oil tends to stay with the refrigerant throughout the system. Miscible refrigerants include Refrigerants 11, 12, 21, and 113.

On the other hand, many refrigerants are immiscible with oil. If oil is picked up by the refrigerant vapor, a heterogeneous mixture is formed. In such a case, the oil is mechanically separable from the refrigerant. Refrigerants which are immiscible with oil include ammonia, sulfur dioxide, carbon dioxide, and Refrigerants 13 and 14.

A few refrigerants are intermediate in their action with oil. Such a refrigerant may be fully miscible with oil above some critical solution temperature but only partially miscible at lower temperatures. Examples are Refrigerants 22 and 114.

The problem of oil miscibility is primarily one of system design. A system using an immiscible refrigerant must be equipped with an efficient oil separator after the compressor. Systems using a miscible refrigerant must be designed for sufficient velocity of flow in evaporators and suction lines. The greatest problem of design occurs with refrigerants which are intermediate in miscibility. Oil logging of the evaporator may occur unless special provisions are made.

5

ABSORPTION REFRIGERATION

5.1. INTRODUCTION

Refrigeration by a mechanical vapor-compression system may be an efficient method. However, the energy input is shaft work, which is high-grade energy and therefore expensive. The relatively large amount of work is required because in compression the vapor undergoes a large change in specific volume.

If means were available for raising the pressure of the refrigerant without appreciably altering its volume, the amount of work required could be greatly reduced. Example 5.1 shows how this may be done by absorption of the refrigerant vapor by a liquid.

> **Example 5.1.** In Example 3.1, let the ammonia vapor leaving the evaporator be absorbed by water, producing a liquid solution with a concentration x of 0.4 lb of ammonia per lb of solution. Assume that the specific volume of the liquid solution is 0.0184 cu ft per lb. Compare the theoretical work required to raise the pressure of the liquid solution from 30.42 to 180.6 psia with that required for vapor compression.
>
> SOLUTION: We will assume that the liquid solution is incompressible. The work required per pound of ammonia is given by

$$W = \frac{\Delta P\, v}{xJ} = \frac{(180.6 - 30.4)(144)(0.0184)}{(0.4)(778)} = 1.28 \text{ Btu/lb of ammonia}$$

By Example 3.1, the theoretical work required for vapor compression is

$$_3W_4 = h_4 - h_3 = 112.2 \text{ Btu/lb of ammonia}$$

Thus, the work requirement for the liquid solution is approximately one per cent of that required for vapor compression.

Example 5.1 demonstrates the principal advantage of the absorption refrigeration cycle. Only a small amount of work is needed. However, we will find that a heat input many times greater than the work input of the mechanical vapor-compression cycle is required. If heat is sufficiently cheap, the absorption cycle may be attractive economically.

Before discussing absorption refrigeration cycles, we will first study some fundamental characteristics of binary mixtures. Our understanding of absorption refrigeration is greatly improved if we are thoroughly acquainted with the behavior of the working fluids.

5.2. ELEMENTARY PROPERTIES OF BINARY MIXTURES

Mixtures are formed by combining two or more pure substances. A mixture may be classified as either *homogeneous* or *heterogeneous*. A *homogeneous* mixture is uniform in composition, regardless of how small the particles are. The various properties such as density, pressure, and temperature are uniform throughout the mixture. A homogeneous mixture cannot be separated into its constituents by pure mechanical means such as settling or centrifuging. Practically all gaseous mixtures are homogeneous because it is the nature of gases to diffuse into each other. An example is moist air which is a homogeneous mixture of dry air and water vapor.

A *heterogeneous* mixture is non-uniform in composition. It can be separated by ordinary mechanical means. An example of a heterogeneous mixture is a fog or cloud, which is a mixture of liquid water and saturated air.

In general, substances may be combined to form heterogeneous or homogeneous mixtures in solid, liquid, or vapor phases. In this chapter we will be concerned only with *homogeneous liquid* and *homogeneous vapor* mixtures, and we will limit our study to binary mixtures. Bosnjakovic* has extensively covered thermodynamics of binary mixtures.

The thermodynamic state of a mixture cannot be established by pressure and temperature alone as may be done with a pure substance. In our more general treatment of binary mixtures in this chapter, we will consider the quantitative composition

---

*Fran Bosnjakovic, *Technical Thermodynamics*, trans. Perry L. Blackshear, Jr. (New York: Holt, Rinehart & Winston, 1965).

in terms of the *concentration* x which is the mass of one arbitrary constituent divided by the mass of the mixture. Knowledge of p, t, and x enables us to establish the thermodynamic state of the mixture.

An important characteristic of a mixture is its *miscibility*. A mixture is *miscible* if through any arbitrary range of concentration values, a homogeneous mixture is formed. Thus, a *non-miscible* mixture is a heterogeneous mixture. In Chapter 4, we briefly considered mixtures of oil and refrigerants. Thus, an oil-ammonia mixture is heterogeneous or non-miscible, whereas an oil-Refrigerant 12 mixture is homogeneous or miscible. Some mixtures are miscible under certain conditions but otherwise non-miscible. Miscibility is materially influenced by temperature. Thus at low temperatures an oil-Refrigerant 22 mixture is non-miscible, while at higher temperatures the mixture becomes miscible.

Binary mixtures suitable for an absorption system must be completely miscible in both the liquid and vapor phases. There can be no *miscibility gap* or range of concentration values where a heterogeneous mixture would exist.

Two important phenomena occurring with the mixing of two liquids are the change of volume and change of temperature of the constituents during mixing. Figure 5.1(a) shows schematically a divided container holding x lb of Liquid A and

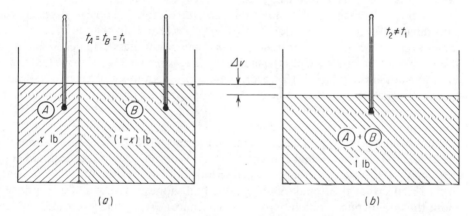

Figure 5.1 Schematic illustrations of change of volume and change of temperature in the mixing of two liquids.

$(1 - x)$ lb of Liquid B. Each liquid is at the same temperature t_1. The volume of the pound of constituents is

$$v_1 = x v_A + (1 - x) v_B$$

We assume that the dividing wall is removed and the two liquids are thoroughly mixed. We generally observe that $v_2 \neq v_1$. If we repeated the experiment with different liquids and with various concentration values, we would find that, in some cases, contraction occurred. In other cases, Δv would be positive. We would find that no general rule was followed, and that experimentation was necessary to find the result.

Our other important observation in the experiment of Fig. 5.1 would be that, in general, $t_2 \neq t_1$. In some cases, we would find that a warming effect occurred; in others, we might observe a decrease of temperature. The effect may be expressed in terms of a *heat of solution* ΔH_x. In the experiment we could measure the heat which we had to remove or add in order to keep the temperature constant. If the mixing were carried out at *constant pressure*, ΔH_x would be strictly related to the enthalpy of the mixture. For the original components, we have

$$h_1 = xh_A + (1 - x)h_B$$

and after the mixture

$$h_2 = h_1 + \Delta H_x = xh_A + (1 - x)h_B + \Delta H_x \tag{5.1}$$

Equation (5.1) allows calculation of the specific enthalpy in Btu per lb of mixture for a solution of known concentration at some fixed temperature and pressure, if the enthalpies of the pure components and the isothermal heat of solution are known.

The pressure-temperature relationships for a boiling-liquid binary mixture or a condensing-vapor binary mixture are of especial importance in absorption refrigeration. We will discuss these relationships through another imaginary experiment where we again restrict ourselves to using only homogeneous solutions.

Figure 5.2 schematically shows the experiment and the observed results on t-x coordinates. We start with a liquid solution (State 1) and slowly add heat to it, at all times keeping the pressure constant. (We may imagine a glass cylinder so that we may view the solution inside.) Until the temperature t_2 is reached we observe that the solution remains entirely liquid. However, upon further addition of heat we find that the piston rises above the liquid indicating that vaporization occurs. If we would stop the experiment at the condition shown in Fig. 5.2(b), and chemically analyze the vapor and liquid components, we would make an interesting discovery. The concentration of the liquid and vapor would be different, and furthermore, both concentrations would be different from the original concentration x_1. We would find that the liquid concentration $x_3 < x_1$, while the vapor concentration $x_4 > x_1$. If we added more heat, we would notice that the liquid would gradually disappear. We would also notice that both the liquid concentration and the vapor concentration had decreased. When we reached the state-point (5), only vapor would remain, and here we would observe that $x_5 = x_1$. Further heating would result in superheating of the vapor at constant concentration.

If we repeated the experiment with different initial concentration values, but with the same pressure, our experimental results would give the equilibrium-boiling and equilibrium-condensing lines shown in Fig. 5.2(c). If we repeated the experiments but with different pressures, we would obtain the results shown in Fig. 5.2(d). If we reversed the experiments, started with a superheated vapor, and then removed heat, we would have the results shown by Fig. 5.2(e).

Thus, binary mixtures, as contrasted with pure substances, do not have a single boiling temperature or condensing temperature for a given pressure. The equilibrium

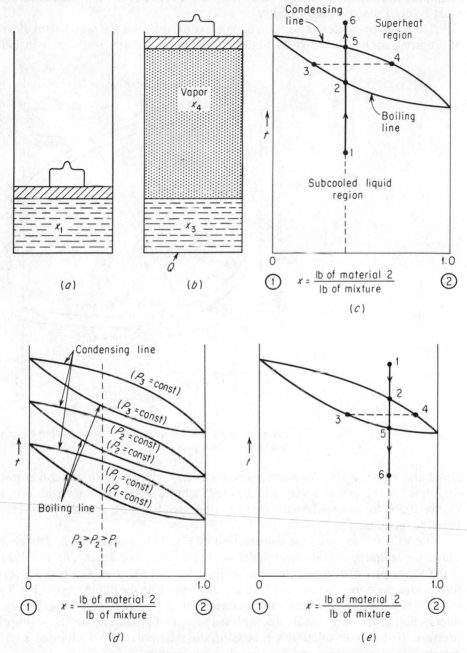

Figure 5.2 Evaporation and condensation characteristics for a homogeneous binary mixture.

or saturation temperature is also dependent on the concentration. These relationships must be determined experimentally.

The enthalpy-concentration diagram (*h-x* coordinates) is the most useful diagram of properties for a binary mixture. Figure 5.3 shows a schematic *h-x* diagram including

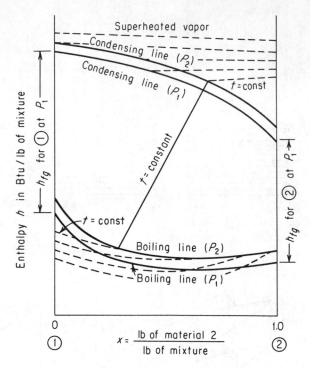

Figure 5.3 Schematic *h-x* diagram for liquid and vapor regions for a homogeneous binary mixture.

liquid and vapor regions for a homogeneous binary mixture. Enthalpy-concentration diagrams for two actual binary mixtures are included in the envelope with this text. Figure E-3 is for ammonia-water combinations, while Fig. E-4 is for water-lithium bromide mixtures.

We will now examine the construction of Fig. E-3 in some detail. The chart is drawn for saturated or equilibrium solutions of ammonia and water. The liquid region is the lower part of the diagram. Equilibrium liquid lines (boiling lines) are shown for various pressures. Isothermal lines are also shown in the liquid region. If a liquid solution is known to be saturated, the state-point may be located on Fig. E-3 by the intersection of the temperature line with the saturated liquid line for the appropriate pressure. If the liquid solution is subcooled, the state-point may be located approximately if the temperature and concentration are known.

No temperature lines are shown in the vapor region of Fig. E-3. However, a saturated vapor state in equilibrium with a known saturated liquid state may be located through use of the equilibrium construction lines. The procedure may be illustrated by an example for saturation conditions of 100 psia and 260 F. A vertical line is drawn upward from the saturated liquid point to the equilibrium construction line for 100 psia. From this intersection, a horizontal line is drawn to the saturated vapor line for 100 psia. This intersection is the vapor state-point. Thus the properties of the saturated vapor state are known ($t = 260$ F, $p = 100$ psia, $h = 906$ Btu per lb, $x = 0.63$ lb of NH_3 per lb of mix). Furthermore, a straight line drawn from the saturated liquid state to saturated vapor state is an isotherm for the liquid-vapor region.

5.3. ELEMENTARY STEADY-FLOW PROCESSES WITH BINARY MIXTURES

In industrial systems using binary homogeneous mixtures, several types of thermodynamic processes may be involved. Some of the common processes are adiabatic mixing of fluid streams, mixing of fluid streams with heat exchange, vaporization and condensation, and throttling. Several specific steady-flow processes will be considered in this section. Study of these separate processes will help us later to understand more clearly how the absorption refrigeration system functions. The examples will also illustrate how helpful the h-x diagram may be in the solution of problems.

(a) *Adiabatic mixing of two streams.* Figure 5.4(a) shows schematically a mixing chamber where two fluid streams are brought together adiabatically. In the following analysis, as well as in all of our subsequent processes, we will use the following

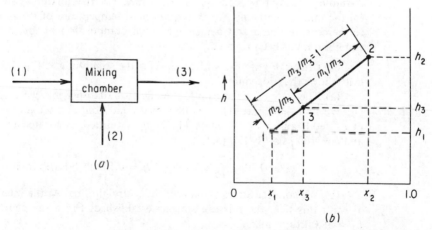

Figure 5.4 Adiabatic mixing of two binary fluid streams under steady-flow conditions.

symbols: specific enthalpy h in Btu per lb of mixture, concentration x in lb of one constituent per lb of mixture, and mass rate of flow m in lb of mixture per unit time.

We may write three fundamental equations for the system of Fig. 5.4. These are

$$m_1 h_1 + m_2 h_2 = m_3 h_3$$

$$m_1 + m_2 = m_3$$

$$m_1 x_1 + m_2 x_2 = m_3 x_3$$

By elimination of m_3, we obtain

$$\frac{m_1}{m_2} = \frac{h_2 - h_3}{h_3 - h_1} = \frac{x_2 - x_3}{x_3 - x_1} \tag{5.2}$$

Equation (5.2) defines a straight line on the h-x diagram as shown by Fig. 5.4(b). Thus, State (3) must lie on a straight line connecting States (1) and (2) on the h-x diagram. The location of State (3) on the line is determined by the mass ratios. We may also deduce that

$$x_3 = x_1 + \frac{m_2}{m_3}(x_2 - x_1) \tag{5.3}$$

$$h_3 = h_1 + \frac{m_2}{m_3}(h_2 - h_1) \tag{5.4}$$

In the adiabatic mixing of two binary-liquid mixtures, determination of the mixture temperature using the h-x diagram may be somewhat involved, since partial vaporization may occur during mixing.

Example 5.2. A stream of liquid aqua-ammonia ($m = 10$ lb per min, $x = 0.7$ lb NH_3 per lb mix, $t = 60$ F, $p = 100$ psia) is adiabatically mixed with a stream of saturated liquid aqua-ammonia ($m = 5$ lb per min, $t = 200$ F, $p = 100$ psia). Assuming steady-flow conditions, determine (a) the mixture concentration, (b) the mixture enthalpy, (c) the equilibrium temperature of the mixture, and (d) percentage liquid and percentage vapor composition of the mixture after equilibrium has been reached.

SOLUTION: Figure E-3 will be used for the solution, and Fig. 5.5 schematically shows the procedure.

(a) The state of the stream for the flow of 10 lb per min we will call State (1). We find that State (1) is a subcooled condition, and we locate the point at $t = 60$ F and $x = 0.7$. State (2), being saturated, is located at $t = 200$ F and $p = 100$ psia. By Eq. (5.3),

$$x_3 = 0.70 + \frac{5}{15}(0.26 - 0.70) = 0.553 \text{ lb } NH_3/\text{lb mix}$$

(b) We connect States (1) and (2) by a straight line. At the intersection of x_3 with this line, the mixture state is established. From the chart, we read $h_3 = 38.0$ Btu/lb mix.

(c) We know that $P_3 = 100$ psia. We observe that State (3) lies above the equilibrium liquid line for 100 psia. Thus, we know that the state is a mechanical

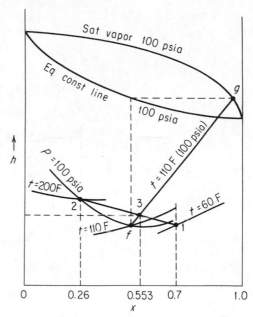

Figure 5.5 Schematic h-x diagram for Example 5.2.

mixture of saturated liquid and saturated vapor. With a straight-edge, we find by trial the line $\overline{f3\,g}$ of Fig. 5.5. This line is the isotherm for the liquid-vapor region connecting States (f) and (g). At (f), we read $t_f = t_g = t_3 = 110$ F.

(d) Since State (3) is a mixture of liquid and vapor, the usual mixing equations apply. Thus,

$$\frac{m_g}{m_3} = \frac{x_3 - x_f}{x_g - x_f} = \frac{\overline{f3}}{\overline{fg}} = 0.033$$

Therefore the mixture consists of 96.7 per cent liquid and 3.3 per cent vapor.

(b) Mixing of two streams with heat exchange. In the mixing of binary fluids, it is often necessary to remove heat or add heat in order to obtain the desired final condition. An example process occurs in the absorber of an absorption refrigeration system. Figure 5.6(a) shows the schematic steady-flow problem where heat *removal* is assumed. The fundamental mixing equations are

$$m_1 h_1 + m_2 h_2 = m_3 h_3 + q$$

$$m_1 + m_2 = m_3$$

$$m_1 x_1 + m_2 x_2 = m_3 x_3$$

From the above equations, we find that Eq. (5.3) is still applicable, but instead of Eq. (5.4) we have

$$h_3 = h_1 + \frac{m_2}{m_3}(h_2 - h_1) - \frac{q}{m_3} \tag{5.5}$$

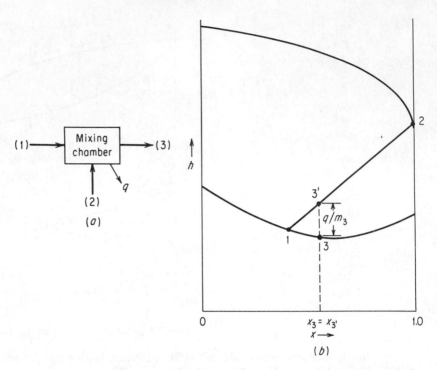

Figure 5.6 Mixing of two binary fluid streams under steady-flow conditions with heat exchange.

With heat removal during mixing, State (3) lies at a vertical distance q/m_3 below the state-point obtained for adiabatic mixing. Figure 5.6(b) shows a schematic h-x diagram for the process.

Example 5.3. One lb per min of saturated ammonia vapor ($x = 1.0$) at 30 psia is mixed with 10 lb per min of saturated liquid aqua-ammonia at 30 psia and 100 F. The final mixture state is saturated liquid at 30 psia. Find (a) the concentration, temperature, and enthalpy of the mixture state, and (b) the heat removal in Btu per min.

SOLUTION: Figure 5.6(b) shows the schematic solution.
(a) By Eq. (5.3),

$$x_3 = 0.345 + \frac{1}{11}(1.0 - 0.345) = 0.405 \text{ lb NH}_3/\text{lb mix}$$

We draw the straight line $\overline{1\,2}$ and locate State (3'). We then proceed vertically downward to the saturated liquid line for 30 psia and locate State (3). We read $t_3 = 80$ F, $h_3 = -24.0$ Btu/lb mix.
(b) By Figure 5.6(b), we see that

$$q/m_3 = h_{3'} - h_3 = 54 - (-24) = 78 \text{ Btu/lb mix}$$

Thus

$$q = (11)(78) = 858 \text{ Btu/min}$$

(*c*) *Simple heating and cooling processes.* Processes of vaporization are important in all absorption refrigeration systems. In the case of some binary mixtures such as aqua-ammonia, rectification of the vapor is required. We will now consider two simple processes—heating and cooling. We will further see the usefulness of the *h-x* diagram in solving problems.

Figure 5.7(a) shows a simple arrangement of heat exchangers by which we may produce a high-purity vapor from an initially weak-liquid solution. In order to simplify

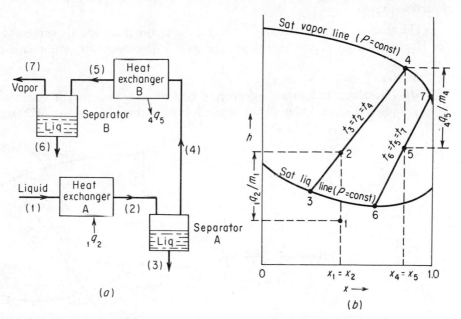

Figure 5.7 Simple heating and cooling processes for steady-flow conditions.

our analysis, we assume that liquid and vapor components are separated after each heat exchanger. For Heat Exchanger A, we obtain for steady-flow conditions

$$_1q_2 = m_1(h_2 - h_1)$$

$$m_1 = m_2$$

$$x_1 = x_2$$

For Separator A, we have

$$m_2h_2 = m_3h_3 + m_4h_4$$

$$m_2 = m_3 + m_4$$

$$m_2x_2 = m_3x_3 + m_4x_4$$

and thus

$$\frac{m_3}{m_2} = \frac{x_4 - x_2}{x_4 - x_3} = \frac{h_4 - h_2}{h_4 - h_3} \tag{5.6}$$

$$\frac{m_4}{m_2} = \frac{x_2 - x_3}{x_4 - x_3} = \frac{h_2 - h_3}{h_4 - h_3} \tag{5.7}$$

Figure 5.7(b) shows the state-points for Heat Exchanger A and Separator A on the h-x diagram. Equations (5.6) and (5.7) show that the fractional components for Separator A may be obtained directly from the diagram. Thus $m_3/m_2 = \overline{2\,4/3\,4}$ and $m_4/m_2 = \overline{3\,2/3\,4}$. Also, the heat requirement $_1q_2/m_1$ may be graphically represented as shown.

The analyses for Heat Exchanger B and Separator B are exactly similar to those for Heat Exchanger A and Separator A. Figure 5.7(b) shows the state-points on the h-x diagram.

(d) *Throttling.* The throttling process occurs in a majority of systems using binary liquids. Figure 5.8(a) shows schematically a throttling valve. We recognize

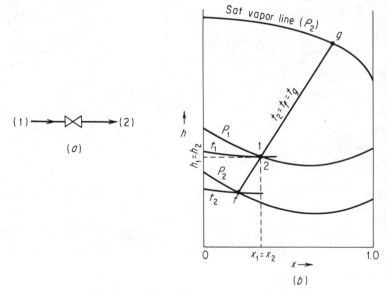

Figure 5.8 Throttling of a binary liquid mixture under steady-flow conditions.

that $x_1 = x_2$ and that $h_1 = h_2$. On the h-x diagram, Points (1) and (2) are identical. However, in the throttling of a binary liquid, vaporization may occur. Figure 5.8(b) shows the procedure where we assume that the liquid and vapor components are in equilibrium. With a straight-edge we find by trial the isotherm $\overline{f\,2\,g}$. We find that t_2

may be considerably less than t_1. The fractional components of liquid and vapor may be obtained from the segment ratios of the line $\overline{f\,2\,g}$.

5.4. RECTIFICATION OF A BINARY MIXTURE

In the previous section we discussed simple steady-flow processes of heating and cooling. In Fig. 5.7, Heat Exchanger A performs the function of a *generator*, while Heat Exchanger B performs the function of a *dephlegmator*. Through the combination shown, a vapor of arbitrarily high purity may be produced at the exit of the dephlegmator.

The combination of Fig. 5.7, however, is inadequate for the separation of a binary mixture. To achieve efficient separation we must introduce a *rectifying column* between the generator and the dephlegmator. In this section we will discuss in some detail the fundamental principles of a rectifying column. This discussion will enable us to understand better the aqua-ammonia absorption system and also to understand better in Chapter 7 how liquefied-gas mixtures are separated.

Figure 5.9(a) shows schematically a rectifying column between a generator and a dephlegmator. The rectifying column includes many capped plates or perforated

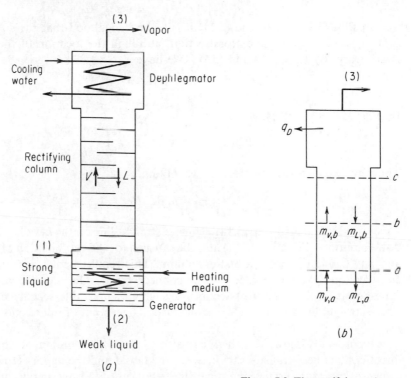

Figure 5.9 The rectifying column.

plates. A vapor solution V ascends through the column and a liquid solution L descends. The purpose of the plates is to bring the liquid and vapor solutions into intimate direct contact. The vapor rising through the tower has some liquid condensed from it, while the liquid trickling downward has some vapor evaporated from it. With regard to composition, the ascending vapor develops a progressively higher concentration while the descending liquid solution develops a progressively weaker concentration.

In Fig. 5.9(b) the dephlegmator and an arbitrary part of the column are shown. Between any two cross-sections such as a and b, for steady-state conditions we have

$$m_{V,a} + m_{L,b} = m_{V,b} + m_{L,a}$$

or

$$m_{V,a} - m_{L,a} = m_{V,b} - m_{L,b} \tag{5.8}$$

Equation (5.8) must hold for every part of the column. Thus, we have, for any cross-section,

$$m_V - m_L = \text{a constant} = m_3 \tag{5.9}$$

In a similar fashion, we may show that

$$m_V x_V - m_L x_L = m_3 x_3 \tag{5.10}$$

$$m_V h_V - m_L h_L = m_3 h_3 + q_D \tag{5.11}$$

In Eqs. (5.9), (5.10), and (5.11) it is to be understood that the subscripts V and L always refer to the same cross-section, although the location of the cross-section is arbitrary. By Eqs. (5.9) and (5.10), we have

$$\frac{m_V}{m_3} = \frac{x_3 - x_L}{x_V - x_L} \tag{5.12}$$

By Eqs. (5.9) and (5.11), we have

$$\frac{m_V}{m_3}(h_V - h_L) = (h_3 - h_L) + \frac{q_D}{m_3} \tag{5.13}$$

Thus, by Eqs. (5.12) and (5.13), we obtain

$$\frac{(x_3 - x_L)}{(x_V - x_L)}(h_V - h_L) = (h_3 - h_L) + \frac{q_D}{m_3} \tag{5.14}$$

As shown by Fig. 5.10, Eq. (5.14) defines a straight line on the h-x diagram for steady-flow conditions. Equation (5.14) must be satisfied for the combination of dephlegmator and any connected portion of the column. This requirement leads to the graphical construction shown by Fig. 5.11. If we connect, by straight lines, the vapor and liquid state-points for the same cross-section, all such lines must intersect at a common point. These straight lines are called *operating lines*. Their point of intersection P_1 is called a *pole*.

In order to know what happens at a plate of the column, we must know the direct-contact heat transfer rate between the liquid and the vapor. However, we may draw an important conclusion regarding rectification by making an analysis using a

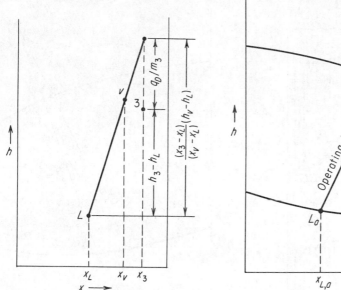

Figure 5.10 Schematic representation of Eq. (5.14) on h-x diagram.

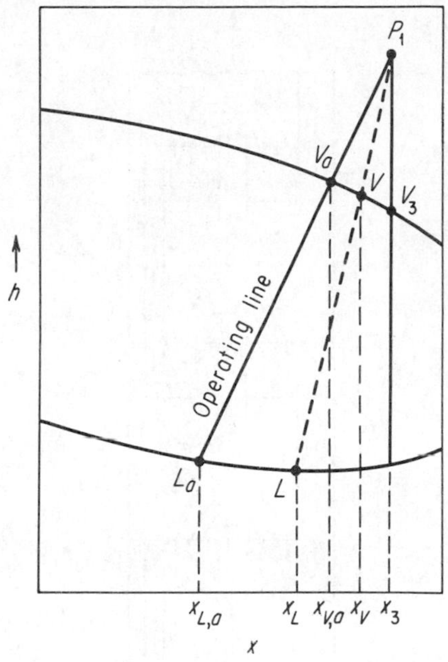

Figure 5.11 Rectification pole and operating lines.

diagram similar to Fig. 5.11. In Fig. 5.12(a), let us assume that Sections a and b represent a portion of a column containing one contact plate. In the limiting case, the vapor passing Section b may be brought to the same temperature as the liquid passing Section a. In a finite case, the temperature of the vapor at Section b would be greater than the temperature of the liquid at Section a. This would require that the vapor state V_b be located to the left of the intersection of the isotherm $t_{L,a}$ and the saturated vapor line. *Thus, for separation to occur in a rectifying column, the operating line for each cross-section must be steeper than the corresponding mixture-region isotherm.* The pole of the operating lines must be located high enough for this requirement to be met for each and every part of the column.

The separation arrangement of Fig. 5.9(a) may be improved upon substantially if the strong liquid, State (1), is introduced part of the way up the column instead of directly into the generator. Such an arrangement combines an *exhausting column* with the generator. We will examine the functioning of a simple exhausting column.

Figure 5.13(a) schematically shows an exhausting column. In a procedure identical to that used for the rectifying column, we may show that for any cross-section of the column

$$m_L - m_V = m_2 \tag{5.15}$$

$$m_L x_L - m_V x_V = m_2 x_2 \tag{5.16}$$

$$m_L h_L - m_V h_V = m_2 h_2 - q_G \tag{5.17}$$

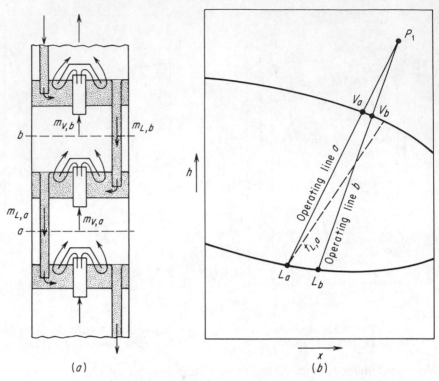

Figure 5.12 Relation of operating lines to mixture-region isotherms.

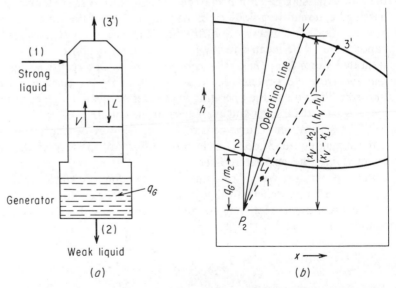

Figure 5.13 The exhausting column.

From Eqs. (5.15)–(5.17), we obtain

$$\frac{(x_V - x_2)}{(x_V - x_L)}(h_V - h_L) = (h_V - h_2) + \frac{q_G}{m_2} \tag{5.18}$$

Equation (5.18) must be satisfied for every cross-section of the column. This requirement leads to the graphical construction shown in Fig. 5.13(b). All cross-section operating lines must intersect at a common point P_2 located at a distance q_G/m_2 below the State-point (2).

The heat requirement q_G/m_2 for the generator may be substantially decreased if the strong liquid is preheated in a heat exchanger by the weak liquid solution. The concentration x of each solution remains constant throughout the heat exchanger. Point (1) in Fig. 5.13(b) is raised, thus also raising the pole P_2 vertically, and thereby decreasing the quantity q_G/m_2.

Figure 5.14(a) schematically shows a double rectifying column. The double column is formed by coupling the exhausting column and the simple rectifying column discussed previously. Vapor boiled off in the generator rises through the exhausting column and is purified to a condition (3′). The vapor is further concentrated in the rectifying column and dephlegmator to a desired final condition (3). Strong liquid

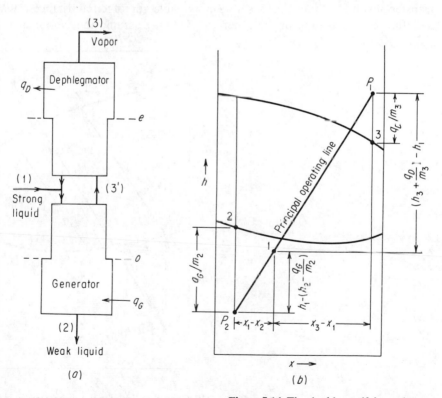

Figure 5.14 The double rectifying column.

solution, State (1), is mixed with liquid leaving the upper column and this mixture is stripped of some material in the exhausting column.

We have previously analyzed the separate columns, and will now examine the combination. We may write the following fundamental equations:

$$m_1 = m_2 + m_3 \tag{5.19}$$

$$m_1 x_1 = m_2 x_2 + m_3 x_3 \tag{5.20}$$

$$m_1 h_1 + q_G = m_2 h_2 + m_3 h_3 + q_D \tag{5.21}$$

By Eqs. (5.19)–(5.21), we may show that

$$\frac{\left(h_3 + \dfrac{q_D}{m_3}\right) - h_1}{x_3 - x_1} = \frac{h_1 - \left(h_2 - \dfrac{q_G}{m_2}\right)}{x_1 - x_2} \tag{5.22}$$

Equation (5.22) requires that the two poles P_1 and P_2 and State-point (1) lie on the same straight line on the h-x diagram. This requirement is evident from Fig. 5.14(b). This straight line is called the *principal operating line*.

The location of the operating lines on the h-x diagram for the exhausting column and the rectifying column depends upon the apparatus and upon the rates of heat transfer involved. Figure 5.15 shows a possible set of circumstances. Because of agitation due to boiling in the generator, we may assume that vapor at a state V_o

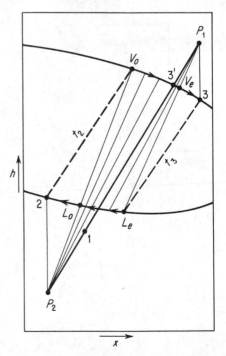

Figure 5.15 Operating lines for the double rectifying column.

leaves the generator in equilibrium with the weak liquid. The liquid state L_o leaving the exhausting column and entering the generator would be given on the h-x diagram by the intersection of the operating line $\overline{P_2\,V_o}$ with the saturated liquid line. Bosnjakovic* has shown that the least number of plates are required in the two columns if the vapor state (3') leaving the exhausting column lies on the principal operating line $\overline{P_2\,1\,P_1}$. Thus all operating lines for the exhausting column would lie between the lines $\overline{P_2\,L_o\,V_o}$ and $\overline{P_2\,1\,P_1}$.

In the upper column it is more difficult to estimate the state of the liquid leaving the dephlegmator and entering the rectifying column. This state-point is dependent upon the type of heat exchanger used as the dephlegmator. If we assume that liquid enters the rectifying column in equilibrium with the vapor state (3), we would have the circumstances shown in Fig. 5.15. All operating lines for the upper column would lie between the lines $\overline{P_2\,1\,P_1}$ and $\overline{L_e\,V_e\,P_1}$.

The position of the principal operating line $\overline{P_2\,1\,P_1}$ fixes the generator heat requirement q_G/m_2 and the dephlegmator cooling requirement q_D/m_3 for fixed values of x_2 and x_3. Both quantities decrease as the principal operating line becomes more flat. However, the principal operating line cannot be drawn arbitrarily. The poles P_1 and P_2 must be located so that all operating lines are steeper than the mixture region isotherms for the corresponding cross-sections.

5.5. SIMPLE THEORETICAL ABSORPTION REFRIGERATION SYSTEM

Having some knowledge of the characteristics of binary mixtures, we will now consider the absorption refrigeration system. Figure 5.16 shows a schematic arrangement of equipment. For simplicity we assume that the absorbent does not vaporize in the generator. Thus only pure refrigerant flows through the condenser and evaporator, and these two components may be identical to those used in a mechanical vapor-compression system. The vapor leaving the evaporator is mixed with a weak liquid solution in the absorber. We observed in Example 5.3 that by the removal of heat the refrigerant vapor could be absorbed by the weak liquid, producing a liquid solution stronger in the refrigerant. The pressure of the liquid solution is then raised to the generator pressure by the pump. By the addition of heat in the generator, refrigerant vapor is driven out of the liquid solution. A heat exchanger is placed in the liquid solution circuit between the generator and absorber. The generator and condenser are on the high-pressure side of the system, while the evaporator and absorber are on the low-pressure side.

An absorption refrigeration system may be properly called a vapor compression system. We see that several components are required to perform the function of the compressor in a mechanical vapor-compression system.

*Bosnjakovic, *Technical Thermodynamics*, p. 182.

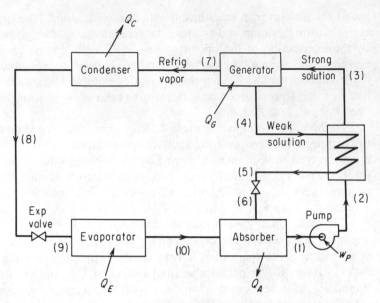

Figure 5.16 Simple absorption refrigeration system.

The system shown in Fig. 5.16 is not intended to represent an ideal type of absorption system. The ideal system would be completely reversible; this is not theoretically possible in Fig. 5.16. In Chapter 3 we found that the Carnot cycle gave the maximum coefficient of performance for a mechanical vapor-compression system. We will use Bosnjakovic's* method to establish the maximum attainable C.O.P. for an absorption system. Figure 5.17 shows the energy transfers to and from the fluids of an absorption system. The generator heating medium adds heat Q_G to the system. The pump adds work W_P. The substance to be cooled in the evaporator adds heat Q_E to the absorption system. The system rejects heat to the environment (cooling water or atmospheric air) in the absorber (Q_A) and in the condenser (Q_C). We will combine these last quantities into the waste heat

$$Q_o = Q_A + Q_C$$

By the first law of thermodynamics

$$Q_o = Q_G + Q_E + W_P \tag{5.23}$$

We will assume that the generator heating medium temperature T_G, the refrigerated substance temperature T_E, and the environmental temperature T_o are constants.

The fluids within the absorption system circulate in a closed cycle. For steady-state operation, the total entropy change of the fluids is zero. The changes of entropy for the overall system are only those occurring externally to the fluids of the absorption system. The entropy change for the generator heating medium is $\Delta S_G = -Q_G/T_G$,

*Bosnjakovic, *Technical Thermodynamics*, pp. 268–270.

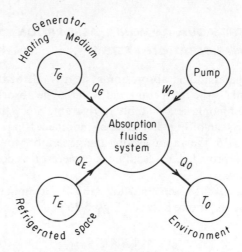

Figure 5.17 External energy transfers for an absorption refrigeration system.

for the refrigerated substance $\Delta S_E = -Q_E/T_E$, and for the environment $\Delta S_o = Q_o/T_o$. By the second law of thermodynamics, we must have

$$\Delta S = \Delta S_G + \Delta S_E + \Delta S_o \geq 0$$

or

$$\Delta S = -\frac{Q_G}{T_G} - \frac{Q_E}{T_E} + \frac{Q_o}{T_o} \geq 0 \tag{5.24}$$

By Eqs. (5.23) and (5.24), we obtain

$$Q_G\left(\frac{T_G - T_o}{T_G}\right) \geq Q_E\left(\frac{T_o - T_E}{T_E}\right) - W_P$$

If we assume that W_P is negligible, we have

$$\text{C.O.P.} = \frac{Q_E}{Q_G} \leq \frac{T_E(T_G - T_o)}{T_G(T_o - T_E)}$$

and for a completely reversible system

$$(\text{C.O.P.})_{\text{max}} = \frac{T_E(T_G - T_o)}{(T_o - T_E)T_G} \tag{5.25}$$

Equation (5.25) shows an interesting result: the maximum attainable coefficient of performance for an absorption system is equal to the coefficient of performance for a Carnot refrigerating cycle working between the temperatures T_E and T_o multiplied by the efficiency of a Carnot engine working between the temperatures T_G and T_o. Equation (5.25) also shows that for a given environmental temperature T_o, the C.O.P. will increase with increase of the generator heating medium temperature T_G and with increase of temperature of the refrigerated region T_E. In practice, however, the C.O.P. is much less than that given by Eq. (5.25).

5.6. THE AQUA-AMMONIA ABSORPTION REFRIGERATION SYSTEM

The aqua-ammonia absorption system is one of the oldest methods of refrigeration. Ammonia is the refrigerant and water is the absorbent. The system may be applicable where ammonia is a suitable refrigerant. Since the absorbent, water, is volatile, the simple system of Fig. 5.16 must be modified. If aqua-ammonia were used in the system of Fig. 5.16, the vapor leaving the generator would contain too much water. Means must be provided to rectify the generator vapor and increase the ammonia concentration.

Figure 5.18 schematically shows an industrial aqua-ammonia system. The principal difference between Figs. 5.16 and 5.18 is that a double rectifying column and

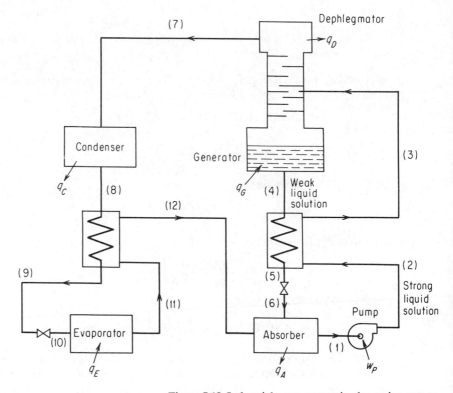

Figure 5.18 Industrial aqua-ammonia absorption system.

a dephlegmator are added in Fig. 5.18. A second heat exchanger is also added. The vapor leaving the generator may contain from five to ten per cent of water vapor. Through the use of the rectifying equipment, the concentration of the vapor entering the condenser can be made almost equal to unity.

The rectifying equipment operates exactly like that described in Section 5.4. Other parts of the system have also been adequately discussed. In Section 5.3 we treated several of the processes involved in Fig. 5.18. We will now consider the solution of a complete system problem.

Example 5.4. The following data are known for the system of Fig. 5.18: condensing pressure, 200 psia; evaporating pressure, 30 psia; generator temperature, 240 F; temperature of vapor leaving dephlegmator, 130 F; and temperature of strong solution entering column, 200 F. The temperature of liquid leaving the condenser is reduced 10 F in the heat exchanger.

Assume equilibrium (saturated) conditions for States 1, 3, 4, 7, 8, and 12. Neglect pressure drop in components and lines. Assume that the system produces 100 tons of refrigeration. Determine (a) thermodynamic properties p, t, x, and h for all state-points of system, (b) mass rate of flow in lb per min for all parts of system, (c) horsepower required for the pump, if a mechanical efficiency of 75 per cent is assumed, (d) system coefficient of performance, (e) system refrigerating efficiency, (f) comparison of coefficient of performance with that of a theoretical vapor-compression ammonia cycle, and (g) an energy balance for entire system in Btu per min.

SOLUTION: It will be necessary to work Parts (a) and (b) concurrently. Table 5.1 shows a tabulation of thermodynamic properties and flow rates for the problem. Figure 5.19 shows the processes on a schematic h-x diagram. The procedure will now be discussed in detail.

All pressures are established from the given data. The temperatures at 3, 4, and 7 are given. Since States 3, 4, and 7 are equilibrium states, we may locate the state-points on Fig. E-3 and read concentration and enthalpy values.

TABLE 5.1

THERMODYNAMIC PROPERTIES AND FLOW RATES FOR EXAMPLE 5.4

State-Point	Pressure P psia	Temperature t F	Concentration x lb NH$_3$/lb mix	Enthalpy h Btu/lb mix	Flow Rate lb mix/min
1	30	79	0.408	−25	262.8
2	200	79	0.408	−24.41	262.8
3	200	200	0.408	109	262.8
4	200	240	0.298	159	221.4
5	200	97	0.298	0	221.4
6	30	97	0.298	0	221.4
7	200	130	0.996	650	41.4
8	200	97	0.996	148	41.4
9	200	86	0.996	137	41.4
10	30	0	0.996	137	41.4
11	30	40	0.996	620	41.4
12	30	57	0.996	631	41.4

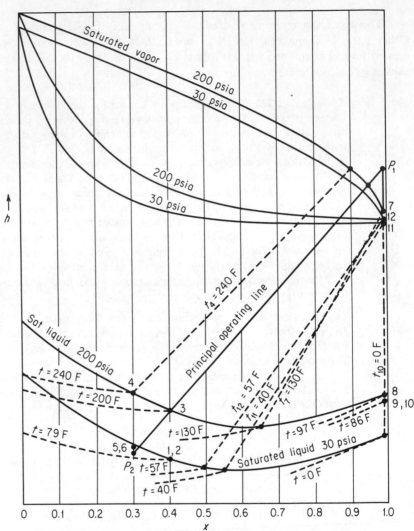

Figure 5.19 Schematic h-x diagram for Example 5.4.

We know that for Points 7 through 12, the concentration is the same. We also know that $x_1 = x_2 = x_3$ and that $x_4 = x_5 = x_6$. Since States 1, 8, and 12 are equilibrium states, we may establish the points on Fig. E-3 from the known pressure and concentration values. Since $t_9 = t_8 - 10$, we may read h_9 at the known values of t and x.

For the combination of evaporator and liquid-suction heat exchanger, we have

$$m_8 h_8 + (100)(200) = m_8 h_{12}$$

$$m_8 = \frac{20,000}{h_{12} - h_8} = \frac{20,000}{483} = 41.4 \text{ lb/min}$$

For the absorber, we have

$$m_{12} + m_6 = m_1$$

$$m_{12}x_{12} + m_6x_6 = m_1x_1$$

Thus

$$m_6 = m_{12}\frac{(x_{12} - x_1)}{(x_1 - x_6)} = (41.4)\frac{(0.588)}{(0.110)} = 221.4 \text{ lb/min}$$

$$m_1 = m_{12} + m_6 = 262.8 \text{ lb/min}$$

Thus, all of the flow rates of the system are known. For the pump, we have

$$h_2 = h_1 + \frac{(P_2 - P_1)v_1}{J}$$

The specific volume v_1 may be found by the empirical relation

$$v_1 = (1 - x_1)v_{\text{H}_2\text{O}} + (0.85)x_1v_{\text{NH}_3}$$

Using steam and ammonia tables, we find

$$v_1 = (0.597)(0.01607) + (0.85)(0.403)\frac{1}{37.53} = 0.0187 \text{ cu ft/lb}$$

and

$$h_2 = -25 + \frac{(200 - 30)(144)(0.0187)}{778} = -24.41 \text{ Btu/lb}$$

For the solution heat exchanger, we have

$$h_5 = h_4 - \frac{m_2}{m_4}(h_3 - h_2) = 159 - \frac{(262.8)(134)}{(221.4)} = 0 \text{ Btu/lb}$$

The temperature at (5) may be read from Fig. E-3 from the known values of x_5 and h_5. Points (5) and (6) are coincidental on Fig. E-3. Since (6) is a subcooled state, its temperature is the same as at (5).

States (9) and (10) are coincidental on the h-x diagram. State (10) is a mechanical mixture of liquid and vapor. Its temperature (0 F) may be found by the method described for Example 5.2(c). The enthalpy at State (11) may be found from the energy balance for the evaporator. State (11) also is a mechanical-mixture state. We find its temperature to be 40 F.

(c) We may find the pump horsepower by

$$\text{Hp} = \frac{m_1(h_2 - h_1)}{42.4\eta_m} = \frac{(262.8)(0.59)}{(42.4)(0.75)} = 4.88$$

(d) Before calculating the coefficient of performance we must first establish the generator heat requirement. The quantities q_G/m_4 and q_D/m_7 may be immediately found once the principal operating line is drawn. However, additional assumptions regarding the column must be made. Referring to the discussion shown in Sec. 5.4 and Fig. 5.15, let us assume that the state of the vapor in the column lies on the principal operating line at the location where the strong solution enters the column. Let us further assume that this vapor state is saturated with a temperature 10 F higher than that of the strong solution which is entering (t_3). These assumptions allow the principal operating line to be constructed. By Fig. E-3, we obtain

$$q_G/m_4 = 173 \text{ Btu/lb} \qquad \text{and} \qquad q_D/m_7 = 121 \text{ Btu/lb}$$

Thus, we have

$$q_G = (221.4)(173) = 38,302 \text{ Btu/min}$$

$$q_D = (41.4)(121) = 5009 \text{ Btu/min}$$

If we neglect the pump work,

$$\text{C.O.P.} = \frac{q_E}{q_G} = \frac{20,000}{38,302} = 0.522$$

(e) The ideal C.O.P. may be calculated by Eq. (5.25). If we assume an environmental temperature of 79 F, then for $t_G = 240$ F and $t_E = 0$ F

$$(\text{C.O.P.})_{\max} = \frac{(460)(161)}{(79)(700)} = 1.34$$

and

$$\eta_R = \frac{\text{C.O.P.}}{(\text{C.O.P.})_{\max}} = \frac{0.522}{1.34} = 0.39$$

(f) For a theoretical ammonia vapor-compression cycle operating with an evaporating temperature of 0 F, and a condensing temperature of 96 F, we have, by Eq. (3.6), that the C.O.P. = 3.82, which is about seven times the C.O.P. for the absorption system.

(g) Calculations for an energy balance are shown in Table 5.2.

TABLE 5.2

ENERGY BALANCE CALCULATIONS FOR EXAMPLE 5.4

Component	Calculations	Btu/min	
		Gains	Losses
Absorber	$q_A = m_6 h_6 + m_{12} h_{12} - m_1 h_1 = $ . .	...	32,694
Pump	$W_P = m_1(h_2 - h_1) = $	155	...
Generator	$q_G = m_4 q_G / m_4 = $	38,302	...
Dephlegmator	$q_D = m_7 q_D / m_7 = $	...	5009
Condenser	$q_C = m_7(h_7 - h_8) = $	...	20,783
Evaporator	Given	20,000	...
Totals	. . . .	58,457	58,486

5.7. THE LITHIUM BROMIDE-WATER ABSORPTION SYSTEM

In recent years the lithium bromide-water absorption system has become prominent in refrigeration for air conditioning. Water is the refrigerant while the absorbent is lithium bromide. Pure lithium bromide is a solid, but when it is mixed with sufficient water, homogeneous liquid solutions are formed.

The outstanding feature of the system is the non-volatility of the lithium bromide. In the generator, only water vapor is driven off. Therefore no rectifying equipment is required. Compared to the aqua-ammonia system, the lithium bromide-water system is simpler and operates with a higher coefficient of performance. The primary disadvantage of the system is its limitation to relatively high evaporating temperatures since the refrigerant is water.

Example 5.5. The following data are known for a lithium bromide-water system of the type shown in Fig. 5.16: condensing temperature = 100 F, evaporating temperature = 40 F, temperature of strong solution leaving absorber = 100 F, temperature of strong solution entering generator = 180 F, generator temperature = 200 F.

Assume saturated conditions for States 3, 4, 8, and 10. Neglect pressure drop in components and lines. Assume a system capacity of one ton of refrigeration. Determine (a) thermodynamic properties p, t, x, and h for all necessary state-points of system, (b) mass rate of flow in lb per min for each part of system, (c) system coefficient of performance, (d) system refrigerating efficiency, and (e) steam consumption for the generator in lb per (hr) (ton), if saturated steam at 220 F is used.

SOLUTION: Parts (a) and (b) will be worked concurrently. Table 5.3 shows a partial tabulation of thermodynamic properties and flow rates for the problem. Properties of pure water were obtained from Table A.1 while solution properties were read from Fig. E-4. It is important to note that Fig. E-4 expresses concentration in terms of the absorbent lithium bromide.

TABLE 5.3

THERMODYNAMIC PROPERTIES AND FLOW RATES FOR EXAMPLE 5.5

State-Point	Pressure p mm. Hg.	Temperature t F	Concentration x lb Li Br/lb mix	Enthalpy h Btu/lb mix	Flow Rate m lb mix/(min)(ton)
1	6.3	100	0.60	...	2.580
2	49.1	...	0.60	...	2.580
3	49.1	180	0.60	−35	2.580
4	49.1	200	0.65	−27	2.380
5	49.1	...	0.65	...	2.380
6	6.3	...	0.65	...	2.380
7	49.1	200	0.00	1151	0.198
8	49.1	100	0.00	68	0.198
9	6.3	40	0.00	68	0.198
10	6.3	40	0.00	1079	0.198

The two pressures are found from the given saturation temperatures at (8) and (10). Enthalpies at (8) and (10) are found from Table A.1. The enthalpy at (7) may be calculated by Eq. (1.32). Since States (3) and (4) are equilibrium

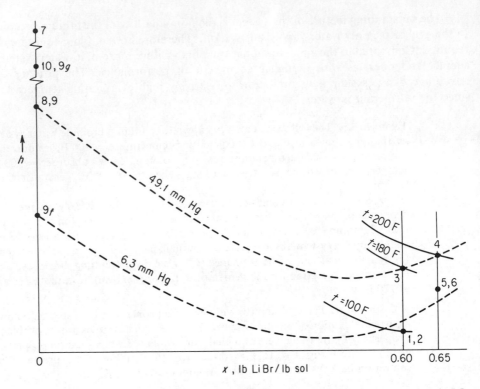

Figure 5.20 Schematic h-x diagram for Example 5.5.

states, we may locate the state-points on Fig. E-4 and read enthalpy and concentration values. Table 5.3 shows the necessary thermodynamic properties to solve the problem. Blank locations may be filled in as a student exercise. Figure 5.20 shows the state-points on a schematic h-x diagram.

For the evaporator, we have

$$m_9 = \frac{200}{h_{10} - h_9} = \frac{200}{1011} = 0.198 \text{ lb/(min)(ton)}$$

For the absorber, we have

$$m_6 = m_{10} \frac{(x_1 - x_{10})}{(x_6 - x_1)} = \frac{(0.198)(0.60)}{(0.05)} = 2.38 \text{ lb/(min)(ton)}$$

$$m_1 = m_6 + m_{10} = 2.58 \text{ lb/(min)(ton)}$$

(c) For the generator, we have

$$q_G = m_4 h_4 + m_7 h_7 - m_3 h_3$$
$$= (2.38)(-27) + (0.198)(1151) - (2.58)(-35)$$
$$= 253.8 \text{ Btu/(min)(ton)}$$

Neglecting the pump work, we have

$$\text{C.O.P.} = \frac{q_E}{q_G} = \frac{200}{253.8} = 0.788$$

(d) For an environmental temperature of 100 F, a generator temperature of 200 F, and an evaporating temperature of 40 F, we have, by Eq. (5.25),

$$\text{(C.O.P.)}_{\max} = \frac{(500)(100)}{(60)(660)} = 1.263$$

and

$$\eta_R = \frac{\text{C.O.P.}}{\text{(C.O.P.)}_{\max}} = \frac{0.788}{1.263} = 0.624$$

(e) We will assume that saturated water at 220 F leaves the steam coil. Thus

$$m_s = \frac{60 q_G}{h_{fg,\,s}} = \frac{(60)(253.8)}{(965.2)} = 15.78 \text{ lb/(hr)(ton)}$$

5.8. COMPARISON OF ABSORPTION SYSTEMS WITH MECHANICAL VAPOR-COMPRESSION SYSTEMS

Example 5.4 showed that the coefficient of performance for an aqua-ammonia absorption system was about one-seventh of that for a theoretical single-stage ammonia system operating at approximately the same evaporating and condensing temperatures. Example 5.5 showed that the C.O.P. for a lithium bromide-water absorption system was somewhat higher than that of an aqua-ammonia system.

It is unfair to the absorption system to make a direct comparison of its C.O.P. with that of a mechanical-compression system. A C.O.P. calculated for an absorption system, as in Example 5.4, is more realistic than a C.O.P. calculated for a theoretical vapor-compression cycle where isentropic compression is assumed. In order to make a comparison, we must multiply the *actual* C.O.P. of the mechanical compression system by the thermal efficiency of the power plant. We must also include the transmission losses up to the compressor. When we do this, we find that a mechanical-compression system has little thermodynamic superiority over an absorption system.

It is worthwhile to examine more closely Eq. (5.25) which gives the maximum attainable C.O.P. for an absorption system. This formula shows that, for a reversible system, the C.O.P. increases with increase of temperature of the generator heating medium. However, in an actual system we cannot choose an arbitrary combination of operating temperatures. For a given refrigerant-absorbent mixture, the necessary generator temperature depends upon the condensing pressure which is determined by the temperature of available cooling water and rate of heat transfer in the condenser. The generator temperature is also related to the low-pressure side of the system. In the aqua-ammonia system, an increase of generator temperature will cause a decrease in absorber pressure, if other operating conditions are maintained constant. Thus, in an absorption system, operating temperatures are limited to those values which are satisfactory in actual practice.

Apart from C.O.P. considerations, the absorption system has some practical advantages over mechanical-compression systems. The absorption system is less subject to wear and maintenance. The absorption system may operate at reduced evaporating pressure with little decrease in refrigerating capacity. Liquid carry-over from the evaporator causes no difficulties.

The choice between an absorption system and a mechanical-compression system is largely determined by economic factors. The absorption water-chilling machine used for air conditioning may be economically attractive where low-cost fuel is available, where electricity rates are high, where heating boiler capacity is idle or partially idle during summer months, where waste steam is available, and where existing electric facilities are inadequate for installing electric motor-driven mechanical compressors.

PROBLEMS

5.1. Saturated water vapor at 50 F is mixed in a steady-flow chamber with a saturated lithium bromide-water solution having a concentration of 0.60 lb Li Br per lb mix. The mass of liquid solution mixed is five times the mass of the water vapor mixed. The mixing process occurs at constant pressure. Determine (a) the concentration of the resulting mixture, and (b) the heat which must be removed in Btu per lb of the final mixture, if a saturated liquid solution is produced.

5.2. A saturated liquid aqua-ammonia solution at 220 F and 200 psia is throttled to a pressure of 10 psia. Determine (a) the equilibrium temperature after the throttling process, and (b) the relative portions of liquid and vapor coexisting after throttling.

5.3. Perform the complete derivation of Eq. (5.22).

5.4. An aqua-ammonia system similar to that in Fig. 5.18 operates as follows: high-side pressure = 200 psia; $t_3 = 190$ F; $t_7 = 140$ F; $t_4 = 210$ F; $m_7 = 100$ lb per min. Assume equilibrium conditions for States 3, 4, and 7. Find (a) the lb per min of strong solution leaving the absorber, and (b) the lb per min of cooling water required for the dephlegmator if the water temperature rise is 15 F. You may make the same assumptions with regard to the rectifying column as made in Part (d) of Example 5.4.

5.5. An aqua-ammonia system similar to that in Fig. 5.18 operates as follows: high-side pressure = 220 psia; low-side pressure = 20 psia; $t_4 = 210$ F; $t_8 = 80$ F; $t_{12} = 40$ F; $m_4 = 1000$ lb per min; $m_{12} = 100$ lb per min. Assume equilibrium states at 1, 3, 4, and 12. Determine (a) the concentration of the strong solution leaving the heat exchanger, (b) the heat removed in the absorber in Btu per min, and (c) the tons of refrigeration produced.

5.6. A lithium bromide-water system of the type shown in Fig. 5.16 operates with a condensing temperature of 110 F, evaporating temperature of 38 F, temperature of solution leaving absorber of 100 F, temperature of solution entering generator of 180 F, and temperature of solution leaving generator of 210 F. Assume saturated conditions for States 3, 4, 8, and 10. Neglect pressure drops in components and lines. Warm water from the load returns to the machine at a temperature of 52 F and at a rate of flow of 600 GPM. Chilled water leaves the machine at a temperature of 44 F. Saturated steam at 25 psia enters the generator and leaves as saturated water. Calculate the required rate of flow of steam in lb per hr.

5.7. Compare the cost in dollars per (hr)(ton) for generator steam in Example 5.5 with cost of electricity in dollars per (hr)(ton) for a Refrigerant 12 compressor system. You may assume that steam costs 40 cents per 1000 lb and that electricity cost is 2 cents per kwhr. Assume that the Refrigerant 12 system is operated at 40 F evaporating temperature and 100 F condensing temperature. You may obtain an estimate for the Hp/ton requirement of the Refrigerant 12 system by extrapolation of Fig. 3.15. Assume an electric motor efficiency of 80 per cent.

SYMBOLS USED IN CHAPTER 5

C.O.P.	Coefficient of performance, dimensionless.
Hp	Horsepower.
ΔH_x	Isothermal heat of solution, Btu per lb.
h	Specific enthalpy, Btu per lb.
$h_{fg, s}$	Latent heat of condensation for steam, Btu per lb.
J	Mechanical equivalent of heat, 778 ft-lb per Btu.
kwhr	Kilowatt hr.
L	Designation for liquid state on h-x diagram.
m	Mass rate of flow, lb per min or lb per (min)(ton); m_L for liquid; m_V for vapor; m_s for steam, lb per (hr)(ton).
P	Pressure, psia, psfa, or mm Hg.
P_1, P_2	Rectification poles on h-x diagram.
Q	Heat transfer, Btu.
q	Rate of heat transfer, Btu per min or Btu per (min)(ton); q_D for dephlegmator, q_G for generator.
S	Entropy, Btu per R.
T	Absolute temperature, R.
t	Temperature, F.
v	Specific volume, cu ft per lb.
V	Designation for vapor state on h-x diagram.
W	Work transfer, Btu per lb or Btu.
x	Concentration; for ammonia-water solutions, lb ammonia per lb mix; for water-lithium bromide solutions, lb lithium bromide per lb mix.
η_m	Mechanical efficiency, dimensionless.
η_R	Refrigerating efficiency, dimensionless.

6

MISCELLANEOUS
REFRIGERATION PROCESSES

6.1. GENERAL REMARKS

Several methods besides mechanical vapor compression systems and absorption systems may be employed to produce a refrigerating effect. Chilled water may be obtained by flash cooling or partial vaporization of warmer water. Chilled water at 40 F to 45 F is widely used in cooling and dehumidifying atmospheric air in comfort air conditioning systems.

Although of lesser importance than in earlier years, water ice is still important commercially. Practically all ice used today is artificially frozen by mechanical refrigeration systems. When calcium chloride or sodium chloride is mixed with ice, a melting temperature less than 32 F is attained. Dry ice or solid carbon dioxide has found a growing market. Since dry ice sublimes at approximately −110 F at atmospheric pressure, it may be used for producing relatively low temperatures. Mechanical compression systems utilizing gaseous air as the refrigerant are important in cooling of aircraft cabins. Through expansion of compressed air in a turbine, air sufficiently cold for refrigeration purposes may be obtained. Thermoelectric cooling offers a novel and potentially important means of refrigeration. By utilizing the reverse principle of the thermocouple, a temperature difference may be established between two junctions by a direct electric current.

In this chapter we will study the principles of several refrigerating processes. We will direct most of our attention to thermoelectric cooling.

6.2. INTRODUCTION TO THERMOELECTRIC COOLING

In 1822 a Prussian, Seebeck, reported experiments which led to the discovery of the thermocouple. Seebeck's experiments showed that an electric current occurs in a closed circuit formed of dissimilar conductors at different junction temperatures. But Seebeck misunderstood his findings, for he thought he had proved that magnetism was caused by a difference in temperature. In 1834 a Frenchman, Peltier, observed a thermal effect at the junction of two dissimilar conductors carrying a current. However, Peltier also misunderstood his results.

In 1857 William Thomson (Lord Kelvin) proved by thermodynamic analysis that the Seebeck effect and Peltier effect were related. He also discovered a third thermoelectric phenomenon which became known as the Thomson effect. But it has only been in recent years, since development of semiconductors, that thermoelectric cooling has become of practical importance. Much of the important modern work was done in the Soviet Union under the direction of A. F. Ioffe.*

6.3. THERMOELECTRIC EFFECTS

Five effects are observed when a current is passed through a thermocouple whose junctions are at different temperatures. These phenomena are the *Seebeck effect*, the *Joulean effect*, the *conduction effect*, the *Peltier effect*, and the *Thomson effect*.

The Seebeck effect defines the principle of the thermocouple. As illustrated by Fig. 6.1(a), an electromotive force is produced when two junctions formed by dis-

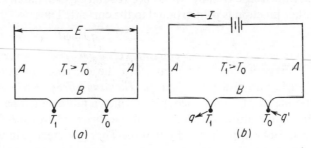

Figure 6.1 Illustration of (a) the Seebeck effect and (b) the Peltier effect.

*A. F. Ioffe, *Semiconductor Thermoelements and Thermoelectric Cooling* (trans. A. Gelbtuch) (London: Infosearch Limited, 1957).

similar conductors A and B are maintained at different temperatures. Experiments have shown that the Seebeck effect is reversible. The electromotive force may be expressed thus:

$$E = \alpha(T_1 - T_0) \qquad (6.1)$$

and for a constant cold junction temperature T_0

$$\alpha = \frac{dE}{dT} \qquad (6.2)$$

where α is called the *thermoelectric power* or the *Seebeck coefficient*. The units are volts per degree Kelvin.

If a closed circuit was formed in Fig. 6.1(a), a current flow I would be caused by the Seebeck effect. As a result, an irreversible generation of heat called the *Joulean effect* would occur in the conductors. The magnitude would be

$$q_j = I^2 R \qquad (6.3)$$

where q_j is the Joulean heat, watts, I is the current, amperes, and R is the total circuit resistance, ohms.

When one junction of a thermocouple is maintained at a higher temperature than the other junction, heat will flow by conduction from the warm junction to the cold junction. This irreversible effect, called the *conduction effect*, may be expressed as

$$q_c = U(T_1 - T_0) \qquad (6.4)$$

where q_c is the rate of heat transfer, watts, U is the overall thermal conductance, watts per K, and T_1, T_0 are the junction temperatures, K.

Figure 6.1(b) shows a thermocouple circuit with a battery added. Experiments show that when a current is passed through a thermocouple whose junctions were initially at the same temperature, the junction temperatures will change. A certain quantity of heat will be released at one junction while a different quantity of heat will be absorbed at the other junction. If the current is reversed, the heating effects are also reversed. This phenomenon, called the *Peltier effect*, forms the basis for thermoelectric cooling. It has been found experimentally that the heat transfer effect at either junction is proportional to the current. Thus

$$q = \phi I \qquad (6.5)$$

where ϕ is the *Peltier coefficient*, volts.

Experiments have established the existence of still another reversible thermoelectric phenomenon called the *Thomson effect*. When a current is passed through a conductor of a thermocouple in which a uniform temperature gradient initially exists, the temperature distribution will be distorted in excess of the Joulean effect. It has been found experimentally that the Thomson heat q_τ in watts per cm is given by

$$q_\tau = \tau I \frac{dT}{dx} \qquad (6.6)$$

where τ is the *Thomson coefficient*, volts per K, I the current, amp, and dT/dx the temperature gradient in the conductor, K per cm.

Thermodynamic analysis shows that the Seebeck coefficient α, the Peltier coefficient ϕ, and the Thomson coefficient τ are related. The quantities α and ϕ are in reality differential terms since they depend upon properties of both arms of the thermocouple. We may use the expression $\alpha = \alpha_A - \alpha_B$ for a thermocouple formed of materials A and B.

Both Ioffe[*] and Zemansky[†] have proved, by fundamental thermodynamics, the relations

$$\phi = (\alpha_A - \alpha_B)T \tag{6.7}$$

$$\frac{\tau_A - \tau_B}{T} = -\frac{d(\alpha_A - \alpha_B)}{dT} \tag{6.8}$$

Both Eqs. (6.7) and (6.8) have been verified experimentally also.

6.4. ANALYSIS OF THERMOELECTRIC COOLING

With a proper choice of materials the Peltier effect may be utilized for refrigeration purposes. Figure 6.2 shows a schematic system. Two dissimilar conductors p and n are joined to form two junctions. The conductors are chosen such that p has a *positive*

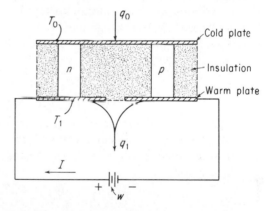

Figure 6.2 Schematic thermoelectric cooling system.

thermoelectric power α_p and n has a *negative* thermoelectric power α_n. The cold junction is attached to a metal plate or some other type of heat-transfer surface. This surface is exposed to the space or substance to be refrigerated and serves the same purpose as the evaporator of a vapor compression system. The hot junction is attached to some type of surface which allows rejection of heat to atmospheric air or some other

[*]Ioffe, *Semiconductor Thermoelements and Thermoelectric Cooling* (trans. A. Gelbtuch) (London: Infosearch Limited, 1957) pp. 8–11.

[†]M. W. Zemansky, *Heat and Thermodynamics*, 4th ed. (New York: McGraw-Hill Book Company, 1957) pp. 305–309.

medium. This surface performs the same function as the condenser of a vapor-compression system. An external battery provides for circulation of a direct current through the circuit. The battery may be compared functionally to the compressor of a mechanical vapor-compression system.

When a current is passed through the circuit of Fig. 6.2, the five thermoelectric effects discussed in Sec. 6.3 will occur. Because of the Peltier effect, the cold plate will be cooled and the warm plate will be heated. Heat will flow from the warm plate to the cold plate by conduction. Heat will be generated in each conductor and at each junction because of the Joulean effect; part of the Joulean heat will flow to each junction. Because of the Seebeck effect, an electromotive force will be produced which opposes that of the battery.

The Thomson effect would also occur in the system shown in Fig. 6.2. However, in a practical problem, it is necessary to use average values for the thermoelectric power $(\alpha_p - \alpha_n)$. According to Eq. (6.8), the net Thomson coefficient $(\tau_p - \tau_n)$ becomes zero if $(\alpha_p - \alpha_n)$ is considered constant. Ioffe* has proved that, to a first approximation, the use of a mean thermoelectric power gives results equivalent to those obtained when the Thomson effect is included. It is usual to assume that one-half of the Joulean heat is transferred to each junction. Ioffe* has shown that, to a first approximation, this assumption is valid.

In addition to the assumptions discussed before, we will assume that heat absorption and heat rejection occur only at the junctions and that all material property values are constants. Under steady-state conditions, we may write the following equations for the system of Fig. 6.2:

$$q_0 = (\alpha_p - \alpha_n)T_0 I - U(T_1 - T_0) - \tfrac{1}{2}I^2 R \qquad (6.9)$$

$$q_1 = (\alpha_p - \alpha_n)T_1 I - U(T_1 - T_0) + \tfrac{1}{2}I^2 R \qquad (6.10)$$

where $(\alpha_p - \alpha_n)$ is the mean thermoelectric power in the temperature range T_0 to T_1, U is the effective thermal conductance between the two junctions, and R is the total electrical resistance (conductors plus contact resistance at junctions).

By Eq. (6.9)

$$T_1 - T_0 = \frac{(\alpha_p - \alpha_n)T_0 I - \tfrac{1}{2}I^2 R - q_0}{U} \qquad (6.11)$$

Equation (6.11) shows that the maximum temperature difference $(T_1 - T_0)$ occurs when the refrigerating effect q_0 is zero.

The power input by the battery w must compensate for the power loss of the Joulean effect and counteract the generation of power by the Seebeck effect. Thus

$$w = (\alpha_p - \alpha_n)(T_1 - T_0)I + I^2 R \qquad (6.12)$$

The coefficient of performance of the system as a refrigerating device is

$$\text{C.O.P.} = \frac{q_0}{w}$$

*Ioffe, *Semiconductor Thermoelements and Thermoelectric Cooling*, pp. 113–114.

By Eqs. (6.9) and (6.12)

$$\text{C.O.P.} = \frac{(\alpha_p - \alpha_n)T_0 I - U(T_1 - T_0) - \frac{1}{2}I^2 R}{(\alpha_p - \alpha_n)(T_1 - T_0)I + I^2 R} \tag{6.13}$$

For a completely reversible thermoelectric system (no Joulean effects and no conduction effects), Eq. (6.13) becomes

$$\text{C.O.P.} = \frac{T_0}{T_1 - T_0} \tag{6.14}$$

which is the Carnot cycle value.

Equations (6.9), (6.11), and (6.13) show that q_0, $(T_1 - T_0)$, and C.O.P. are functions of the current I. Each quantity may be maximized by equating the first partial derivative with respect to I to zero in the appropriate equation and solving for an optimum current. For *maximum temperature difference* and *maximum refrigerating capacity*, we obtain

$$I_{\text{opt}} = \frac{(\alpha_p - \alpha_n)T_0}{R} \tag{6.15}$$

By Eqs. (6.11) and (6.15) with $q_0 = 0$, we have

$$(T_1 - T_0)_{\max} = \frac{1}{2}ZT_0^2 \tag{6.16}$$

where

$$Z = \frac{(\alpha_p - \alpha_n)^2}{UR} \tag{6.17}$$

By Eqs. (6.9), (6.15), and (6.17)

$$q_{0,\,\max} = U\left[\frac{ZT_0^2}{2} - (T_1 - T_0)\right] \tag{6.18}$$

For *maximum* C.O.P., the optimum current is

$$I_{\text{opt}} = \frac{(\alpha_p - \alpha_n)(T_1 - T_0)}{R(\sqrt{1 + ZT_m} - 1)} \tag{6.19}$$

By Eqs. (6.13) and (6.19)

$$(\text{C.O.P.})_{\max} = \frac{\left(\dfrac{T_0}{T_1 - T_0}\right)\left(\sqrt{1 + ZT_m} - \dfrac{T_1}{T_0}\right)}{\sqrt{1 + ZT_m} + 1} \tag{6.20}$$

where $T_m = (T_1 + T_0)/2$.

Equations (6.16), (6.18), and (6.20) show that the performance of a thermoelectric cooling system is a function of the parameter Z given by Eq. (6.17). This parameter is called the *figure of merit*. It has units of the reciprocal of temperature and is a function of the materials of the system only.

Figure 6.3, calculated by Eq. (6.16), shows the maximum temperature difference for a thermoelectric system where the hot junction temperature is 300 K (80.6 F). Figure 6.4 shows the maximum C.O.P. as calculated by Eq. (6.20), where the hot-junction temperature is held constant at 300 K (80.6 F).

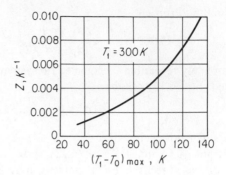

Figure 6.3 Maximum theoretical temperature difference for a thermoelectric element. (Reproduced by permission from R. L. Eichorn, "Thermoelectric Refrigeration," *Refrigerating Engineering*, Vol. 66, June 1958, p. 32.)

Figures 6.3 and 6.4 show that the figure of merit Z is decisive in determining the performance of a cooling couple. Equation (6.17) shows that for a large value of Z, the couple must have a large thermoelectric power $(\alpha_p - \alpha_n)$, small thermal conductance U, and small electrical resistance R.

The electrical resistance R is made up principally of the resistances of the elements and the contact resistances resulting from attachment of the elements to the copper connecting strips. In the following procedure, we will assume equal length L for the elements and negligible thermal contact resistance, but we will allow for an electrical contact resistance of r ohm-cm^2 at each junction. For each couple, we may write

$$U = U_p + U_n = k_p\frac{A_p}{L} + k_n\frac{A_n}{L}$$

$$R = R_p + R_n = \rho_p\frac{L}{A_p} + \frac{2r}{A_p} + \rho_n\frac{L}{A_n} + \frac{2r}{A_n}$$

We obtain

$$UR = (k_p y + k_n)\left[\frac{\rho_p}{y}\left(1 + \frac{2r}{\rho_p L}\right) + \rho_n\left(1 + \frac{2r}{\rho_n L}\right)\right] \tag{6.21}$$

where $y = A_p/A_n$.

We may maximize Z by finding a minimum value for UR. For UR to be a minimum with respect to y, we must have

$$y = \sqrt{\frac{k_n\rho_p\left(1 + \dfrac{2r}{\rho_p L}\right)}{k_p\rho_n\left(1 + \dfrac{2r}{\rho_n L}\right)}} \tag{6.22}$$

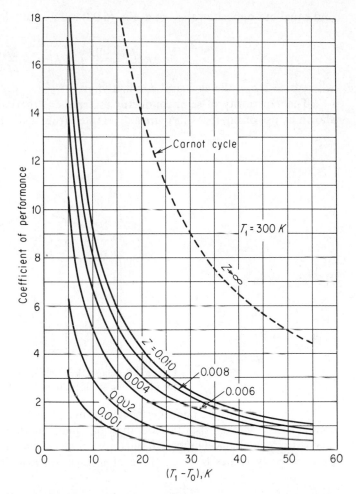

Figure 6.4 Maximum theoretical coefficient of performance for a thermoelectric element. (Reproduced by permission from R. L. Eichorn, "Thermoelectric Refrigeration," *Refrigerating Engineering*, Vol. 66, June 1958, p. 32.)

By Eqs. (6.17), (6.21), and (6.22)

$$Z_{max} = \left[\frac{(\alpha_p - \alpha_n)}{\sqrt{k_p \rho_p \left(1 + \dfrac{2r}{\rho_p L}\right)} + \sqrt{k_n \rho_n \left(1 + \dfrac{2r}{\rho_n L}\right)}} \right]^2 \tag{6.23}$$

For the special case where each element of a couple has equal values of k and ρ

and equal but opposite values of α, Eq. (6.23) reduces to

$$Z_{\max} = \frac{\alpha^2}{k\rho\left(1 + \dfrac{2r}{\rho L}\right)} = \frac{\alpha^2\sigma}{k\left(1 + \dfrac{2r}{\rho L}\right)} \tag{6.24}$$

where $\sigma = 1/\rho = $ the electrical conductivity.

The superiority of semi-conductor materials as thermoelectric elements may be shown by use of Eq. (6.24). Figure 6.5, adapted from a similar graph given by Ioffe,*

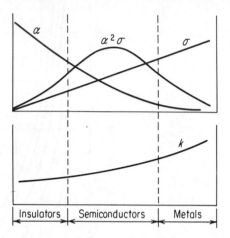

Figure 6.5 Schematic comparative thermoelectric characteristics for insulators, semiconductors, and metals.

shows that insulators are poor thermoelectric materials because of their small electrical conductivity. Metals also fail because of small thermoelectric power and large thermal conductivity. The best results are obtained with the semi-conductors which have property values between those of metals and insulators.

6.5. DESIGN AND APPLICATION OF THERMOELECTRIC COOLING SYSTEMS

According to the *ASHRAE Handbook of Fundamentals*,† materials used for the elements of thermoelectric cooling couples include alloys of bismuth, tellurium, and antimony for p-type elements, and alloys of bismuth, tellurium, and selenium for n-type elements. Typical parameters applicable to both p-type and n-type elements are shown in Table 6.1.

Electrical contact resistance has an important influence upon the actual perfor-

*Ioffe, *Semiconductor Thermoelements and Thermoelectric Cooling*, p. 100.

†*ASHRAE Handbook of Fundamentals* (New York: American Society of Heating, Refrigerating and Air Conditioning Engineers, 1967) p. 29.

TABLE 6.1*

TYPICAL PARAMETERS FOR THERMOELECTRIC ELEMENTS

Parameter	Value
Thermoelectric power α, volt per K	0.00021
Thermal conductivity k, watt per (cm)(K)	0.015
Electrical resistivity ρ, ohm-cm	0.001
Electrical contact resistance r, ohm-cm^2	0.00001 to 0.0001

ASHRAE Handbook of Fundamentals, pp. 26–27.

mance of a thermoelectric couple. Table 6.2 shows values of Z_{max} calculated by Eq. (6.24) for various element lengths and for values of r of 10^{-5} and 10^{-4} ohm-cm^2. Table 6.2 shows that the length may not be made arbitrarily small without possible serious decrease in Z_{max}.

TABLE 6.2

INFLUENCE OF ELECTRICAL CONTACT RESISTANCE
UPON FIGURE OF MERIT

L cm	r ohm-cm^2	$1 + \dfrac{2r}{\rho L}$	Z_{max} K^{-1}
—	0	1.00	0.00294
1.0	0.00001	1.02	0.00288
0.5	0.00001	1.04	0.00283
1.0	0.0001	1.20	0.00245
0.5	0.0001	1.40	0.00210

Figure 6.6 adapted from a similar diagram in the *ASHRAE Handbook of Fundamentals** shows a schematic thermoelectric module. The alternate *p-type* and *n-type* elements may be soldered to the heat transfer surfaces with the couples connected in series electrically. The separating spaces are filled with insulation. The external construction of the heat transfer surfaces depends upon the particular problem. Where air is the heat source or heat sink, it is advantageous to attach fins to the surface. A difficult problem is to find a satisfactory electrical insulation which is also a good thermal conductor. The *ASHRAE Handbook of Fundamentals** states that beryllia, anodized aluminum, filled epoxy coatings, and alumina ceramic wafers have been used as electrical insulations.

Two important variables, the current I and the ratio of element cross-sectional area to length A/L, must be considered when designing a thermoelectric module. The current needed for maximum coefficient of performance increases as A/L increases.

ASHRAE Handbook of Fundamentals, p. 29.

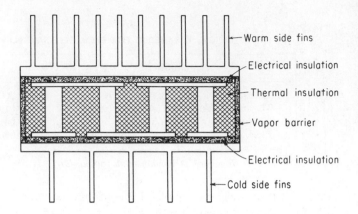

Warm side fins

Electrical insulation

Thermal insulation

Vapor barrier

Electrical insulation

Cold side fins

Figure 6.6 Schematic thermoelectric module.

As A/L increases, the maximum refrigerating capacity increases but correspondingly higher currents are required.

Example 6.1. A thermoelectric cooling system is to be designed to maintain a small insulated chamber at 40 F when the ambient temperature is 90 F. The estimated cooling load is 100 Btu per hr. Each thermoelectric element will be cylindrical with a length of 1.25 cm and a diameter of 1.00 cm. Thermoelectric properties are given in Table 6.1. Determine (a) the number of couples required, (b) the rate of heat rejection from the heat dissipator in Btu per hr, (c) the C.O.P., and (d) the overall voltage drop and watts capacity for the direct-current power source.

SOLUTION: (a) We will assume the cold junction temperature to be 30 F and the warm junction temperature to be 100 F. Thus, $T_0 = 272.1$ K and $T_1 = 311.0$ K. By Table 6.1, we will assume $r = 0.0001$ ohm-cm² at each junction. For each element

$$A = \frac{\pi(1.00)^2}{4} = 0.7854 \text{ cm}^2$$

$$\frac{L}{A} = \frac{1.25}{0.7854} = 1.592 \text{ cm}^{-1}$$

For each couple

$$R = 2\left(\rho\frac{L}{A} + \frac{2r}{A}\right) = 2\left[(0.001)(1.592) + \frac{(2)(0.0001)}{0.7854}\right] = 0.00369 \text{ ohm}$$

$$U = 2k\frac{A}{L} = \frac{(2)(0.015)}{1.592} = 0.0188 \text{ watt/K}$$

$$Z = \frac{4\alpha^2}{UR} = \frac{(4)(0.00021)^2}{(0.0188)(0.00369)} = 0.00254 \text{ K}^{-1}$$

We will calculate the current for maximum C.O.P. By Eq. (6.19)

$$I = \frac{(2)(0.00021)(38.9)}{(0.00369)[\sqrt{1 + (0.00254)(291.5)} - 1]} = 13.87 \text{ amp}$$

By Eq. (6.9), each couple would have a cooling capacity of

$$q_0 = (2)(0.00021)(272.1)(13.87) - (0.0188)(38.9) - \frac{(13.87)^2(0.00369)}{2}$$
$$= 0.499 \text{ watt}$$

and

$$\text{Number of couples} = \frac{100}{(3.413)(0.499)} = 59$$

The 59 couples would be connected in series as shown schematically in Fig. 6.6.
    (b) By Eq. (6.10)

$$q_1 = (59)(3.413)\Big[(2)(0.00021)(311.0)(13.87) - (0.0188)(38.9)$$
$$+ \frac{(13.87)^2(0.00369)}{2}\Big] = 289 \text{ Btu/hr}$$

(c) Equation (6.13) reduces to

$$\text{C.O.P.} = \frac{q_0}{q_1 - q_0} = \frac{100}{189} = 0.529$$

(d) The overall voltage drop would be

$$E = (59)[(\alpha_p - \alpha_n)(T_1 - T_0) + IR]$$
$$- (59)[(2)(0.00021)(38.9) + (13.87)(0.00369)] = 3.98 \text{ volt}$$

The d-c power required would be

$$w = EI = (3.98)(13.87) = 55.2 \text{ watt}$$

Thermoelectric cooling offers many advantages when compared to other methods of refrigeration. These include (a) ease of interchanging the cooling and heating functions, (b) no wear and noise from moving parts, (c) no problem in containment of refrigerant, (d) ease of miniaturization for very small capacity systems, (e) ease of controlling capacity by varying applied voltage, and (f) ability to operate under zero gravity or many g's of gravity and in any position. Major disadvantages include high cost of couples and low coefficient of performance unless a way is found to significantly increase the figure of merit, Z.

Thermoelectric cooling has been applied to numerous special situations where cooling capacities were relatively small. For some applications, thermoelectric cooling offers the only practical method.

6.6. FLASH COOLING

Flash cooling is important commercially in obtaining chilled water and in the manufacture of dry ice. Figure 6.7 shows a flash chamber maintained under an extremely low absolute pressure by a compressor. Water admitted to the chamber is partially vaporized so that the remaining water is cooled to the saturation temperature corresponding to the chamber pressure. Example 6.2 shows the solution of a typical problem.

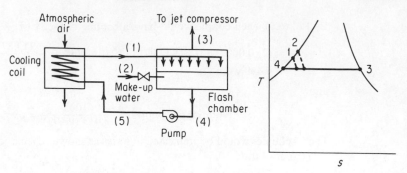

Figure 6.7 Schematic flash cooling process for obtaining chilled water.

Example 6.2. In Fig. 6.7, the flash chamber is maintained under a pressure of 0.1216 psia. Warm water returns from the cooling coil at 50 F. Make-up water temperature is 80 F. The pump circulates 150 lb per min. Dry-and-saturated vapor enters the jet compressor. Ignore the work input to the pump. Determine (a) the tons of refrigeration produced, and (b) the volume of flash vapor entering the compressor in cu ft per min.

SOLUTION: By Table A.1, we obtain

$$h_1 = 18.07 \text{ Btu/lb}, \qquad h_2 = 48.05 \text{ Btu/lb}, \qquad t_3 = t_4 = 40 \text{ F},$$

$$h_3 = 1078.68 \text{ Btu/lb}, \qquad v_3 = 2445.4 \text{ cu ft/lb}, \qquad h_4 = 8.04 \text{ Btu/lb}$$

(a) For the cooling coil

$$\text{Tons} = \frac{m_5(h_1 - h_5)}{200} = \frac{m_5(h_1 - h_4)}{200}$$

$$= \frac{(150)(10.03)}{200} = 7.52$$

(b) For the flash chamber, we have

$$m_1 h_1 + m_2 h_2 = m_3 h_3 + m_4 h_4$$

$$m_1 = m_4$$

$$m_2 = m_3$$

Thus

$$m_3 = m_1 \frac{(h_1 - h_4)}{(h_3 - h_2)} = \frac{(150)(10.03)}{1030.63} = 1.46 \text{ lb/min}$$

and

$$\text{Volume of flash vapor} = m_3 v_3 = (1.46)(2445.4) = 3570 \text{ cu ft/min}$$

Example 6.2 shows that a relatively enormous volume of flash vapor must be handled by the compressor. A steam-jet compressor is commonly employed. Steam-jet-water-vapor systems possess the advantage of few moving parts and low maintenance, use of a cheap, non-toxic refrigerant (water vapor), and minimum power requirements. They have the disadvantages of requiring relatively large quantities of motive steam and condensing water and are limited to flash chamber temperatures higher than about 40 F. They may be more economical than mechanical compression systems for chilling water if cheap steam and cheap condensing water are available.

The principle of flash cooling is also utilized in the manufacture of dry ice. Figure 6.8 shows a schematic three-stage CO_2 system for making dry ice. The system

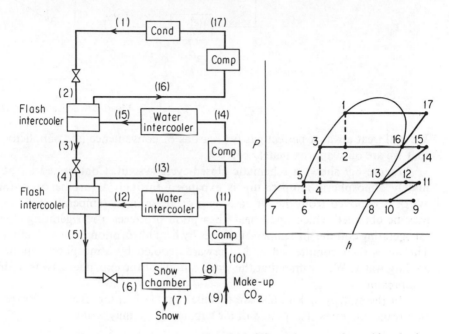

Figure 6.8 Schematic system for making dry ice.

is basically a mechanical compression refrigeration system. The triple point of CO_2 occurs at -69.9 F and 75.1 psia. If liquid CO_2 is expanded to a lower pressure, the equilibrium products are solid CO_2 and flash vapor. Solid CO_2 so produced may be compressed into cakes of dry ice.

Dry ice is of commercial importance where relatively low temperatures are needed. It is particularly adaptable to infrequently operated facilities and in the transport of expensive commodities. The economics of dry ice systems are relatively low initial costs but relatively high operating costs.

6.7. GASEOUS AIR CYCLE SYSTEMS

Dry air may serve as a refrigerant in a mechanical compression system. The air remains gaseous throughout the system. Although most air cycle systems used today are not closed systems, a closed thermodynamic cycle will be first considered here.

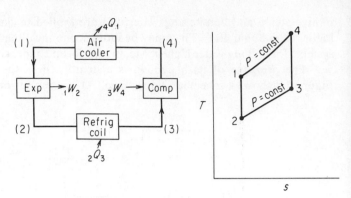

Figure 6.9 Schematic air cycle system.

We will treat air as a perfect gas for reasons of convenience and simplicity, realizing that we are only approximately correct.

Figure 6.9 shows a schematic closed-cycle system. Compressed air at approximately atmospheric temperature is expanded through a turbine or piston device. Work is removed from the air causing a decrease in its temperature. Thus, cold air may be obtained which upon heat absorption produces a refrigerating effect. In the refrigerating coil the air temperature rises by heat absorption from a product or space. The air is then compressed and afterwards cooled by atmospheric air or ordinary cooling water. We assume that the work output of the expander assists in driving the compressor.

In the system of Fig. 6.9, we assume that both compression and expansion are isentropic processes. Per pound of air circulated, we may write

$$_1W_2 = h_1 - h_2 = c_p(t_1 - t_2)$$

$$_2Q_3 = h_3 - h_2 = c_p(t_3 - t_2)$$

$$_3W_4 = h_4 - h_3 = c_p(t_4 - t_3)$$

$$_4Q_1 = h_4 - h_1 = c_p(t_4 - t_1)$$

The coefficient of performance is given by

$$\text{C.O.P.} = \frac{_2Q_3}{_3W_4 - {_1}W_2} \tag{6.25}$$

The refrigerating efficiency is given by

$$\eta_R = \frac{\text{C.O.P.}\,(T_1 - T_3)}{T_3} \tag{6.26}$$

Example 6.3. An air cycle system similar to Fig. 6.9 produces 10 tons of refrigeration. High-side pressure is 60 psia and low-side pressure is 20 psia. Air enters the expander at 120 F and enters the compressor at 0 F. Assume isentropic compression and expansion. Determine (a) the coefficient of performance,

(b) the refrigerating efficiency, (c) piston displacement of compressor in cu ft per min, and (d) piston displacement of expander in cu ft per min.

SOLUTION: We must first find the air temperatures at States (2) and (4). Since $\gamma = c_p/c_v = 1.40$ for air, we have

$$T_2 = \frac{T_1}{\left(\frac{P_1}{P_2}\right)^{(\gamma-1)/\gamma}} = \frac{580}{(3.0)^{0.286}} = 423.7 \text{ R}$$

$$t_2 = -36.3 \text{ F}$$

$$T_4 = T_3\left(\frac{P_4}{P_3}\right)^{(\gamma-1)/\gamma} = 460(3.0)^{0.286} = 629.7 \text{ R}$$

$$t_4 = 169.7 \text{ F}$$

(a) By Eq. (6.25),

$$\text{C.O.P.} = \frac{c_p(t_3 - t_2)}{c_p(t_4 - t_3) - c_p(t_1 - t_2)} = \frac{t_3 - t_2}{(t_4 - t_3) - (t_1 - t_2)}$$

$$= \frac{36.3}{169.7 - 156.3} = 2.71$$

(b) By Eq. (6.26),

$$\eta_R = \frac{(2.71)(120)}{460} = 0.707$$

(c) Assuming 100 per cent volumetric efficiency for the compressor, we have

$$\text{P.D.} = mv_3$$

$$m = \frac{(10)(200)}{c_p(t_3 - t_2)} = \frac{2000}{(0.240)(0 + 36.3)} = 229.4 \text{ lb/min}$$

$$v_3 = \frac{RT_3}{P_3} = \frac{(53.34)(460)}{(20)(144)} = 8.52 \text{ cu ft/lb}$$

and

$$\text{P.D.} = (229.4)(8.52) = 1954 \text{ cu ft/min}$$

(d) Assuming 100 per cent volumetric efficiency for the expander, we have

$$\text{P.D.} = mv_2 = (229.4)(7.85) = 1800 \text{ cu ft/min}$$

where

$$v_2 = \frac{RT_2}{P_2} = \frac{(53.34)(423.7)}{(20)(144)} = 7.85 \text{ cu ft/lb}$$

The outstanding disadvantage of the air cycle system is its low coefficient of performance. This statement is particularly true when we consider thermodynamic losses for the compressor and expander. If in Example 6.3 we assume a mechanical efficiency of 50 per cent for both expander and compressor, we have

$$\text{C.O.P.} = \frac{{}_2Q_3}{({}_3W_4/0.5) - (0.5 \, {}_1W_2)} = \frac{t_3 - t_2}{[(t_4 - t_3)/0.5] - [0.5(t_1 - t_2)]} = 0.139$$

or about five per cent of the theoretical value.

We may make a more practical analysis of the air cycle system just as we did for the vapor cycle in Sec. 3.5. The equations of Sec. 3.5 are directly applicable to the air cycle system. Since we are assuming air to be a perfect gas, Eq. (3.14) may be written as

$$\eta_v = \left[1 + C - C\left(\frac{P_c}{P_b}\right)^{1/n} \right] \frac{T_3 P_b}{T_b P_3} \tag{6.27}$$

Equation (3.18) may be written as

$$W = \frac{n}{(n-1)} \frac{R}{J} (t_c - t_b) \tag{6.28}$$

with W in Btu/lb.

For conventional refrigeration requirements, the air cycle system has too low a coefficient of performance to compete with the vapor compression system. In the cooling of aircraft cabins, the air cycle system has some particular advantages. Aircraft cooling is the main application of the air cycle system.

Aircraft cabins require cooling to offset heat gains from passengers, electrical and mechanical equipment, solar radiation, and heat transmission through the walls of the plane. In addition, the ram temperature rise resulting from adiabatic stagnation of air taken into the plane and pressurization of the cabin, further contribute to cooling requirements. Air cycle refrigeration systems for aircraft may be either open systems or semi-closed systems. Air circulated through the refrigeration equipment is also circulated through the cabin. Advantages claimed for air cycle refrigeration systems compared to vapor compression systems in aircraft include (1) less weight per ton of refrigeration, (2) a refrigeration unit which may be easily removed and repaired, and (3) unimportance of refrigerant leakage.

Air cycle systems for military and commercial aircraft involve various equipment arrangements. Figure 6.10 shows an elementary type of system for a jet fighter plane. All air which circulates through the cabin is exhausted.

In aircraft air cycle systems the refrigerant air is not perfectly dry, as was assumed in Example 6.3. Some condensation of moisture occurs in the turbine which reduces

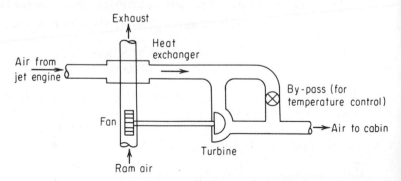

Figure 6.10 Schematic air cycle cooling system for a jet fighter plane.

the refrigerating capacity of the unit. A moisture separator must be installed after the turbine to minimize carry-over of water droplets into the cabin.

PROBLEMS

6.1. Beginning with Eq. (6.13), show the complete derivation for Eq. (6.20).

6.2. Show that when the refrigerating capacity of a thermoelectric system is a maximum, the coefficient of performance is given by

$$\text{C.O.P.} = \frac{T_0}{2T_1} - \frac{(T_1 - T_0)}{ZT_0T_1}$$

6.3. Rework Example 6.1 based upon the current for maximum refrigerating capacity.

6.4. A thermoelectric cooling system is to be designed to maintain an experimental food storage cabinet at 35 F when the ambient temperature is 90 F. The estimated cooling load is 500 Btu per hr. You may assume an element length of 1.00 cm, $A/L = 0.5$ cm, and electrical contact resistance $r = 0.00005$ ohm-cm^2. You may use $t_0 = 25$ F and $t_1 = 100$ F. The system is to be designed for maximum coefficient of performance. Determine (a) the number of couples required, and (b) the cost in cents per hr for operating the refrigerator if electricity costs 3 cents per kwhr and if the power demand of the complete system (including rectifier, etc.) is 10 per cent greater than the d-c power demand of the thermoelectric circuit alone.

6.5. A two-stage thermoelectric cooling system is to be designed to maintain a small insulated chamber at 0 F when ambient temperature is 90 F. You may assume a cold-plate temperature of -15 F, intermediate-plate temperature of 60 F, and a warm-plate temperature of 105 F. Estimated cooling load is 120 Btu per hr. You may assume cylindrical elements having a length of 1.00 cm and a diameter of 1.00 cm. Electrical contact resistance r may be assumed as 0.0001 ohm-cm^2. Each stage is to be designed for maximum refrigerating capacity. Determine (a) the number of couples required for each stage, and (b) the required output capacity of the d-c power supply unit.

6.6. A steam-jet-water-vapor system operates at 45 F flash chamber temperature. Warm water enters the spray nozzles at 60 F. Make-up water enters at 70 F. Flash vapor enters the steam ejector in a dry-and-saturated state at a rate of 7000 cu ft per min. Calculate the tons of refrigeration produced.

6.7. Show why a throttling valve may not be used for the expansion process in a gaseous air cycle system.

6.8. A closed air cycle refrigeration system produces 10 tons of refrigeration. Air enters the compressor at 65 psia and 20 F and is compressed polytropically ($Pv^{1.35} = C$) to 225 psia. The temperature of the air leaving the air cooler is 80 F. Expansion is polytropic with $n = 1.30$. Assume frictionless flow. Calculate the coefficient of performance, assuming that the expander helps drive the compressor.

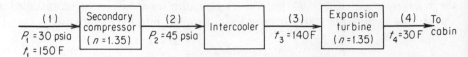

Figure 6.11 Schematic air cycle system for Problem 6.9.

6.9. Figure 6.11 shows part of an aircraft air cycle refrigeration system. Assume frictionless flow. Assume a mechanical efficiency of 80 per cent for both the compressor and the expansion turbine. Will the power output of the expansion turbine be adequate for driving the secondary compressor?

6.10. In Fig. 6.10, assume that air is bled from the jet engine at 620 F and 100 psia at a rate of 18 lb per min. Air enters the turbine at 215 F. For the turbine assume a polytropic exponent $n = 1.20$ and a mechanical efficiency of 85 per cent. Cabin pressure is 13.50 psia and the air is exhausted from the cabin at 75 F. Assume that no air is by-passed around the turbine. Determine (a) the turbine power output, and (b) the tons of refrigeration produced.

SYMBOLS USED IN CHAPTER 6

C	Cylinder clearance fraction, dimensionless.
C.O.P.	Coefficient of performance, dimensionless.
c_p	Specific heat at constant pressure, Btu per (lb) (F).
E	Electromotive force, volt.
h	Specific enthalpy, Btu per lb.
I	Electric current, amp.
K	Deg Kelvin.
k	Thermal conductivity, watt per (cm²) (K per cm).
L/A	Ratio of length of thermoelectric element to cross-sectional area, cm⁻¹.
m	Mass rate of flow, lb per min.
n	Polytropic exponent, dimensionless.
P	Pressure, psfa.
P.D.	Piston displacement, cu ft per min.
Q	Heat transfer, Btu per lb.
q	Rate of heat transfer, watt.
R	Electrical resistance, ohm.
R	Constant in perfect gas law, ft lb per (lb) (R).
r	Electrical contact resistance, ohm-cm².
T	Absolute temperature, R.
t	Temperature, F.
U	Overall thermal conductance, watt per K.
v	Specific volume, cu ft per lb.
W	Work transfer, Btu per lb.
Z	Figure of merit given by Eq. (6.17), K⁻¹.
α	Thermoelectric power, volt per K.

γ	c_p/c_v, dimensionless.
η_R	Refrigerating efficiency, dimensionless.
ρ	Electrical resistivity, ohm-cm.
σ	$1/\rho$, ohm^{-1} cm^{-1}.
τ	Thomson coefficient, volt per K.
ϕ	Peltier coefficient, volt.

7

ULTRA-LOW-TEMPERATURE
REFRIGERATION: CRYOGENICS

7.1. INTRODUCTION

In the previous chapters we studied a variety of refrigerating methods. With these methods, the lowest temperatures may be produced by the cascade vapor compression arrangement. Table 3.2 shows that a cascade system using Refrigerant 14 as the low-side refrigerant, should be capable of attaining a minimum temperature of about −250 F. However, there remains an approximate range of 210 F between this temperature and absolute zero. In this range of temperatures we must use liquefied gases to achieve refrigeration. The term *cryogenics* is used to describe such applications.

Table 7.1 shows boiling temperature (14.696 psia), freezing temperature, and critical temperature and pressure for several cryogenic fluids. Figure E-5 (in the envelope) gives general thermodynamic properties for dry air.

Oxygen, nitrogen, argon, neon, xenon, and krypton are constituents of air. Each element may be obtained by liquefaction and separation of atmospheric air. Pure hydrogen may be obtained by separation from coke-oven gas and from other sources. Pure helium is commonly obtained by separation from helium-bearing natural gas.

The subject of cryogenics has become increasingly important in recent years. Many laboratories are now equipped with ultra-low-temperature systems which allow

TABLE 7.1

PRESSURE AND TEMPERATURE DATA FOR VARIOUS CRYOGENIC FLUIDS

Substance	Boiling Temperature (14.696 psia) F	Freezing Temperature F	Critical Temperature F	Critical Pressure psia
Xenon	−162.4	−169.6	61.9	852
Krypton	−243.8	−251.0	−82.8	798
Oxygen	−297.4	−361.8	−181.8	737
Argon	−302.6	−308.9	−188.3	705
Air	−312.7[a] −318.1[b]	−351.2[c] −357.2[d]	−221.3	546
Nitrogen	−320.4	−346.0	−232.8	492
Neon	−411.0	−415.5	−379.7	395
Hydrogen	−423.0	−434.6	−399.8	188
Helium	−452.1	...	−450.2	33

[a]Condensing temperature of saturated vapor.
[b]Boiling temperature of saturated liquid.
[c]Freezing temperature of liquid.
[d]Melting temperature of solid.

realization of temperatures within a fraction of a degree of absolute zero. The most important commercial application is the separation of oxygen and nitrogen from the atmosphere. Pure oxygen has many industrial uses, one of the foremost being in steel making. Liquid oxygen is also used as an oxidant in rocket propulsion. (Liquid hydrogen is used as the fuel in some rocket engines.)

The subject of cryogenics is a broad field of study, the liquefaction of gases being an important aspect. Adiabatic demagnetization is used to realize temperatures close to absolute zero. Special problems arise in thermometry. Elaborate insulation techniques are necessary. In this chapter we will only touch upon some of the more elementary aspects of cryogenics. More complete information may be obtained from cryogenic books such as *Cryogenic Engineering* by Scott.*

7.2. MINIMUM WORK REQUIRED TO LIQUEFY A GAS

In order to liquefy a gas, heat must be withdrawn from the gas and rejected to an environment at higher temperature. Such a process requires an expenditure of work. It is of thermodynamic importance to know the theoretical minimum work required. The minimum work input occurs when the processes involved are reversible. Figure

*R. B. Scott, *Cryogenic Engineering* (Princeton, New Jersey: D. Van Nostrand Company, Inc., 1959).

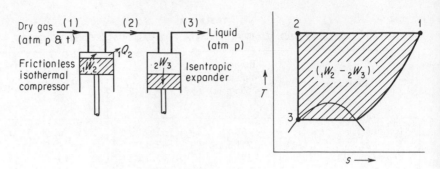

Figure 7.1 Schematic diagrams illustrating a reversible arrangement for liquefying a gas.

7.1 shows one reversible arrangement which we will use for our analysis. The pure gas at State (1) is isothermally compressed to State (2). An isentropic expansion allows liquefaction such that at State (3) pure liquid exists. For the system shown in Fig. 7.1, for steady-flow conditions, we may write

$$_1W_2 = (h_2 - h_1) + T_1(s_1 - s_2)$$
$$_2W_3 = h_2 - h_3$$

Thus

$$W_{net} = {}_1W_2 - {}_2W_3 = T_1(s_1 - s_3) - (h_1 - h_3) \qquad (7.1)$$

Equation (7.1) may be generalized in the form

$$W_{z,min} = T_o(s_o - s_f) - (h_o - h_f) \qquad (7.2)$$

where $W_{z,\,min}$ is the minimum work required per pound of liquid produced. The subscript o refers to the gas state, and the subscript f refers to the liquid state.

In practice the work required for gas liquefaction is many times more than that given by Eq. (7.2).

7.3. COOLING OF A GAS BY EXPANSION

In order to liquefy a gas, heat must be withdrawn from it. Ordinary refrigeration methods are of course inadequate although they may be used for precooling of the gas. Two methods of expansion may be used for reducing the gas temperature: (1) a restrained expansion where work is performed by the gas, and (2) an unrestrained expansion or throttling of the gas.

In Fig. 7.1 an isentropic expansion was theoretically employed. The system of Fig. 7.1 would not be practical since the pressure at State (2) would be prohibitively great. Because of the very low temperature, irreversibilities in the expander would be large, so that in practice the gas could not be liquefied by a restrained expansion alone. However, restrained expansion using a turbine or piston device is an efficient means of gas cooling. Such expansion devices are often employed in gas liquefaction apparatus.

Cooling of a gas by throttling is of major importance, and therefore we will consider this process in some detail. The *Joule-Thomson coefficient,*

$$\mu = \left(\frac{\partial T}{\partial P}\right)_h \tag{7.3}$$

has particular significance. If $\mu = 0$, the temperature of the gas remains constant with throttling. If $\mu > 0$, the temperature of the gas decreases with throttling. If $\mu < 0$, the temperature of the gas increases with throttling. Thus, in cooling of a gas by throttling we require that the gas show a large positive value of μ.

The Joule-Thomson coefficient for a gas is not a constant, and is a function of both pressure and temperature. We will now derive a functional relationship for the coefficient. We begin with the general property relation

$$T\,ds = dh - \frac{v\,dP}{J}$$

Thus

$$ds = \frac{1}{T}\,dh - \frac{v}{TJ}\,dP$$

Since *ds* is an exact differential, we must have

$$\left[\frac{\partial\left(\frac{1}{T}\right)}{\partial P}\right]_h = -\frac{1}{J}\left[\frac{\partial\left(\frac{v}{T}\right)}{\partial h}\right]_P$$

But

$$\left[\frac{\partial\left(\frac{1}{T}\right)}{\partial P}\right]_h = -\frac{1}{T^2}\left[\frac{\partial T}{\partial P}\right]_h$$

and

$$-\frac{1}{J}\left[\frac{\partial\left(\frac{v}{T}\right)}{\partial h}\right]_P = -\frac{1}{J}\left[\frac{\partial\left(\frac{v}{T}\right)}{\partial T}\right]_P\left(\frac{\partial T}{\partial h}\right)_P = -\frac{1}{J}\frac{\left[T\left(\frac{\partial v}{\partial T}\right)_P - v\right]}{T^2}\left(\frac{\partial T}{\partial h}\right)_P$$

Since

$$\frac{1}{c_p} = \left(\frac{\partial T}{\partial h}\right)_P$$

we have

$$\mu = \frac{1}{Jc_p}\left[T\left(\frac{\partial v}{\partial T}\right)_P - v\right] \tag{7.4}$$

Equation (7.4) is general and is valid for liquids as well as gases. In order to calculate the Joule-Thomson coefficient, we must know the equation of state for the substance. For a *perfect gas* $(Pv = RT)$, we have

$$\left(\frac{\partial v}{\partial T}\right)_P = \frac{R}{P} = \frac{v}{T}$$

and the Joule-Thomson coefficient is always equal to zero.

The magnitude of the Joule-Thomson coefficient is a measure of the imperfection of a gas or of its deviation from perfect-gas behavior. For real gases, μ may have either positive or negative values depending upon the thermodynamic state. The temperature at which μ equals zero is called the *inversion temperature* for a given

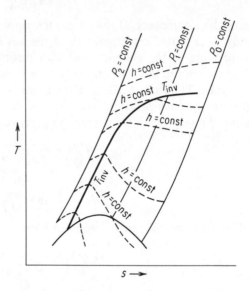

Figure 7.2 Schematic illustration of inversion-temperature line.

pressure. The inversion temperature of a gas may be easily determined from a diagram of properties such as Fig. E-5 for air. An inversion-temperature line may be drawn by connecting the peaks of the constant enthalpy lines. Figure E-5 shows that the inversion temperature for air decreases with increase of pressure. Other gases behave similarly in this respect. Figure 7.2 shows schematically an illustration of an inversion-temperature line on $T\text{-}s$ coordinates.

7.4. LINDE AIR LIQUEFACTION CYCLE

Air liquefaction is important in the recovery of oxygen, nitrogen, and other gases from the atmosphere. In this section we will treat air liquefaction cycles in some detail. Our fundamental analyses will be applicable to the liquefaction of other gases as well.

The most elementary air liquefaction method is the *simple Linde* cycle. Figure 7.3 shows the method. Equipment includes a compressor, a heat exchanger, and a separator. The heat exchanger and separator must be extremely well insulated. Figure 7.3 also shows a $T\text{-}s$ diagram for operation after the system has adjusted itself to steady-state conditions. The compressor discharges air at a relatively high pressure

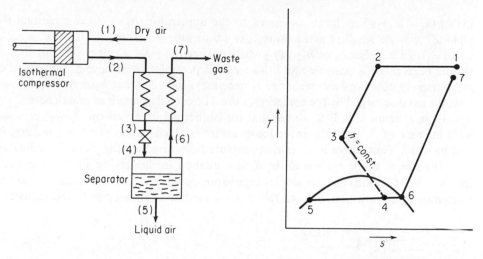

Figure 7.3 Simple Linde cycle for liquefying air.

(50 to 200 atm). It is necessary to cool the air as much as possible during compression. In the heat exchanger the high-pressure air is cooled sufficiently so that liquid air can result from a throttling process. The system can pull itself down to operating conditions after start-up from a warm condition if the temperature of the air leaving the compressor is less than the inversion temperature.

Two performance quantities are of particular interest in gas liquefaction systems. These are (1) the *yield Z* in pounds of liquid produced per pound of gas compressed, and (2) the *specific work requirement W_z* in Btu per pound of liquid produced. We will now derive expressions for Z and W_z for the simple Linde cycle.

For the combination of separator and heat exchanger, we have for steady-flow conditions:

$$m_2 h_2 = m_s h_s + m_7 h_7$$

$$m_2 = m_s + m_7$$

Eliminating m_7, we have

$$Z_L = \frac{m_s}{m_2} = \frac{h_7 - h_2}{h_7 - h_s} \tag{7.5}$$

Assuming isothermal compression, we may write for each pound of air passing through the compressor

$$_1W_2 = T_1(s_1 - s_2) - (h_1 - h_2)$$

The specific work requirement $W_{z,L}$ becomes

$$W_{z,L} = \frac{m_2(_1W_2)}{m_s} = \frac{_1W_2}{Z_L} = [T_1(s_1 - s_2) - (h_1 - h_2)]\frac{(h_7 - h_s)}{(h_7 - h_2)} \tag{7.6}$$

It is useful to examine Eq. (7.5) closely. This equation shows that the yield for the simple Linde cycle is dependent upon States (2), (5), and (7) only. Since the denominator $(h_7 - h_5)$ is very large compared to the numerator $(h_7 - h_2)$, we see that the yield Z_L will be small. Furthermore, the numerator $(h_7 - h_2)$ is decisive, since a small change in either h_7 or h_2 may greatly change the yield Z_L. It is obvious that h_7 should be as large as possible and h_2 as small as possible. To achieve a high value of h_7, extremely effective heat exchange is necessary. In order that h_2 may be relatively low, the gas discharged by the compressor must be cooled as much as practicable.

If we examine Fig. E-5, we see that for isothermal compression, h_2 will decrease with increase of P_2 until the inversion pressure is reached. Beyond this pressure, h_2 will increase. Thus, there is a limiting pressure beyond which the yield Z_L decreases.

The simple Linde cycle is an inefficient method for liquefying air for essentially the same basic reasons that the single-stage vapor-compression cycle is inefficient at low evaporating temperatures. Figure 7.4 shows a modified Linde cycle which requires a

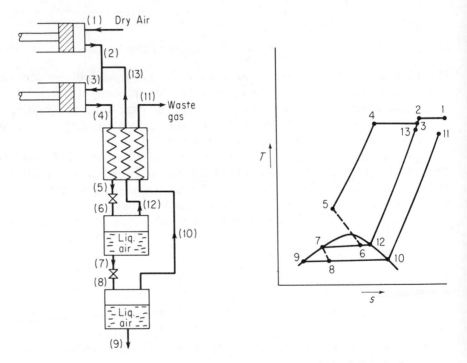

Figure 7.4 Dual-pressure Linde air liquefaction system.

smaller specific work requirement than the simple cycle. Air is liquefied at an intermediate pressure and the flash vapor formed in the first throttling process (5 to 6) is compressed in the high-pressure compressor only. Such an arrangement considerably reduces the throttling loss of the simple Linde cycle.

Both the simple Linde cycle and the dual-pressure Linde cycle may be improved upon if the compressed air is precooled by a mechanical vapor-compression system.

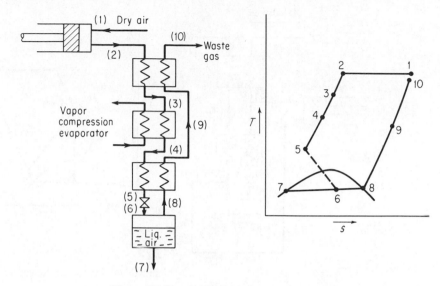

Figure 7.5 Simple Linde air liquefaction system with precooling by mechanical-compression refrigeration system.

Figure 7.5 shows a simple Linde system equipped with a vapor-compression system evaporator for precooling. Since the vapor-compression cycle using conventional refrigerants such as ammonia or Refrigerant 12 is thermodynamically superior to the Linde system itself, the specific work requirement may be considerably reduced. The best performance is obtained when the compressed air is precooled to a relatively low temperature. Thus a multistage or cascade vapor-compression system may be advantageous.

7.5. CLAUDE AIR LIQUEFACTION CYCLE

The elementary Claude air liquefaction system differs from the simple Linde cycle by the addition of an expansion engine and a second heat exchanger. Figure 7.6 shows the Claude cycle. Because of friction and heat conduction losses, the expansion process (3-8) deviates greatly from the theoretical isentropic case (3-A). For the combination of two heat exchangers and separator, we may write

$$m_2 h_2 + m_8 h_8 = m_8 h_3 + m_6 h_6 + m_{11} h_{11}$$

$$m_2 = m_6 + m_{11}$$

Eliminating m_{11}, we have

$$Z_c = \frac{m_6}{m_2} = \frac{h_{11} - h_2}{h_{11} - h_6} + \frac{m_8(h_3 - h_8)}{m_2(h_{11} - h_6)} \tag{7.7}$$

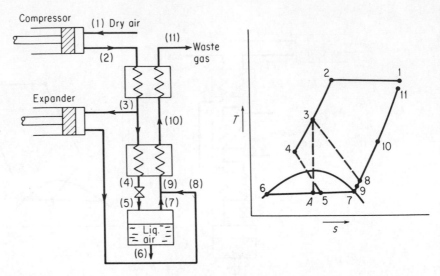

Figure 7.6 Claude air liquefaction system.

The quantity $(h_{11} - h_2)/(h_{11} - h_6)$ represents the yield of the simple Linde cycle. Equation (7.7) may be written as

$$Z_C = Z_L + \frac{m_8(h_3 - h_8)}{m_2(h_{11} - h_6)} \tag{7.8}$$

Thus the yield of the Claude cycle is greater than that of the simple Linde cycle.

Assuming isothermal compression, we have

$${}_1W_2 = T_1(s_1 - s_2) - (h_1 - h_2)$$

The *specific work requirement* $W_{z,c}$ in Btu per pound of liquid air produced is

$$W_{z,c} = \frac{m_2({}_1W_2) - m_8({}_3W_8)}{m_6} = \frac{{}_1W_2}{Z_C} - \frac{m_8}{m_6}({}_3W_8)$$

Thus

$$W_{z,c} = \frac{T_1(s_1 - s_2) - (h_1 - h_2)}{Z_C} - \frac{m_8}{m_6}({}_3W_8) \tag{7.9}$$

The quantity $[T_1(s_1 - s_2) - (h_1 - h_2)]$ may also be equal to the work W_L of the simple Linde cycle. Thus,

$$W_{z,c} = \frac{W_L}{Z_L + \dfrac{m_8(h_3 - h_8)}{m_2(h_{11} - h_6)}} - \frac{m_8}{m_6}({}_3W_8) \tag{7.10}$$

Thus the specific work requirement for the Claude cycle is less than that for the simple Linde cycle.

The main advantage of the expander in the Claude cycle is for cooling of the compressed air rather than for work recovery. The expander work output is often wasted.

Table 7.2 shows a comparison of the specific work requirement in various air-liquefaction cycles.

TABLE 7.2*

SPECIFIC WORK REQUIREMENT FOR VARIOUS AIR LIQUEFACTION CYCLES

System	Work Required Btu/lb Liquid Air
Reversible process (calculated)	324
Simple Linde (observed)	4440
Simple Linde with precooling to −49 F (observed)	2390
Dual-pressure Linde (observed)	2370
Dual-pressure Linde with precooling to −49 F (observed)	1370
Claude (observed)	1370

*Scott, *Cryogenic Engineering* (Princeton, New Jersey: D. Van Nostrand Company, Inc., 1959) p. 17.

7.6. SEPARATION OF AIR

Air liquefaction is of great importance in the production of the pure components of dry air—oxygen, nitrogen, argon, and the rare gases neon, krypton, and xenon. The recovery of oxygen, nitrogen, and other gases from the atmosphere requires the separation of liquefied air. Such applications date from the year 1902 when Linde demonstrated that by rectification of liquid air the separate gases, oxygen and nitrogen, could be obtained.

In this section we will study in an elementary manner how dry air may be separated into its constituents. Bosnjakovic* has covered this topic in detail. Ruhemann† has also treated the subject extensively, including separation of the rare gases. We will emphasize methods for the production of oxygen and nitrogen of commercial purity.

Table 8.2 shows approximately the composition of dry air. The constituents neon, krypton, and xenon have been ignored. These gases appear in dry air only minutely as shown by the following approximate volumetric percentages: neon, 0.0018; krypton, 0.0001; and xenon, 0.000008. For practical problems involving the commercial production of oxygen and nitrogen, it is customary to consider dry air as a binary mixture of 21 per cent oxygen and 79 per cent nitrogen on a volume basis.

Figure E-6 (in the envelope) is an enthalpy-concentration diagram for nitrogen-oxygen mixtures. Coordinates are h in Btu per lb mole and mol-fraction x in moles

*Fr. Bosnjakovic, *Technische Thermodynamik*, Zweiter Teil (Dresden und Leipzig: Theodor Steinkopff, 1937) pp. 142–151.

†M. Ruhemann, *The Separation of Gases* (London: Oxford University Press, 1949).

of nitrogen per mole of mixture. Figure E-6 may be used in exactly the same manner as described in Secs. 5.2–5.4.

When we consider air as a binary mixture, rectification of liquid air follows the same basic principles discussed in Sec. 5.4. Although the principles are the same, there are some important differences between rectifying liquid air and rectifying a binary mixture such as ammonia and water. The system shown in Fig. 5.14 employs external energy transfers. The heat q_G added to the generator comes from an external source such as condensing steam. The heat q_D rejected in the dephlegmator is absorbed by an external medium such as cooling water. These procedures are suitable to rectification of aqua-ammonia but are not permissible in rectification of liquid air. Since very low temperatures are involved, there is obviously no cheap medium available to reject heat to in the dephlegmator. Moreover, any introduction of external heat to the generator would necessitate a vastly increased expenditure of work. In the liquefaction and rectification of air every effort should be made to minimize the addition of any external heat. Thus the rectification column must operate as nearly adiabatic as possible, and all energy transfers must be of an internal nature.

Before considering the double column used for separation of oxygen and nitrogen we will study the more elementary single Linde column. Figure 7.7 shows an arrange-

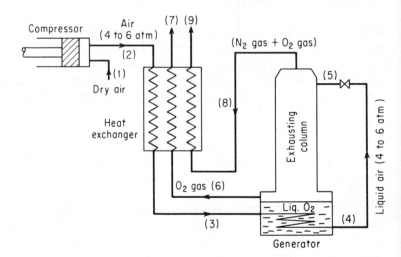

Figure 7.7 Linde single-column apparatus for separation of oxygen.

ment in which a simple exhausting column is used. Figure 7.8 shows a schematic *h-x* diagram for the exhausting column. Dry compressed air at a pressure of four to six atmospheres is cooled in the heat exchanger and is used as a heat source in the generator. Here the air is liquefied (State 4) by transfer of heat to liquid oxygen boiling at atmospheric pressure. The liquid air is then throttled to atmospheric pressure and is admitted to the top of the exhausting column as a mechanical mixture of liquid and vapor (States 5f and 5g).

The exhausting column operates similarly to that described in Sec. 5.4 for the system of Fig. 5.13. One difference is that in Figs. 7.7 and 7.8, the weak component (oxygen) is removed as a gas instead of as a liquid. All operating lines for the exhausting column in Fig. 7.8 are contained between the lines $\overline{P_2\,6}$ and $\overline{P_2\,58}$.

The striking disadvantage of the single Linde column is the loss of oxygen with the nitrogen gas. Figure 7.8 shows that each mole of gas withdrawn from the top of the

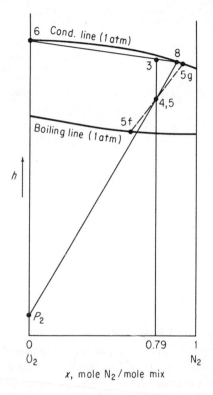

Figure 7.8 Schematic h-x diagram for exhausting column of Fig. 7.7.

column (State 8) will contain $(1 - x_8)$ moles of oxygen. In the limiting case the gas would contain $(1 - x_{5g})$ moles of oxygen, which would represent an oxygen content of about nine per cent. Since air contains only 21 per cent oxygen by volume, the system of Fig. 7.7 could recover only somewhat less than two-thirds of the available oxygen.

In order to obtain a more complete yield of oxygen from air, a double rectifying column is needed. This method was introduced by Linde in 1910. Figure 7.9 schemati-

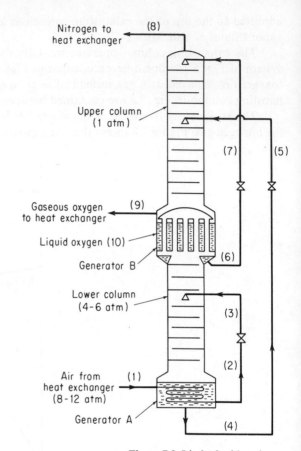

Nitrogen to heat exchanger (8)

Upper column (1 atm)

(7)   (5)

Gaseous oxygen (9) to heat exchanger

Liquid oxygen (10)

Generator B

(6)

Lower column (4-6 atm)

(3)

(2)

Air from (1) heat exchanger (8-12 atm)

Generator A

(4)

Figure 7.9 Linde double column.

cally shows the Linde double column. Such an apparatus may produce both oxygen and nitrogen of arbitrary purity. Compressed air, precooled in the heat exchanger, is liquefied in the coil of Generator A by heat exchange with the boiling liquid in the bottom of the lower column. The liquid air (State 2) is then throttled to the pressure P_L of the lower column and admitted to the column at an intermediate location. In a manner similar to that described in Sec. 5.4 for the system of Fig. 5.14, the vapor rising in the lower column is enriched with nitrogen, while the descending liquid acquires a greater oxygen content. Almost pure nitrogen vapor at the top of the lower column is condensed in Generator B (dephlegmator for lower column). Part of this liquid falls back into the lower column, while the remainder is trapped and used as reflux for the upper column.

     Liquid (State 4) is removed from the bottom of the lower column at an intermediate concentration. This liquid is throttled to the pressure P_U of the upper column and enters the upper column as a mechanical mixture of liquid and vapor. Almost pure nitrogen liquid (State 6) is likewise throttled and admitted to the top of the upper column. Almost pure liquid oxygen (State 10) boils in the bottom of the upper column (Generator B). The oxygen vapor (State 9) is removed at a location above the boiling liquid. Nitrogen vapor (State 8) is removed from the top of the upper column.

Operation of the Linde double column is better understood by consideration of the processes on the h-x diagram. Construction of Fig. 7.10 will now be explained by the following itemized statements. Many of these statements are evident from the discussion in Sec. 5.4.

1. The desired product States (8) and (9) may be located. We assume each may be of arbitrary purity.

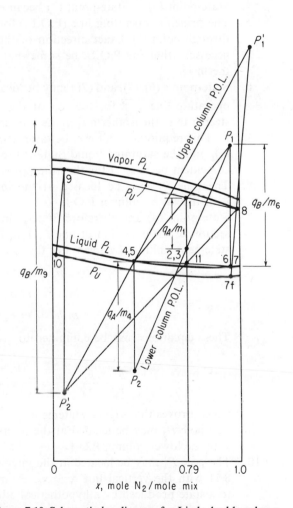

Figure 7.10 Schematic h-x diagram for Linde double column.

2. The required State (1) of the inlet air may be located on the straight line $\overline{8\ 9}$ at $x = 0.79$. This statement is valid if we assume no heat leakage into the column from the surroundings. The statement is proved by the solution of the following equations:

$$m_1 = m_8 + m_9$$
$$m_1 x_1 = m_8 x_8 + m_9 x_9$$
$$m_1 h_1 = m_8 h_8 + m_9 h_9$$

3. The State-point (2) may be located on the boiling line (saturated liquid line) for the inlet pressure at $x_2 = x_1 = 0.79$.

4. State-point (3) = State-point (2) because of the throttling process.

5. The principal operating line (P.O.L.) for the lower column may be drawn through point (3). Exact direction of the line is somewhat arbitrary. It is necessary that the P.O.L. be somewhat steeper than mixture region isotherm t_3.

6. State-points (6), (7), and (7f) may be located in the following manner: The operating line $\overline{7f\ 8}$ for the top of the upper column must be somewhat steeper than the isotherm t_7 in the mixture region, since t_7 must be less than t_8. This requirement allows the saturated liquid Point 7f, the mixture State (7), and the saturated liquid State (6) to be located. States (6) and (7) are coincident because of the throttling process.

7. The pole P_1 may be located at the intersection of vertical line $x = x_6$ with the lower column P.O.L.

8. State-point (4) and therefore State-point (5) may be located at the intersection of the extension of line $\overline{P_1\ 1}$ with the P_L boiling line. Proof of this statement will now be given. For the lower column, we may write

$$m_1 = m_4 + m_6$$
$$m_1 x_1 = m_4 x_4 + m_6 x_6$$
$$m_1 h_1 = m_4 h_4 + m_6 h_6 + q_B$$

These equations may be combined to give the relation

$$\frac{h_1 - h_4}{x_1 - x_4} = \frac{\dfrac{q_B}{m_6} - (h_4 - h_6)}{x_6 - x_4}$$

which proves the original statement.

9. The pole P_2 may be located at the intersection of the vertical line $x = x_4$ with the lower column P.O.L.

10. The pole P_2' may be located at the intersection of the extension of the line $\overline{8\ 11}$ with the vertical line $x = x_9$. Point 11 is a fictitious state equivalent to a state produced by a hypothetical adiabatic mixing of streams 4 and 6.

Thus

$$m_{11} = m_4 + m_6 = m_1$$
$$m_1 x_{11} = m_4 x_4 + m_6 x_6$$
$$m_1 h_{11} = m_4 h_4 + m_6 h_6$$

For the upper column, we may write

$$m_6 + m_4 = m_8 + m_9$$
$$m_6 x_6 + m_4 x_4 = m_8 x_8 + m_9 x_9$$
$$m_6 h_6 + m_4 h_4 + q_B = m_8 h_8 + m_9 h_9$$

Combining the two sets of equations gives the relation

$$\frac{h_8 - h_{11}}{x_8 - x_{11}} = \frac{\dfrac{q_B}{m_9} - (h_9 - h_{11})}{x_{11} - x_9}$$

which proves the original statement.

11.   The upper column P.O.L. may be drawn through Points P_2' and 5.
12.   The pole P_1' may be located at the intersection of the upper column P.O.L. with the extension of the operating line $\overline{7f\ 8}$.
13.   State-point 10 for the liquid oxygen may be located by the assumption that $t_{10} = t_9$.

In an actual problem, construction of the h-x diagram would have to be checked to prove that all operating lines for each column were steeper than the corresponding mixture-region isotherms, or, stated in a different way, that no negative temperature differences existed. If this requirement were not satisfied, the compressed-air pressure and possibly the lower column pressure would need to be adjusted.

Figure 7.11 shows a contemporary type of double column for production, of gaseous oxygen. Such a column has both theoretical and practical advantages over the Linde double column of Fig. 7.9. Partially liquefied compressed air is admitted to the lower column at a pressure of four to six atmospheres. In the lower column the rising vapor is enriched with nitrogen. High-purity nitrogen vapor is condensed to liquid in the generator by transfer of heat to liquid oxygen boiling at a pressure of approximately one atmosphere. Part of this liquid nitrogen is used as reflux for the upper column. Liquid (State 2) is removed from the bottom of the lower column at an intermediate concentration, throttled to the pressure of the upper column, and admitted to the upper column as a mechanical mixture of liquid and vapor. The upper column is also supplied with air vapor (State 6) from an expansion turbine. High-purity oxygen vapor (State 7) is removed at a location above the boiling liquid in the generator. High-purity nitrogen vapor (State 8) is removed from the top of the upper column.

Figure 7.12 shows a schematic h-x diagram for the column of Fig. 7.11. States (10), (11), and (12) are all fictitious and are introduced for convenience in construc-

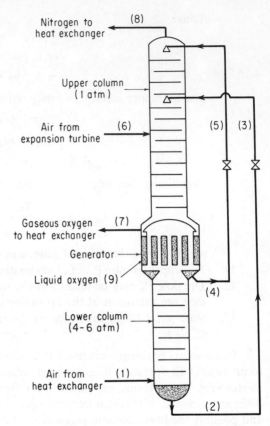

Figure 7.11 Contemporary double column.

tion of the diagram. State (10) is the adiabatic equivalent (resulting from a hypothetical adiabatic mixing) of streams 2 and 4. State (11) is the adiabatic equivalent of streams 3, 5, and 6, while State (12) is the adiabatic equivalent of streams 1 and 6, or of 7 and 8. Construction of Fig. 7.12 should be obvious from the detailed discussion given earlier for Fig. 7.10. In Fig. 7.12 we may also show that several mass-flow ratios are given by the graphical construction. These include the relations

$$\frac{m_2}{m_4} = \frac{\overline{4\,10}}{\overline{10\,2}} = \frac{\overline{P_1\,1}}{\overline{1\,2}}$$

$$\frac{m_6}{m_1} = \frac{\overline{12\,1}}{\overline{6\,12}} = \frac{\overline{11\,10}}{\overline{6\,11}}$$

$$\frac{m_7}{m_8} = \frac{\overline{8\,12}}{\overline{12\,7}} = \frac{\overline{8\,11}}{\overline{11\,P_2'}}$$

The column of Fig. 7.11 is superior to the Linde column of Fig. 7.9. Only two pressure levels are needed instead of the three involved with the Linde double column.

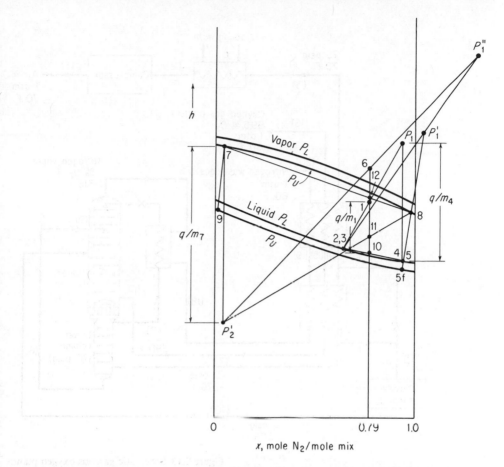

Figure 7.12 Schematic h-x diagram for column of Fig. 7.11.

Thus a lower supply air pressure and therefore a lower compressor power input is needed for a system with the column of Fig. 7.11.

An analysis utilizing the second law of thermodynamics is of benefit to a study of air-separation systems. Although we will not include a second-law analysis here,* we may observe that such a study would show that fewer irreversibilities result in a system employing the column of Fig. 7.11 than in a system using the Linde double column. This circumstance also leads to a lower power requirement for a system with the column of Fig. 7.11.

The column of Fig. 7.11 has further advantages compared to the Linde double column. Generator A is eliminated. A lesser heat-transfer rate is required in the generator of Fig. 7.11 than in Generator B of Fig. 7.9.

---

*For further reading see H. Bliss and B. F. Dodge, "Oxygen Manufacture, Part I," *Chemical Engineering Progress*, Vol. 45, No. 1, January 1949, pp. 56–64.

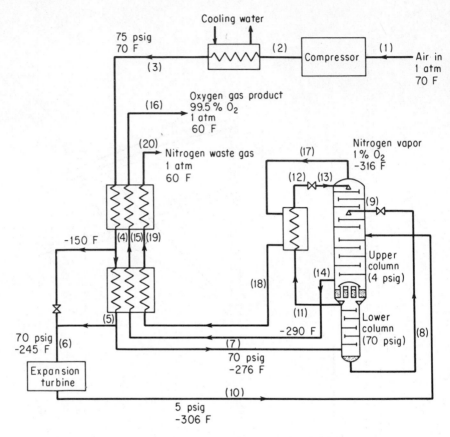

Figure 7.13 Schematic gaseous oxygen plant.

Figures 7.13 and 7.14 show schematic diagrams of contemporary type plants for producing oxygen. Moisture removal and air purification equipment are not shown. The diagrams and associated operating conditions have been adapted from a paper by Zenner.* Figure 7.13 shows a schematic gaseous oxygen plant. The rectifying column is identical to that of Fig. 7.11. Figure 7.14 shows a schematic liquid oxygen plant. A very high compressed-air pressure is required. Zenner states that maximum yield of liquid oxygen is obtained when about 50 per cent of the high-pressure air is expanded in the engine. Zenner also states that power costs for the gaseous oxygen product are about one-third those for liquid oxygen product.

We will only briefly consider how the rare constituents of air may be separated. Figure 7.15 shows the column of Fig. 7.11 equipped with devices which allow separa-

*G. H. Zenner, "Cryogenics and Mechanical Engineering," *Mechanical Engineering*, Vol. 83, September 1961, pp. 39–43.

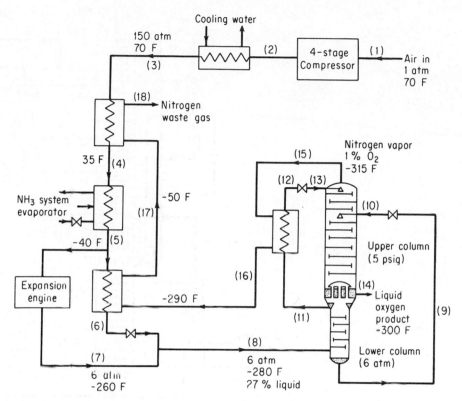

Figure 7.14 Schematic liquid oxygen plant.

tion of argon, krypton and xenon, and neon and helium. The boiling point of argon is but slightly lower than that of oxygen. Argon is most concentrated at about the middle of the upper column. Vapor withdrawn at this point may contain from 10 to 15 per cent of argon while the remainder is mostly oxygen. The mixture is rectified in the argon column where most of the oxygen is condensed and withdrawn as liquid from the bottom of the column while crude argon vapor is removed at the top. The argon vapor may be further purified by selective adsorption.

Both neon and helium appear in dry air in very small amounts. Both gases are much more volatile than nitrogen. In air separation these gases form as non-condensibles in the dome of the dephlegmator of the lower column. As shown in Fig. 7.15, gas (non-condensibles plus nitrogen, primarily) may be withdrawn from the dome of the lower column and passed through a rectifier which is refrigerated by liquid nitrogen. Most of the nitrogen vapor present with the helium and neon may be condensed and separated. Neon may be separated from the mixture of neon and helium by a selective adsorption arrangement.

Both krypton and xenon have higher boiling temperatures than oxygen. Both

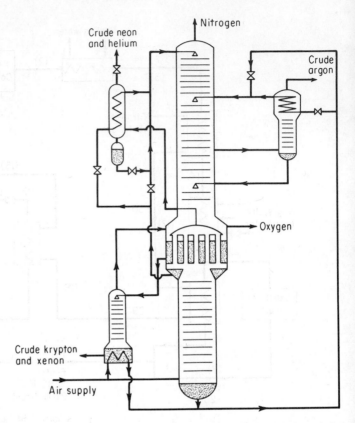

Figure 7.15 Schematic apparatus for separation of air.

remain in the liquid oxygen in the bottom of the upper column. As shown in Fig. 7.15 liquid may be withdrawn from the bottom of the upper column and admitted to the top of an auxiliary exhausting column. Here the oxygen portion may be distilled from the mixture. The crude krypton and xenon may be further purified and separated by a selective adsorption system.

7.7. LIQUEFACTION OF HYDROGEN AND HELIUM

Hydrogen and helium are the most difficult gases to liquefy because of their extremely low liquefaction temperatures. All other substances freeze at temperatures higher than the boiling points (14.696 psia) of either element.

Figure 7.16 shows a schematic arrangement of equipment for liquefying hydrogen. Pure hydrogen gas at about 100 atmospheres pressure is precooled in a divided heat exchanger arrangement (Heat Exchangers A and B) to about −300 F. In Heat Exchanger C, the hydrogen vapor is further cooled to about −340 F by nitrogen boiling

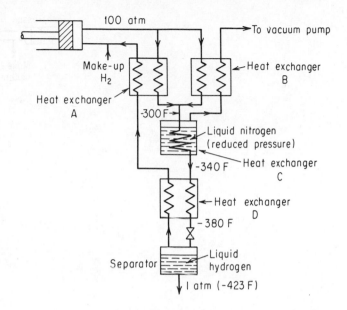

Figure 7.16 Schematic hydrogen liquefier.

under reduced pressure. In Heat Exchanger D, the hydrogen vapor is still further cooled to about −380 F by the low-pressure hydrogen vapor returning from the Separatoι. By throttling to atmospheric pressure, liquid hydrogen may then be produced.

Helium is the most difficult of all gases to liquefy. At 14.696 psia it boils at approximately −452 F. Its maximum inversion temperature is approximately −390 F. Helium was first liquefied by H. K. Onnes of the University of Leiden in 1908. Helium may be liquefied by an arrangement similar to that in Fig. 7.16, where both liquid nitrogen and liquid hydrogen are used for precooling. Disadvantages of this method include the high cost and hazardous nature of liquid hydrogen.

Helium may also be liquefied through use of the Claude principle, where expansion engines are used for producing refrigeration. Figure 7.17 shows a system developed by Collins.* It illustrates the principle of the *Collins Cryostat* manufactured by Arthur D. Little, Inc. Helium at approximately 12 atm pressure is supplied to the liquefier by a four-stage compressor. Part of the helium is precooled by liquid nitrogen. Through use of the combination of heat exchangers and expansion engines, the high pressure helium gas may be cooled to about −430 F. Liquid helium is then produced by throttling of the vapor to atmospheric pressure. Collins has reported, for one such plant, a liquefaction rate of 25 to 32 liters per hour with a power requirement of 45 kw.

*S. C. Collins, "Helium Liquefier," *Science*, Vol. 116, September 19, 1952, pp. 289–294.

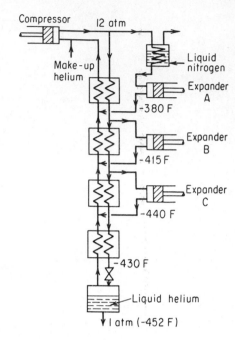

Figure 7.17 Schematic helium liquefier.

7.8. *APPROACH TO ABSOLUTE ZERO BY THE ADIABATIC DEMAGNETIZATION OF A PARAMAGNETIC SALT*

Through the lowering of pressure over liquid helium, it is possible to attain a temperature of about −458 F. Lower temperatures may be achieved by adiabatic demagnetization of certain paramagnetic salts.

The atoms of a paramagnetic salt may be considered as tiny magnets. When the salt is not magnetized, the atoms are oriented in a random manner such that the magnetic forces are in balance. If such a substance is exposed to a strong magnetic field, the atoms attempt to align themselves in the direction of the field. The realignment of the atoms requires work. This work is converted into heat and causes a temperature rise unless the heat is removed by some form of cooling. When the magnetic field is removed, the atoms readjust their positions to the original random arrangement. Such readjustment requires that the salt atoms perform work. In the absence of external heat exchange, the internal energy of the salt decreases. Consequently the salt must cool itself.

Figure 7.18 shows a schematic arrangement for adiabatic demagnetization of a paramagnetic salt. The inner chamber containing the salt specimen is initially filled with gaseous helium. This chamber is surrounded by a bath of liquid helium which

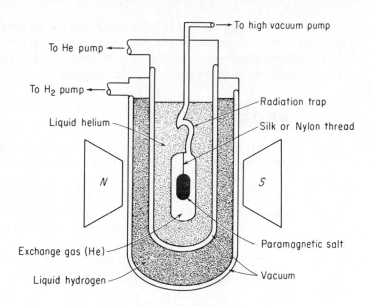

Figure 7.18 Schematic adiabatic demagnetization apparatus.

in turn is surrounded by a bath of liquid hydrogen. External insulation is not shown in Fig. 7.18.

The salt is first cooled to about −458 F by reducing the pressure over the liquid helium. Next, the salt is exposed to a strong magnetic field of about 25,000 gauss. Heat produced by magnetization of the salt is transferred to the liquid helium without causing an increase of the salt temperature. With the magnetic field still present, the inner chamber containing the salt is evacuated of gaseous helium. The salt is then almost completely thermally isolated. Upon release of the magnetic field, the salt temperature decreases in an almost perfectly isentropic way. Temperatures of the salt as low as approximately 0.001 K have been reported.

An interesting and important problem in adiabatic demagnetization is the determination of the very low temperatures produced. In the neighborhood of absolute zero, all ordinary means of temperature measurement fail. Scott* has given a thorough review of temperature measurement in cryogenic work.

Close to absolute zero, temperature cannot be measured but must be calculated. The temperature may be calculated approximately by the *Curie relation*

$$x = \frac{C}{T}$$

where x is the magnetic susceptibility of the salt, T the absolute temperature, and C a constant. Through magnetic measurements the absolute temperature may be calculated.

*Scott, *Cryogenic Engineering*, pp. 109–141.

The most accurate method of calculating the final temperature produced in a salt through adiabatic demagnetization directly utilizes the second law of thermodynamics. The demagnetization process is almost perfectly isentropic. By means of heat measurements, the temperature may be calculated.

PROBLEMS

7.1. Determine the theoretical minimum work required in Btu per lb to produce liquid air at 14.696 psia from dry air initially at 70 F and 14.696 psia.

7.2. Dry air at 70 F and 14.696 psia is to be liquefied by the simple Linde method. Assume the air is isothermally compressed at 70 F to 2500 psia. Also assume that the waste gas leaves the system at 70 F. Determine (a) the yield of liquid air in lb per lb of air compressed, and (b) the specific work requirement in Btu per lb of liquid air.

7.3. For the dual-pressure Linde cycle of Fig. 7.4, prove that the specific work requirement $W_{Z,\,2L}$ in Btu per lb of liquid air is

$$W_{Z,\,2L} = \frac{(h_{10} - h_9)}{(h_{10} - h_8)}(_1W_2) + \left[\frac{(h_{13} - h_9)}{(h_{13} - h_4)} - \frac{(h_8 - h_9)(h_{11} - h_{13})}{(h_{10} - h_8)(h_{13} - h_4)}\right](_3W_4)$$

where

$$_1W_2 = T_1(s_1 - s_2) - (h_1 - h_2)$$
$$_3W_4 = T_3(s_3 - s_4) - (h_3 - h_4)$$

7.4. Dry air at 70 F and 14.696 psia is to be liquefied at 14.696 psia by the dual-pressure Linde method. Intermediate pressure is 200 psia and high-side pressure is 2500 psia. Assume that air is isothermally compressed at 70 F in both stages. Also assume that the waste gas leaves the system at 70 F. Determine (a) the yield of liquid air m_9/m_1 in lb per lb of air compressed in the low-pressure compressor, and (b) the specific work requirement in Btu per lb of liquid air.

7.5. Dry air at 70 F and 14.696 psia is to be liquefied by the Claude method. Assume the air is isothermally compressed at 70 F to 2500 psia. Assume that 80 per cent of the total mass of air compressed passes through the expander. Assume that the temperature of the air entering the expander is -110 F while the temperature of the waste gas leaving the system is 70 F. Assume that the enthalpy drop across the expander is 50 per cent of the isentropic value. Determine (a) the yield of liquid air in lb per lb of air compressed, (b) the specific work requirement in Btu per lb of liquid air if the enthalpy drop across the expander is fully recovered as work to help drive the compressor, and (c) the specific work requirement in Btu per lb of liquid air if the expander work output is wasted.

7.6. Gaseous oxygen (99% pure, molar basis) is to be produced at a rate of 350 cu ft per hr in a Linde single-column plant similar to the one depicted in Fig. 7.7. The column operates at a pressure of 1.0 atm. Air leaves the compressor at a pressure of 10.0 atm. The nitrogen product is to have 91% purity. Dry compressed air enters the heat exchanger at 70 F and the outgoing products leave the heat exchanger at 60 F. You may assume that the inlet air and the gaseous products behave as ideal gases near room temperature and at atmospheric pressure. Determine (a) the volume of air in cu ft per hr which must be compressed, (b) the oxygen yield (ratio of moles of oxygen produced to moles of oxygen available), and (c) the temperature at which the air must leave the heat exchanger.

7.7. A Linde double-column apparatus processes 200,000 cu ft of air ($t = 70$ F, $p = 14.696$ psia) per hr. The air is compressed and cooled and passed through the boiler (Generator A) at a pressure of 8.0 atm. The oxygen product has a concentration of 0.01 (molar basis) and the nitrogen product has a concentration of 0.95. The upper column operates at a pressure of 1.0 atm. Determine (a) the heat exchange rate in Btu per hr in Generator A if the air at 8.0 atm pressure leaves as a saturated liquid, and (b) the percentage of incoming air which is exhausted as oxygen product (molar basis).

7.8. Dry air is to be separated in a Linde double-column apparatus into oxygen (99.5% purity, molar basis) and nitrogen (99.0% purity). The precooled compressed air leaves the generator as a saturated liquid at 6.0 atm pressure. The upper column operates at 1.0 atm pressure and the lower column at 4.0 atm. Assume equilibrium exists between nitrogen product and reflux at top of upper column ($t_7 = t_8$). The concentration x_4 of the oxygen-rich liquid leaving Generator A is 0.59. (a) Plot the entire h-x diagram and, in a table, show values of p, t, h, and x for all ten principal locations shown in Fig. 7.9. (b) Determine the h-x coordinates for each of the four poles. (c) Determine the heat transferred in Generator A per mole of oxygen-rich liquid removed at (4). (d) Determine the heat transferred in Generator B per mole of oxygen product removed at (9). (e) Determine the oxygen yield (ratio of moles of oxygen produced to moles of oxygen available).

7.9. Analyze the schematic h-x diagram of Fig. 7.12 for the double column of Fig. 7.11. Develop a set of itemized statements on construction of the diagram similar to those shown in Sec. 7.6 for the Linde double column. Show details and fully justify each statement.

7.10. The gaseous oxygen plant of Fig. 7.13 is to be designed to produce 10 tons of oxygen product per hr. For the compressor, you may assume polytropic compression ($Pv^{1.35} = C$) and a mechanical efficiency of 70 per cent. Assume sea-level pressure and neglect pressure drop in the water-cooled heat exchanger following the compressor. Calculate the required shaft horsepower input for the compressor.

7.11. Develop a general schematic h-x diagram analysis similar to Fig. 7.12, but for the double column of Fig. 7.14. Ignore specific property data shown in Fig. 7.14. Also ignore heat exchanger between nitrogen vapor and reflux from lower column.

SYMBOLS USED IN CHAPTER 7

c_p Specific heat at constant pressure, Btu per (lb)(F).

h Specific enthalpy, Btu per lb, or Btu per lb mole.

J Mechanical equivalent of heat, 778 ft-lb per Btu.

m Mass rate of flow, lb per min, or lb moles per min.

P Pressure, psia or psfa.

q Rate of heat transfer, Btu per min.

R Constant in perfect-gas law, ft-lb per (lb)(R).

s Specific entropy, Btu per (lb)(R.).

T Absolute temperature, R.

t Temperature, F.

v Specific volume, cu ft per lb.

W Work transfer, Btu per lb; W_L for Linde cycle; W_C for Claude cycle.

W_Z Specific work requirement, Btu per lb of liquid produced; $W_{Z,L}$ for Linde cycle; $W_{Z,C}$ for Claude cycle.

x Concentration of nitrogen in mixture of nitrogen and oxygen, moles N_2 per mole mix.

Z Yield of liquid, lb liq per lb of gas compressed; Z_L for Linde cycle; Z_C for Claude cycle.

μ Joule-Thomson coefficient $(\partial T/\partial P)_h$, F per psf.

Part III

PSYCHROMETRICS

8

THERMODYNAMIC PROPERTIES
OF MOIST AIR

8.1. ATMOSPHERIC AIR

The earth's atmosphere is a mixture of several gases including nitrogen, oxygen, argon, carbon dioxide, water vapor, and traces of other gases. Atmospheric air usually contains various particulate matter. Additional vapors are often present. Dust particles and condensible vapors such as water vapor are usually concentrated in the atmosphere only within a few thousand feet of the earth's surface. Above an altitude of about 20,000 ft, the atmosphere consists essentially of dry air.

Barometric pressure is the force per unit area due to the weight of the atmosphere. Standard sea-level pressure is 14.696 psia. As one proceeds vertically into the atmosphere, pressure decreases. Temperature also decreases until the tropopause is reached.

The *U.S. Standard Atmosphere* provides a reference standard with respect to barometric pressure for an air-conditioning engineer. The *ASHRAE Handbook of Fundamentals** has given the following information which forms the definition of the U.S. Standard Atmosphere:

*ASHRAE Handbook of Fundamentals (New York: American Society of Heating, Refrigerating and Air Conditioning Engineers, 1967) pp. 103–104.

1. The atmosphere consists of dry air which behaves as a perfect gas; thus

$$Pv = RT \tag{8.1}$$

2. Gravity is constant at 32.174 ft per sec².
3. At sea level, pressure is 29.921 in. Hg and temperature is 59 F.
4. Temperature t decreases linearly with altitude z up to the lower limit of the isothermal atmosphere according to the relation

$$t = t_0 - 0.003566z \tag{8.2}$$

with t and t_0 in F and z in ft. The isothermal atmosphere (-66 F) begins at 35,332 ft.

Equations (8.1) and (8.2) may be combined to give the relation

$$P = P_o \left(1 - \frac{0.003566z}{T_o}\right)^{5.256} \tag{8.3}$$

Table 8.1 shows variation of pressure and temperature for the U.S. Standard Atmosphere for altitudes from 1000 ft below sea level to 35,000 ft above sea level.

TABLE 8.1*

VARIATION OF PRESSURE AND TEMPERATURE WITH ALTITUDE
FOR U. S. STANDARD ATMOSPHERE

Altitude z ft	Pressure P in. Hg	Temperature t F
$-1,000$	31.02	$+62.6$
-500	30.47	$+60.8$
0	29.921	$+59.0$
$+500$	29.38	$+57.2$
$+1,000$	28.86	$+55.4$
5,000	24.90	$+41.2$
10,000	20.58	$+23.4$
15,000	16.89	$+5.5$
20,000	13.76	-12.3
25,000	11.12	-30.0
30,000	8.90	-47.8
35,000	7.06	-65.6

*Abstracted by permission from *ASHRAE Handbook of Fundamentals*, p. 104.

Although the actual atmosphere above a locality would not correspond precisely to the U.S. Standard Atmosphere, Table 8.1 provides a convenient means for estimating barometric pressure for a given altitude above sea level.

8.2. FUNDAMENTAL DISCUSSION OF MOIST AIR

The composition of atmospheric air is variable, particularly with regard to amounts of water vapor and particulate matter. Before we can discuss thermodynamic properties, the substance must be precisely defined. The working substance in air conditioning problems is called *moist air*. Moist air is defined as a *binary mixture of dry air and water vapor*. Goff,* in a final report of the Working Subcommittee, International Joint Committee on Psychrometric Data, has defined dry air as shown in Table 8.2.

TABLE 8.2*

COMPOSITION OF DRY AIR

Substance	Molecular Weight	Mol-fraction Composition in Dry Air	Partial Molecular Weight in Dry Air
Oxygen (O_2)	32.000	0.2095	6.704
Nitrogen (N_2)	28.016	0.7809	21.878
Argon (A)	39.944	0.0093	0.371
Carbon Dioxide (CO_2)	44.01	0.0003	0.013
		1.0000	28.966

*Reprinted by permission from *Trans. ASHVE*, Vol. 55, p. 463.

Although somewhat arbitrary, this composition is regarded as exact, by definition. The molecular weights for dry air and water vapor are 28.966 and 18.016, respectively.

Moist air may contain variable amounts of water vapor from zero (dry air) to that of saturated moist air. Goff† has defined *saturation of moist air* as that condition where moist air may coexist in neutral equilibrium with associated condensed water, presenting a flat surface to it.

For a pure substance, such as water vapor or dry air, the virial equation of state can be used. Thus

$$P\bar{v} = \bar{R}T + A_2P + A_3P^2 + \ldots \qquad (8.4)$$

where $\bar{v}$ and $\bar{R}$ are molar values. The coefficients A_2 and A_3 are the second and third virial coefficients, respectively, and are sole functions of temperature. They account for forces of attraction and/or repulsion between molecules. The form of Eq. (8.4) is predicted by theory, but experimentation is needed for evaluating the virial coefficients.

*J. A. Goff, "Standardization of Thermodynamic Properties of Moist Air," *Trans. ASHVE*, Vol. 55, 1949, pp. 463–464.

†*Trans. ASHVE*, Vol. 55, pp. 473–474.

Formulations for specific enthalpy and specific entropy can be derived from Eq. (8.4) through use of appropriate identical relations of thermodynamics. For specific enthalpy, we may show that

$$\bar{h} = \bar{h}^0 + B_2 P + \tfrac{1}{2} B_3 P^2 + \ldots \qquad (8.5)$$

where $\bar{h}$ and $\bar{h}^0$ are molar enthalpies with $\bar{h}^0$ taken at zero pressure. The coefficients B_2 and B_3 are, respectively, the second and third virial coefficients for enthalpy. They are functions of the respective virial coefficients A_2 and A_3 in Eq. (8.4).

Let us now consider a container filled with moist air. Molecules may react among themselves in various ways. Like molecules in any numerical combination may react, while unlike molecules in all possible combinations may interact.

For moist air, statistical mechanics predicts that

$$P\bar{v} = \bar{R}T - [x_a^2 A_{aa} + 2x_a x_w A_{aw} + x_w^2 A_{ww}]P$$
$$- [x_a^3 A_{aaa} + 3x_a^2 x_w A_{aaw} + 3x_a x_w^2 A_{aww} + x_w^3 A_{www}]P^2 - \ldots \qquad (8.6)$$

where x_a is mol-fraction of dry air and x_w is mol-fraction of water vapor. A_{aa} is a virial coefficient for one molecule of dry air reacting with another molecule of dry air; A_{aaa} for three molecules of dry air, etc. A_{aw} is an interaction coefficient for one molecule of dry air interacting with one molecule of water vapor, etc.

Equation (8.6) and identical relations of thermodynamics lead to the following relation for specific enthalpy:

$$\bar{h} = x_a \bar{h}_a^0 + x_w \bar{h}_w^0 - [x_a^2 B_{aa} + 2x_a x_w B_{aw} + x_w^2 B_{ww}]P$$
$$- \tfrac{1}{2}[x_a^3 B_{aaa} + 3x_a^2 x_w B_{aaw} + 3x_a x_w^2 B_{aww} + x_w^3 B_{www}]P^2 - \ldots \qquad (8.7)$$

where $\bar{h}_a^0$ and $\bar{h}_w^0$ are specific enthalpies at zero pressure for dry air and water vapor, respectively. Meanings of other terms should be obvious.

For convenience in calculations, the properties enthalpy h and volume v are based upon unit weight of dry air. The conversion of units is accomplished by dividing $\bar{v}$ in Eq. (8.6) and $\bar{h}$ in Eq. (8.7) by $28.966x_a$. Likewise, the property expressing composition is also based upon unit weight of dry air. By definition, the *humidity ratio W* is that weight of water vapor associated with unit weight of dry air. Humidity ratio is related to the mol-fraction of water vapor by the expression

$$W = \frac{18.016}{28.966} \frac{x_w}{(1 - x_w)} = 0.622 \frac{x_w}{1 - x_w} \qquad (8.8)$$

For humidity ratio at saturation W_s, Goff* has given the relation

$$W_s = 0.622 \frac{f_s P_{w,s}}{P - f_s P_{w,s}} \qquad (8.9)$$

where $P_{w,s}$ is the saturation pressure of pure water (temperature function only) and f_s is a coefficient which is a function of temperature and pressure. Table 8.3 shows values of f_s for sea-level pressure and for 0 to 125 F.

Three moist-air properties are associated with temperature. *Dry-bulb temperature*

Trans. ASHVE, Vol. 55, p. 474.

t is the true temperature of moist air at rest. *Dew-point temperature t_d* is defined as the solution $t_d(P, W)$ of the equation

$$W_s(P, t_d) = W \tag{8.10}$$

In words, dew-point temperature is the saturation temperature (temperature of incipient condensation) corresponding to the humidity ratio and pressure of a given moist-air state.

TABLE 8.3*

COEFFICIENT f_s IN EQ. (8.9) FOR 14.696 PSIA

Temperature F	f_s	Temperature F	f_s
0	1.0048	70	1.0045
10	1.0046	80	1.0047
20	1.0046	90	1.0048
30	1.0045	100	1.0050
40	1.0044	110	1.0053
50	1.0044	120	1.0055
60	1.0044	125	1.0057

*Reprinted by permission from *Heating, Ventilating, Air Conditioning Guide*, Vol. 37, p. 14.

Thermodynamic wet-bulb temperature t^* is separately considered in Sec. 8.3.

Two measures of relative saturation are commonly used. *Degree of saturation μ* is defined by the relation

$$\mu = \frac{W}{W_s} \tag{8.11}$$

where W_s is the humidity ratio at saturation for the same temperature and pressure as those of the actual state.

Relative humidity ϕ is defined by the relation

$$\phi = \frac{x_w}{x_{w,s}} \tag{8.12}$$

where $x_{w,s}$ is the mol-fraction of water vapor at saturation for the same temperature and pressure as those of the actual state.

We may convert Eq. (8.12) to the forms

$$\phi = \frac{1 + \dfrac{0.622}{W_s}}{1 + \dfrac{0.622}{W}} = \mu \frac{(0.622 + W_s)}{(0.622 + W)} \tag{8.13}$$

It is important to observe that neither μ nor ϕ are defined when the temperature of moist air exceeds the saturation temperature of pure water corresponding to the moist-air pressure. For sea-level pressure, W_s approaches infinity at 212 F. Thus for 14.696 psia, μ and ϕ are undefined for temperatures higher than 212 F.

8.3. THERMODYNAMIC WET-BULB TEMPERATURE

Adiabatic saturation temperature is that temperature at which water, by evaporating into air, can bring the air to saturation adiabatically at the same temperature. It is unnecessary to inject practical details into a discussion of adiabatic saturation since the results of the process are given by definition. However, to help us understand it better, we may consider how such a process could be approached.

Figure 8.1 schematically shows a device which may provide adiabatic saturation. The chamber could be indefinitely long and perfectly insulated. The total quantity

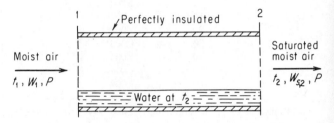

Figure 8.1 Schematic adiabatic saturation device.

of water present could be large compared to that added to the air in a given length of time. We may assume no temperature gradients within the water body. Regardless of the initial temperature of the water, we would expect that after sufficient time the water would assume a constant temperature. This limiting temperature of water should be less than the entering air dry-bulb temperature, but greater than the entering air dew-point temperature.

Referring to Fig. 8.1 and assuming steady-state conditions, it is convenient to write the steady-flow energy equation on a basis of one pound of dry air passing through the chamber. Thus

$$h_1 + (W_{s,2} - W_1)h_{f,2} = h_{s,2} \tag{8.14}$$

Since the leaving air is saturated, and since we assume a constant pressure P, the quantities $W_{s,2}$, $h_{s,2}$, and $h_{f,2}$ are sole functions of temperature t_2. We may then deduce that t_2 is a function of h_1, W_1, and P, or that t_2 is a function of State (1). Therefore, t_2 is a thermodynamic property of State (1). We call this property the *thermodynamic wet-bulb temperature* $t^\star$. Equation (8.14) may be written as

$$h + (W_s^\star - W)h_f^\star = h_s^\star \tag{8.15}$$

Thus, for given values of h, W, and P (given moist-air state), the thermodynamic wet-bulb temperature $t^\star$ is that value of temperature which satisfies Eq. (8.15). There are numerous practical problems where the concept of thermodynamic wet-bulb temperature is useful.

8.4. THE GOFF AND GRATCH TABLES FOR MOIST AIR

By applying fundamental procedures of statistical mechanics, Goff and Gratch* calculated accurate thermodynamic properties of moist air for standard sea-level pressure. These properties for dry air and for saturated moist air are shown in Table A.5 of the Appendix. In the following list, brief explanations of the data in Table A.5 are shown.

W_s = humidity ratio of saturated air, lb of water vapor per lb of dry air.
v_a = specific volume of dry air under 14.696 psia pressure, cu ft per lb.
v_s = volume of saturated air, cu ft per lb of dry air.
$v_{as} = v_s - v_a$, cu ft per lb of dry air.
h_a = specific enthalpy of dry air, Btu per lb. Zero enthalpy for dry air is taken at 0 F.
h_s = enthalpy of saturated air, Btu per lb of dry air.
$h_{as} = h_s - h_a$, Btu per lb of dry air.
s_a = specific entropy of dry air, Btu per (lb)(R). Zero entropy for dry air is taken at 0 F.
s_s = entropy of saturated air, Btu per (lb of dry air)(R).
$s_{as} = s_s - s_a$, Btu per (lb of dry air)(R).

Calculations for volume, enthalpy, and entropy of unsaturated moist-air states are closely given by the relations

$$v = v_a + \mu v_{as} + \bar{\bar{v}} \tag{8.16}$$

$$h = h_a + \mu h_{as} + \bar{\bar{h}} \tag{8.17}$$

$$s = s_a + \mu s_{as} + \bar{\bar{s}} \tag{8.18}$$

where

$$\bar{\bar{v}} = \frac{\mu(1-\mu)A}{1 + 1.608\mu W_s} \tag{8.19}$$

$$\bar{\bar{h}} = \frac{\mu(1-\mu)B}{1 + 1.608\mu W_s} \tag{8.20}$$

$$\bar{\bar{s}} = \frac{\mu(1-\mu)C}{1 + 1.608\mu W_s} \tag{8.21}$$

The constants A, B, and C are given in Table 8.4. For temperatures less than 96 F, they may be taken as zero.

*J. A. Goff and S. Gratch, "Thermodynamic Properties of Moist Air," *Trans. ASHVE*, Vol. 51, 1945, pp. 125–164.

TABLE 8.4*

CONSTANTS A,B, AND C FOR EQS. (8.19)–(8.21)

Temperature F	A cu ft/lb$_a$	B Btu/lb$_a$	C Btu/(lb$_a$)(R)
96	0.0018	0.0268	0.00004
112	0.0042	0.0650	0.00009
128	0.0096	0.1439	0.00020
144	0.0215	0.3149	0.00042
160	0.0487	0.6969	0.00091
176	0.1169	1.636	0.00207
192	0.3363	4.608	0.00567

*Abstracted by permission from *ASHRAE Handbook of Fundamentals*, p. 100.

Example 8.1. Moist air exists at 80 F dry-bulb temperature, 60 F dew-point temperature, and 14.696 psia pressure. Determine (a) the humidity ratio, lb$_w$ per lb$_a$, (b) degree of saturation, (c) relative humidity, (d) enthalpy, Btu per lb$_a$, and (e) the volume in cu ft per lb$_a$.

SOLUTION: (a) Since sea-level pressure exists, Table A.5 will be used. By Eq. (8.10), $W = W_s$ at 60 F. Thus $W = 0.01108$ lb$_w$/lb$_a$.

(b) At 80 F, $W_s = 0.02233$ lb$_w$/lb$_a$. By Eq. (8.11),

$$\mu = 0.01108/0.02233 = 0.496 \text{ or } 49.6 \text{ per cent}$$

(c) By Eq. (8.13)

$$\phi = \frac{(0.496)(0.622 + 0.02233)}{(0.622 + 0.01108)} = 0.505 \text{ or } 50.5 \text{ per cent}$$

(d) We may calculate the enthalpy by Eq. (8.17) with $\bar{h} = 0$. Using Table A.5, we have

$$h = 19.221 + (0.496)(24.47) = 31.36 \text{ Btu/lb}_a$$

(e) We may calculate the volume by Eq. (8.16) with $\bar{v} = 0$. Using Table A.5, we have

$$v = 13.601 + (0.496)(0.486) = 13.84 \text{ cu ft/lb}_a$$

Example 8.2. Moist air exists at 160 F dry-bulb temperature and 30 per cent degree of saturation. Pressure is 14.696 psia. Determine (a) the enthalpy, Btu per lb$_a$, and (b) the volume, cu ft per lb$_a$.

SOLUTION: (a) By Table A.5, $W_s = 0.2990$ lb$_w$/lb$_a$. By Table 8.4, $B = 0.6969$ Btu/lb$_a$. By Eq. (8.20),

$$\bar{h} = \frac{(0.30)(0.70)(0.6969)}{1 + (1.608)(0.30)(0.2990)} = 0.128 \text{ Btu/lb}_a$$

By Eq. (8.17),

$$h = 38.472 + (0.30)(337.8) + 0.128 = 139.94 \text{ Btu/lb}_a$$

(b) By Table 8.4, $A = 0.0487$ cu ft/lb$_a$. By Eq. (8.19),

$$\bar{\bar{v}} = \frac{(0.30)(0.70)(0.0487)}{1 + (1.608)(0.30)(0.2990)} = 0.009 \text{ cu ft/lb}_a$$

By Eq. (8.16),

$$v = 15.622 + (0.30)(7.446) + 0.009 = 17.87 \text{ cu ft/lb}_a$$

Example (8.2) shows that terms $\bar{h}$ and $\bar{v}$ could have been neglected with only a small error.

Example 8.3. Moist air exists at 80 F dry-bulb temperature, 60 F thermodynamic wet-bulb temperature, and 14.696 psia pressure. Through use of Table A.5, determine (a) the degree of saturation, and (b) the enthalpy.

SOLUTION: (a) By Eqs. (8.17) and (8.15), and with $\bar{h} = 0$,

$$h_a + \mu h_{as} = h_s^\star - (W_s^\star - W)h_f^\star = h_s^\star - W_s^\star h_f^\star + \mu W_s h_f^\star$$

Thus

$$\mu = \frac{h_s^\star - W_s^\star h_f^\star - h_a}{h_{as} - W_s h_f^\star} = \frac{26.46 - (0.01108)(28.08) - 19.22}{24.47 - (0.02233)(28.08)} = 0.284$$

(b) By Eq. (8.17),

$$h = 19.22 + (0.284)(24.47) = 26.17 \text{ Btu/lb}_a$$

8.5. PERFECT-GAS RELATIONSHIPS FOR APPROXIMATE CALCULATIONS

The methods of Sec. 8.4 allow accurate calculation of moist-air properties through use of Table A.5. However, Table A.5 is restricted to a pressure of 14.696 psia. Basic relationships shown in Sec. 8.2 may be applied for any existing pressure, but expressions such as Eqs. (8.6) and (8.7) are too complicated for ordinary calculations.

Equation (8.6) reduces to the perfect-gas form for very low pressures. At ordinary atmospheric pressures, $P\bar{v} = \bar{R}T$ may be an excellent approximation to Eq. (8.6). Use of perfect-gas relations (Dalton's Rule) is highly convenient compared to complete solution of formulae such as Eqs. (8.6) and (8.7). In this section, we will assume that circumstances permit assumption of Dalton's Rule. Formulations for this special situation will be developed from the more exact expressions of Sec. 8.2.

In Eq. (8.8), x_w reduces to P_w/P for the case of Dalton's Rule. Thus

$$W = 0.622 \frac{P_w}{P - P_w} \tag{8.22}$$

In Eq. (8.9), f_s is unity for the case of Dalton's Rule. Thus

$$W_s = 0.622 \frac{P_{w,s}}{P - P_{w,s}} \tag{8.23}$$

For enthalpy, Eq. (8.7) reduces to the zero-pressure form

$$\bar{h} = x_a \bar{h}_a^0 + x_w \bar{h}_w^0 \tag{8.24}$$

when Dalton's Rule is applied. Division of Eq. (8.24) by $28.966 x_a$ gives the moist-air enthalpy in units of Btu per pound of dry air. Thus

$$h = h_a^0 + W h_w^0 \tag{8.25}$$

where the specific enthalpies h_a^0 and h_w^0 are functions of temperature only. It is customary to make h_a^0 zero at 0 F. For water vapor, standard procedure is to make the enthalpy of saturated liquid water zero at 32 F. With these procedures, the following expressions (see Sec. 1.14) are suitable within the accuracy of Dalton's Rule:

$$h_a^0 = 0.240t$$

$$h_w^0 = h_g = 0.45t + 1061$$

where h_g is the specific enthalpy of saturated water vapor, Btu per lb, at the dry-bulb temperature t. Thus

$$h = 0.240t + W h_g \tag{8.26}$$

or

$$h = c_{p,a} t + 1061 W \tag{8.27}$$

where

$$c_{p,a} = 0.240 + 0.45 W \tag{8.28}$$

The quantity $c_{p,a}$ is the specific heat of moist air in units of Btu per (lb of dry air) (F).

A Dalton's Rule expression for the volume of moist air per pound of dry air may be found by dividing $\bar{v} = \bar{R}T/P$ by $28.966 x_a$. Thus

$$v = \frac{R_a T}{P - P_w} \tag{8.29}$$

By Eqs. (8.13), (8.22), and (8.23)

$$\phi = \frac{P_w}{P_{w,s}} \tag{8.30}$$

when Dalton's Rule is assumed.

By Eqs. (8.10) and (8.22), we may deduce that the dew-point temperature t_d is equal to the saturation temperature corresponding to the vapor pressure P_w when Dalton's Rule is accepted.

Equation (8.15) may be altered through use of Dalton's Rule relations. By Eqs. (8.15), (8.27), and (8.28)

$$(W_s^\star - W) h_{fg}^\star = c_{p,a}(t - t^\star) \tag{8.31}$$

By Eqs. (8.15) and (8.26)

$$W = \frac{W_s^\star h_{fg}^\star - 0.240(t - t^\star)}{h_g - h_f^\star} \tag{8.32}$$

As stated earlier, equations of this section are only approximate. Figure 8.2 shows per cent error in calculation of humidity ratio W_s', enthalpy h_s', and volume v_s' by Eqs. (8.23), (8.26), and (8.29) respectively, for saturated air at 14.696 psia. Correct

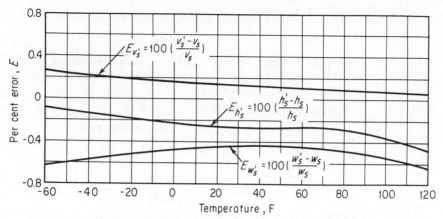

Figure 8.2 Error of perfect-gas relations in calculation of humidity ratio, enthalpy, and volume of saturated air at 14.696 psia pressure.

values W_s, h_s, and v_s were taken from Table A.5. Figure 8.2 shows that the error in calculation of humidity ratio by Eq. (8.23) is less than about 0.6 per cent in the range of -50 F to 110 F. Except for temperatures above about 100 F, the error in calculation of enthalpy by Eq. (8.26) is less than about 0.4 per cent. For the temperature range of -30 F to 120 F, the volume may be calculated by Eq. (8.29) with an error of less than 0.2 per cent.

Example 8.4. Rework Example 8.1, using perfect-gas relations and steam table data.

SOLUTION: (a) We may deduce from Eq. (8.22) that the dew-point temperature t_d is the saturation temperature corresponding to the partial pressure of water vapor P_w. By Table A.1, at 60 F, $P_w = 0.25618$ psi. By Eq. (8.22),

$$W = \frac{(0.622)(0.256)}{14.696 - 0.256} = 0.01103 \text{ lb}_w/\text{lb}_a$$

(b) By Table A.1, at 80 F, $P_{w,s} = 0.50701$ psia. By Eq. (8.23),

$$W_s = \frac{(0.622)(0.507)}{14.696 - 0.507} = 0.02223 \text{ lb}_w/\text{lb}_a$$

By Eq. (8.11),

$$\mu = \frac{0.01103}{0.02223} = 0.496$$

(c) By Eq. (8.30),

$$\phi = \frac{0.25618}{0.50701} = 0.505$$

(d) By Table A.1, at 80 F, $h_g = 1096.12$ Btu/lb$_w$. By Eq. (8.26),

$$h = (0.240)(80) + (0.01103)(1096.12) = 31.29 \text{ Btu/lb}_a$$

(e) By Eq. (8.29),

$$v = \frac{(53.35)(540)}{(14.44)(144)} = 13.85 \text{ cu ft/lb}_a$$

We observe that the answers of Example 8.4 differ but slightly from those of Example 8.1.

Example 8.5. Moist air exists at 90 F dry-bulb temperature, 40 per cent relative humidity, and 12.56 psia barometric pressure. Determine (a) the dew-point temperature, and (b) the enthalpy.

SOLUTION: (a) By Table A.1, at 90 F, $P_{w,s} = 0.69838$ psi. By Eq. (8.30),

$$P_w = (0.40)(0.69838) = 0.27935 \text{ psi}$$

Interpolation in Table A.1 gives $t_d = 62.4$ F.

(b) By Eq. (8.22),

$$W = \frac{(0.622)(0.2794)}{12.56 - 0.279} = 0.01415 \text{ lb}_w/\text{lb}_a$$

By Table A.1, at 90 F, $h_g = 1100.44$ Btu/lb$_w$. By Eq. (8.26),

$$h = (0.240)(90) + (0.01415)(1100.44) = 37.17 \text{ Btu/lb}_a$$

Example 8.6. Moist air exists at 100 F dry-bulb temperature, 80 F thermodynamic wet-bulb temperature, and 13.86 psia pressure. Determine (a) the humidity ratio, and (b) the relative humidity.

SOLUTION: (a) By Table A.1, at 80 F, $P_{w,s}^\star = 0.50701$ psi. By Eq. (8.23),

$$W_s^\star = \frac{(0.622)(0.507)}{13.86 - 0.507} = 0.02362 \text{ lb}_w/\text{lb}_a$$

By Table A.1, at 80 F, $h_{fg}^\star = 1048.07$ Btu/lb$_w$, $h_f^\star = 48.05$ Btu/lb$_w$. At 100 F, $h_g = 1104.74$ Btu/lb$_w$. By Eq. (8.32),

$$W = \frac{(0.02362)(1048.07) - (0.240)(20)}{1104.74 - 48.05} = 0.01888 \text{ lb}_w/\text{lb}_a$$

(b) Equation (8.22) may be changed to the form

$$P_w = \frac{1.608 PW}{1 + 1.608 W}$$

Thus

$$P_w = \frac{(1.608)(13.86)(0.01888)}{1 + (1.608)(0.01888)} = 0.408 \text{ psi}$$

By Table A.1, at 100 F, $P_{w,s} = 0.94959$ psi. By Eq. (8.30),

$$\phi = \frac{0.408}{0.950} = 0.430 \text{ or } 43.0 \text{ per cent}$$

PROBLEMS

8.1. Moist air exists at 80 F dry-bulb temperature, 0.0150 lb$_w$ per lb$_a$ humidity ratio, and 14.696 psia pressure. Through the use of Table A.5 and fundamental relations, determine (a) the dew-point temperature, (b) relative humidity, (c) volume in cu ft per lb$_a$, and (d) the enthalpy in Btu per lb$_a$.

8.2. Calculate values of humidity ratio, enthalpy, and volume for saturated air at 14.696 psia pressure using perfect-gas relations and Table A.1, for temperatures of (a) 70 F, and (b) −20 F. Compare your results with those shown in Table A.5.

8.3. The atmosphere within a room is at 70 F dry-bulb temperature, 50 per cent degree of saturation, and 14.696 psia pressure. The inside surface temperature of the windows is 40 F. Will moisture condense out of the air upon the window glass?

8.4. Assume that the dimensions of the room of Problem 8.3 are 30 ft by 15 ft by 8 ft high. Calculate the number of pounds of water vapor in the room.

8.5. Moist air exists at a dry-bulb temperature of 100 F, relative humidity of 20 per cent, and 14.696 psia pressure. Find the enthalpy in Btu per lb dry air. Base solution on Table A.5 and fundamental relations. Do not use Dalton's Rule expressions.

8.6. Moist air exists at a dew-point temperature of 65 F, a relative humidity of 60.3 per cent, and a pressure of 14.00 psia. Determine (a) the humidity ratio in lb_w per lb_a, and (b) the volume in cu ft per lb_a.

8.7. Determine the relative humidity and dew-point temperature of moist air at 95 F dry-bulb temperature, 80 F thermodynamic wet-bulb temperature, and 13.20 psia pressure.

8.8. Calculate the enthalpy in Btu per lb_a of moist air at 70 F thermodynamic wet-bulb temperature, 34 F dew-point temperature, and 14.696 psia pressure.

8.9. Calculate the dry-bulb temperature of moist air at 80 F thermodynamic wet-bulb temperature, 0.01250 lb_w per lb_a humidity ratio, and 13.00 psia pressure.

8.10. Develop the complete derivation for Eq. (8.3).

8.11. Through the use of basic definitions and perfect-gas relations, derive the following equations:

(a) $v = \dfrac{R_a T}{P}(1 + 1.608W)$

(b) $\phi = 1.608\dfrac{P}{P_{w,s}}\left(\dfrac{W}{1 + 1.608W}\right)$

8.12. Through the use of basic definitions and Dalton's Rule expressions, show that

$$v = v_a + \mu v_{as}$$

reduces to

$$v = \frac{R_a T}{P - P_w}$$

SYMBOLS USED IN CHAPTER 8

A Coefficient in Eq. (8.19), given by Table 8.4, cu ft per lb dry air.
B Coefficient in Eq. (8.20), given by Table 8.4, Btu per lb dry air.

C Coefficient in Eq. (8.21), given by Table 8.4, Btu per (lb dry air) (R).

$c_{p,a}$ Specific heat of moist air at constant pressure, Btu per (lb dry air) (F).

f_s Coefficient in Eq. (8.9), dimensionless.

h Enthalpy of moist air, Btu per lb dry air; h_s for air saturated at t; $h_s^\star$ for air saturated at $t\star$.

h_a Specific enthalpy of dry air, Btu per lb; h_a^0 for case of zero pressure.

h_{as} $h_s - h_a$, Btu per lb dry air.

h_f Specific enthalpy of liquid water, Btu per lb; $h_f^\star$ evaluated at $t\star$ and P.

h_g Specific enthalpy of saturated water vapor, Btu per lb.

$h_{fg}^\star$ $h_g^\star - h_f^\star$, Btu per lb.

h_w^0 Specific enthalpy of water vapor at zero pressure, Btu per lb.

$\bar{h}$ Specific enthalpy, Btu per lb mole; $\bar{h}^0$ for case of zero pressure.

$\bar{h}_a^0$ Specific enthalpy of dry air at zero pressure, Btu per lb mole.

$\bar{h}_w^0$ Specific enthalpy of water vapor at zero pressure, Btu per lb mole.

$\bar{\bar{h}}$ Correction factor for enthalpy given by Eq. (8.20), Btu per lb dry air.

P Pressure of moist air, psia, psfa, or in. Hg.

P_w Pressure of water vapor, psia or psfa; $P_{w,s}$ for saturated water vapor.

R Constant in perfect-gas equation of state, ft-lb per (lb) (R); R_a for dry air.

$\bar{R}$ Constant in perfect-gas equation of state, equals 1545 ft-lb per (lb mole) (R).

s Entropy of moist air, Btu per (lb dry air) (R); s_a for dry air; s_s for saturated moist air at t; $s_{as} = s_s - s_a$.

$\bar{s}$ Correction factor for entropy given by Eq. (8.21), Btu per (lb dry air) (R).

T Absolute dry-bulb temperature, R.

t Dry-bulb temperature, F.

t_d Dew-point temperature, F.

$t\star$ Thermodynamic wet-bulb temperature, F.

v Volume of moist air, cu ft per lb dry air; v_s for saturated moist air.

v_a Specific volume of dry air, cu ft per lb.

v_{as} $v_s - v_a$, cu ft per lb dry air.

$\bar{v}$ Specific volume, cu ft per lb mole.

$\bar{\bar{v}}$ Correction factor for volume given by Eq. (8.19), cu ft per lb dry air.

W Humidity ratio, lb water vapor per lb dry air; W_s for air saturated at t; $W_s^\star$ for air saturated at $t\star$.

x_a Mol-fraction of dry air, moles dry air per mole moist air.

x_w Mol-fraction of water vapor, moles water vapor per mole moist air; $x_{w,s}$ for case of saturated air at t.

z Altitude, ft.

μ Degree of saturation, dimensionless.

ϕ Relative humidity, dimensionless.

9

PSYCHROMETRIC CHARTS AND
ELEMENTARY APPLICATIONS

9.1. INTRODUCTION

In Chapter 8 we considered various thermodynamic properties of moist air and the equations relating them. With these relations we may accurately solve problems concerning moist air. However, the calculations are tedious and time-consuming.

It is of considerable advantage to plot the relations to give a nomograph called a *psychrometric chart*. Such a chart not only allows graphical reading of the various properties but also provides for convenient graphical solutions to many process problems.

A thermodynamic state for moist air is uniquely fixed if the barometric pressure and two independent properties are known. A psychrometric chart may be constructed for some single value of barometric pressure. Traditionally, standard sea-level pressure has been used. The choice of coordinates is, of course, arbitrary. Most psychrometric charts used in the United States have employed dry-bulb temperature and humidity ratio as the basic coordinates. In 1923, Richard Mollier* of Dresden,

*Richard Mollier, "Ein neues Diagram für Dampfluftgemische," *ZVDI*, Vol. 67, September 8, 1923, pp. 869–872.

Germany introduced a chart using enthalpy and humidity ratio as the coordinates. This chart received wide acceptance in Europe.

The use of enthalpy and humidity ratio as basic coordinates presents many advantages. Thermodynamic wet-bulb temperature lines are identically straight. A majority of the common psychrometric processes appear as straight lines on h-W coordinates. In general, the Mollier type of chart allows the most fundamentally consistent treatment of air conditioning problems with a minimum of approximations.

9.2. CONSTRUCTION OF THE PSYCHROMETRIC CHART

Through use of the psychrometric relations of Chapter 8, we may readily construct an h-W chart. Experience has shown that the best intersections result when enthalpy is used as an oblique coordinate and humidity ratio as a rectangular coordinate.

The construction method used here has been adapted from Goodman's* procedure. Figure 9.1 shows the basic geometry. The enthalpy lines are inclined at an

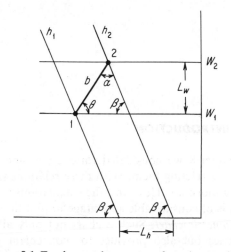

Figure 9.1 Fundamental geometry of psychrometric chart.

angle β to the horizontal humidity ratio lines. The line $\overline{1\,2}$ may represent any straight line. The vertical scalar distance representing $(W_2 - W_1)$ is L_W and the *horizontal* scalar distance representing $(h_2 - h_1)$ is L_h. By the law of sines, we may write

$$\frac{L_h}{\sin \alpha} = \frac{b}{\sin \beta} \tag{9.1}$$

*William Goodman, *Air Conditioning Analysis* (New York: The Macmillan Company, 1943) pp. 271–276.

Also

$$\alpha = 180 - (\theta + \beta)$$
$$\sin \alpha = \sin(\theta + \beta) = \sin \theta \cos \beta + \cos \theta \sin \beta \tag{9.2}$$

and

$$b = \frac{L_w}{\sin \theta} \tag{9.3}$$

By Eqs. (9.1)–(9.3), we have

$$\cot \beta + \cot \theta = \frac{L_h}{L_w} \tag{9.4}$$

We may define an enthalpy scale factor s_h in Btu per (lb$_a$)(inch) by

$$s_h = \frac{(h_2 - h_1)}{L_h}$$

and a humidity ratio scale factor s_W in lb$_w$ per (lb$_a$)(inch) by

$$s_W = \frac{(W_2 - W_1)}{L_W}$$

Then

$$\frac{L_h}{L_W} = \frac{s_W(h_2 - h_1)}{s_h(W_2 - W_1)} = \frac{q'}{S} \tag{9.5}$$

where

$$S = \frac{s_h}{s_W} = \text{chart scale factor in Btu per lb}_w$$

and

$$q' = \frac{(h_2 - h_1)}{(W_2 - W_1)} = \text{enthalpy-moisture ratio in Btu/lb}_w$$

By Eqs. (9.4) and (9.5),

$$\cot \theta + \cot \beta = \frac{q'}{S} \tag{9.6}$$

Equation (9.6) is the general equation used for constructing the various straight lines on the psychrometric chart. Before any lines may be drawn, the scale factor S and the angle β must be established. Each may be arbitrarily fixed, but their choice greatly influences the appearance and usability of the resulting chart. A large scale factor S is desirable so that the angle θ may vary more uniformly with uniform changes in the enthalpy-moisture ratio q'. On the other hand, the enthalpy scale factor s_h can be made too large for accurate reading of enthalpy values.

The inclination angle β of the enthalpy lines may be fixed by choosing some property line and considering it to be vertical. Usually some one of the dry-bulb temperature lines is chosen. By Eq. (8.25), and noting that h_a^0 and h_w^0 are constants, it is true for a line of constant dry-bulb temperature that

$$h_2 - h_1 = (W_2 - W_1)h_w^0$$

or

$$q' = h_w^0 = h_g \tag{9.7}$$

Thus the dry-bulb temperature lines are straight but not parallel, since h_g varies with temperature. If we arbitrarily choose some dry-bulb temperature line t to be vertical, by Eqs. (9.6) and (9.7), we have

$$\tan \beta = \frac{S}{h_{g,t}} \tag{9.8}$$

Equation (9.8) establishes the inclination of the enthalpy lines. The grid of h and W lines may then be constructed.

The remaining procedure for constructing a complete psychrometric chart will now be reviewed.

Saturation curve. The saturation curve is a locus of points representing saturated air. For standard barometric pressure, values of h_s and W_s at various temperatures may be read from Table A.5 and the points plotted. For other barometric pressures, Eq. (8.23) and Eq. (8.26) may be used.

Dry-bulb temperature lines. We have already shown that, within the accuracy of the perfect-gas approximation, these lines are straight. By Eqs. (9.6) and (9.7),

$$\cot \theta = \frac{h_g}{S} - \cot \beta$$

For various dry-bulb temperatures we may determine h_g and calculate $\cot \theta$. One point on each line may be conveniently located at $W = 0$ and the line then may be drawn through knowledge of the angle θ.

Thermodynamic wet-bulb temperature lines. By Eq. (8.15),

$$q' = h_f^\star = \frac{h_s^\star - h}{W_s^\star - W}$$

For a line of constant wet-bulb temperature, Eq. (9.6) becomes

$$\cot \theta = \frac{h_f^\star}{S} - \cot \beta$$

Thus thermodynamic wet-bulb temperature lines are identically straight. For various values of $t^\star$, we may determine θ and extend the lines through one known point. Locations are known on the saturation curve. For unsaturated air, a point may be calculated by solving Eq. (8.15) or Eq. (8.32).

Volume lines. Lines of constant volume in cu ft per lb_a are not strictly straight, but their curvature is so slight that they may be drawn as straight lines. For a line of constant volume, with the perfect-gas approximation, we may show that

$$\frac{dh}{dW} = 0.001201 \frac{Pv}{(1 + 1.608W)^2} + 854$$

where P is in lb per sq ft, v in cu ft per lb_a, and W in lb_w per lb_a. An average value of

W may be used with little error. For example, with $W = 0.01$,

$$q' = 0.001163Pv + 854$$

and by Eq. (9.6),

$$\cot \theta = \frac{0.001163Pv + 854}{S} - \cot \beta$$

For a chosen v, a value of dry-bulb temperature may be found at $W = 0$. The volume line may then be extended through knowledge of θ.

Relative humidity lines. By Eqs. (8.22) and (8.30), we have

$$W = 0.622 \frac{\phi P_{w,s}}{P - \phi P_{w,s}}$$

For chosen values of ϕ and t, we may calculate values of W and plot points for a line of constant relative humidity.

Enthalpy-moisture ratio protractor. A convenient aid in many psychrometric chart problems is a protractor showing the direction of straight lines for various values of the enthalpy-moisture ratio q'. Such a protractor may be directly calculated from Eq. (9.6).

9.3. THE ASHRAE PSYCHROMETRIC CHARTS

Three psychrometric charts are included in the envelope accompanying this text. These are the three ASHRAE sea-level pressure charts. They are of the Mollier type, each having basic coordinates of enthalpy and humidity ratio. Figure E-7 is a psychrometric chart for low temperatures covering the range of -40 to 50 F. Figure E-8 is for the normal range of temperatures from 32 to 120 F where most air conditioning problems occur. Figure E-9 is for the high temperature range of 60 to 250 F. The three charts were constructed using data from the Goff and Gratch tables for moist air. Palmatier* has discussed construction of the normal-temperature-range chart.

The three ASHRAE psychrometric charts are similar in format. We will limit further discussion in this section to Fig. E-8. Humidity-ratio lines are horizontal and are shown for the range from zero (dry air) to 0.03 lb moisture per lb dry air. Enthalpy lines are obliquely drawn across the chart in intervals of 5 Btu per lb dry air. All enthalpy lines are precisely parallel. Edge scales for enthalpy are shown above the saturation curve and at the bottom and right-hand margins for intervals of 0.2 Btu per lb dry air.

Dry-bulb temperature lines are shown in one deg F intervals. The dry-bulb temperature lines are drawn straight, are inclined slightly from the vertical position, and are not strictly parallel to one another. Thermodynamic wet-bulb temperature lines are obliquely drawn across the chart in intervals of one deg F. Their directions

*Palmatier, E. P., "Construction of the Normal Temperature ASHRAE Psychrometric Chart," *ASHRAE Trans.*, Vol. 69, 1963, pp. 7–12.

differ but slightly from that of the enthalpy lines. The thermodynamic wet-bulb temperature lines are exactly straight, but are not strictly parallel to each other.

Relative humidity lines are shown in intervals of 10 per cent from zero ($W = 0$) to 100 per cent (saturation curve). Volume lines are obliquely drawn straight lines in intervals of 0.5 cu ft per lb dry air. The volume lines are not strictly parallel to one another.

A narrow region above the saturation curve has been developed for fog conditions. A fog is a mechanical mixture of saturated moist air and water droplets, both at the same temperature. A fog is equivalent to the condition produced by taking saturated moist air and adiabatically supersaturating it with water at the same temperature. Thus an isotherm in the fog region is an extension of a thermodynamic wet-bulb temperature line.

A protractor and a nomograph are shown to the left of the chart body. The protractor shows two scales—one for the ratio of enthalpy difference to humidity-ratio difference ($q' = \Delta h / \Delta W$) and one for the sensible-total heat ratio. The nomograph provides an alternate method for finding moist air enthalpy and also allows a direct reading of the enthalpy ($\Delta W h_f$) in Btu per lb dry air for liquid water added or rejected in a process.

Example 9.1. Moist air exists at a condition of 100 F dry-bulb temperature, 65 F thermodynamic wet-bulb temperature, and 14.696 psia pressure. Determine (a) the humidity ratio, (b) enthalpy, (c) dew-point temperature, (d) relative humidity, and (e) the volume.

SOLUTION: The state-point may be located on Fig. E-8 at the intersection of the 100 F dry-bulb temperature line and the 65 F thermodynamic wet-bulb temperature line.

(a) Read $W = 0.00523$ lb water vapor per lb dry air.

(b) The enthalpy may be found by two methods. Through use of two triangles, draw a line parallel to nearest enthalpy line (30 Btu per lb dry air) through the state-point to the nearest edge scale. Read $h = 29.80$ Btu per lb dry air.

An alternate method for determining enthalpy will now be described. By Eq. (8.15)

$$h = h_s^\star - (W_s^\star - W)h_f^\star = h_s^\star + D$$

where $h_s^\star$ is the enthalpy of saturated moist air at the thermodynamic wet-bulb temperature and D is the enthalpy deviation given by the nomograph. At 65 F, read $h_s^\star = 30.06$ Btu per lb dry air. By the nomograph, at $W = 0.00523$ and $t^\star = 65$, read $D = -0.26$ Btu per lb dry air. Thus, $h = 30.06 - 0.26 = 29.80$ Btu per lb dry air.

(c) The dew-point temperature may be read at the intersection of $W = 0.00523$ lb water vapor per lb dry air with the saturation curve. Thus, $t_d = 40.1$ F.

(d) Read $\phi = 13$ per cent.

(e) The volume may be accurately found by linear interpolation between the volume lines for 14.0 and 14.5 cu ft per lb dry air. Thus, $v = 14.22$ cu ft per lb dry air.

9.4. ELEMENTARY PSYCHROMETRIC PROCESSES

Many of the problems in processing moist air with various apparatus result in rather complex process lines on a psychrometric chart. Psychrometric processing with actual apparatus will be covered in later chapters. At this time we will consider a number of types of fundamental problems which are independent of heat- and/or mass-transfer rates in equipment. These problems will further illustrate some of the many uses for a psychrometric chart.

All of the processes considered here will be for steady-flow conditions. The total or barometric pressure will be assumed constant throughout the process. This assumption is valid in almost all psychrometric processing problems since even in actual apparatus such as finned coils or spray chambers, the pressure drop is typically less than one inch of water.

Sensible heating or cooling of moist air. If heat is added to moist air with no addition of moisture, then we speak of the process as one of *sensible* heating. Such a process may occur if moist air is passed across a heated surface such as a bundle of finned tubes where a medium such as hot water or steam circulates inside the tubes.

Sensible cooling is the reverse of sensible heating and may occur if air is passed across a cool surface. To restrict the process to only sensible cooling, the surface temperature must be higher than the air dew-point temperature. Figure 9.2(a) shows a

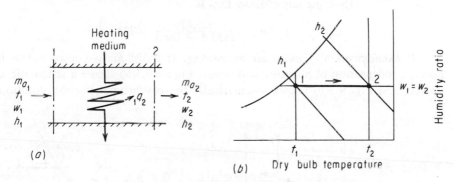

Figure 9.2 Schematic illustration of sensible heating of moist air.

schematic device for heating air and Fig. 9.2(b) shows the process on a schematic psychrometric chart. The humidity ratio remains constant.

The steady-flow energy and material balance equations are

$$m_{a,1} h_1 + {}_1 q_2 = m_{a,2} h_2$$

$$m_{a,1} = m_{a,2}$$

$$m_{a,1} W_1 = m_{a,2} W_2$$

Thus for sensible heating

$$_1q_2 = m_a(h_2 - h_1) \tag{9.9}$$

Since the humidity ratio remains constant, it is closely true by Eqs. (8.27), (8.28), and (9.9) that

$$_1q_2 = m_a(0.24 + 0.45W)(t_2 - t_1) \tag{9.10}$$

Example 9.2. Moist air enters a steam-heating coil at 40 F dry-bulb temperature and 36 F thermodynamic wet-bulb temperature, at a rate of 2000 cu ft per min. Barometric pressure is 14.696 psia. The air leaves the coil at a dry-bulb temperature of 140 F. Determine the lb per hr of saturated steam at 220 F required, if the condensate leaves the coil at 200 F.

SOLUTION: By Fig. E-8, we find that $W_1 = 0.00359$ lb$_w$/lb$_a$, $h_1 = 13.47$ Btu/lb$_a$, and $v_1 = 12.66$ cu ft/lb$_a$. By Fig. E-9 at $t_2 = 140$ F and $W_2 = 0.00359$ lb$_w$/lb$_a$, we read $h_2 = 37.70$ Btu/lb$_a$. Also,

$$m_a = \frac{(2000)}{12.66}(60) = 9479 \text{ lb}_a/\text{hr}$$

By Eq. (9.9),

$$_1q_2 = (9479)(37.70 - 13.47) = 229{,}676 \text{ Btu/hr}$$

or by Eq. (9.10),

$$_1q_2 = (9479)[0.24 + (0.45)(0.00359)](140 - 40) = 229{,}013 \text{ Btu/hr}$$

For the steam, by Table A.2, h_g (at 220 F) $= 1153.4$ Btu/lb. For the condensate by Table A.1, h_f(at 200 F) $= 168.10$ Btu/lb.

Thus, the rate of steam flow is

$$m_s = \frac{229{,}676}{1153.4 - 168.1} = 233.1 \text{ lb/hr}$$

Dehumidification of moist air by cooling. If moist air is cooled below its dew point, condensation of moisture will occur. Figure 9.3(a) shows a schematic cooling device, and Fig. 9.3(b) shows schematically the psychrometric chart solution when

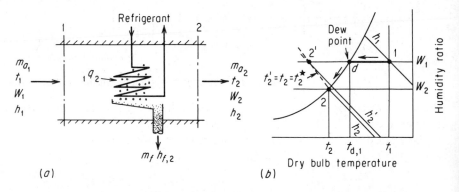

Figure 9.3 Schematic illustration of dehumidification by cooling.

moist air is cooled below its initial dew-point temperature. We assume that the air is uniformly and perfectly contacted, which is an idealized situation. Moisture will be separated at a variable temperature ranging from the initial dew point to the final saturation temperature. Let us assume that all of the condensed moisture is brought to the final temperature t_2 before it drains from the system.

The process may be interpreted in two ways. For the moist-air system alone, the process proceeds at constant humidity ratio from (1) to (d) and down the saturation curve to (2). For the total system of air and condensed moisture, the process is one of constant humidity ratio from (1) to the fog condition (2'). Thus the steady-flow energy and material-balance equations are

$$m_{a,1}h_1 = m_{a,2}h_2 + {}_1q_2 + m_fh_{f,2}$$

or

$$m_{a,1}h_1 = m_{a,2}h_{2'} + {}_1q_2$$

and

$$m_{a,1} = m_{a,2}$$
$$m_{a,1}W_1 = m_{a,2}W_2 + m_f$$

Thus

$${}_1q_2 = m_a(h_1 - h_2) - m_fh_{f,2} \qquad (9.11)$$

or

$${}_1q_2 = m_a(h_1 - h_{2'}) \qquad (9.12)$$

and

$$m_f = m_a(W_1 - W_2) \qquad (9.13)$$

Example 9.3. Moist air enters a refrigeration coil at 80 F dry-bulb temperature and 50 per cent relative humidity at a rate of 200 lb$_a$ per min. Barometric pressure is 14.696 psia. The air leaves saturated at 50 F. Calculate the tons of refrigeration required.

SOLUTION: Using Fig. E-8 for the necessary properties, by Eqs. (9.11) and (9.13), we find

$${}_1q_2 = (200)[31.20 - 20.30) - (0.01096 - 0.00766)(18.07)] = 2168 \text{ Btu/min}$$

or by Eq. (9.12),

$${}_1q_2 = (200)(31.20 - 20.35) = 2170 \text{ Btu/min}$$

Thus

$$\text{Tons of refrigeration} = \frac{2168}{200} = 10.8$$

Combined heating and humidification of moist air. In winter, atmospheric air may be heated and humidified in processing equipment prior to its introduction into an air conditioned space. In summer, air passing through a conditioned space absorbs the heat and moisture gains of the space and is heated and humidified.

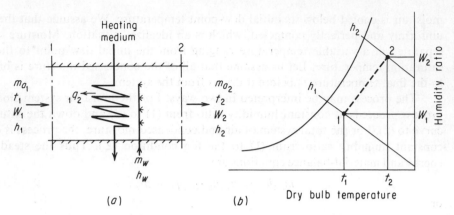

Figure 9.4 Schematic illustration of addition of heat and moisture to moist air.

Figure 9.4(a) shows schematically a device for heating and humidifying moist air in steady flow. The steady-flow energy and material-balance equations are

$$m_{a,1}h_1 + m_w h_w + {}_1q_2 = m_{a,2}h_2$$

$$m_{a,1} = m_{a,2}$$

$$m_{a,1}W_1 + m_w = m_{a,2}W_2$$

Thus

$${}_1q_2 = m_a(h_2 - h_1) - m_w h_w \tag{9.14}$$

$$m_w = m_a(W_2 - W_1) \tag{9.15}$$

By dividing Eq. (9.14) by Eq. (9.15), we have

$$q' = \frac{h_2 - h_1}{W_2 - W_1} = \frac{{}_1q_2}{m_w} + h_w \tag{9.16}$$

Equation (9.16) gives the value of the enthalpy-moisture ratio which fixes the direction of a straight line connecting the initial and final state points for the combined process.

The analysis for heating and humidifying and Eqs. (9.14)–(9.16) may be generalized to apply to any process where heat and/or moisture are added or removed. Thus for sensible heating alone, where m_w is zero, Eq. (9.14) reduces to Eq. (9.9). Equations (9.14) and (9.15) with ${}_1q_2$ and m_w negative, agree with Eqs. (9.11) and (9.13).

> **Example 9.4.** Moist air enters a chamber at 40 F dry-bulb and 36 F thermodynamic wet-bulb temperature, at a rate of 3000 cu ft per min. Barometric pressure is 14.696 psia. In passing through the chamber, the air absorbs sensible heat at a rate of 153,000 Btu per hr and picks up 83 lb per hr of saturated steam at 230 F. Determine the dry-bulb and thermodynamic wet-bulb temperatures of the leaving air.

SOLUTION: The initial state point is identical to that for Example 9.2. Thus, $W_1 = 0.00359$ lb$_w$/lb$_a$, $h_1 = 13.47$ Btu/lb$_a$, and $v_1 = 12.66$ cu ft/lb$_a$ and

$$m_a = \frac{3000}{12.66} = 237.0 \text{ lb}_a/\text{min}$$

By Eq. (9.15),

$$W_2 = W_1 + \frac{m_w}{m_a} = 0.00359 + \frac{83}{(237.0)(60)} = 0.00943 \text{ lb}_w/\text{lb}_a$$

By Eq. (9.14),

$$h_2 = h_1 + \frac{1q_2}{m_a} + \frac{m_w}{m_a} h_w = 13.47 + \frac{153,000}{(237.0)(60)} + \frac{(83)(1157.02)}{(237.0)(60)} = 30.98 \text{ Btu/lb}_a$$

State (2) is established by the intersection of h_2 and W_2. Thus $t_2 = 85.8$ F and $t_2^* = 66.4$ F.

An alternate procedure using the protractor on the chart may be used to establish State (2). Figure 9.5 shows the procedure. By Eq. (9.16),

$$q' = \frac{153,000}{83} + 1157.02 = 3000 \text{ Btu/lb}_w$$

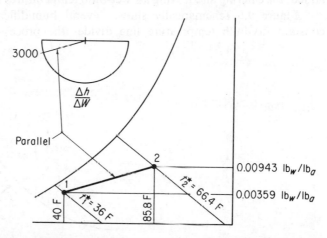

Figure 9.5 Schematic illustration of use of protractor in Example 9.4.

Through the use of two straight edges draw a line through State (1) parallel to the line on the protractor for $q' = 3000$ Btu/lb$_w$. The intersection of this line with $W_2 = 0.00943$ lb$_w$/lb$_a$ locates State (2). We find $t_2 = 85.8$ F and $t_2^* = 66.4$ F.

Humidification of moist air. A frequent psychrometric process is that of adding only moisture to air passing through a chamber. No other energy is added to the air and the moisture may be either in vapor or liquid form. (It may also be solid although this case occurs infrequently.) We assume that all moisture added in the chamber is retained by the air passing through.

The process is a special case of heating and humidifying. We may imagine that the heating coil in Fig. 9.4(a) is removed. Equations (9.14)–(9.16) apply with $_1q_2$ equal to zero. Thus

$$m_w h_w = m_a(h_2 - h_1)$$
$$m_w = m_a(W_2 - W_1)$$

and

$$q' = \frac{h_2 - h_1}{W_2 - W_1} = h_w \qquad (9.17)$$

The direction of the condition line connecting States (1) and (2) depends on the enthalpy of the moisture added. Two unique cases will be mentioned. Each has been considered previously.

Equation (9.7) states that when air is humidified at constant dry-bulb temperature, the steam added must have a specific enthalpy equal to that of saturated steam at the air dry-bulb temperature. If water at the air thermodynamic wet-bulb temperature is added, the entering and leaving air wet-bulb temperatures must be identical.

Figure 9.6 schematically shows several humidification condition lines. The constant dry-bulb temperature line divides the processes into two categories. If

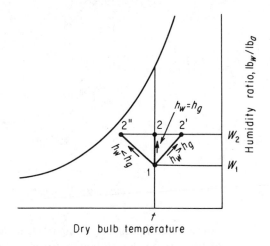

Figure 9.6 Schematic illustration of humidification processes.

$h_w > h_g$, then air may be sensibly heated as well as humidified with a moisture spray. If $h_w < h_g$, the air may be sensibly cooled during the process of humidification.

Adiabatic mixing of two streams of moist air. In almost every air conditioning system, the mixing of two or more air streams may occur. Usually, such mixing processes occur under essentially adiabatic conditions. Figure 9.7(a) schematically shows a

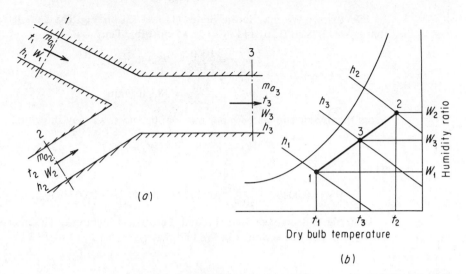

Figure 9.7 Schematic adiabatic mixing of two streams of moist air.

mixture of two air streams. The fundamental equations applying to the process are

$$m_{a,1}h_1 + m_{a,2}h_2 = m_{a,3}h_3$$

$$m_{a,1} + m_{a,2} = m_{a,3}$$

$$m_{a,1}W_1 + m_{a,2}W_2 = m_{a,3}W_3$$

By elimination of $m_{a,3}$, we obtain

$$\frac{m_{a,1}}{m_{a,2}} = \frac{h_2 - h_3}{h_3 - h_1} = \frac{W_2 - W_3}{W_3 - W_1} \tag{9.18}$$

Equation (9.18) tells us that the resulting State (3) must lie on a straight line connecting States (1) and (2) on the psychrometric chart. Furthermore, segments of this line are proportional to the masses of dry air mixed. In Fig. 9.7(b), we may write

$$\frac{m_{a,1}}{m_{a,2}} = \frac{\overline{32}}{\overline{13}} \quad \text{or} \quad \frac{m_{a,1}}{m_{a,3}} = \frac{\overline{32}}{\overline{12}} \quad \text{or} \quad \frac{m_{a,2}}{m_{a,3}} = \frac{\overline{13}}{\overline{12}} \tag{9.19}$$

There are several choices in the solution of a mixing problem. We may solve a problem algebraically using the mixing equations, or we may use a graphical solution on the psychrometric chart. The following example illustrates one convenient method of solution.

Example 9.5. One stream of moist air (1000 cu ft per min, 60 F dry-bulb temperature, 56 F thermodynamic wet-bulb temperature) is mixed with a second stream (400 cu ft per min, 80 F dry-bulb temperature and 67 F thermodynamic wet-bulb temperature). Barometric pressure is 14.696 psia. Determine the dry-bulb and wet-bulb temperatures of the resulting mixture.

SOLUTION: We may locate States (1) and (2) on Fig. E-8 and determine that $v_1 = 13.28$ cu ft/lb$_a$ and $v_2 = 13.85$ cu ft/lb$_a$. Thus,

$$m_{a,1} = \frac{1000}{13.28} = 75.3 \text{ lb}_a/\text{min}$$

$$m_{a,2} = \frac{400}{13.85} = 28.9 \text{ lb}_a/\text{min}$$

From the dry-air and water-vapor material balances, we may show that

$$W_3 = W_1 + \frac{m_{a,2}}{m_{a,3}}(W_2 - W_1)$$

Thus

$$W_3 = 0.00867 + \frac{28.9}{104.2}(0.01123 - 0.00867) = 0.00938 \text{ lb}_w/\text{lb}_a$$

On Fig. E-8, connect States (1) and (2) with a straight line. The intersection of W_3 with this line is State (3). We find that $t_3 = 65.4$ F, $t_3^* = 59.3$ F.

PROBLEMS

9.1. Moist air exists under conditions of 85 F dry-bulb temperature, 40 per cent relative humidity, and 14.696 psia pressure. By the psychrometric chart, determine (a) the dew-point temperature, (b) thermodynamic wet-bulb temperature, (c) humidity ratio, (d) enthalpy, and (e) the volume.

9.2. It is planned to construct an h-W diagram for sea-level pressure from Table A.5. Scale factor S for the diagram is to be 1000 Btu per lb water. The 80 F dry-bulb temperature line is to be vertical. Determine the slope (tan θ) of the 60 F thermodynamic wet-bulb temperature line.

9.3. For a line of constant volume in cu ft per lb$_a$ on the Mollier psychrometric chart, and assuming that moist air is a perfect-gas mixture, prove that

$$\frac{dh}{dW} = 0.001201 \frac{Pv}{(1 + 1.608W)^2} + 854$$

where h is in Btu per lb$_a$, W in lb$_w$ per lb$_a$, P in lb per sq ft, and v in cu ft per lb$_a$.

9.4. Construct a Mollier-type psychrometric chart for a pressure of 28 inches of mercury. The following lines are to be shown in addition to the basic coordinates of enthalpy and humidity ratio: (a) saturation curve, (b) dry-bulb temperature lines for 30, 50, 70, 90, 110, and 130 F, (c) thermodynamic wet-bulb temperature lines for 30, 50, 70, and 90 F, (d) degree of saturation lines for $\mu = 0.3$, 0.5, and 0.7, and (e) volume lines for 13.0, 14.0, and 15.0 cu ft per lb$_a$. At some location on the paper show a protractor for $q' = -10,000$, -1000, 0, 500, 1000, 2000, 3000, and 10,000 Btu per lb$_w$. Use a scale factor of $S = 1250$ Btu per lb$_w$, and incline the enthalpy lines so that the 70 F dry-bulb temperature line is vertical.

9.5. Moist air at 84 F dry-bulb temperature and 70 F thermodynamic wet-bulb temperature enters a perfect contact refrigeration coil at a rate of

3500 cu ft per min. The air leaves the coil at 54 F. Assume 14.696 psia pressure. Determine the tons of refrigeration required.

9.6. Moist air enters a refrigeration coil at 89 F dry-bulb temperature and 65 F thermodynamic wet-bulb temperature, at a rate of 1400 cu ft per min. The surface temperature of the coil is 55 F. If 3.5 tons of refrigeration are available, find the dry-bulb and wet-bulb temperatures of the air leaving the coil. Assume sea-level pressure.

9.7. Saturated steam at a pressure of 25 psia is sprayed into a stream of moist air. The initial condition of the air is 55 F dry-bulb temperature and 35 F dew-point temperature. The mass rate of air flow is 2000 lb_a per min. Barometric pressure is 14.696 psia. Determine (a) how much steam must be added in lb per min to produce a saturated air condition, and (b) the resulting temperature of the saturated air.

9.8. Moist air at 70 F dry-bulb temperature and 45 per cent relative humidity is recirculated from a room and mixed with outdoor air at 97 F dry-bulb temperature and 83 F thermodynamic wet-bulb temperature. Determine the mixture state dry-bulb and wet-bulb temperatures if the volume of recirculated

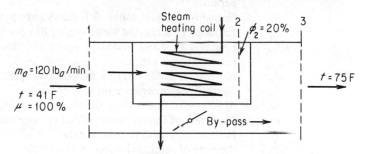

Figure 9.8 Schematic system for Problem 9.9.

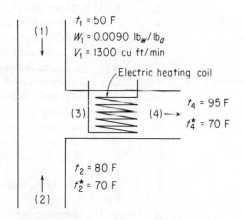

Figure 9.9 Schematic system for Problem 9.10.

air (cu ft per min) is three times the volume of outdoor air. Assume sea-level pressure.

9.9. Moist air is heated by steam condensing inside the tubes of a heating coil as shown by Fig. 9.8. Part of the air passes through the coil and part is by-passed around the coil. Barometric pressure is 14.696 psia. Determine (a) the lb_a per min which by-pass the coil, and (b) the heat added by the coil in Btu per hr.

9.10. Figure 9.9 schematically shows part of a winter-type air conditioning system. Barometric pressure is 14.696 psia. Determine (a) the temperature t_3 of the mixed air entering the heating coil, and (b) the rate of heat addition to the air by the heating coil in Btu per hr.

SYMBOLS USED IN CHAPTER 9

h — Enthalpy of moist air, Btu per lb_a; h_s for air saturated at dry-bulb temperature t; $h_s^\star$ for air saturated at thermodynamic wet-bulb temperature $t^\star$.

h_f — Enthalpy of liquid water, Btu per lb; $h_f^\star$ for water at $t^\star$.

h_g — Enthalpy of saturated water vapor, Btu per lb.

h_w — Enthalpy of water added to moist air, Btu per lb; $h_w = h_g$ for low-pressure water vapor.

L_h — Horizontal scalar distance on h-W chart, in.

L_W — Vertical scalar distance on h-W chart, in.

m_a — Mass rate of flow of dry air, lb_a per hr or lb_a per min.

m_f — Mass rate of flow of water added or removed, lb per hr.

P — Barometric pressure, psia or psfa.

$P_{w,s}$ — Pressure of saturated water, psia or psfa.

q — Rate of heat transfer, Btu per hr.

q' — $(h_2 - h_1)/(W_2 - W_1)$, Btu per lb of water.

R_a — 53.35 ft lb per (lb) (R).

S — Psychrometric chart scale factor s_h/s_W, Btu per lb_w.

s_h — Enthalpy scale factor, Btu per $(lb_a)(in.)$.

s_W — Humidity ratio scale factor, lb_w per $(lb_a)(in.)$.

T — Absolute dry-bulb temperature, R.

t — Dry-bulb temperature, F.

$t^\star$ — Thermodynamic wet-bulb temperature, F.

v — Volume of moist air, cu ft per lb of dry air.

W — Humidity ratio of moist air, lb_w per lb_a; W_s for air saturated at t; $W_s^\star$ for air saturated at $t^\star$.

α — $180 - (\theta + \beta)$, deg (see Fig. 9.1).

β — Inclination angle of enthalpy lines (see Fig. 9.1).

θ — Inclination angle of any straight line on psychrometric chart (see Fig. 9.1).

μ — W/W_s, dimensionless

ϕ — Relative humidity, dimensionless.

10

THE PSYCHROMETER
AND HUMIDITY MEASUREMENT

10.1. INTRODUCTION

In Chapter 8, we discussed various thermodynamic properties of moist air and their relations to each other. In Chapter 9, we found that the psychrometric chart is a useful graphical aid in solution of moist-air problems. However, to use the psychrometric equations or the psychrometric chart, we must know the thermodynamic state of the air. This requires knowledge of the barometric pressure and two other independent properties.

Barometric pressure and dry-bulb temperature may be measured easily and precisely. Certain properties such as enthalpy in Btu per lb_a, volume in cu ft per lb_a, and thermodynamic wet-bulb temperature are incapable of measurement. The primary problem in psychrometric measurements is determination of some property related to the air moisture content. Measurement of air humidity is difficult and most of the techniques available are subject to error.

The wet-bulb thermometer is one of the most convenient devices available for humidity measurement. It is a reliable instrument if properly applied. In this chapter, we will study the wet-bulb thermometer in detail. We will also study briefly some of the other techniques for measurement of air humidity.

One of the most intriguing problems in psychrometrics is the relationship between thermodynamic wet-bulb temperature and psychrometer wet-bulb temperature. We will closely examine that relationship in this chapter.

10.2. MASS TRANSFER AND THE EVAPORATION OF WATER INTO MOIST AIR

The theory of the wet-bulb thermometer involves the mechanism of evaporation of water into moist air. Some of our problems in later chapters will also deal with simultaneous heat transfer and water-vapor transfer. We will now consider, in some detail, the fundamental problem of evaporation from a free-water surface.

Figure 10.1 will serve as our schematic model. A free-water surface at temperature t_w is exposed to a moving stream of moist air. Adjacent to the water surface is a bound-

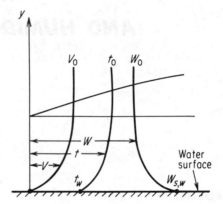

Figure 10.1 Schematic illustration of boundary-layer problem for the evaporation of water into moist air.

ary layer of air whose velocity varies from zero at the water surface to the main air-stream velocity V_o at its outer edge. Temperature increases within the boundary layer from t_w at the water surface to t_o of the main air stream. Air immediately adjacent to the water surface is assumed to be saturated. Humidity ratio decreases in the y direction from $W_{s,w}$ at the water surface to W_o of the main air stream. The thickness of the boundary layer is assumed identical for the three gradients of velocity, temperature, and humidity ratio.

The transfer processes are heat transfer from the air to the water surface and water-vapor transfer from the water surface to the air. Heat transfer through the boundary layer is by the combined processes of convection and conduction. Vapor transfer occurs by the combined processes of convection and diffusion.

For heat transfer, we may write

$$h_c(t_o - t_w) = k\left(\frac{\partial t}{\partial y}\right)_{y=0}$$

By defining a dimensionless temperature $t' = (t - t_w)/(t_o - t_w)$, and a dimensionless length $y' = y/L$, where L is a reference length, we obtain

$$\frac{h_c L}{k} = \left(\frac{\partial t'}{\partial y'}\right)_{y=0}$$

(10.1)

Eckert* has shown that the solution of Eq. (10.1) has the form

$$\frac{h_c L}{k} = f(\text{Re}, \text{Pr})$$

(10.2)

where $h_c L/k$ is the dimensionless *Nusselt* number, $\text{Re} = LV\rho/\mu$ is the dimensionless *Reynolds* number and $\text{Pr} = c_p\mu/k$ is the dimensionless *Prandtl* number.

The basic concept of diffusion is given by Fick's law which may be written as

$$m_w = -D\rho_a\, dW/dy$$

where m_w = the mass flow of vapor, lb per (hr) (sq ft); D = the vapor diffusivity or diffusion coefficient, sq ft per hr; ρ_a = the air density, mass of dry air per unit volume, lb_a per cu ft; W = the humidity ratio, lb_w per lb_a; and y = the diffusion length, ft.

A mass transfer coefficient h_D for the transfer of water vapor through the boundary layer of Fig. 10.1 may be defined by the equation

$$m_w = h_D(W_{s,w} - W_o)$$

We may then write

$$h_D(W_{s,w} - W_o) = -D\rho_a\left(\frac{\partial W}{\partial y}\right)_{y=0}$$

By use of the dimensionless quantities $W' = (W_{s,w} - W)/(W_{s,w} - W_o)$ and $y' = y/L$, we obtain

$$\frac{h_D L}{\rho_a D} = \left(\frac{\partial W'}{\partial y'}\right)_{y=0}$$

(10.3)

Eckert† has also shown that the solution of Eq. (10.3) has the form

$$\frac{h_D L}{\rho_a D} = f(\text{Re}, \text{Sc})$$

(10.4)

and has stated that the function in Eqs. (10.2) and (10.4) may be assumed as identical for turbulent flow of air over wetted flat plates, around cylinders and spheres, and through packed beds. In Eq. (10.4), $\text{Sc} = \mu/\rho D$ is the dimensionless *Schmidt* number.

Equations (10.2) and (10.4) may be expressed also as

$$\frac{h_c L}{k} = a\left(\frac{LV\rho}{\mu}\right)^b\left(\frac{c_p\mu}{k}\right)^c$$

$$\frac{h_D L}{\rho_a D} = a\left(\frac{LV\rho}{\mu}\right)^b\left(\frac{\mu}{\rho D}\right)^c$$

*E.R.G. Eckert and R. M. Drake, Jr., *Heat and Mass Transfer* (New York: McGraw-Hill Book Company, 1959) pp. 469–471.

†Eckert and Drake, *Heat and Mass Transfer*, pp. 471–472.

We then obtain

$$\frac{h_c}{h_D} = \frac{k}{D\rho_a}\left(\frac{D}{\alpha}\right)^c \qquad (10.5)$$

where $\alpha = k/\rho c_p$, the *thermal diffusivity*, sq ft per hr. By dividing both sides of Eq. (10.5) by $c_{p,a}$, Btu per (lb_a) (F), we obtain

$$\frac{h_c}{h_D c_{p,a}} = \left(\frac{\alpha}{D}\right)^{1-c} \qquad (10.6)$$

The dimensionless term $h_c/(h_D)(c_{p,a})$ is called the *Lewis number* Le. Kusuda* has made a review of available correlations for calculating the Lewis number. For forced convection air flow, he recommends the relation

$$\text{Le} = \left(\frac{\alpha}{D}\right)^{2/3} \qquad (10.7)$$

For the case of natural convection, Kusuda recommends the same form of equation but with an exponent of 0.48 instead of $\frac{2}{3}$. In the same paper, Kusuda made a study of data available on transport properties of dry and saturated moist air. Table 10.1 shows his values of α, D, and α/D for temperatures from 50 F to 140 F. In applying these values to evaporation problems, Kusuda recommends that the properties be evaluated for saturated moist air at the water surface temperature.

10.3. THEORY OF THE PSYCHROMETER

The wet-bulb thermometer or some variation has been in use for more than a century. In its simplest form, a wet-bulb thermometer consists of an ordinary thermometer whose sensing bulb is covered by a moistened cloth wick. The thermometer is ventilated by whirling it in a calm atmosphere or by exposing it to forced air circulation. When both a dry-bulb thermometer and a wet-bulb thermometer are included in the same instrument, the device is called a *psychrometer*.

Figure 10.2 shows a schematic psychrometer exposed to a moving stream of moist air. In the following analyses we will assume that conduction heat-transfer effects along the thermometer stems are negligible. We will also assume that air velocities are low enough so that impact influences upon the thermometers are negligible. If the temperature t_s of surrounding surfaces is different from the air dry-bulb temperature t, the dry-bulb thermometer will indicate a value t_{db} somewhere between the values of t_s and t. The moistened bulb at temperature t_{wb} may receive heat by convection transfer from the air stream and by radiation transfer from surrounding surfaces. Some water may evaporate from the moistened bulb exerting a cooling

*Tamami Kusuda, "Calculation of the Temperature of a Flat-Plate Wet Surface under Adiabatic Conditions with Respect to the Lewis Relation," in *Humidity and Moisture, Volume 1: Principles and Methods of Measuring Humidity in Gases*, edited by Robert E. Ruskin (New York: Reinhold Publishing Corporation, 1965) pp. 16–32.

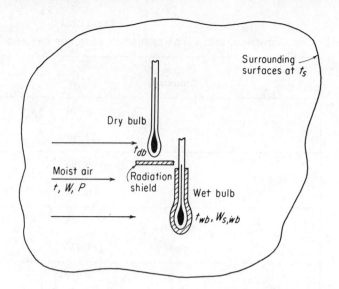

Figure 10.2 Schematic psychrometer.

effect which depresses the reading of the thermometer below that of the air dry-bulb temperature.

Assuming the steady state for the system of Fig. 10.2, we may write for the dry-bulb thermometer

$$h_c(t_{db} - t) = h_R(t_s - t_{db})$$

or

$$t = t_{db} - \left(\frac{h_R}{h_c}\right)_{db}(t_s - t_{db}) \tag{10.8}$$

Equation (10.8) allows calculation of the air dry-bulb temperature.

For the wet-bulb thermometer of Fig. 10.2, we may write

$$h_D(W_{s,wb} - W)h_{fg,wb} = h_c(t - t_{wb}) + h_R(t_s - t_{wb})$$

Substitution of $h_D - h_c/(\mathrm{Le})(c_{p,a})$ gives

$$W - W_{s,wb} - K(t - t_{wb}) \tag{10.9}$$

where

$$K = \frac{\mathrm{Le}\,c_{p,a}}{h_{fg,wb}}\left[1 + \frac{h_R(t_s - t_{wb})}{h_c(t - t_{wb})}\right] \tag{10.10}$$

In our procedure, we will evaluate the specific heat $c_{p,a}$ at the arithmetic-mean value of humidity ratio. Thus, by Eq. (8.28)

$$c_{p,a} = 0.240 + 0.45\frac{(W + W_{s,wb})}{2} \tag{10.11}$$

TABLE 10.1*

THERMAL AND VAPOR DIFFUSIVITY DATA FOR DRY AND SATURATED MOIST AIR

Temperature F	Degree of Saturation	α ft^2/hr	D ft^2/hr	α/D
50	0	0.770	0.901	0.855
	1	0.769		0.854
60	0	0.799	0.936	0.854
	1	0.797		0.852
70	0	0.828	0.971	0.853
	1	0.826		0.850
80	0	0.858	1.007	0.852
	1	0.854		0.848
90	0	0.888	1.044	0.851
	1	0.883		0.846
100	0	0.919	1.081	0.850
	1	0.911		0.843
110	0	0.949	1.119	0.848
	1	0.938		0.838
120	0	0.981	1.157	0.848
	1	0.963		0.832
130	0	1.012	1.196	0.846
	1	0.985		0.823
140	0	1.044	1.235	0.845
	1	1.003		0.812

*Adapted by permission from Tamami Kusuda, "Calculation of the Temperature of a Flat-Plate Wet Surface under Adiabatic Conditions with Respect to the Lewis Relation," in *Humidity and Moisture, Volume 1: Principles and Methods of Measuring Humidity in Gases*, edited by Robert E. Ruskin (New York: Reinhold Publishing Corporation, 1965) p. 29.

By Eqs. (10.9)–(10.11), we obtain

$$K = \frac{0.240 + 0.45W_{s,wb}}{\dfrac{h_{fg,wb}}{\text{Le}\left[1 + \dfrac{h_R(t_s - t_{wb})}{h_c(t - t_{wb})}\right]} + 0.225(t - t_{wb})} \tag{10.12}$$

Equations (10.9) and (10.12) allow calculation of the air humidity ratio W. The humidity ratio $W_{s,wb}$ of saturated moist air at the wet-bulb temperature may be calculated by Eq. (8.23) or read from Table A.5 for sea-level pressure. The *wet-bulb coefficient* K must be separately evaluated. Equation (10.12) allows calculation of K for the general case. Two special cases exist. For the common situation where the mean temperature of surfaces surrounding the wet-bulb may be assumed equal to the air dry-bulb temperature ($t_s = t$), Eq. (10.12) reduces to

$$K = \frac{0.240 + 0.45W_{s,wb}}{\dfrac{h_{fg,wb}}{\text{Le}\left(1 + \dfrac{h_{R,t}}{h_c}\right)} + 0.225(t - t_{wb})} \tag{10.13}$$

where $h_{R,t}$ means that h_R is evaluated for the condition of $t_s = t$. If the wet-bulb is perfectly shielded from radiation effects

$$K = \frac{0.240 + 0.45W_{s,wb}}{\dfrac{h_{fg,wb}}{\text{Le}} + 0.225(t - t_{wb})} \tag{10.14}$$

Evaluation of the wet-bulb coefficient K is the principal problem in psychrometry. The Lewis number Le may be found from Eq. (10.7). The radiation coefficient h_R may be calculated by Eq. (2.18) which for the wet-bulb thermometer reduces to

$$h_R = 0.1713\epsilon_{wb} \frac{\left[\left(\dfrac{T_s}{100}\right)^4 - \left(\dfrac{T_{wh}}{100}\right)^4\right]}{(t_s - t_{wb})} \tag{10.15}$$

where ϵ_{wb} is the surface emissivity of the wet-bulb.

The convection coefficient h_c may be calculated from equations given by McAdams.* For the flow of air normal to single wires or cylinders, McAdams recommends the relations

$$\frac{h_c d}{k_f} = 0.615\left(\frac{d V \rho_f}{\mu_f}\right)^{0.466} \quad \text{for } 40 < \text{Re} < 4000 \tag{10.16}$$

$$\frac{h_c d}{k_f} = 0.174\left(\frac{d V \rho_f}{\mu_f}\right)^{0.618} \quad \text{for } 4000 < \text{Re} < 40{,}000 \tag{10.17}$$

where d is the wet-bulb diameter. The properties k_f, ρ_f, and μ_f are evaluated at a *mean film-temperature* t_f.

Figure 10.3 shows the ratio $h_{R,t}/h_c$ for a wet-bulb diameter of 0.3 in., calculated by Eqs. (10.15)–(10.17), for the *special case of the temperature of surrounding surfaces* t_s

*W. H. McAdams, *Heat Transmission* (New York: McGraw-Hill Book Company, 1954) p. 260.

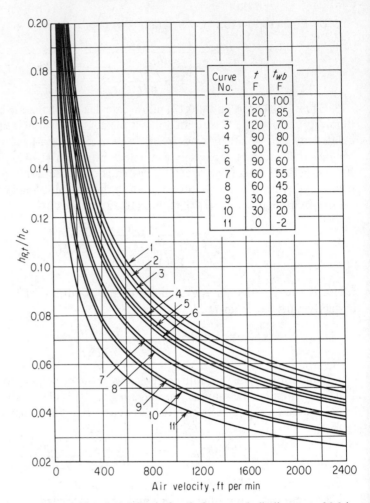

Figure 10.3 Ratio $h_{R,t}/h_c$ for a wet-bulb diameter of 0.3 in.

equal to the air dry-bulb temperature t. A value of $\epsilon_{wb} = 0.9$ was used in Eq. (10.15). Figure 10.4 is similar to Fig. 10.3 except that it is for a wet-bulb diameter of 0.1 in. Figures 10.3 and 10.4 show that $h_{R,t}/h_c$ decreases with decrease of both dry-bulb temperature and wet-bulb temperature. In addition, $h_{R,t}/h_c$ increases rapidly with decrease of air velocity for velocities below about 500 ft per min. At velocities above about 1000 ft per min, $h_{R,t}/h_c$ is less dependent upon air velocity, particularly for a wet-bulb diameter of 0.1 in.

Figures 10.3 and 10.4 apply directly only to calculation of the wet-bulb coefficient by Eq. (10.13). For the general case, where the surrounding-surfaces temperature t_s is different from the air dry-bulb temperature t, we may obtain K by the relation

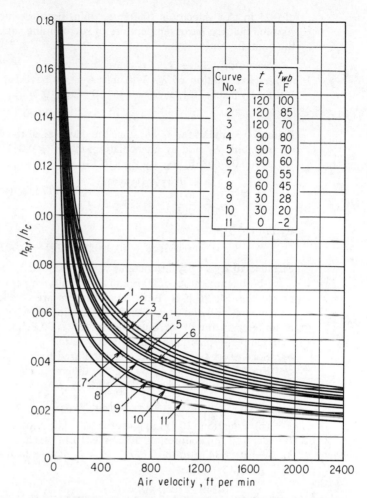

Curve No.	t F	t_{wb} F
1	120	100
2	120	85
3	120	70
4	90	80
5	90	70
6	90	60
7	60	55
8	60	45
9	30	28
10	30	20
11	0	-2

Figure 10.4 Ratio $h_{R,t}/h_c$ for a wet-bulb diameter of 0.1 in.

$$K = \cfrac{0.240 + 0.45W_{s,wb}}{\cfrac{h_{fg,wb}}{\mathrm{Le}\left\{1 + \cfrac{h_{R,t}}{h_c}\left[\cfrac{\left(\frac{T_s}{100}\right)^4 - \left(\frac{T_{wb}}{100}\right)^4}{\left(\frac{T}{100}\right)^4 - \left(\frac{T_{wb}}{100}\right)^4}\right]\right\}} + 0.225(t - t_{wb})} \tag{10.18}$$

where $h_{R,t}/h_c$ is read from Fig. 10.3 or from Fig. 10.4.

Example 10.1. Measurements with a psychrometer indicate a wet-bulb tempera-ture of 50 F and a dry-bulb temperature of 75 F. Barometric pressure is 14.696 psia. Both thermometers are mercury-in-glass types with a bare-bulb diameter of 0.25 in. Air velocity is 1000 ft per min. Both thermometers are unshielded. Assume that the mean temperature of surrounding surfaces is 75 F. Determine the humidity ratio.

SOLUTION: Since $t_s = t_{db}$, we have $t = t_{db} = 75$ F. Equation (10.13) applies for solution of K. By Table A.1, we have for 50 F, $h_{fg, wb} = 1064.99$ Btu/lb. By Table 10.1, $\alpha/D = 0.854$. By Eq. (10.7)

$$\text{Le} = (0.854)^{2/3} = 0.900$$

Figure 10.3 applies for $h_{R,t}/h_c$ since the diameter of the wet-bulb wick would be about 0.3 in. We find by interpolation, $h_{R,t}/h_c = 0.063$. By Table A.5, at 50 F, $W_{s, wb} = 0.007658$ lb$_w$/lb$_a$. By Eq. (10.13)

$$K = \frac{0.240 + (0.45)(0.007658)}{\dfrac{1064.99}{(0.90)(1.063)} + (0.225)(25)} = 0.002175 \text{ lb}_w/(\text{lb}_a)(\text{F})$$

By Eq. (10.9)

$$W = 0.007658 - (0.0002175)(25) = 0.00222 \text{ lb}_w/\text{lb}_a$$

Example 10.2. A psychrometer, with unshielded thermometers, indicates a dry-bulb temperature of 70 F and a wet-bulb temperature of 60 F. Barometric pressure is 14.696 psia. Both thermometers are of the mercury-in-glass type with a bare-bulb diameter of 0.25 in. Air velocity is 200 ft per min. Assume that mean temperature of surrounding surfaces is 90 F. Determine (a) the true dry-bulb temperature, and (b) the humidity ratio.

SOLUTION: (a) Figure 10.3 may be used for finding $(h_R/h_c)_{db}$ if we replace t of Fig. 10.3 by t_s and t_{wb} by t_{db}. The difference in diameter of 0.05 in. should not introduce much error. By Fig. 10.3, $(h_R/h_c)_{db} = 0.146$. By Eq. (10.8)

$$t = 70 - (0.146)(90 - 70) = 67.1 \text{ F}$$

(b) Equation (10.18) applies for calculation of the wet-bulb coefficient K. By identical procedures of Example 10.1, we find $W_{s, wb} = 0.01108$ lb$_w$/lb$_a$, $h_{fg, wb} = 1059.34$ Btu/lb$_w$, $\alpha/D = 0.852$, Le $= 0.898$, and $h_{R, t}/h_c = 0.134$. By Eq. (10.18)

$$K = \frac{0.240 + (0.45)(0.01108)}{\dfrac{1059.34}{0.898\left[1 + (0.134)\dfrac{(5.5^4 - 5.2^4)}{(5.27^4 - 5.2^4)}\right]} + (0.225)(7.1)} = 0.000335 \text{ lb}_w/(\text{lb}_a)(\text{F})$$

By Eq. (10.9)

$$W = 0.01108 - (0.000335)(67.1 - 60) = 0.00870 \text{ lb}_w/\text{lb}_a$$

Example 10.2 shows that radiation effects may cause considerable error in the dry-bulb thermometer reading and may also significantly influence the wet-bulb coefficient K when the air velocity is relatively low. Since the mean temperature of surrounding surfaces t_s is rarely known accurately, much more reliable results are obtained with a psychrometer when the air velocity is moderately high, about 1000 ft per min. The most reliable results, of course, are obtained when both thermometers are shielded from radiation effects. For this case, the simpler relation, Eq. (10.14), applies for the wet-bulb coefficient. Since perfect radiation shields are not possible, it is still desirable to use a moderately high velocity.

There will always be, of course, some uncertainty in calculation of the wet-bulb coefficient K. The influence of an error in K upon W in Eq. (10.9) is most severe at low relative humidities. When W is very small compared to both $W_{s,wb}$ and $K(t - t_{wb})$, the psychrometer may yield an unreliable result. However, almost all of the commonly available techniques for measuring humidity are affected in a similar adverse manner.

10.4. PRACTICAL USE OF A PSYCHROMETER

From our discussion of the psychrometer in Sec. 10.3, we recognize that several factors may affect the readings of the thermometers. Careful application is necessary to obtain reliable results. The use of a psychrometer is covered in detail in *NBS Circular 512*.*

Two types of psychrometers are commonly used. Each comprises two thermometers with the bulb of one covered by a moistened wick. It is necessary to separate the two sensing bulbs so that radiation heat exchange between them is negligible. The *sling psychrometer* is widely used for measurements involving room air or other applications where the rate of air movement is small. Air circulation is obtained by whirling the psychrometer. The *aspiration psychrometer* uses two stationary thermometers and is ventilated by a motor-driven blower. Unventilated psychrometers are unreliable and should not be used. Most psychrometers use mercury-in-glass thermometers. However, resistance thermometers, thermocouples, and bimetallic elements may be used.

The function of the wick is to provide a thin film of water on the wet-bulb. Cotton or linen cloth of a soft mesh is satisfactory. Any factors which prevent a continuous film of water on the wet-bulb may cause an erroneous reading. Thus, the wick material should have no sizing or encrustations, should be clean, and should fit snugly. Wicks should be replaced frequently and *only distilled water* should be used for saturating. It is desirable for the wick to extend for one or two inches beyond the sensing bulb to help reduce heat conduction along the stem. The wick should be maintained in a fully saturated condition, since a partially dry wick may cause an erroneous reading. The temperature of water used for saturating the wick should be close to the wet-bulb temperature. Otherwise, sufficient time must be allowed for the wick to reach the wet-bulb temperature.

Wile† has discussed application of wet-bulb thermometers for temperatures below freezing. Here it is desirable to discard the wick and to freeze a layer of ice

---

*Arnold Wexler and W. B. Brombacher, *Methods of Measuring Humidity and Testing Hygrometers*, U. S. Department of Commerce, National Bureau of Standards, Circular 512 (Washington, D. C.: Government Printing Office, 1951).

†D. D. Wile, "Psychrometry in the Frost Zone," *Refrigerating Engineering*, Vol. 48, October 1944, pp. 291–301.

directly on the wet-bulb. Some uncertainty may exist as to whether ice or subcooled water is in equilibrium with the wet-bulb. The wet-bulb thermometer is less reliable and less convenient to use for temperatures below freezing.

10.5. CORRELATION OF PSYCHROMETER WET-BULB TEMPERATURE WITH THERMODYNAMIC WET-BULB TEMPERATURE

In Chapter 8, we discussed thermodynamic wet-bulb temperature or the temperature of adiabatic saturation. In preceding sections of this chapter, we analyzed wet-bulb temperature as found from a thermometer.

Some confusion arises because there are two types of wet-bulb temperature. We should realize that there is a *distinct difference* between thermodynamic wet-bulb temperature and ordinary wet-bulb temperature which we may read from a thermometer. Thermodynamic wet-bulb temperature is a hypothetical temperature which, strictly speaking, can only be approached in a limiting case, and cannot be measured directly. We should emphasize that only thermodynamic wet-bulb temperature is a thermodynamic property. A wet-bulb temperature as read from a thermometer is influenced by heat and mass transfer rates and is therefore not a sole function of the air state to which the thermometer is exposed. Thus in psychrometric equations and psychrometric charts where wet-bulb temperature appears, it is always *thermodynamic wet-bulb temperature* which is considered.

An interesting problem is the relationship between psychrometer wet-bulb temperature t_{wb} and thermodynamic wet-bulb temperature $t^\star$. Any air stream with the properties t, W, etc. to which we expose a wet-bulb thermometer also must have some value of $t^\star$. Our problem is to study how t_{wb} may differ from $t^\star$.

We may rewrite Eq. (8.31), obtained from an analysis of the adiabatic saturation process, as

$$W = W_s^\star - K^\star(t - t^\star) \tag{10.19}$$

where

$$K^\star = \frac{c_{p,a}}{h_{fg}^\star} \tag{10.20}$$

The quantity $K^\star$ is analogous to a wet-bulb coefficient for the adiabatic saturation process. Comparison of Eqs. (10.9), (10.10), (10.19), and (10.20) shows that a shielded wet-bulb thermometer will indicate a temperature *less* than the thermodynamic wet-bulb temperature since the Lewis number Le is less than one.

However, t_{wb} may be equal to $t^\star$, providing that in Eq. (10.10)

$$\text{Le} \left[1 + \frac{h_R(t_s - t_{wb})}{h_c(t - t_{wb})} \right] = 1$$

Thus, radiation heat transfer to the wet-bulb may compensate for the Lewis number being less than unity.

We will now derive a general relationship between psychrometer wet-bulb temperature t_{wb} and thermodynamic wet-bulb temperature $t^\star$. By Eqs. (10.9) and (10.19),

$$(t_{wb} - t^\star) + \frac{(W_{s,wb} - W_s^\star)}{K^\star} = \left(\frac{K}{K^\star} - 1\right)(t - t_{wb}) \qquad (10.21)$$

Since we find that t_{wb} differs but slightly from $t^\star$, for a small interval of temperature, we may write

$$W_{s,wb} = A + Bt_{wb}$$
$$W_s^\star = A + Bt^\star$$

where A and B are constants. We obtain

$$W_{s,wb} - W_s^\star = B(t_{wb} - t^\star) \qquad (10.22)$$

By Eqs. (10.21) and (10.22),

$$\frac{t_{wb} - t^\star}{t - t_{wb}} = \frac{(K/K^\star) - 1}{1 + (B/K^\star)} \qquad (10.23)$$

Equation (10.23) expresses the deviation $(t_{wb} - t^\star)$ in terms of the *wet-bulb depression* $(t - t_{wb})$. Evaluation of the psychrometer wet-bulb coefficient K was discussed in Section 10.3. Figure 10.5 shows the quantity B in Eq. (10.23) for barometric pressures of 12.00 and 14.696 psia.

For the special case when the mean temperature of the surfaces surrounding the wet-bulb is equal to the air dry-bulb temperature $(t_s = t)$, Eq. (10.23) reduces to

$$\frac{t_{wb} - t^\star}{t - t_{wb}} = \frac{\text{Le}(1 + h_{R,t}/h_c) - 1}{1 + (B/K^\star)} \qquad (10.24)$$

since $h_{fg,wb}$ is almost equal to $h_{fg}^\star$.

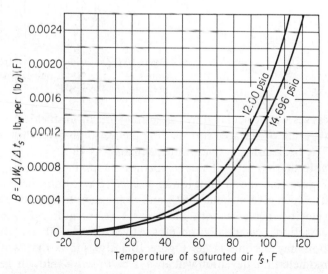

Figure 10.5 Rate of change of humidity ratio with temperature for saturated air.

For the special case of a shielded wet-bulb thermometer, Eq. (10.23) reduces to

$$\frac{t_{wb} - t^\star}{t - t_{wb}} = \frac{Le - 1}{1 + (B/K^\star)} \tag{10.25}$$

Figure 10.6 shows deviation of the psychrometer wet-bulb temperature from the thermodynamic wet-bulb temperature for a wet-bulb diameter of 0.3 in. The solid-line

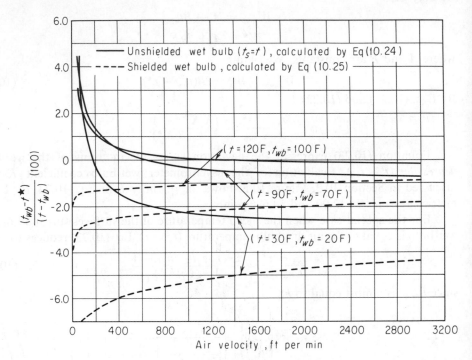

Figure 10.6 Deviation $(t_{wb} - t^\star)$ in per cent of the wet-bulb depression $(t - t_{wb})$ for a wet-bulb diameter of 0.3 in., and a barometric pressure of 14.696 psia.

curves were calculated by Eq. (10.24) for an unshielded wet-bulb for the case of temperature of surfaces surrounding the wet-bulb equal to the air dry-bulb temperature. The broken-line curves were calculated by Eq.(10.25) for the case of a shielded wet-bulb thermometer. Three temperature combinations (t, t_{wb}) are shown for each case. Figure 10.6 shows that, for an *unshielded* wet-bulb thermometer, there is some velocity which gives equality of t_{wb} and $t^\star$ for each temperature combination. The required velocity increases with increase of temperature. Beyond a certain velocity, the deviation is essentially constant. At velocities less than about 100 ft per min, the deviation may become large. Figure 10.6 also shows that with a *shielded* wet-bulb thermometer, the deviation will always be *negative* and, in general, larger than for the unshielded wet-bulb where $t_s = t$. Figure 10.7 shows results for conditions similar to those in Fig. 10.6 except for a wet-bulb diameter of 0.1 in. Equality of t_{wb} and $t^\star$ occurs at much lower velocities than when the wet-bulb diameter is 0.3 in.

Although Figs. 10.6 and 10.7 show results for only three temperature combinations, several general observations may be made. For atmospheric temperatures above freezing, where the wet-bulb depression does not exceed about 20 F, and where no

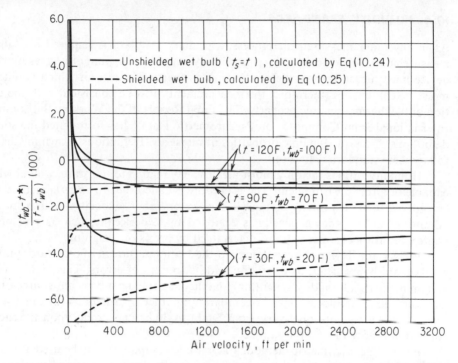

Figure 10.7 Deviation $(t_{wb} - t^\star)$ in per cent of the wet-bulb depression $(t - t_{wb})$ for a wet-bulb diameter of 0.1 in., and a barometric pressure of 14.696 psia.

unusual radiation circumstances exist, t_{wb} should differ from $t^\star$ by less than about 0.5 F for an unshielded mercury-in-glass wet-bulb thermometer as long as the air velocity exceeds about 100 ft per min. When a thermocouple is used as a wet-bulb thermometer, similar accuracy exists, except that somewhat lower air velocities are permissible.

Thus we conclude that in a majority of engineering problems, a wet-bulb temperature obtained from a properly operated unshielded psychrometer may be used directly as the thermodynamic wet-bulb temperature. The moist-air state may then be obtained directly from psychrometric tables or from a psychrometric chart. When the wet-bulb depression is large but the mean temperature of surrounding surfaces is known to differ very slightly from the air dry-bulb temperature, a more accurate procedure with an unshielded psychrometer is to find the moist-air state through the use of Eqs. (10.9) and (10.13). When the mean temperature of surrounding surfaces is believed substantially different from the air dry-bulb temperature, or when temperatures in direct sunshine are to be measured, both thermometers of the psychrometer should be shielded.

10.6. HUMIDITY STANDARDS

As in any other measurement process, a primary standard is required for humidity measurement. To be acceptable as a primary standard for humidity measurement, a device must measure some thermodynamic property of moist air which is related to moisture content. Furthermore, the measurement must be consistent with the definition of the thermodynamic property. The Final Report of the Working Subcommittee, International Joint Committee on Psychrometric Data,* has formulated the standard definitions of thermodynamic properties of moist air. Definitions shown in Chapter 8 follow these standard definitions.

Humidity ratio is the only moist-air property related to moisture content which is capable of direct measurement. The procedure is called the *gravimetric* method and is considered to be a primary standard. The water vapor associated with a measured volume of air of known density is removed by a desiccant such as phosphorous pentoxide. Mass of water vapor is determined by precision weighing.

Wexler and Hyland† have described the National Bureau of Standards gravimetric standard hygrometer, its operation and its sources of errors. This highly complex system includes a humidity generator capable of supplying a constant source of moist air over a long period of time. In comparing another instrument with this standard, the time for a test run may vary from 5 min to 30 hr depending upon the moisture content and flow rate of the sample.

Besides the gravimetric method, an *atmosphere producer* may be used as a humidity standard. An atmosphere producer is essentially a precise air-conditioning system which may produce a defined set of atmospheric conditions in a test-chamber section. Most atmosphere producers operate on the principle of altering the condition of a saturated atmosphere in a defined and calculable way. Precision of the device is primarily dependent upon how precisely the saturated atmosphere may be attained.

Amdur and White‡ have described a two-pressure type of atmosphere producer. Compressed moist air is brought to saturation at a known temperature. The air is then throttled to atmospheric pressure in a test chamber. Relative humidity in the

*J. A. Goff, "Standardization of Thermodynamic Properties of Moist Air," *ASHVE Transactions*, Vol. 55, 1949, pp. 459–484.

†Arnold Wexler and Richard W. Hyland, "The NBS Standard Hygrometer," in *Humidity and Moisture, Volume 3: Fundamentals and Standards*, edited by Arnold Wexler and William A. Wildhack (New York: Reinhold Publishing Corporation, 1965) pp. 389–432.

‡Elias J. Amdur and Robert W. White, "Two-pressure Relative Humidity Standards," in *Humidity and Moisture, Volume 3: Fundamentals and Standards*, edited by Arnold Wexler and William A. Wildhack (New York: Reinhold Publishing Corporation, 1965) pp. 445–459.

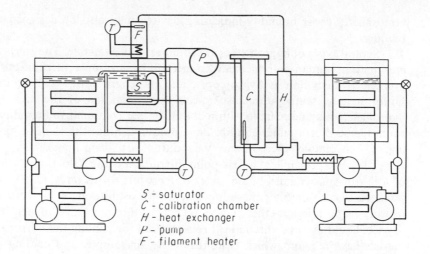

S - saturator
C - calibration chamber
H - heat exchanger
P - pump
F - filament heater

Figure 10.8 Schematic illustration of two-temperature, recirculation-type atmosphere producer. (Reprinted by permission from C. E. Till and G. O. Handegord, "Proposed Humidity Standard," *ASHRAE Journal*, Vol. 2, June 1960, p. 45.)

test chamber is determined through precise measurements of the two air pressures.

Figure 10.8 shows the atmosphere producer proposed by Till and Handegord. There are two controlled-temperature liquid sources which maintain the calibration chamber at a higher temperature than the saturator chamber. Air is recirculated in the system. It is assumed that air is brought to saturation in equilibrium with the temperature of the water in the saturator. The air is then reheated in the two heat exchangers, *F* and *H*, to a known temperature in the calibration chamber. The system of Fig. 10.8 was designed to allow calibration of electrical-resistance hygrometers within an accuracy of 0.2 per cent.

10.7. OTHER METHODS FOR MEASURING AIR HUMIDITY*

Besides the psychrometer, many other techniques are available for measurement of air humidity. Most of these devices are much more complicated than the psychrometer and, unfortunately, many of them are less reliable. A *hygrometer* is a device that gives a direct reading of relative humidity. A *dew-point indicator* allows direct determination of the dew-point temperature. Other humidity-measurement devices are available, some of which allow an absolute determination of the amount of water vapor in the

*For further information see *Humidity and Moisture, Volume 1, Methods of Measuring Humidity in Gases*, edited by Robert E. Ruskin (New York: Reinhold Publishing Corporation, 1965).

atmosphere. Most humidity-indicating devices may also be designed to record the readings.

Several types of hygrometers are commercially available. The *mechanical hygrometer* uses a human-hair element connected by a simple mechanical linkage to a pointer. Human hair is hygroscopic, and its length varies with relative humidity. Unfortunately, temperature also affects the elongation of the hair element. Hair hygrometers may be reliable within about ±3 per cent relative humidity for ordinary room temperatures under equilibrium conditions. Since the element has a large time lag, it is unsuitable where humidity conditions are inconstant.

The basic principle of a dew-point instrument is the reduction of the temperature of a mirror surface until liquid water or frost just forms on it. One type uses a thin, polished silver thimble containing ether, with aspiration of air across the thimble causing evaporation of ether and cooling of the thimble. Other models have employed dry ice, liquid air, and mechanical refrigeration for cooling the mirror surface. Some models have a pump which compresses the air sample, and cooling is effected by expansion to atmospheric pressure. Although measurement of the dew point may appear to be a fundamental method, completely reliable results are somewhat difficult to obtain. It is difficult to measure the temperature of the mirror surface, and the exact point of incipient condensation is uncertain. The temperature of the mirror surface is affected by rates of heat transfer and water-vapor transfer.

The Dunmore electric hygrometer* depends upon the hygroscopic and electrical characteristics of an aqueous salt solution. A change in air relative humidity can be determined directly by the change in electrical resistance of a film of an unsaturated solution of lithium chloride. The Dunmore cell is available as a commercial instrument from several manufacturers. The device is highly sensitive, responds rapidly to changes in relative humidity, and is well suited to remote sensing.

PROBLEMS

10.1. A psychrometer indicates a dry-bulb temperature of 90 F and a wet-bulb temperature of 70 F. Barometric pressure is 13.00 psia. Air velocity is 600 ft per min. Assume that both thermometers are shielded from radiation effects. Find the relative humidity of the air stream.

10.2. Readings for an unshielded psychrometer for an air velocity of 1200 ft per min are 120 F dry-bulb temperature and 75 F wet-bulb temperature. Barometer reading is 27.56 in. Hg. The wet-bulb diameter is 0.3 in. Assume that surrounding surfaces are at 120 F. Find the humidity ratio of the air stream.

10.3. An unshielded-type psychrometer indicates a dry-bulb temperature of 90 F and a wet-bulb temperature of 60 F. Air velocity past the thermometers

*F. W. Dunmore, "An Improved Electric Hygrometer," *Journal of Research, National Bureau of Standards*, Vol. 23, 1939, p. 701.

is 500 ft per min. Diameter of the dry-bulb is 0.25 in.; diameter of the wet-bulb is 0.3 in. Barometric pressure is 14.696 psia. The mean temperature of surrounding surfaces is estimated to be 120 F. Determine the dry-bulb temperature, humidity ratio, and thermodynamic wet-bulb temperature of the air stream.

10.4. Estimate the reading of the dry-bulb thermometer in Problem 10.3 if the sensing bulb was tightly wrapped with a metal foil having an emissivity of 0.05.

10.5. A psychrometer with unshielded thermometers indicates a dry-bulb temperature of 250 F and a wet-bulb temperature of 100 F. Barometric pressure is 14.696 psia. The mean temperature of surfaces surrounding the sensing bulbs is 250 F. Air velocity is 900 ft per min. Diameter of the wet-bulb is 0.3 in. Determine the thermodynamic wet-bulb temperature of the air stream.

10.6. A psychrometer is exposed to an air stream having true properties of 100 F dry-bulb temperature, 70 F thermodynamic wet-bulb temperature, and 14.00 psia barometric pressure. Surfaces surrounding the sensing bulbs are at 100 F. Determine the required air velocity, ft per min, such that the wet-bulb thermometer reading will be equal to the thermodynamic wet-bulb temperature if (a) the diameter of the wet-bulb is 0.1 in., and (b) the diameter of the wet-bulb is 0.5 in.

10.7. Compressed moist air at saturation conditions under a pressure P_1 is throttled to a lower pressure P_2. Assuming perfect-gas behavior, prove that the relative humidity ϕ_2 is given by the relation

$$\phi_2 = \frac{P_2}{P_1}$$

Note: This problem demonstrates the idealized principle of the two-pressure atmosphere producer.

SYMBOLS USED IN CHAPTER 10

A	Cross-sectional area, sq ft.
A	Coefficient in the equation $W_{s,wb} = A + Bt_{wb}$, lb_w per lb_a.
a	Constant, dimensionless.
B	Coefficient in the equation $W_{s,wb} = A + Bt_{wb}$, lb_w per (lb_a) (F).
b	Constant, dimensionless.
c	Constant, dimensionless.
c_p	Specific heat at constant pressure, Btu per (lb) (F).
$c_{p,a}$	Specific heat at constant pressure, Btu per (lb of dry air) (F).
D	Water vapor diffusivity, sq ft per hr.
d	Diameter, ft.
h_c	Convection heat transfer coefficient, Btu per (hr) (sq ft) (F).
h_D	Convection mass transfer coefficient, lb_w per (hr) (sq ft) (lb_w per lb_a).
h_{fg}	Latent heat of vaporization for water, Btu per lb; $h_{fg,wb}$ evaluated at t_{wb}; $h_{fg}^\star$ evaluated at $t^\star$.

h_R Radiation heat transfer coefficient, Btu per (hr) (sq ft) (F); $h_{R,t}$ evaluated for $t_s = t$.

K Psychrometer wet-bulb coefficient, lb_w per (lb_a) (F).

$K^\star$ Equivalent psychrometer wet-bulb coefficient for adiabatic saturation process, lb_w per (lb_a) (F).

k Thermal conductivity of moist air, Btu per (hr) (sq ft) (F per ft); k_f evaluated at mean film temperature, t_f.

L Reference length, ft.

Le   Lewis number, $h_c/h_D c_{p,a}$, dimensionless.

m_w Mass flow of water vapor, lb per (hr) (sq ft).

Pr   Prandtl number, $c_p \mu / k$, dimensionless.

q Rate of heat transfer, Btu per (hr) (sq ft).

Re   Reynolds number, $LV\rho/\mu$, dimensionless.

Sc   Schmidt number, $\mu/\rho D$, dimensionless.

T.   Absolute dry-bulb temperature of air, R.

T_s Absolute temperature of surfaces surrounding psychrometer, R.

T_{wb} Absolute temperature as indicated by wet-bulb thermometer, R.

t Dry-bulb temperature of moist air, F.

$t^\star$ Thermodynamic wet-bulb temperature of moist air, F.

t' $(t - t_w)/(t_o - t_w)$, dimensionless (see Fig. 10.1).

t_{db} Temperature indicated by dry-bulb thermometer, F.

t_f Mean film temperature, F.

t_o Temperature of bulk-air stream, F (see Fig. 10.1).

t_s Temperature of surfaces surrounding psychrometer, F.

t_{wb} Temperature indicated by wet-bulb thermometer, F.

V Air velocity, ft per hr; V_o for bulk stream (see Fig. 10.1).

W Humidity ratio of moist air, lb_w per lb_a; W_o for bulk-air stream in Fig. 10.1.

W_s Humidity ratio of saturated moist air, lb_w per lb_a; $W_{s,w}$ evaluated at t_w of wet surface; $W_{s,wb}$ evaluated at t_{wb}; $W_s^\star$ evaluated at $t^\star$.

W' $(W_{s,w} - W)/(W_{s,w} - W_o)$, dimensionless (see Fig. 10.1).

y Coordinate length, ft.

y' y/L, dimensionless.

α Thermal diffusivity, $k/\rho c$, sq ft per hr.

ϵ_{wb} Emissivity of wet-bulb, dimensionless.

θ Time, hr.

μ Absolute viscosity, lb per (ft) (hr); μ_f evaluated at t_f.

ρ Density, lb per cu ft; ρ_f evaluated at t_f.

ρ_a Density of moist air, lb of dry air per cu ft.

11

DIRECT CONTACT
TRANSFER PROCESSES BETWEEN
MOIST AIR AND WATER

11.1. INTRODUCTION

In Chapter 9 we discussed the problem of adding moisture to air. We assumed that all moisture in contact with the air was retained by the air stream. We found that the condition line on the psychrometric chart was a function solely of the enthalpy of the added moisture.

In this chapter we will consider problems involving direct contact of air and water different from those covered in Chapter 9. We will now study processes where a relatively large flow of water contacts moist air. The rate of addition (or withdrawal) of moisture to the air stream will be extremely small compared to the flow rate of the water entering the apparatus. Depending upon the moist-air state and the temperature of the water, it is possible to have a variety of results. Air may be heated and humidified, cooled and humidified, or cooled and dehumidified by direct contact with water.

In this chapter we will be concerned with thermal processes only. However, in practical apparatus, atmospheric air may be partially cleansed of dust particles and water-soluble vapors when it is washed by water.

11.2. CONTACT OF MOIST AIR BY DIRECTLY RECIRCULATED SPRAY WATER: THE AIR WASHER

Spray devices using directly recirculated water have been in use for many years. Such a device is called an *air washer*. Figure 11.1 shows plan and elevation views of an air washer with one bank of spray nozzles. Water is withdrawn from the sump by

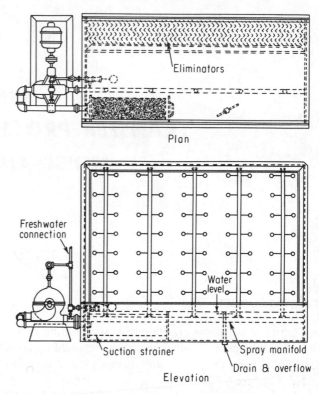

Figure 11.1 Air washer with single bank of spray nozzles. (Reprinted by permission from *Heating, Ventilating, Air Conditioning Guide*, Vol. 28, 1950, p. 710.)

an external pump and sprayed into the chamber in fine droplets by a bank of nozzles. At the air-outlet end, staggered metal baffles called *eliminator plates* minimize physical carry-over of water droplets with the air stream.

Figure 11.2 shows a schematic diagram that we will use for our analysis. We will assume that the rate at which make-up water is added to the sump is negligibly small compared to the rate of water flow through the nozzles. We will assume that heat transfer through the walls of the chamber from the ambient surroundings may be ignored. We will further assume that the small addition of energy to the water by the

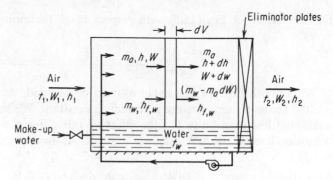

Figure 11.2 Schematic diagram of air washer using directly recirculated spray water.

pump has a negligible effect upon the water temperature. Transfer processes in the chamber involve evaporation of water droplets and convection heat transfer from the air to the water. For steady-flow conditions, and for a differential volume dV, we have

$$m_a \, dh = m_a \, dW \, h_{f,w}$$

or

$$q' = \frac{dh}{dW} = h_{f,w} \tag{11.1}$$

Since the water temperature remains constant, the condition line on the psychrometric chart would be straight with a direction of $q' = h_{f,w}$.

By Eqs. (8.26)–(8.28), we obtain

$$dh = (0.240 + 0.45W) \, dt + (1061 + 0.45t) \, dW = c_{p,a} \, dt + h_{g,t} \, dW \tag{11.2}$$

By Eqs. (11.1) and (11.2)

$$\frac{dt}{dW} = \frac{-(h_{g,t} - h_{f,w})}{c_{p,a}} \tag{11.3}$$

Heat transfer for evaporation of water added must come from convection cooling of the air stream. Thus

$$h_D A_V \, dV \, (W_{s,w} - W) h_{fg,w} = h_c A_V \, dV \, (t - t_w) \tag{11.4}$$

where h_D is a mass transfer coefficient, lb_w per (hr) (sq ft) (lb_w per lb_a); A_V is surface area of the water droplets, sq ft per cu ft; V is contact volume, cu ft; $W_{s,w}$ and W are respectively the humidity ratio of saturated moist air in equilibrium with the water and of the air stream, lb_w per lb_a; $h_{fg,w}$ is the latent heat of vaporization of the water, Btu per lb_w; h_c is a convection heat transfer coefficient, Btu per (hr) (sq ft) (F); and t and t_w are respectively the moist air dry-bulb temperature and the water temperature, F.

Substitution of $Le = h_c/h_D c_{p,a}$ in Eq. (11.4) gives

$$(W_{s,w} - W)h_{fg,w} = Le \, c_{p,a}(t - t_w) \tag{11.5}$$

Differentiation of Eq. (11.5) with respect to W (assuming Le and t_w as constants) gives

$$\frac{dt}{dW} = \frac{-(h_{g,t} - h_{g,w} + h_{fg,w}/\text{Le})}{c_{p,a}} \qquad (11.6)$$

Since both Eqs. (11.3) and (11.6) must be satisfied, we conclude that Le must be unity if the water temperature is to remain constant. We know by Sec. 10.2 that Le is somewhat less than unity but the effect of this difference upon the direction of the condition line on the h-W diagram is small. As an approximation, we will write Eq. (10.5) as

$$(W_{s,w} - W)h_{fg,w} = c_{p,a}(t - t_w) \qquad (11.7)$$

When compared to Eq. (8.31), Eq. (11.7) shows that the water temperature must be equal to the thermodynamic wet-bulb temperature $t_1^\star$. Furthermore, the condition line on the psychrometric chart must coincide with a line of constant thermodynamic wet-bulb temperature. Figure 11.3 shows a schematic case. Experiments on actual air

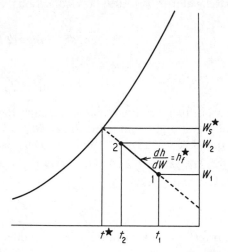

Figure 11.3 Schematic condition line for moist air passing through air washer of Fig. 11.2.

washers using directly recirculated spray water show that these conclusions are closely true.

The rate of evaporation for the volume element dV in Fig. 11.2 may be written as

$$m_a\, dW = h_D A_V\, dV\, (W_s^\star - W)$$

Assuming m_a, $h_D\, A_V$, and t_w as constants, we obtain

$$\frac{W_s^\star - W_2}{W_s^\star - W_1} = e^{-z} \qquad (11.8)$$

where

$$Z = \frac{h_D A_V V}{m_a}$$

The *air washer efficiency* η_w is defined by the relation

$$\eta_w = \frac{W_2 - W_1}{W_s^{\star} - W_1} \tag{11.9}$$

By Eqs. (11.8) and (11.9), we have

$$\eta_w = 1 - e^{-z} \tag{11.10}$$

If we make the approximation that $c_{p,a}$ is constant in Eq. (8.31), we find that

$$\eta_w = \frac{t_1 - t_2}{t_1 - t^{\star}} \tag{11.11}$$

Equation (11.9) is useful in evaluating the performance of an air washer as a *humidification* device. However, a dry-bulb temperature reduction occurs in the process. In hot, relatively dry climates an air washer using directly recirculated spray water (or other devices employing the same principle) may be beneficial for reducing the dry-bulb temperature of outdoor air admitted to ventilation systems. Equation (11.11) is useful in evaluating the performance of an air washer as an *evaporative cooling* device.

> **Example 11.1.** Moist air enters an air washer similar to that of Fig. 11.2 at 90 F dry-bulb temperature and 60 F thermodynamic wet-bulb temperature at a rate of 5000 cu ft per min. Barometric pressure is 14.696 psia. It is desired that the humidity ratio of the air leaving the washer be 0.0080 lb_w per lb_a. Face velocity of the entering air is 500 ft per min. It is estimated that the mass transfer coefficient $h_D A_V$ has a value of 300 lb_w per (hr) (cu ft) (lb_w per lb_a). Determine (a) the dry-bulb temperature of the air leaving the washer, (b) the air washer efficiency, and (c) the required length of the washer.
>
> SOLUTION: (a) Since $t_2^{\star} = t_1^{\star}$, we read $t_2 = 73.4$ F from Fig. E-8.
> (b) By Eq. (11.9),
>
> $$\eta_w = \frac{0.0080 - 0.00425}{0.01108 - 0.00425} = 0.549$$
>
> (c) We have $v_1 = 13.95$ cu ft/lb_a. Thus
>
> $$m_a = \frac{(5000)(60)}{13.95} = 21,505 \ lb_a/hr$$
>
> By Eq. (11.10), $e^{-z} = 0.451$ and $Z = 0.795$. Thus,
>
> $$V = \frac{m_a Z}{h_D A_V} = \frac{(21,505)(0.795)}{300} = 57.0 \text{ cu ft}$$
>
> The face area of the washer is 5000/500 = 10 sq ft. Thus, the required length is 5.70 ft.

11.3. COUNTER-FLOW CONTACT OF MOIST AIR BY HEATED SPRAY WATER: THE COOLING TOWER

Probably the most important device utilizing direct contact between water and atmospheric air is the *cooling tower*. Here the objective is not the processing of the air but cooling of the spray water. Cooling towers may be used thermally to reclaim cir-

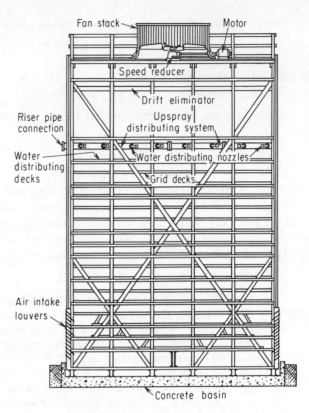

Figure 11.4 Illustration of a counter-flow, induced-draft cooling tower. (Reprinted by permission from *Heating, Ventilating, Air Conditioning Guide*, Vol. 37, 1959, p. 582.)

culating water for re-use in refrigerant condensers, power-plant condensers, and other heat exchangers.

Figure 11.4 shows a sectional view of a counter-flow, induced-draft cooling tower. Atmospheric air is circulated upward through the tower by a fan. The warm water is admitted in the upper part of the tower and falls downward in counter-flow to the air. Most towers contain a *fill* of some type, such as wood slats or latticework. The fill retards the rate of water fall and increases the water surface exposed to the air. Eliminator plates at the top of the tower minimize *drift* or carry-over of liquid water in the exhaust air. Water is lost from the tower by *evaporation*, *drift*, and *blowdown*. Blowdown, or wasting of some of the sump water, prevents undue concentration of solids. Total water loss depends upon the water-cooling range, but is usually three per cent or less. More detailed practical discussion of cooling towers is given in the ASHRAE Guide and Data Book.*

*ASHRAE Guide and Data Book: Systems and Equipment (New York: American Society of Heating, Refrigerating and Air Conditioning Engineers, 1967) pp. 223–236.

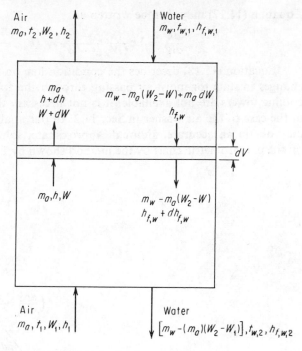

Figure 11.5 Schematic diagram of counter-flow cooling tower.

Figure 11.5 shows a schematic diagram of a counter-flow cooling tower. We will ignore water loss by drift. We will also ignore heat transfer through the walls of the tower. For steady-flow conditions, for the differential volume element we have

$$m_a\, dh = -[m_w - m_a(W_2 - W)]dh_{f,w} + m_a\, dW\, h_{f,w} \qquad (11.12)$$

or approximately,

$$m_a\, dh = -m_w\, dh_{f,w} + m_a\, dW\, h_{f,w} \qquad (11.13)$$

We may also write

$$-m_w\, dh_{f,w} = h_c A_V\, dV(t_w - t) + h_D A_V\, dV\,(W_{s,w} - W)h_{fg,w} \qquad (11.14)$$

$$m_a\, dW = h_D A_V\, dV\,(W_{s,w} - W) \qquad (11.15)$$

By substitution of Le $= h_c/h_D c_{p,a}$ in Eq. (11.14), we obtain

$$-m_w\, dh_{f,w} = h_D A_V\, dV[\text{Le }c_{p,a}(t_w - t) + (W_{s,w} - W)h_{fg,w}] \qquad (11.16)$$

By Eqs. (11.13), (11.15), and (11.16),

$$\frac{dh}{dW} = \text{Le }c_{p,a}\frac{(t_w - t)}{(W_{s,w} - W)} + h_{g,w} \qquad (11.17)$$

Using the approximation of constant $c_{p,a}$ in Eq. (8.27), we have

$$h_{s,w} - h = c_{p,a}(t_w - t) + 1061(W_{s,w} - W)$$

Equation (11.17) may then be written as

$$\frac{dh}{dW} = \text{Le}\,\frac{(h_{s,w} - h)}{(W_{s,w} - W)} + (h_{g,w} - 1061\,\text{Le}) \tag{11.18}$$

Equation (11.18) describes the condition line on the psychrometric chart for the changes in state for moist air passing through the tower. Although data for Le for cooling towers are not available, it is not necessary that Le be restricted to unity as in the case of the air washer in Sec. 11.2. In our analysis, we will use Eq. (10.7). We may obtain an accurate, although approximate, solution to Eq. (11.18) graphically on the psychrometric chart by the method shown by Fig. 11.6. (We may also program

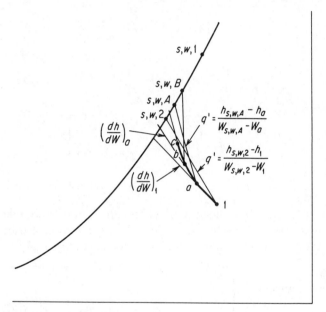

Figure 11.6 Graphical solution of Eq. (11.18) on the psychro-metric chart.

the solution on a digital computer.) We assume that the water temperatures $t_{w,1}$ and $t_{w,2}$, the water-flow rate m_w, the air-flow rate m_a, and the entering air state are known. With this information, we may solve Eq. (11.18) for the direction dh/dW of the condition line at State (1). With the chart protractor, we draw a line segment with this direction through State (1). At a short distance from State (1) on this line we arbitrarily locate a new State (a). Equation (11.13) may be written as

$$-\Delta t_w = \frac{m_a}{m_w c_w}(\Delta h - \Delta W\,h_{f,w}) \tag{11.19}$$

With Eq. (11.19), we may calculate the water temperature $t_{w,A}$ corresponding to State (a). Equation (11.18) may then be solved for $(dh/dW)_a$ and the procedure continued until the complete condition line is drawn and the final air state determined. The accuracy of the method depends upon the extent of the assumed incremental changes of air state.

For the entire cooling tower, by Fig. 11.5 we have

$$m_a h_1 + m_w h_{f,w,1} = m_a h_2 + [m_w - m_a(W_2 - W_1)]h_{f,w,2}$$

or

$$h_2 = h_1 + \frac{m_w c_w}{m_a}(t_{w,1} - t_{w,2}) + (W_2 - W_1)h_{f,w,2} \qquad (11.20)$$

Assuming that the complete condition line is already drawn on the psychrometric chart, the outlet air condition may be checked by Eq. (11.20).

With the condition line and the outlet-air state known, and with the average mass transfer coefficient $h_D A_V$ known, the required tower volume may be obtained from Eq. (11.15). Thus

$$V = \frac{m_a}{h_D A_V} \int_{W_1}^{W_2} \frac{dW}{W_{s,w} - W} \qquad (11.21)$$

The integral in Eq. (11.21) may be solved by a numerical method.

A convenient method for solving integrals of the type in Eq. (11.21) is by use of the *Stevens diagram*. Figure 11.7 shows the diagram where a dimensionless factor f is presented as a function of two dimensionless variables y_m/y_1 and y_m/y_2. Using the integral of Eq. (11.21) to typify a general example, we have

$$y_1 = W_{s,w,1} - W_2 \qquad (11.22)$$

$$y_2 = W_{s,w,2} - W_1 \qquad (11.23)$$

$$y_m = W_{s,w,m} - W_m \qquad (11.24)$$

where $W_m = (W_1 + W_2)/2$ and $W_{s,w,m}$ is evaluated at the same location where W_m exists. The solution of the integral is given by

$$\int_{W_1}^{W_2} \frac{dW}{W_{s,w} - W} = \frac{W_2 - W_1}{f y_m} \qquad (11.25)$$

where f is given by Fig. 11.7.

The Stevens diagram was constructed on the premise that the difference function y varies along a second-degree parabola fitted to the three known values y_1, y_m, and y_2.

> **Example 11.2.** A cooling tower is to be designed to cool 1500 GPM of water from 100 F to 85 F when the outside air is at 95 F dry-bulb temperature and 75 F thermodynamic wet-bulb temperature. Barometric pressure is 14.696 psia. The ratio of water flow to dry-air flow (m_w/m_a) is 1.00. Assume a constant air mass velocity of 1400 lb_a per (hr) (sq ft). It is estimated that the average mass transfer coefficient $h_D A_V$ is 120 lb_w per (hr) (cu ft) (lb_w per lb_a). (a) Construct the complete condition line on the psychrometric chart; (b) determine the dry-bulb temperature, thermodynamic wet-bulb temperature, and humidity ratio of the air leaving the tower; and (c) calculate the required tower volume in cu ft.

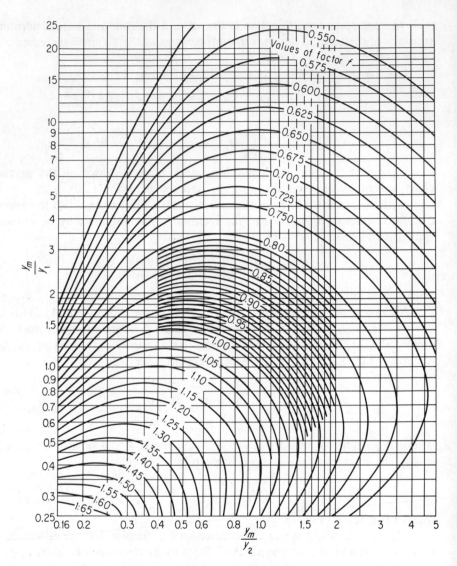

Figure 11.7 The Stevens diagram. (Reprinted from W. F. Carey and G. J. Williamson, "Gas Cooling and Humidification: Design of Packed Towers from Small-Scale Tests," *Proceedings of the Institution of Mechanical Engineers*, Vol. 163, 1950, p. 49.)

SOLUTION: By Table A.1 at 100 F, $v_{f,w} = 0.01613$ cu ft/lb$_w$. Thus

$$m_w = \frac{(1500)(231)(60)}{(1728)(0.01613)} = 746,000 \text{ lb}_w/\text{hr}$$

Since $m_w/m_a = 1.00$, $m_a = 746,000$ lb$_a$/hr.

(a) In order to construct the condition line, both the normal temperature psychrometric chart (Fig. E-8) and the high temperature psychrometric chart (Fig. E-9) are required. We begin the construction at the air-inlet end of the tower using Fig. E-8. We may locate State (1) of the inlet air and State (s,w,2) of saturated air in equilibrium with the leaving water. By Table A.1, we have $h_{f,w} = 53.0$ Btu/lb$_w$ and $h_{g,w} = 1098$ Btu/lb$_w$. By Eq. (10.7), we find Le = 0.895. With the chart protractor, we obtain

$$(h_{s,w,2} - h_1)/(W_{s,w,2} - W_1) = 890 \text{ Btu/lb}_w$$

By Eq. (11.18),

$$\left(\frac{dh}{dW}\right)_1 = (0.895)(890) + 1098 - (0.895)(1061) = 945 \text{ Btu/lb}_w$$

With the chart protractor, we draw a short line of direction $(dh/dW)_1$ through State (1). We arbitrarily locate State (a) at the intersection of $h = 40.00$ Btu/lb$_a$ with this line. By Eq. (11.19)

$$t_{w,A} - t_{w,2} = \frac{(1)}{(1)}[(h_a - h_1) - (W_a - W_1)h_{f,w,2}]$$

We obtain $t_{w,A} = 86.52$ F. Next we locate the state of saturated air at this temperature on the psychrometric chart. We then repeat the procedure.

Table 11.1 shows a tabulation of calculations for the complete condition line. Figure 11.8 shows a plot of the condition line on the psychrometric chart.

(b) The leaving air state was determined as the final point on the condition line. We read $t_2 = 94.8$ F, $t_2^* = 88.8$ F, and $W_2 = 0.0285$ lb$_w$/lb$_a$. State 2 may be checked by Eq. (11.20). At $W_2 = 0.0285$ lb$_w$/lb$_a$, we obtain

$$h_2 = 38.39 + \frac{(1)}{(1)}(100 - 85) + (0.0285 - 0.01408)(53) = 54.15 \text{ Btu/lb}_a$$

which closely agrees with $h_2 = 54.26$ Btu/lb$_a$ obtained by the graphical procedure.

(c) We may obtain the tower volume by graphical integration of Eq. (11.21). All necessary numerical data are known from Part (a). Table 11.2 shows a tabulation of calculations for $f(W) = 1/(W_{s,w} - W)$ for various values of W. Figure 11.9 shows a plot of $f(W)$. The area under the curve is the value of the integral. We obtain

$$F(W) = \int_{W_1}^{W_2} \frac{dW}{W_{s,w} - W} = 1.162$$

and

$$V = \left(\frac{m_a}{h_D A_V}\right)F(W) = \frac{(746,000)(1.162)}{120} = 7220 \text{ cu ft}$$

Since an air-mass velocity of 1400 lb$_a$/(hr) (sq ft) was used, the tower cross-sectional area would be $746,000/1400 = 533$ sq ft. The required tower height would be $7220/533 = 13.5$ ft.

The Stevens diagram may be used also for solving the integral of Eq. (11.21). We have $W_m = (W_1 + W_2)/2 = 0.02129$ lb$_w$/lb$_a$. By Table 11.1, we may estimate $t_{w,m} = 91.83$ F at the location where $W_m = 0.02129$ lb$_w$/lb$_a$ occurs. Thus, $W_{s,w,m} = 0.03305$ lb$_w$/lb$_a$. By Eqs. (11.22)–(11.24), we obtain $y_1 = 0.0147$

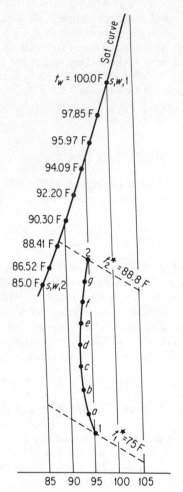

Figure 11.8 Air-process line for Example 11.2.

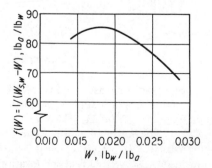

Figure 11.9 Plot of $f(W) = 1/(W_{s,w} - W)$ for Example 11.2.

lb_w/lb_a, $y_2 = 0.01234$ lb_w/lb_a, and $y_m = 0.01176$ lb_w/lb_a. Thus, $y_m/y_1 = 0.80$, and $y_m/y_2 = 0.953$. By Fig. 11.7, $f = 1.045$.

By Eq. (11.25)

$$F(W) = \frac{W_2 - W_1}{fy_m} = \frac{(0.01442)}{(1.045)(0.01176)} = 1.173$$

which closely checks the value of 1.162 obtained before.

TABLE 11.1
CALCULATIONS FOR PART (a) OF EXAMPLE 11.2

Air State	t F	h Btu/lb$_a$	W lb$_w$/lb$_a$	t_w F	$h_{f,w}$ Btu/lb$_w$	Le	$\dfrac{h_{s,w} - h}{W_{s,w} - W}$ Btu/lb$_w$	$h_{g,w}$ Btu/lb$_w$	$\dfrac{dh}{dW}$ Btu/lb$_w$
1	95.0	38.39	0.01408	85.00	53.0	0.895	890	1098	945
a	93.8	40.00	0.01585	86.52	54.6	0.895	940	1099	990
b	93.0	42.00	0.01782	88.41	56.4	0.895	1000	1100	1045
c	92.6	44.00	0.01975	90.30	58.3	0.895	1080	1101	1118
d	92.7	46.00	0.02154	92.20	60.2	0.895	1095	1101	1131
e	93.0	48.00	0.0233	94.09	62.1	0.895	1130	1102	1163
f	93.3	50.00	0.0250	95.97	64.0	0.895	1150	1103	1182
g	94.0	52.00	0.0267	97.85	65.9	0.895	1170	1104	1201
2	94.8	54.27	0.0285	100.00	....	....	....	....	....

TABLE 11.2
CALCULATIONS FOR PART (c) OF EXAMPLE 11.2

Air State	t_w F	W lb$_w$/lb$_a$	$W_{s,w}$ lb$_w$/lb$_a$	$W_{s,w} - W$ lb$_w$/lb$_a$	$f(W) = 1/(W_{s,w} - W)$ lb$_a$/lb$_w$
1	85.00	0.01408	0.02635	0.01227	81.5
a	86.52	0.01585	0.02775	0.01190	84.0
b	88.41	0.01782	0.02952	0.01170	85.5
c	90.30	0.01975	0.03157	0.01182	84.6
d	92.20	0.02154	0.0336	0.01206	82.9
e	94.09	0.0233	0.0358	0.01250	80.0
f	95.97	0.0250	0.0380	0.01300	76.9
g	97.85	0.0267	0.0404	0.01370	73.0
2	100.00	0.0285	0.0432	0.01470	68.0

An interesting result, both academically and practically, is that of determining the minimum possible temperature at which water may be withdrawn from a cooling tower. By Eq. (11.13),

$$\frac{dh}{dW} = -\frac{m_w c_w}{m_a}\frac{dt_w}{dW} + h_{f,w}$$

The minimum water temperature would occur when $dt_w/dW = 0$, or when $dh/dW = h_{f,w}$. The transfer processes would then correspond to those for an air washer using directly recirculated spray water. For the reasons explained in Sec. 11.2, the minimum possible temperature for water leaving a cooling tower would be the *thermodynamic wet-bulb temperature of the inlet air*.

The methods of analysis for a cooling tower so far presented provide an accurate procedure for design purposes. Example 11.2 shows that rather lengthy calculations are required. It is necessary to have both water temperatures known in order to apply the procedure conveniently. In order to analyze tower performance at other than design conditions (given values of tower volume, m_w, m_a, $t_{w,1}$, and inlet-air state, but unknown values of $t_{w,2}$ and outlet-air state), a tedious trial-and-error procedure is necessary.

It is possible to simplify the analysis if certain approximations are made. Noting that $m_w\, dh_{f,w} = m_w c_w\, dt_w$, by Eqs. (11.13) and (11.18), we have

$$-\frac{dt_w}{h_{s,w} - h} = \frac{m_a}{m_w c_w}\left[\text{Le} + \frac{(h_{fg,w} - 1061\,\text{Le})}{(h_{s,w} - h)/(W_{s,w} - W)}\right]\frac{dW}{(W_{s,w} - W)} \qquad (11.26)$$

On the right side of Eq. (11.26), the second term of the group in brackets is small compared to the Lewis number. Example (11.2) shows that Le was constant for a single problem. Thus, we may assign a mean value to the bracketed group with only a small error. By Eqs. (11.21) and (11.26),

$$\int_{t_{w,2}}^{t_{w,1}} \frac{dt_w}{h_{s,w} - h} = \frac{h_D A_V V}{m_w c_w}\left[\text{Le} + \frac{(h_{fg,w} - 1061\,\text{Le})}{(h_{s,w} - h)/(W_{s,w} - W)}\right]_m \qquad (11.27)$$

By Eq. (11.13),

$$\frac{dh}{dt_w} = -\frac{m_w c_w}{m_a} + \frac{dW}{dt_w} h_{f,w}$$

or

$$-\frac{dh}{dt_w} = \frac{m_w c_w/m_a}{1 - h_{f,w}/(dh/dW)} \qquad (11.28)$$

Table 11.1 shows that for Example 11.2, the term $h_{f,w}/(dh/dW)$ was small compared to unity and that it was also essentially constant. We may assign a mean value to this term and thus assume that approximately h varies linearly with t_w.

Solution of Eqs. (11.27) and (11.28) is simplified by use of the Stevens diagram and the analysis shown by Fig. 11.10. Figure 11.10 shows a plot of the air enthalpy h and the enthalpy of saturated air $h_{s,w}$ as a function of the water temperature. The *operating line* showing the variation of h with t_w is drawn by Eq. (11.28). Through use of the Stevens diagram, we have

$$\int_{t_{w,2}}^{t_{w,1}} \frac{dt_w}{h_{s,w} - h} = \frac{t_{w,1} - t_{w,2}}{fy_m} \qquad (11.29)$$

The quantities y_1, y_2, and y_m for use with Fig. 11.7 are shown on Fig. 11.10.

> **Example 11.3.** Determine the required tower volume for Example 11.2 by the approximate method given by Eqs. (11.27)–(11.29).

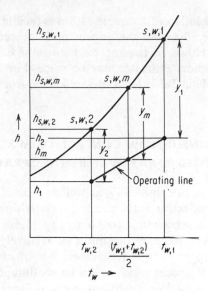

Figure 11.10 Schematic variation of $(h_{s,w} - h)$ for a cooling tower.

SOLUTION: By Table 11.1, using Air State (d),

$$[h_{f,w}/(dh/dW)]_m = \frac{60.2}{1131} = 0.053$$

By Eq. (11.28),

$$h_2 = h_1 - (dh/dt_w)(t_{w,1} - t_{w,2}) = 38.39 + 15/0.947 = 54.23 \text{ Btu/lb}_a$$

$$h_m = h_1 - (dh/dt_w)[(t_{w,1} - t_{w,2})/2] = 38.39 + 7.5/0.947 = 46.31 \text{ Btu/lb}_a$$

At $t_{w,1} = 100$ F, $h_{s,w,1} = 71.73$ Btu/lb$_a$; at $t_{w,m} = 92.5$ F, $h_{s,w,m} = 59.52$ Btu/lb$_a$; at $t_{w,2} = 85$ F, $h_{s,w,2} = 49.43$ Btu/lb$_a$. Thus

$$y_1 = h_{s,w,1} - h_2 = 71.73 - 54.23 = 17.50 \text{ Btu/lb}_a$$

$$y_2 = h_{s,w,2} - h_1 = 49.43 - 38.39 = 11.04 \text{ Btu/lb}_a$$

$$y_m = h_{s,w,m} - h_m = 59.52 - 46.31 = 13.21 \text{ Btu/lb}_a$$

and $y_m/y_1 = 0.755$, $y_m/y_2 = 1.20$. By Fig. 11.7, $f = 1.01$. By Table 11.1, using Air State (d),

$$\left[\text{Le} + \frac{(h_{fg,w} - 1061 \text{ Le})}{(h_{s,w} - h)/(W_{s,w} - W)}\right]_m = 0.895 + \frac{[1041 - (1061)(0.895)]}{1095} = 0.978$$

By Eqs. (11.27) and (11.29),

$$V = \frac{m_w c_w}{h_D A_V} \frac{(t_{w,1} - t_{w,2})/fy_m}{\left[\text{Le} + \dfrac{(h_{fg,w} - 1061 \text{ Le})}{(h_{s,w} - h)/(W_{s,w} - W)}\right]_m} = \frac{(746,000)(1)(15)}{(120)(0.978)(1.01)(13.21)}$$

$$= 7140 \text{ cu ft}$$

which is 1.1 per cent lower than the answer obtained for Example 11.2.

Our solution of Example 11.3 was facilitated by known calculations in Example 11.2. Generally, with the approximate procedure, such calculations would not be available. However, the approximations in Eqs. (11.27) and (11.29) are not critical, and satisfactory estimates may be made. For example, if we base the solution upon Air State (1) in Table 11.1 (all data known from given conditions), we find $V = 7020$ cu ft.

11.4. COUNTER-FLOW CONTACT OF MOIST AIR BY CHILLED SPRAY WATER: THE SPRAY DEHUMIDIFIER

It is possible to dehumidify, as well as to cool, moist air by direct contact with cold water. For effective heat exchange, counter-flow must be employed.

From a schematic standpoint, Fig. 11.5 is applicable for an analysis of a spray dehumidifier. We may imagine that relatively cold well water, or else refrigerated water, is supplied to the chamber. Air which leaves the contact chamber may be circulated to spaces which are to be conditioned.

For steady-flow conditions, we have for a differential volume

$$-m_a\, dh = m_w\, dh_{f,w} - m_a\, dW\, h_{f,w} \tag{11.30}$$

$$m_w\, dh_{f,w} = m_w c_w\, dt_w = h_c A_V\, dV\,(t - t_w) + h_D A_V\, dV\,(W - W_{s,w})(h_{g,t} - h_{f,w}) \tag{11.31}$$

$$-m_a\, dW = h_D A_V\, dV\,(W - W_{s,w}) \tag{11.32}$$

By procedures completely analogous to those for the cooling tower, we obtain

$$\frac{dh}{dW} = \text{Le}\,\frac{(h - h_{s,w})}{(W - W_{s,w})} + (h_{g,t} - 1061\,\text{Le}) \tag{11.33}$$

$$h_2 = h_1 - \frac{m_w c_w}{m_a}(t_{w,2} - t_{w,1}) - (W_1 - W_2)h_{f,w,2} \tag{11.34}$$

$$V = \frac{m_a}{h_D A_V} \int_{W_2}^{W_1} \frac{dW}{W - W_{s,w}} \tag{11.35}$$

Comparison of Eqs. (11.30)–(11.35) with Eqs. (11.12)–(11.21) shows that the analysis of a counter-flow spray dehumidifier is the same as that for the cooling tower. Using the psychrometric chart, we may construct the condition line and determine the outlet-air state. The required chamber volume may be found from Eq. (11.35).

Figure 11.11 shows a schematic representation of Eq. (11.33) on the psychrometric chart for the air-inlet end of the chamber. Because of the magnitude of the term $(h - h_{s,w})/(W - W_{s,w})$, the condition line of direction dh/dW may be steeper than the line of direction $q' = (h - h_{s,w})/(W - W_{s,w})$.

Spray dehumidifiers have rather infrequent practical application for cooling and dehumidifying atmospheric air compared to heat exchangers using extended surfaces. In Chapter 12 we will study in detail the cooling and dehumidifying of moist air through the use of finned-tube coils.

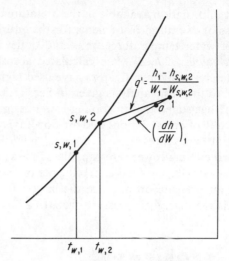

Figure 11.11 Schematic solution of Eq. (11.33) on psychrometric chart.

11.5. MASS TRANSFER COEFFICIENTS FOR WATER-AIR DIRECT CONTACT DEVICES

In design calculations on air washers, cooling towers, and spray dehumidifiers, knowledge of the mass transfer coefficient h_n and the surface-area-to-volume ratio A_v are needed. Conventional procedure has been to combine the two factors into one coefficient $h_D A_v$ since it is difficult to separate them.

A large amount of performance information has been published on laboratory cooling towers. Less information is available on commercial cooling towers. Very few mass transfer data are available on air washers or spray dehumidifiers.

Simpson and Sherwood* have reported results of experimental studies on several small cooling towers. They found that the coefficient $h_D A_v$ was a function of the dry-air flow rate, water-flow rate, and inlet water temperature. Lichtenstein† has discussed the performance and selection of cooling towers including some correlations for the mass transfer coefficient.

Caution must be observed in applying mass transfer coefficient information given in the literature on cooling towers. Because of wide variation in types of packing or fill, one type of design may show a very different performance from another type. Most reported information has come from small laboratory towers where the ratio of wall surface to volume is much larger than in commercial towers.

---

*W. M. Simpson and T. K. Sherwood, "Performance of Small Mechanical Draft Cooling Towers," *Refrigerating Engineering*, Vol. 52, December 1946, pp. 535–543, 574–576.

†J. Lichtenstein, "Performance and Selection of Mechanical Draft Cooling Towers," *A.S.M.E. Trans.*, Vol. 65, 1943, pp. 779–787.

Most information available in the literature on the coefficient $h_D A_V$ was obtained from gross or overall measurements for the entire tower. Based upon measured inlet and outlet water temperatures, inlet and outlet air states, air-flow rate, and water-flow rate, values of $h_D A_V$ were calculated according to some analytical procedure. Unfortunately, many investigators have used highly approximate methods of analysis compared to the basic analysis given in Sec. 11.3.

Kern* has stated that in cooling-tower design, water-flow rates of 500 to 2000 lb$_w$ per (hr) (sq ft of cross-section) and air-flow rates of 1300 to 1800 lb$_a$ per (hr) (sq ft of cross-section) are representative of commercial practice. Kern further states that few commercial cooling towers are capable of showing an average coefficient $h_D A_V$ much greater than 100 lb$_w$ per (hr) (cu ft) (lb$_w$ per lb$_a$) without exceeding the usual allowable limits on air-pressure drop. Since the largest factor in tower operating cost is fan power, relatively small air-pressure drops of 2 in. of water or less are usually permitted.

PROBLEMS

11.1. An air washer using directly recirculated spray water operates under steady-state conditions. Air enters the washer at 100 F dry-bulb temperature, 65 F thermodynamic wet-bulb temperature, and 14.696 psia barometric pressure with a velocity of 400 ft per min. The efficiency of the washer is 75 per cent. Determine the dry-bulb and thermodynamic wet-bulb temperatures of the air leaving the washer.

11.2. A system including a preheating coil and an air washer (directly recirculated spray water) is to be designed to process 10,000 cu ft per min of outdoor air. Barometric pressure is 14.696 psia. Outdoor air saturated at 30 F enters the preheater where heat is supplied by steam condensing inside the tubes of the coil. The air then passes through the air washer. The air state at the exit of the washer is 70 F dry-bulb temperature and 60 per cent degree-of-saturation. Average face velocity of the air through the washer is 500 ft per min. Use $h_D A_V$ equal to 400 lb$_w$ per (hr) (cu ft) (lb$_w$ per lb$_a$). Determine (a) the dry-bulb and thermodynamic wet-bulb temperatures of the air entering the washer, (b) the required capacity of the preheating coil in Btu per hr, (c) the quantity of make-up water required for the air washer in GPM, and (d) the necessary contact volume of the air washer in cu ft.

11.3. A cooling tower operates under steady-state conditions as shown by Fig. 11.12. Assume that one per cent of the inlet water is lost from the tower as drift (liquid carry-over with the exhaust air). Assume that rate of overflow from the sump is 20 per cent of the flow rate of make-up water. Determine (a) the volume of moist air entering the fan in cu ft per min, and (b) the GPM of make-up water required.

*Donald Q. Kern, *Process Heat Transfer* (New York: McGraw-Hill Book Company, 1950) p. 601.

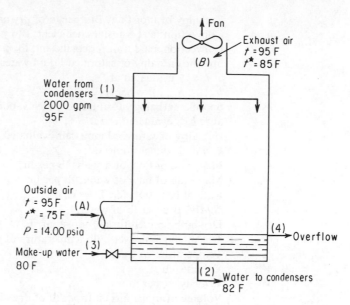

Figure 11.12 Schematic cooling tower for Problem 11.3.

11.4. Determine the required tower volume for Example 11.2 if water leaves the tower at 80 F. You may use the approximate method given by Eqs. (11.27)–(11.29).

11.5. A counter-flow spray dehumidifier is to process moist air having an inlet condition of 85 F dry-bulb temperature and 70 F thermodynamic wet-bulb temperature. Barometric pressure is 14.696 psia. The rate of air flow entering the dehumidifier is 5000 cu ft per min. Water at 45 F enters the dehumidifier at a rate of 50 GPM. The temperature of the leaving water is 55 F. Assume an average air velocity of 500 ft per min. Construct the complete condition line and determine the dry-bulb temperature and thermodynamic wet-bulb temperature of the air leaving the chamber.

11.6. By use of Fig. 11.7 determine the contact volume in cu ft required for Problem 11.5. Assume a mass transfer coefficient $h_D A_V$ of 600 lb_w per (hr) (cu ft) (lb_w per lb_a).

SYMBOLS USED IN CHAPTER 11

A_V	Surface area of water droplets, sq ft per cu ft.
$c_{p,a}$	Specific heat at constant pressure of moist air, Btu per (lb of dry air) (F).
c_w	Specific heat of water, Btu per (lb)(F).
G_a	Mass velocity of dry air, lb per (min)(sq ft).

h	Enthalpy of moist air, Btu per lb of dry air.
h_c	Convection heat transfer coefficient, Btu per (hr)(sq ft)(F).
h_D	Convection mass transfer coefficient, lb_w per (hr)(sq ft)(lb_w per lb_a).
h_f	Specific enthalpy of saturated liquid water, Btu per lb; $h_{f,w}$ evaluated at t_w; $h_f^\star$ evaluated at $t^\star$.
$h_{fg,w}$	$h_{g,w} - h_{f,w}$, Btu per lb.
h_g	Specific enthalpy of saturated water vapor, Btu per lb; $h_{g,t}$ evaluated at t; $h_{g,w}$ evaluated at t_w.
$h_{s,w}$	Enthalpy of saturated moist air evaluated at t_w, Btu per lb of dry air.
Le	$h_c/h_D c_{p,a}$, dimensionless.
m_a	Mass rate of flow of dry air, lb per hr.
m_w	Mass rate of flow of water, lb per hr.
q	Rate of heat transfer, Btu per hr.
q'	dh/dW, Btu per lb of water.
t	Dry-bulb temperature of moist air, F.
$t^\star$	Thermodynamic wet-bulb temperature of moist air, F.
t_w	Temperature of water, F.
V	Volume, cu ft.
V	Velocity, ft per min.
v	Volume of moist air, cu ft per lb of dry air.
W	Humidity ratio of moist air, lb of water vapor per lb of dry air.
$W_{s,w}$	Humidity ratio of saturated moist air evaluated at t_w, lb of water vapor per lb of dry air.
Z	$h_D A_V V/m_a$, dimensionless.
η_w	Air washer efficiency, dimensionless.

12

HEATING AND COOLING
OF MOIST AIR BY
EXTENDED SURFACE COILS

12.1. GENERAL REMARKS

Almost every thermal environmental system involves heating and/or cooling of atmospheric air. In winter heating is a major function, while in summer heating is often required in air conditioning systems where air humidity is controlled. In summer air conditioning systems, cooling and dehumidification are prime functions. In this chapter we will be concerned primarily with heat transfer problems where forced-convection, turbulent air flow occurs.

Atmospheric air may be heated or cooled in duct coils which are banks of bare tubes or banks of tubes which have finned or extended surfaces. Practically all modern coils are of the extended-surface type. Figure 12.1 shows two types of finned tubing. Figure 12.1(a) employs spiral fins while Fig. 12.1(b) shows continuous flat-plate fins. The heating or cooling medium passes through the tubes while moist air flows across the tubes and through the fins. The tubes are commonly made of copper, aluminum, or red brass; the secondary surface is made of aluminum or copper. The fins are usually mechanically bonded to the tubes.

In contrast to bare-pipe coils of the same capacity, finned coils are much more compact, have a much smaller weight, and are usually less expensive. The secondary surface area of a finned coil may be from 10 to 30 times or more that of the bare tubes.

235

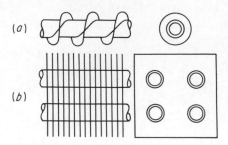

Figure 12.1 Schematic illustrations of finned tubing.

Heat transfer in finned coils is complicated. The simple expression for the overall heat transfer coefficient for a bare pipe derived in Sec. 2.7 must be modified. In addition, the fluids are typically in some type of cross-flow arrangement and the logarithmic mean temperature difference (Sec. 2.7) may also not be applicable.

The analysis of cooling coils is more involved than for heating coils since mass transfer (dehumidification) may occur simultaneously with heat transfer. In this chapter, we will first analyze heat transfer where the fins are dry. We will find that some of our results may be extended to the case of wet coils as well.

12.2. TRUE MEAN TEMPERATURE DIFFERENCE FOR HEAT EXCHANGERS OF THE CROSS-FLOW, FINNED-TUBE TYPE

For equal surface areas and for the same value of overall heat transfer coefficient, a counter-flow heat exchanger such as shown in Fig. 12.2 provides the maximum rate of heat transfer between two fluids. Such an arrangement gives the highest mean temperature difference Δt_m between the fluids. In Sec. 2.7 it was shown that for pure counter-flow

$$\Delta t_{m,cf} = \frac{(T_2 - t_1) - (T_1 - t_2)}{\ln\left(\dfrac{T_2 - t_1}{T_1 - t_2}\right)} \tag{12.1}$$

which is the *logarithmic mean temperature difference.*

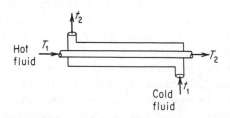

Figure 12.2 Schematic counter-flow heat exchanger.

In a more general study of heat exchangers, it is desirable to follow the plan of Bowman, Mueller, and Nagle* and express Δt_m in terms of the quantities

$$R = \frac{T_1 - T_2}{t_2 - t_1} \tag{12.2}$$

$$P = \frac{t_2 - t_1}{T_1 - t_1} \tag{12.3}$$

By Eqs. (12.1)–(12.3), we obtain for pure counter-flow

$$\Delta t_{m, cf} = \frac{(t_2 - t_1)(R - 1)}{\ln\left(\dfrac{1 - P}{1 - RP}\right)} \tag{12.4}$$

In the case where air is one of the fluids, pure counter-flow is generally not practicable. The most economical heat exchanger is usually the finned-tube type employing some form of *cross-flow* arrangement. Figure 12.3 shows two schematic

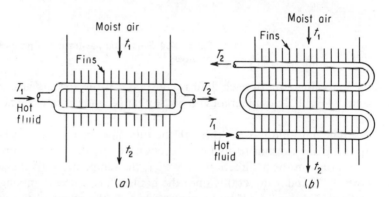

Figure 12.3 Schematic cross-flow arrangements.

arrangements. Figure 12.3(a) shows a *pure cross-flow* heat exchanger with two rows of tubes. This type with one or two rows of tubes is commonly used in steam coils for heating air. Figure 12.3(b) shows a *counter-cross-flow* arrangement with four tube passes. This type with two or more tube passes is commonly used where hot water or chilled water passes through the tubes.

As will be shown later, the logarithmic mean temperature difference is valid for cross-flow heat exchangers only when one fluid temperature remains constant (condensing steam, evaporating refrigerant, etc.). Otherwise, the conditions required for the derivation of the logarithmic mean temperature difference do not exist when cross-flow is employed.

*R. A. Bowman, A. C. Mueller, and W. M. Nagle, "Mean Temperature Difference in Design," *A.S.M.E. Trans.*, Vol. 62, 1940, pp. 283–294.

It is convenient to express the mean temperature difference Δt_m for a cross-flow heat exchanger as

$$\Delta t_m = F \Delta t_{m,\,cf} \tag{12.5}$$

where F is a correction factor and $\Delta t_{m,\,cf}$ is the logarithmic mean temperature difference calculated for *pure counter-flow*. Solutions for Eq. (12.5) have been published for only a limited number of cross-flow arrangements. A derivation for F will now be given for the simple case of pure cross-flow with one row of tubes as shown in Fig. 12.4. We will assume that (1) the overall heat transfer coefficient U_o is constant, (2)

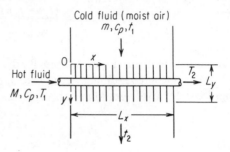

Figure 12.4 Schematic illustration of pure cross-flow with one row of tubes.

the mass-flow rate of each fluid is constant, (3) the specific heat of each fluid is constant, (4) each fluid undergoes no phase change, and (5) external heat losses are negligible.

The hot fluid passing through the tube may be assumed to be thoroughly mixed at any plane normal to its flow direction. Thus, the temperature of the hot fluid T varies only in the x direction. However, the temperature of the cold fluid t varies in both the x and y directions since the baffles (fins) prevent mixing in a plane normal to flow. The final cold-fluid temperature t_2 is the result of the mixture of the many separate streams beyond the heat transfer surface.

We will first consider how the cold-fluid temperature changes along a plane in the y direction for an element of heat transfer surface. We may write that in the y direction

$$U_o \, dA_o \, (T - t) = m \frac{dx}{L_x} c_p \, dt$$

but $dA_o = A_o(dx/L_x)(dy/L_y)$. Thus

$$\int_{t_1}^{t'} \frac{dt}{T - t} = \frac{U_o A_o}{m c_p L_y} \int_0^{L_y} dy \tag{12.6}$$

The solution of Eq. (12.6) for the cold-fluid temperature t' leaving the heat exchanger for a strip of width dx, is

$$t' = t_1 + (T - t_1)(1 - e^{-K_1}) \tag{12.7}$$

where

$$K_1 = U_o A_o / m c_p \tag{12.8}$$

Let us now consider how the fluid temperatures change in the x direction. We may write

$$-MC_P\,dT = m\frac{dx}{L_x}c_p(t' - t_1)$$

With Eq. (12.7),

$$\int_{T_1}^{T_2}\frac{dT}{T - t_1} = -\frac{mc_p}{MC_P}\frac{(1 - e^{-K_1})}{L_x}\int_0^{L_x}dx \tag{12.9}$$

The solution of Eq. (12.9) for the hot-fluid temperature T_2 at the exit of the heat exchanger is

$$T_2 = T_1 - (T_1 - t_1)(1 - e^{-K_2}) \tag{12.10}$$

where

$$K_2 = \frac{mc_p}{MC_P}(1 - e^{-K_1}) \tag{12.11}$$

Since

$$mc_p(t_2 - t_1) = MC_P(T_1 - T_2)$$

we have

$$t_2 = t_1 + \frac{MC_P}{mc_p}(T_1 - t_1)(1 - e^{-K_2}) \tag{12.12}$$

Thus, Eqs. (12.10) and (12.12) allow calculation of the final temperatures of the fluids.

We will now develop an expression for the mean temperature difference Δt_m between the fluids. By Eqs. (12.2), (12.3), (12.10), and (12.11),

$$e^{-K_2} = 1 - RP$$

and

$$K_2 = R(1 - e^{-K_1})$$

Thus

$$e^{K_1} = \frac{R}{R + \ln(1 - RP)} \tag{12.13}$$

But

$$K_1 = \frac{U_o A_o}{mc_p} = \frac{t_2 - t_1}{\Delta t_m} \tag{12.14}$$

By Eqs. (12.13) and (12.14)

$$\Delta t_m = \frac{t_2 - t_1}{\ln\left[\dfrac{R}{R + \ln(1 - RP)}\right]} \tag{12.15}$$

and by Eqs. (12.4), (12.5), and (12.15)

$$F = \frac{\Delta t_m}{\Delta t_{m,cf}} = \frac{\ln\left[\dfrac{1 - P}{1 - RP}\right]}{(R - 1)\ln\left[\dfrac{R}{R + \ln(1 - RP)}\right]} \tag{12.16}$$

Equation (12.16) shows that the correction factor F is a function of only the parameters R and P.

Bowman, Mueller, and Nagle* have presented solutions of Eq. (12.5) for the correction factor F for many heat-exchanger arrangements. Three of their solutions are shown in Figs. 12.5–12.7. Figure 12.5 shows the solution of Eq. (12.16) for the single-pass, cross-flow heat exchanger of Fig. 12.4. Figure 12.6 shows the correction-factor for a single-pass, cross-flow heat exchanger where both fluids are unmixed. Figure 12.6 applies to cases similar to Fig. 12.3(a) having many rows of tubes. Figure 12.7 shows the correction factor for a counter-cross-flow heat exchanger with two

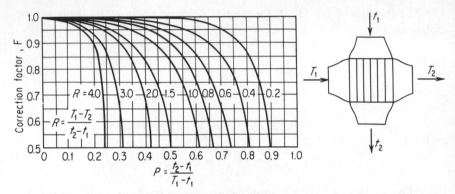

Figure 12.5 Correction-factor $F = \Delta t_m / \Delta t_{m,cf}$ for a single-pass, cross-flow heat exchanger, one fluid (T) mixed, other fluid (t) unmixed. (Reprinted from "Mean Temperature Difference in Design" by R. A. Bowman, A. C. Mueller, and W. M. Nagle, *ASME Trans.*, Vol. 62, p. 289, with permission of the publisher, Amer. Soc. Mech. Engrs.)

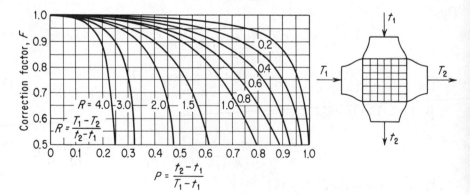

Figure 12.6 Correction-factor $F = \Delta t_m / \Delta t_{m,cf}$ for a single-pass cross-flow heat exchanger, both fluids unmixed. (Reprinted from "Mean Temperature Difference in Design" by R. A. Bowman, A. C. Mueller, and W. M. Nagle, *ASME Trans.*, Vol. 62, 1940, p. 288, with permission of the publisher, Amer. Soc. Mech. Engrs.)

ASME Trans., Vol. 62, pp. 283–294.

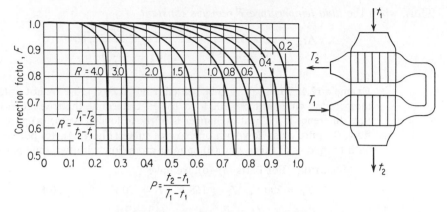

Figure 12.7 Correction-factor $F = \Delta t_m / \Delta t_{m,cf}$ for two-pass, counter-cross-flow heat exchanger one fluid (T) mixed, other fluid (t) unmixed except between passes. (Reprinted from "Mean Temperature Difference in Design" by R. A. Bowman, A. C. Mueller and W. M. Nagle, *ASME Trans.*, Vol. 62, 1940, p. 289, with permission of the publisher, Amer. Soc. Mech. Engrs.)

tube passes, with one fluid mixed and the other unmixed except between passes. Figure 12.7 applies approximately to a heat exchanger of the type shown in Fig. 12.3(b), having two tube passes. In Fig. 12.3(b) the air is unmixed between tube passes and the correction factor F would be slightly higher than that given by Fig. 12.7. For counter-cross-flow heat exchangers having more than two tube passes one may usually estimate F as being between unity and the value given by Fig. 12.7.

We will now show that if one fluid temperature remains constant, the logarithmic mean temperature difference applies regardless of the flow arrangement. We will again analyze the heat exchanger of Fig. 12.4 but assuming the hot-fluid temperature T to be constant. The cold-fluid temperature t would vary only in the y direction. The solution of Eq. (12.6) would be

$$t_2 = t_1 + (T - t_1)(1 - e^{-K_1}) \qquad (12.17)$$

If there were more than one tube pass, we might easily show that Eq. (12.17) would still apply by replacing K_1 for the single pass by NK_1 where N is the number of tube passes. Furthermore, it would be immaterial how the tube passes were arranged (counter, cross-flow, etc.) if the fluid temperature in the tubes remained constant.

By Eq. (12.17),

$$e^{K_1} = \frac{T - t_1}{T - t_2}$$

and

$$K_1 = \frac{U_o A_o}{mc_p} = \frac{t_2 - t_1}{\Delta t_m} = \ln\left(\frac{T - t_1}{T - t_2}\right)$$

Thus, when the *fluid temperature T remains constant*

$$\Delta t_m = \frac{t_2 - t_1}{\ln\left(\dfrac{T - t_1}{T - t_2}\right)} = \frac{(T - t_1) - (T - t_2)}{\ln\left(\dfrac{T - t_1}{T - t_2}\right)} \tag{12.18}$$

Example 12.1. Moist air is to be heated from 70 F to 150 F by hot water whose temperature changes from 200 F to 168 F. Determine the true mean temperature difference if the heat exchanger is of the following types: (a) pure counter-flow, (b) pure cross-flow with one row of tubes (Fig. 12.4), (c) pure cross-flow with four rows of tubes, and (d) counter-cross-flow with two tube passes.

SOLUTION: For Parts (a)–(d), we have

$$T_1 = 200 \text{ F}, \quad T_2 = 168 \text{ F}, \quad t_1 = 70 \text{ F}, \quad t_2 = 150 \text{ F}$$

$$R = \frac{T_1 - T_2}{t_2 - t_1} = 0.40, \qquad P = \frac{t_2 - t_1}{T_1 - t_1} = 0.615$$

The solution of the entire problem is shown in Table 12.1.

TABLE 12.1

SOLUTION FOR EXAMPLE 12.1

Case	Method of Solution	F	$\dfrac{\Delta t_m}{F}$
(a) Pure counter-flow	Eq. (12.1)	1.00	71.3
(b) Pure cross-flow, one tube pass	Fig. 12.5	0.93	66.3
(c) Pure cross-flow, four tube passes . . .	Fig. 12.6	0.94	67.1
(d) Counter-cross-flow, two tube passes . .	Fig. 12.7	0.99	70.6

12.3. THE EFFICIENCY OF VARIOUS EXTENDED SURFACES

Heat exchangers used for heating or cooling moist air may have fins of various types on the surface contacted by the air. Figure 12.3 shows two schematic arrangements with rectangular-plate fins. Circular-plate fins, bar fins, and various types of spines may also be employed.

The addition of fins to the tubes greatly increases the outer surface area but at the expense of decreasing the mean temperature difference between the surface and the air stream. Whereas the thermal resistance of the bare tube may be negligible, the thermal resistance of the extended surface may be considerable.

A significant quantity in evaluating the thermal effectiveness of fins is the *fin efficiency* ϕ defined as

$$\phi = \frac{t_{F,m} - t}{t_{F,B} - t} = \frac{\Delta t_{F,m}}{\Delta t_{F,B}} \tag{12.19}$$

where $t_{F,m}$ is the mean temperature of the fin, $t_{F,B}$ is the temperature at the base of the fin, and t is the air dry-bulb temperature.

In this section we will study efficiencies of several types of finned surfaces. We will first derive the efficiency for a bar fin which is mathematically the most elementary type. Figure 12.8 schematically shows a bar fin attached to a tube. We will assume (1)

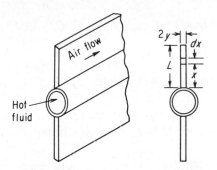

Figure 12.8 Schematic illustration of bar fin.

steady-state heat transfer, (2) constant thermal conductivity for the fin, (3) a constant temperature at the base of the fin, (4) one-dimensional heat conduction in the fin, (5) negligible heat transfer from the outer edge of the fin, (6) a uniform temperature of the air stream, and (7) constant outside surface convection coefficient $h_{c,o}$.

At any cross-section of unit length in Fig. 12.8, we have

$$q_F = -kA\frac{dt_F}{dx} = -2ky\frac{dt_F}{dx}$$

or

$$dq_F = -2ky\frac{d^2t_F}{dx^2}\,dx = -2ky\frac{d^2\Delta t_F}{dx^2}\,dx$$

But

$$dq_F = -2h_{c,o}\,dx\,(t_F - t) = -2h_{c,o}\,dx\,\Delta t_F$$

Thus

$$\frac{d^2\Delta t_F}{dx^2} = \frac{h_{c,o}}{ky}\Delta t_F \tag{12.20}$$

The solution of Eq. (12.20) for the conditions: at $x = 0$, $\Delta t_F = \Delta t_{F,B}$ and at $x = L$, $d\Delta t_F/dx = 0$ is

$$\Delta t_F = \Delta t_{F,B}\left[\frac{e^{p(L-x)} + e^{-p(L-x)}}{e^{pL} + e^{-pL}}\right] \tag{12.21}$$

where $p = \sqrt{h_{c,o}/ky}$ and $\Delta t_{F,B} = t_{F,B} - t$.

We may find the total rate of heat transfer for a unit length of the fin by

$$q_F = 2h_{c,o}\int_0^L \Delta t_F\,dx$$

With Eq. (12.21), we have

$$q_F = \frac{2h_{c,o}\,\Delta t_{F,B}}{p}\,(\tanh pL) \tag{12.22}$$

By definition of the mean fin temperature $t_{F,m}$

$$q_F = h_{c,o}A_F(t_{F,m} - t) = 2h_{c,o}L\,\Delta t_{F,m} \tag{12.23}$$

Thus, by Eqs. (12.19), (12.22), and (12.23), the efficiency of the bar fin is given by

$$\phi = \frac{\tanh pL}{pL} \tag{12.24}$$

Figure 12.9 shows the efficiency of a bar fin as calculated by Eq. (12.24) for values of pL up to 5.0.

Circular-plate fins are more commonly applied to heat exchangers than bar fins. Figure 12.10 shows two schematic circular-plate fins. Fin (a) has a uniform thickness

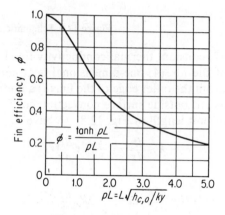

Figure 12.9 Efficiency of a bar fin.

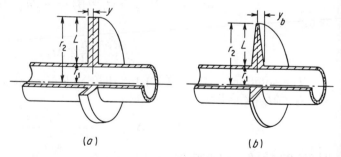

Figure 12.10 Schematic illustrations of circular-plate fins (a) having a uniform thickness, and (b) having a constant cross-sectional area.

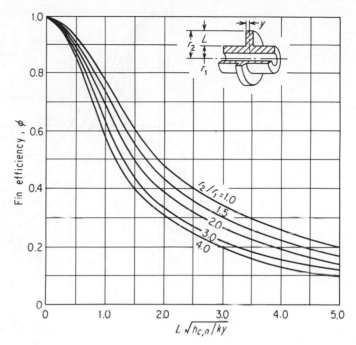

Figure 12.11 Efficiency for a circular-plate fin of uniform thickness. (Reprinted from "Efficiency of Extended Surfaces" by K. A. Gardner, *ASME Trans.*, Vol. 67, 1945, p. 625, with permission of the publisher, Amer. Soc. Mech. Engrs.)

while Fin (b) has a constant cross-sectional area. Gardner* has solved the differential equations for temperature distribution in the two fins shown in Fig. 12.10 and has calculated their efficiencies. Figures 12.11 and 12.12 were reproduced from Gardner's paper.

The rectangular-plate fin of uniform thickness is commonly used in finned coils for heating or cooling air. It is not possible to obtain an exact mathematical solution for the efficiency of such a fin. Carrier and Anderson† have shown that an adequate approximation is to assume that the fin area served by each tube is equivalent in performance to a flat circular-plate fin of equal area. Figure 12.13 shows the method where the equivalent outer radius of the circular fin is determined as

$$r_2 = \sqrt{\frac{ac}{\pi}} \tag{12.25}$$

*K. A. Gardner, "Efficiency of Extended Surfaces," *A.S.M.E. Trans.*, Vol. 67, 1945, pp. 621–631.
†W. H. Carrier and S. W. Anderson, "The Resistance to Heat Flow Through Finned Tubing," *A.S.H.V.E. Trans.*, Vol. 50, 1944, pp. 117–152.

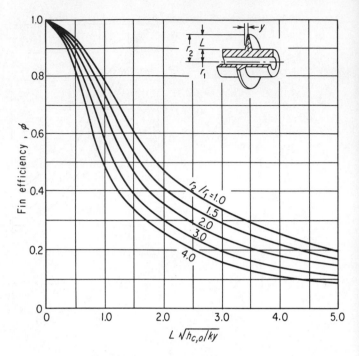

Figure 12.12 Efficiency for a circular-plate fin having a constant cross-sectional area. (Reprinted from "Efficiency of Extended Surfaces" by K. A. Gardner, *ASME Trans.*, Vol. 67, 1945, p. 625, with permission of the publisher, Amer. Soc. Mech. Engrs.)

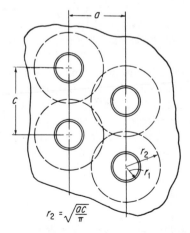

Figure 12.13 Approximation method for treating a rectangular-plate fin of uniform thickness in terms of a flat circular-plate fin of equal area.

After determination of the equivalent outer radius, the fin efficiency may be found from Fig. 12.11.

12.4. OVERALL HEAT TRANSFER COEFFICIENT FOR A DRY FINNED-TUBE HEAT EXCHANGER

Design problems with finned-tube heat exchangers involve solution of the equation

$$q = U_o A_o \Delta t_m \tag{12.26}$$

In Sec. 12.3 we studied calculation of the mean temperature difference Δt_m. We will now investigate calculation of the *overall heat transfer coefficient* U_o where we assume the fin surfaces to be dry.

Figure 12.14 shows a schematic local section of a finned-tube heat exchanger. We will assume (1) steady-state heat transfer, and (2) negligible contact resistance

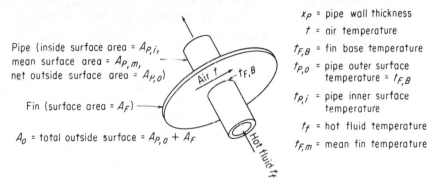

Pipe (inside surface area = $A_{P,i}$, mean surface area = $A_{P,m}$, net outside surface area = $A_{P,o}$)

Fin (surface area = A_F)

A_O = total outside surface = $A_{P,o} + A_F$

Air t — $t_{F,B}$

Hot fluid t_f

x_P = pipe wall thickness
t = air temperature
$t_{F,B}$ = fin base temperature
$t_{P,o}$ = pipe outer surface temperature = $t_{F,B}$
$t_{P,i}$ = pipe inner surface temperature
t_f = hot fluid temperature
$t_{F,m}$ = mean fin temperature

Figure 12.14 Schematic illustration of a finned-tube heat exchanger.

between the base of the fin and the pipe ($t_{F,B} = t_{P,o}$). We may write the following equations for the rate of heat transfer

$$q = h_i A_{P,i}(t_f - t_{P,i}) \tag{12.27}$$

$$q = \frac{k_P A_{P,m}(t_{P,i} - t_{P,o})}{x_P} \tag{12.28}$$

$$q = h_{c,o,P} A_{P,o}(t_{P,o} - t) + h_{c,o,F} A_F(t_{F,m} - t) \tag{12.29}$$

$$q = U_o A_o(t_f - t) \tag{12.30}$$

By Eqs. (12.19) and (12.29) and assuming $h_{c,o,P} = h_{c,o,F} = h_{c,o}$, we have

$$q = h_{c,o}(A_{P,o} + \phi A_F)(t_{P,o} - t) \tag{12.31}$$

By Eqs. (12.27), (12.28), (12.30), and (12.31), we obtain

$$U_o = \frac{1}{\dfrac{A_o}{A_{P,i}h_i} + \dfrac{A_o x_P}{A_{P,m}k_P} + \dfrac{1 - \phi}{h_{c,o}(A_{P,o}/A_F + \phi)} + \dfrac{1}{h_{c,o}}} \tag{12.32}$$

Equation (12.32) shows that, for a heat exchanger of known dimensions, the overall coefficient U_o may be calculated if the heat transfer coefficients h_i and $h_{c,o}$, the pipe thermal conductivity k_P, and the fin efficiency ϕ are known.

It is of considerable interest to study the influences upon U_o of the various quantities in Eq. (12.32). Some of these are shown in Example 12.2 which follows.

> **Example 12.2.** An air heating coil is constructed of rectangular-plate fins bonded to copper tubes. The tubes are arranged in line in adjacent rows. Physical data are as follows: $a = 0.141$ ft, $c = 0.125$ ft, $x_P = 0.00233$ ft, $r_1 = 0.0208$ ft, fin thickness $= 0.000833$ ft, fin spacing $= 8$ per in., $A_{P,i} = 0.116$ sq ft per lineal ft of tube, $A_{P,o} = 0.120$ sq ft per lineal ft of tube, $A_{P,m} = 0.123$ sq ft per lineal ft of tube, $A_F = 3.11$ sq ft per lineal ft of tube, and $A_o = 3.23$ sq ft per lineal ft of tube. (a) Determine the individual thermal resistances and the overall heat transfer coefficient U_o if $h_{c,o} = 10.0$ Btu per (hr) (sq ft) (F), $h_i = 600$ Btu per (hr) (sq ft) (F), and the fin material is aluminum; (b) rework Part (a) but with copper fins; (c) rework Part (a) but with $h_{c,o} = 5.0$ Btu per (hr) (sq ft) (F); and (d) rework Part (a) but with $h_i = 1200$ Btu per (hr) (sq ft) (F).
>
> SOLUTION: All calculations are shown in Tables 12.2 and 12.3. Table 12.2 shows determination of fin efficiency ϕ and fin thermal resistance R_F. Table

TABLE 12.2

DETERMINATION OF FIN EFFICIENCY ϕ AND FIN THERMAL RESISTANCE R_F
FOR EXAMPLE 12.2

Component or Calculation	Part (a)	Part (b)	Part (c)	Part (d)
r_1, ft	0.0208	0.0208	0.0208	0.0208
$r_2 = \sqrt{ac/\pi}$, ft	0.0749	0.0749	0.0749	0.0749
$L = (r_2 - r_1)$, ft	0.0541	0.0541	0.0541	0.0541
$h_{c,o}$, Btu/(hr) (sq ft) (F) . .	10.0	10.0	5.0	10.0
k, Btu/(hr) (sq ft) (F/ft) . .	120	223	120	120
y, ft	0.000417	0.000417	0.000417	0.000417
r_2/r_1	3.60	3.60	3.60	3.60
$L\sqrt{h_{c,o}/ky}$	0.765	0.560	0.541	0.765
ϕ, (Fig. 12.11)	0.73	0.83	0.84	0.73
$A_{P,o}/A_F$	0.0386	0.0386	0.0386	0.0386
$R_F = \dfrac{1}{h_{c,o}}\left(\dfrac{1-\phi}{\phi + A_{P,o}/A_F}\right)$ . .	0.0351	0.0196	0.0364	0.0351

12.3 shows the individual thermal resistances R_i, R_P, R_F, and R_o and their per cent of the total resistance R_t, as well as the overall coefficient U_o.

Table 12.2 shows that the fin thermal resistance R_F is affected strongly by the fin thermal conductivity k but only in a small way by the outside-surface coefficient $h_{c,o}$. Table 12.3 shows that all of the thermal resistances are important *except the pipe wall resistance R_P*, but that the outside-surface resistance R_o

TABLE 12.3

DETERMINATION OF THERMAL RESISTANCES AND U_o FOR EXAMPLE 12.2

Component or Calculation	Part (a)	Part (b)	Part (c)	Part (d)
Inside surface resistance,				
$\quad R_i = A_o/A_{P,i}h_i$	0.0464	0.0464	0.0464	0.0232
Pipe wall resistance,				
$\quad R_P = A_o x_P/(A_{P,m}k_P)$ . .	0.00027	0.00027	0.00027	0.00027
Fin resistance, R_F (Table 12.2)	0.0351	0.0196	0.0364	0.0351
Outside surface resistance,				
$\quad R_o = 1/h_{c,o}$	0.1000	0.1000	0.2000	0.1000
Total resistance,				
$\quad R_t = R_i + R_P + R_F + R_o$	0.18177	0.16627	0.28307	0.15857
R_i in per cent of R_t	25.5	27.9	16.4	14.6
R_P in per cent of R_t	0.2	0.2	0.1	0.2
R_F in per cent of R_t	19.3	11.8	12.9	22.1
R_o in per cent of R_t	55.0	60.1	70.6	63.1
$U_o = 1/R_t$, Btu/(hr) (sq ft) (F)	5.50	6.01	3.53	6.31

is dominant. Part (b) shows that use of copper fins instead of aluminum fins increased U_o by 9.3 per cent. Part (c) shows that a 50 per cent reduction in $h_{c,o}$ (10.0 to 5.0) decreased U_o by 36 per cent. Part (d) shows that a 100 per cent increase in h_i (600 to 1200) increased U_o by 14.7 per cent.

12.5. OVERALL HEAT TRANSFER PROBLEMS INVOLVING DRY FINNED SURFACES

As stated in Sec. 12.4, a practical heat exchanger design problem involves solution of Eq. (12.26). For a practical problem, Eq. (12.32) should be modified to allow for deposit coefficients. A minor deposit on the outside surface of a finned coil generally has little effect upon U_o because of the usually large magnitude of $1/h_{c,o}$. Sometimes an allowance is made for imperfect bonding of the fins to the tubes but this effect is difficult to evaluate and, with good construction, it should be small. It is more important to include a deposit coefficient for the inside surface of the tubes. Example 12.2 shows that the thermal resistance R_P of the tube wall may be neglected with little error. Thus, for most cases

$$U_o = \frac{1}{\dfrac{A_o}{A_{P,i}h_i} + \dfrac{A_o}{A_{P,i}h_{d,i}} + \dfrac{1-\phi}{h_{c,o}(A_{P,o}/A_F + \phi)} + \dfrac{1}{h_{c,o}}} \qquad (12.33)$$

Information was given in Chapter 2 for evaluating the individual coefficients h_i and $h_{d,i}$. Equations shown in Chapter 2 for forced-convection coefficients are usually not applicable for $h_{c,o}$ for finned-tube coils. The type of fins, fin spacing, and other factors affect the value of the coefficient. It is usually necessary to have experimental information for the particular fin arrangement. A rather small amount of basic information is available in published literature.

Table 12.4 shows dimensional data for two plate-fin-and-tube arrangements. The surface consists of aluminum fins bonded to copper tubes. Figure 12.15 shows a

TABLE 12.4

DIMENSIONAL DATA FOR TWO PLATE-FIN-AND-TUBE SURFACE ARRANGEMENTS

Data	Surface I	Surface II
Dimensions (See sketch with Fig. 12.15.)		
A, tube outside diameter, in.	0.402	0.676
B, tube spacing across face, in.	1.00	1.50
C, tube spacing between rows, in.	0.866	1.75
D, spacing of fins, center to center, in.	0.125	0.129
E, thickness of aluminum fins, in.	0.013	0.016
Flow passage hydraulic diameter, $4r_h$, ft	0.01192	0.01268
Area Data		
$A_{o,1}$, sq ft external surface/(sq ft face area) (row)	12.92	22.86
$A_o/A_{P,i}$, sq ft external surface/sq ft internal surface	12.27	19.31
A_c, sq ft net flow area/sq ft face area	0.534	0.497
A_F/A_o, sq ft fin surface/sq ft external surface	0.839	0.905

schematic sketch of the surface and also presents a heat transfer correlation for the external surface. The information in both Table 12.4 and Fig. 12.15 was taken from *Compact Heat Exchangers*, by Kays and London.* Similar information for many other types of extended surface is also given in this reference.

Figure 12.15 provides an estimation for the external surface convection coefficient $h_{c,o}$. The dimensionless grouping $(h_{c,o}/Gc_p)(c_p\mu/k)^{2/3}$ is shown as a function of the dimensionless Reynolds number $(4r_h G/\mu)$. The mass velocity G is based upon the minimum free-flow area A_c. The Reynolds number is based upon a hydraulic diameter $4r_h$ defined by the relation $4r_h/L = 4A_c/A_o$ where L is the flow length of the heat exchanger and A_o is the total external surface area. For a given type of surface, r_h is a constant independent of the depth of the coil. Kays and London recommend that the fluid properties c_p, μ, and k be evaluated at the mean mixed fluid temperature.

> **Example 12.3.** A heat exchanger is to be sized to process 12,000 cu ft per min of saturated air at 20 F to a final temperature of 150 F. Heating medium is dry-and-saturated steam condensing at 5 psig. Barometric pressure is 14.00 psia.

*W. M. Kays and A. L. London, *Compact Heat Exchangers* (Palo Alto, California: The National Press, 1955).

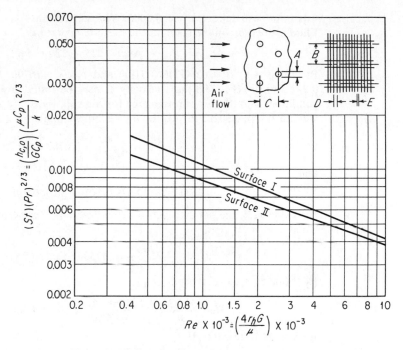

$$Re \times 10^{-3} = \left(\frac{4 r_h G}{\mu}\right) \times 10^{-3}$$

Figure 12.15 Correlated external surface heat transfer data for surfaces of Table 12.4. (Adapted by permission from W. M. Kays and A. L. London, *Compact Heat Exchangers*, The National Press, 1955, p. 177.)

Type II surface shown in Table 12.4 is to be used. Inlet face velocity of the air is to be 500 ft per min. Determine (a) the required coil face area in sq ft, (b) required total outside-surface area in sq ft, (c) the number of rows of tubes required, and (d) the lb per hr of steam required if the steam condensate leaves the coil as saturated liquid.

SOLUTION: (a) From the given data we may calculate the coil face area as $12,000/500 = 24$ sq ft.

(b) We must solve for A_o from Eq. (12.26). We will first solve for Δt_m which is given by the logarithmic mean temperature difference, since the steam temperature is constant at 225.2 F (sat. temp. at 19 psia). By Eq. (12.18),

$$\Delta t_m = \frac{150 - 20}{\ln\left(\dfrac{225.2 - 20}{225.2 - 150}\right)} = 129 \text{ F}$$

To find U_o we must first evaluate the individual quantities in Eq. (12.33). By Table 12.4, $A_o/A_{P,i} = 19.31$ and $A_F/A_o = 0.905$. Thus $A_{P,o}/A_F = A_o/A_F - 1 = 0.105$.

Based on the discussion in Sec. 2.5, we will assume $h_i = 1200$ Btu/(hr) (sq ft) (F). We will assume that the steam is free of oil, and since we have copper tubes, we will take $1/h_{d,i} = 0.0005$. We will estimate $h_{c,o}$ from Fig. 12.15. Fluid properties will be taken at the mean temperature

$$t_m = t_{steam} - \Delta t_m = 225 - 129 = 96 \text{ F}$$

By Table A.6, $\mu = 0.0457 \text{ lb}/(\text{ft})(\text{hr})$ and $c_p\mu/k = 0.706$. We may use $c_p = 0.241 \text{ Btu}/(\text{lb})(\text{F})$. The mass velocity G may be found from the entering air conditions and the per cent free area for the coil. By Eq. (8.23) and Table A.1, we find $W_1 = 0.00225 \text{ lb}_w/\text{lb}_a$. By Eq. (8.29), we find $v_1 = 12.73 \text{ cu ft}/\text{lb}_a$. Thus

$$\rho_1 = \frac{1 + W_1}{v_1} = \frac{1.002}{12.73} = 0.0787 \text{ lb/cu ft}$$

and

$$m = (12,000)(0.0787)(60) = 56,700 \text{ lb/hr}$$

By Table 12.4, the net area between the fins is 49.7 per cent of the face area. Thus,

$$G = \frac{56,700}{(0.497)(24)} = 4750 \text{ lb}/(\text{hr})(\text{sq ft})$$

By Table 12.4, $4r_h = 0.01268$. Thus

$$Re = \frac{4r_h G}{\mu} = \frac{(0.01268)(4750)}{0.0457} = (1.32)(10^3)$$

By Fig. 12.15 (Surface II)

$$\frac{h_{c,o}(c_p\mu/k)^{2/3}}{Gc_p} = 0.0078$$

Thus

$$h_{c,o} = \frac{(0.0078)Gc_p}{(c_p\mu/k)^{2/3}} = \frac{(0.0078)(4750)(0.241)}{(0.706)^{2/3}} = 11.3 \text{ Btu}/(\text{hr})(\text{sq ft})(\text{F})$$

We may next evaluate the fin efficiency. By Eq. (12.25),

$$r_2 = \sqrt{\frac{(1.75)(1.5)}{\pi}} = 0.914 \text{ in.}$$

and

$$\frac{r_2}{r_1} = \frac{0.914}{0.338} = 2.70$$

$$L\sqrt{h_{c,o}/ky} = \frac{0.575}{12} \sqrt{\frac{(11.3)(2)(12)}{(120)(0.016)}} = 0.570$$

By Fig. 12.11, $\phi = 0.86$.

We may calculate U_o by Eq. (12.33). Thus,

$$U_o = \frac{1}{\frac{19.31}{1200} + (19.31)(0.0005) + \frac{0.14}{(11.3)(0.97)} + \frac{1}{11.3}} = 7.87 \text{ Btu}/(\text{hr})(\text{sq ft})(\text{F})$$

We may find the total heat transfer by

$$q = mc_p(t_2 - t_1) = (56,700)(0.241)(150 - 20) = 1,776,400 \text{ Btu/hr}$$

Thus, by Eq. (12.26),

$$A_o = \frac{q}{U_o \, \Delta t_m} = \frac{1,776,400}{(7.87)(129)} = 1740 \text{ sq ft}$$

(c) By Table 12.4, we find that Surface II has an outside surface area of 22.86 sq ft/(sq ft of face area) (row). Thus,

$$\text{Rows required} = \frac{1740}{(22.86)(24)} = 3.17$$

(d) We may calculate the required steam flow by

$$m_s = \frac{q}{h_{fg,\,s}} = \frac{1,776,400}{961.9} = 1840 \text{ lb/hr}$$

Example 12.4. Assume the same problem as Example 12.3 but take the heating medium to be hot water. Assume that water enters the coil at 200 F and leaves at 180 F. Assume a counter-cross-flow arrangement similar to Fig. 12.3(b). Find (a) the required coil face area in sq ft, (b) the GPM of water required, (c) outside surface area required in sq ft, and (d) the number of rows of tubes required.

SOLUTION: (a) The required face area is identical to that calculated for Example 12.3 (24 sq ft). However, in order to find the water velocity later, we must fix the face dimensions. We will use face dimensions of 4 ft in height and 6 ft in width.

(b) The total rate of heat transfer is the same as for Example 12.3. We may calculate the required rate of water flow by

$$m_w = \frac{q}{c_{p,\,w} \, \Delta t_w} = \frac{1,776,400}{(1)(20)} = 88,820 \text{ lb/hr}$$

At 190 F, $\rho_w = 60.4$ lb/cu ft. One gallon of water at 190 F weighs $(231)(60.4)/1728 = 8.08$ lb. Thus, based upon the average water density,

$$\text{GPM} = \frac{m_w}{(8.08)(60)} = \frac{88,820}{(8.08)(60)} = 183$$

(c) Since we know that 3.17 rows were required for Example 12.3, it will be sufficiently accurate to calculate Δt_m as being equal to that for pure counter-flow. Thus, by Eq. (12.1),

$$\Delta t_m = \frac{(180 - 20) - (200 - 150)}{\ln \left(\dfrac{180 - 20}{200 - 150} \right)} = 94.6 \text{ F}$$

Before evaluating h_i, we must solve for the water velocity in the tubes. Since the face height is 48 in., there are 32 vertical tubes per row. From the area ratios in Table 12.4, we may calculate the inside diameter of the tube and find it to be 0.565 in. The internal cross-sectional area of each tube is 0.00174 sq ft. Thus,

$$\text{Water velocity} = \frac{88,820}{(3600)(60.4)(32)(0.00174)} = 7.3 \text{ ft/sec}$$

We may show that Re $>$ 2200. Thus, we may find h_i by Eq. (2.21). We have

$$h_i = (13.5)(190)^{0.54} \frac{(7.3)^{0.8}}{(0.047)^{0.2}} = 2070 \text{ Btu/(hr)(sq ft)(F)}$$

The mean mixed air temperature is approximately

$$t_m = 190 - 95 = 95 \text{ F}$$

Thus $h_{c,o}$ would be the same as for Example 12.3. Furthermore, the fin efficiency would be the same as for Example 12.3. If we use a deposit coefficient $1/h_d$ = 0.0005, by Eq. (12.33) we have

$$U_o = \frac{1}{\frac{19.31}{2070} + (19.31)(0.0005) + \frac{0.14}{(11.3)(0.97)} + \frac{1}{11.3}} = 8.31 \text{ Btu/(hr)(sq ft)(F)}$$

Thus, by Eq. (12.26),

$$A_o = \frac{1,776,400}{(8.31)(94.6)} = 2250 \text{ sq ft}$$

(d) Rows of tubes required $= \dfrac{2250}{(22.86)(24)} = 4.10$

12.6. INTRODUCTION TO HEAT TRANSFER IN WET-SURFACE COOLING COILS

Fin-and-tube surfaces are widely used in applications for cooling atmospheric air. If no moisture is separated from the air, we may use the procedures developed in earlier sections of this chapter. However, in cooling applications it is more common that dehumidification of the air also occurs. With dehumidification, the air-side surface is wetted (liquid water or frost). Besides transfer of sensible heat, there is a transfer of heat because of condensation. Since water-vapor transfer is not dependent upon temperature difference alone, it follows that the analyses presented earlier in this chapter do not suffice.

Figure 12.16 shows schematically a cold surface in contact with a moving stream of moist air. A moving film of water is formed on the surface by condensation of moisture from the air stream. There is a boundary layer of air next to the water surface. In this layer, we assume that air temperature, air humidity ratio, and air velocity vary in a plane perpendicular to bulk motion of the air. Immediately next to the water film, we assume that the air is saturated at the water surface temperature t_w. The transfer processes between the air stream and the water surface are similar to those described in Sec. 11.4 for the spray dehumidifier. For the differential surface area in Fig. 12.16, we have

$$-m_a \, dh = dq - m_a \, dW \, h_{f,w} \tag{12.34}$$

$$dq = h_{c,o} \, dA_o \, (t - t_w) + h_{D,o} \, dA_o \, (W - W_{s,w})(h_{g,t} - h_{f,w}) \tag{12.35}$$

$$-m_a \, dW = h_{D,o} \, dA_o \, (W - W_{s,w}) \tag{12.36}$$

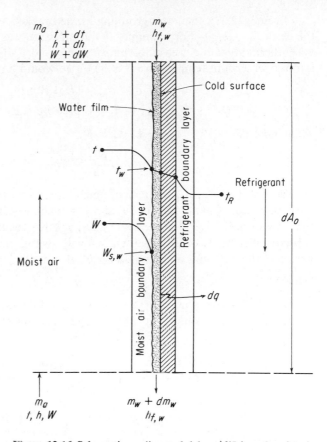

Figure 12.16 Schematic cooling and dehumidifying of moist air.

Using the relation $\text{Le} = h_{c,o}/h_{D,o}c_{p,a}$, Eq. (12.35) may be written as

$$dq = \frac{h_{c,o}\,dA_o}{c_{p,a}}\left[c_{p,a}(t - t_w) + \frac{(W - W_{s,w})(h_{g,t} - h_{f,w})}{\text{Le}}\right]$$

or

$$dq = \frac{h_{c,o}\,dA_o}{c_{p,a}}\left[(h - h_{s,w}) + \frac{(W - W_{s,w})(h_{g,t} - h_{f,w} - 1061\,\text{Le})}{\text{Le}}\right] \qquad (12.37)$$

By Eqs. (12.34), (12.36), and (12.37), we may show that

$$\frac{dh}{dW} = \text{Le}\,\frac{(h - h_{s,w})}{(W - W_{s,w})} + (h_{g,t} - 1061\,\text{Le}) \qquad (12.38)$$

Equation (12.38) describes the process line on the psychrometric chart for the cooling and dehumidifying of moist air by a cold surface. Equation (12.38) is identical to Eq. (11.33) for the spray dehumidifier.

In Eq. (12.37), the latter grouping in the brackets is typically small compared to the term $(h - h_{s,w})$. For example, if the air state was 85 F dry-bulb temperature, 70 F thermodynamic wet-bulb temperature, and 14.696 psia barometric pressure, and the water-film temperature was 60 F, we would have

$$h - h_{s,w} = 7.51 \text{ Btu/lb}_a$$

$$(W - W_{s,w})(h_{g,t} - h_{f,w} - 1061 \text{ Le})/\text{Le} = 0.142 \text{ Btu/lb}_a$$

Thus, approximately

$$dq = \frac{h_{c,o} \, dA_o}{c_{p,a}} (h - h_{s,w}) \tag{12.39}$$

We will find Eq. (12.39) to be of much importance in our studies on cooling coils. Although approximate, Eq. (12.39) allows a much easier analysis of wet cooling coils than does Eq. (12.37). Besides Eq. (12.39), another relation will be repeatedly used in subsequent sections of this chapter. We will assume that over a small range of temperature, the enthalpy of saturated air h_s, Btu per lb_a, may be represented as

$$h_s = a + bt_s \tag{12.40}$$

Figure 12.17 shows that over a narrow range of temperature, such as about 10 F, h_s may be closely given by Eq. (12.40) if the coefficients a and b are average values.

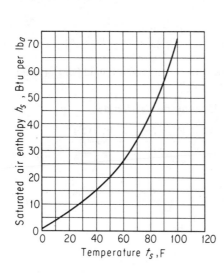

Figure 12.17 Enthalpy of saturated air as a function of temperature for 14.696 psia pressure.

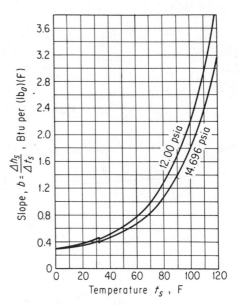

Figure 12.18 Slope $\Delta h_s / \Delta t_s$ for saturated air.

Figure 12.18 shows variation of the coefficient b in Eq. (12.40) for barometric pressures of 12.00 and 14.696 psia.

12.7. THE EFFICIENCY OF VARIOUS EXTENDED SURFACES WHEN BOTH COOLING AND DEHUMIDIFYING OCCUR

The efficiency for a dry bar fin was developed in Sec. 12.3. We will now study the performance of a bar fin when condensation occurs on its surfaces; Figure 12.19 schematically shows the problem. We will make the same assumptions given in Sec.

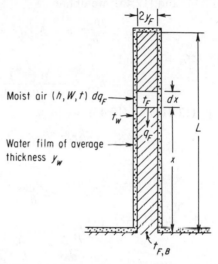

Moist air $(h, W, t)\ dq_F$

t_w

Water film of average thickness y_w

Figure 12.19 Schematic illustration of a bar fin wetted with moisture condensed from moist air.

12.3. We will also assume that heat conduction through the water film occurs in only the y direction. For a unit length of the fin, we have

$$q_F = 2k_F y_F \frac{dt_F}{dx} \tag{12.41}$$

where the subscript F refers to the metal fin. Also,

$$dq_F = -2\frac{k_w}{y_w}(t_w - t_F)\,dx \tag{12.42}$$

where k_w and y_w are respectively the thermal conductivity and thickness of the water film. By Eqs. (12.40) and (12.42),

$$dq_F = -\frac{2k_w}{b_w y_w}(h_{s,w} - a_w - b_w t_F)\,dx$$

But the quantity $(a_w + b_w t_F)$ has the dimensions of *moist air enthalpy*. Let us define a *fictitious air enthalpy* h_F as

$$h_F = a_w + b_w t_F \tag{12.43}$$

where the quantities a_w and b_w are evaluated at the surface temperature t_w. Thus

$$dq_F = -\frac{2k_w}{b_w y_w}(h_{s,w} - h_F)\,dx \tag{12.44}$$

By Eq. (12.39)

$$dq_F = -\frac{2h_{c,o}}{c_{p,a}}(h - h_{s,w})\,dx \tag{12.45}$$

By Eqs. (12.44) and (12.45), we obtain

$$dq_F = -\frac{2h_{o,w}}{b_w}(h - h_F)\,dx = -\frac{2h_{o,w}}{b_w}\Delta h_F\,dx \tag{12.46}$$

where $\Delta h_F = (h - h_F)$ and

$$h_{o,w} = \frac{1}{c_{p,a}/(b_w h_{c,o}) + y_w/k_w} \tag{12.47}$$

By Eqs. (12.41) and (12.43), we have

$$q_F = \frac{2k_F y_F}{b_w}\frac{dh_F}{dx} = -2\frac{k_F y_F}{b_w}\frac{d\,\Delta h_F}{dx}$$

and thus

$$dq_F = -2\frac{k_F y_F}{b_w}\frac{d^2\,\Delta h_F}{dx^2}\,dx \tag{12.48}$$

By Eqs. (12.46) and (12.48)

$$\frac{d^2\,\Delta h_F}{dx^2} = \frac{h_{o,w}}{k_F y_F}\Delta h_F \tag{12.49}$$

The boundary conditions for Eq. (12.49) are $\Delta h_F = \Delta h_{F,B}$ at $x = 0$, *and* $d\,\Delta h_F/dx = 0$ *at* $x = L$.

Equation (12.49) and its boundary conditions are completely analogous to Eq. (12.20) and its boundary conditions. Thus, the solution of Eq. (12.49) has the same form as Eq. (12.20). Furthermore, if we define the efficiency of the wet fin as

$$\phi_w = \frac{h - h_{F,m}}{h - h_{F,B}} = \frac{\Delta h_{F,m}}{\Delta h_{F,B}} \tag{12.50}$$

we find that

$$\phi_w = \frac{\tanh pL}{pL}$$

where

$$p = \sqrt{h_{o,w}/(k_F y_F)}$$

Thus, the solution for efficiency of the wet bar fin is of identical form to that of the dry bar fin. If we analyzed other types of fins and made similar substitutions, we would obtain analogous results. Thus we have the important conclusion that *solutions for efficiency of dry fins also apply for efficiency of wet fins (Eq. 12.50) if we substitute $h_{o,w}$ (Eq. 12.47) for the wet fin in place of $h_{c,o}$ for the dry fin.*

12.8.   OVERALL HEAT TRANSFER COEFFICIENT FOR A WET FINNED-TUBE HEAT EXCHANGER

In Sec. 12.4 we developed an expression for the overall coefficient U_o where the fin surfaces were *dry*. In this section, we will develop an expression for the overall coefficient where the fins are wetted by moisture condensed from the air passing over the outside surface.

We may use Fig. 12.14 but imagine a refrigerant at temperature t_R in the tube instead of the hot fluid. We will assume that the thermal resistance of the tube wall is negligible and that the tube has a uniform temperature t_P. We will also assume that the fin and tube are covered by a thin film of water having an average thickness y_w. The air passing over the surface has an enthalpy h. We may write for the local rate of heat transfer

$$q = h_i A_{P,i}(t_P - t_R) \tag{12.51}$$

By definition, let

$$b'_R = \frac{h_{s,P} - h_{s,R}}{t_P - t_R} \tag{12.52}$$

where $h_{s,P}$ and $h_{s,R}$ are fictitious enthalpies of saturated moist air evaluated at the respective temperatures t_P and t_R. By Eqs. (12.51) and (12.52), we obtain

$$q = \frac{h_i A_{P,i}}{b'_R}(h_{s,P} - h_{s,R}) \tag{12.53}$$

Based upon our development in Sec. 12.7 with $h_{o,w}$ given by Eq. (12.47), we have

$$q = \frac{h_{o,w}}{b_{w,P}} A_{P,o}(h - h_{s,P}) + \frac{h_{o,w}}{b_{w,m}} A_F(h - h_{F,m})$$

where $b_{w,P}$ is evaluated from Fig. 12.18 at the temperature of the surface of the water film on the tube and $b_{w,m}$ is evaluated at the mean surface temperature of the water film on the fin. Making the approximations $b_{w,P} = b_{w,m}$ and $h_{s,P} = h_{F,B}$, with Eq. (12.50) we have

$$q = \frac{h_{o,w}}{b_{w,m}}(A_{P,o} + \phi_w A_F)(h - h_{s,P}) \tag{12.54}$$

By definition of $U_{o,w}$, we may write

$$q = U_{o,w} A_o(h - h_{s,R}) \tag{12.55}$$

We may show by Eqs. (12.53)–(12.55) that

$$U_{o,w} = \frac{1}{\dfrac{b'_R A_o}{A_{P,i} h_i} + \dfrac{b_{w,m}(1 - \phi_w)}{h_{o,w}(A_{P,o}/A_F + \phi_w)} + \dfrac{b_{w,m}}{h_{o,w}}} \tag{12.56}$$

Equation (12.56) is of similar form to Eq. (12.32). Units of $U_{o,w}$ are Btu per (hr) (sq ft of outside surface) (Btu per lb dry air enthalpy difference). In order to calculate $U_{o,w}$ by Eq. (12.56), we must first assume values of the mean water film surface temperature $t_{w,m}$ and of the pipe temperature t_P. These assumptions allow initial approximations to be made for $b_{w,m}$ and b_R' respectively. After calculation of $U_{o,w}$, we must check the assumptions. We will now derive equations for these procedures. By Eqs. (12.51) and (12.55), we have for the pipe temperature

$$t_P = t_R + \frac{U_{o,w}A_o(h - h_{s,R})}{h_i A_{P,i}} \tag{12.57}$$

To establish a procedure for checking $t_{w,m}$, we begin by writing the relation

$$h - h_{F,m} = \phi_w(h - h_{s,P}) = \frac{b_{w,m}h_{c,o}}{h_{o,w}c_{p,a}}(h - h_{s,w,m})$$

By Eqs. (12.53) and (12.55),

$$h - h_{s,P} = \left(1 - \frac{b_R' U_{o,w}A_o}{h_i A_{P,i}}\right)(h - h_{s,R})$$

Thus, we obtain

$$h_{s,w,m}' = h - \frac{c_{p,a}h_{o,w}\phi_w}{b_{w,m}h_{c,o}}\left(1 - \frac{b_R' U_{o,w}A_o}{h_i A_{P,i}}\right)(h - h_{s,R}) \tag{12.58}$$

Equation (12.58) allows determination of $t_{w,m}$ through calculation of the enthalpy of saturated air, $h_{s,w,m}$ at the same temperature.

We will now study the influences upon $U_{o,w}$ of the various quantities in Eq. (12.56). Some of these are shown in Example 12.5 which is a similar problem to Example 12.2.

Example 12.5. The same heat-transfer surface as in Example 12.2 is used for cooling and dehumidifying moist air. Air conditions are 80 F dry-bulb temperature, 68 F thermodynamic wet-bulb temperature, and 14.696 psia barometric pressure. Refrigerant temperature is 44 F. Assume that a film of water of 0.005 in. average thickness covers all of the outside surface. (a) Determine the individual thermal resistances and the overall heat transfer coefficient $U_{o,w}$ if $h_{c,o}$ = 10.0 Btu/(hr)(sq ft)(F), h_i = 600 Btu/(hr)(sq ft)(F), and the fin material is aluminum; (b) rework Part (a) but with copper fins; (c) rework Part (a) but with $h_{c,o}$ = 5.0 Btu/(hr)(sq ft)(F); and (d) rework Part (a) but with h_i = 1200 Btu/(hr)(sq ft)(F).

SOLUTION: All dimensional factors are the same as for Example 12.2. We will first show the entire solution for Part (a). In order to evaluate the inside-surface resistance, we will assume t_P = 54 F. By Eq. (12.52) and with Table A.5, we find

$$b_R' = \frac{22.615 - 17.149}{54 - 44} = 0.547 \text{ Btu/(lb}_a)(F)$$

Thus

$$R_{i,w} = \frac{b_R' A_o}{h_i A_{P,i}} = \frac{(0.547)(27.8)}{600} = 0.0253$$

In order to calculate $R_{F,w}$, we must first assume a mean water-film temperature. Assume $t_{w,m} = 60$ F. By Fig. 12.18, $b_{w,m} = 0.68$ Btu/(lb$_a$)(F). Also, it is satisfactory to write

$$c_{p,a} = 0.24 + 0.45W = 0.24 + (0.45)(0.01) = 0.245 \text{ Btu/(lb}_a)(F)$$

For water, $k_w = 0.33$ Btu/(hr)(sq ft)(F/ft). By Eq. (12.47),

$$h_{o,w} = \cfrac{1}{\cfrac{0.245}{(0.68)(10)} + \cfrac{0.005}{(12)(0.33)}} = \frac{1}{0.0360 + 0.0013} = 26.8 \text{ Btu/(hr)(sq ft)(F)}$$

Thus

$$L\sqrt{h_{o,w}/(k_F y_F)} = 0.0541\sqrt{(26.8)/(120)(0.000417)} = 1.25$$

and we know that $r_2/r_1 = 3.60$. By Fig. 12.11, $\phi_w = 0.49$. Thus

$$R_{F,w} = \frac{b_{w,m}(1 - \phi_w)}{h_{o,w}(A_{P,o}/A_F + \phi_w)} = \frac{(0.68)(0.51)}{(26.8)(0.529)} = 0.0245$$

$$R_{o,w} = \frac{b_{w,m}}{h_{o,w}} = \frac{0.68}{26.8} = 0.0254$$

$$R_{t,w} = R_{i,w} + R_{F,w} + R_{o,w} = 0.0752$$

$$U_{o,w} = 1/R_{t,w} = 13.30 \text{ Btu/(hr)(sq ft)(Btu/lb}_a)$$

We may now check our assumed values of t_P and $t_{w,m}$. By the psychrometric chart, $h = 32.30$ Btu/lb$_a$. By Eq. (12.57),

$$t_P = 44 + \frac{(13.30)(27.8)(32.30 - 17.15)}{600} = 53.3 \text{ F}$$

TABLE 12.5

DETERMINATION OF THERMAL RESISTANCES AND $U_{o,w}$ FOR EXAMPLE 12.5

Component or Calculation	Part (a)	Part (b)	Part (c)	Part (d)
Inside-surface resistance, $R_{i,w} = b'_R A_o/(A_{P,i}h_i)$	0.0253	0.0255	0.0246	0.0121
Fin resistance, $R_{F,w} = \dfrac{b_{w,m}(1 - \phi_w)}{h_{o,w}(A_{P,o}/A_F + \phi_w)}$	0.0245	0.0140	0.0264	0.0245
Outside-surface resistance, $R_{o,w} = b_{w,m}/h_{o,w}$	0.0254	0.0254	0.0498	0.0254
Total resistance, $R_{t,w} = R_{i,w} + R_{F,w} + R_{o,w}$	0.0752	0.0649	0.1008	0.0620
$R_{i,w}$ in per cent of $R_{t,w}$	33.7	39.3	24.4	19.5
$R_{F,w}$ in per cent of $R_{t,w}$	32.6	21.5	26.2	39.5
$R_{o,w}$ in per cent of $R_{t,w}$	33.7	39.2	49.4	41.0
$U_{o,w} = 1/R_{t,w}$, Btu/(hr)(sq ft)(Btu/lb$_a$) . .	13.3	15.4	9.9	16.1

which is close to our assumed value. We may next use Eq. (12.58) to check our assumed $t_{w,m}$. We find

$$h_{s,w,m} = 32.30 - \frac{(0.245)(26.8)(0.49)}{(0.68)(10.0)}\left[1 - \frac{(0.547)(13.30)(27.8)}{600}\right] (15.15)$$

$$= 27.5 \text{ Btu/lb}_a$$

and we find that $t_{w,m}$ is 61.5 F. Thus, our assumption of 60 F should be close enough. Table 12.5 shows results for the complete problem. Compared to Example 12.2 (Table 12.3), Table 12.5 shows that with wet fins, the inside-surface resistance $R_{i,w}$ and the fin resistance $R_{F,w}$ are more significant than when the fins are dry. Compared to aluminum fins, Table 12.5 (Part b) shows that with copper fins, $U_{o,w}$ is increased by about 16 per cent. Part (c) shows that reducing $h_{c,o}$ from 10.0 to 5.0 decreases $U_{o,w}$ by about 26 per cent. Part (d) shows that increasing h_i from 600 to 1200 increases $U_{o,w}$ by about 21 per cent.

12.9. MEAN AIR ENTHALPY DIFFERENCE FOR WET FINNED-TUBE HEAT EXCHANGERS

Equation (12.55) shows that when simultaneous cooling and dehumidifying occur, the *overall heat transfer coefficient*, $U_{o,w}$, is based upon an *air enthalpy difference*. Furthermore, the enthalpy h is the true air enthalpy, but the quantity $h_{s,R}$ is a *fictitious enthalpy of saturated air* calculated at the refrigerant temperature. We now need to develop an expression for the mean enthalpy difference for a cooling and dehumidifying coil. The mean enthalpy difference Δh_m is defined by the equation

$$q = U_{o,w}A_o \Delta h_m \tag{12.59}$$

We may recall that for dry fins where only sensible heat transfer occurs, Δt_m is given by the logarithmic mean temperature difference if the fluid temperature within the tubes remains constant. Furthermore, in Sec. 12.2 we observed that for a counter-cross-flow heat exchanger with more than two tube passes, and where the tube-side fluid temperature changed, it was generally acceptable to calculate Δt_m for *pure counter-flow*.

Two common cases exist with cooling coils. One occurs where the coil serves as the evaporator of a direct-expansion refrigeration system. Here the refrigerant temperature remains essentially constant and we would expect the logarithmic mean enthalpy difference to apply. The other case occurs where the refrigerant (chilled water, brine, etc.) temperature changes. However, counter-flow is always desirable and in almost all cases more than two tube passes are used. Thus, we would expect that in such cases the logarithmic mean enthalpy difference calculated for pure counter-flow would be a sufficiently accurate approximation for Δh_m.

We can show that, with certain approximations, for pure counter-flow, the mean air enthalpy difference is given by

$$\Delta h_m = \frac{(h_1 - h_{s,R,2}) - (h_2 - h_{s,R,1})}{\ln\left(\dfrac{h_1 - h_{s,R,2}}{h_2 - h_{s,R,1}}\right)} \tag{12.60}$$

where h_1 and h_2 are respectively the true enthalpies, Btu/lb$_a$, of the entering and leaving air stream, and $h_{s,R,1}$ and $h_{s,R,2}$ are respectively fictitious enthalpies, Btu/lb$_a$, of saturated air calculated at the entering and leaving refrigerant temperatures. Equation (12.60) is restricted to cases where the refrigerant temperature change is small, since in the derivation it is necessary to assume the quantities a_R and b_R as constants in the relation $h_{s,R} = a_R + b_R t_R$. We must also ignore the term $m_a\, dW\, h_{f,w}$ in Eq. (12.34).

12.10. OVERALL HEAT TRANSFER PROBLEMS INVOLVING WET FINNED SURFACES

Practical cooling coil design problems require solution of Eq. (12.59). As discussed in Sec. 12.5, it is usually necessary to include a deposit coefficient for the inside surface of the tubes in the calculation of the overall heat transfer coefficient. Thus, for most cases, we would use

$$U_{o,w} = \cfrac{1}{\cfrac{b_R' A_o}{A_{P,i} h_i} + \cfrac{b_R' A_o}{A_{P,i} h_{d,i}} + \cfrac{b_{w,m}(1 - \phi_w)}{h_{o,w}(A_{P,o}/A_F + \phi_w)} + \cfrac{b_{w,m}}{h_{o,w}}} \qquad (12.61)$$

Estimation of h_i in Eq. (12.61) generally poses little difficulty except in the case of evaporating refrigerants. As discussed in Sec. 2.6, few correlations for h_i for boiling liquids are available, and we may need to resort to experiments. The wet-surface coefficient $h_{o,w}$ must be calculated by Eq. (12.47). In Eq. (12.47) the term y_w/k_w is usually small, so that an estimate of the water film thickness is not critical. However, in case of frost formation, the term y_w/k_w may be more important.

The convection heat transfer coefficient $h_{c,o}$ in Eq. (12.47) is usually the controlling factor for $h_{o,w}$. As discussed in Sec. 12.5 for dry fins, direct experimental data are needed for an accurate estimate. Little information is available in published literature for $h_{c,o}$ for a wet cooling coil.

Myers[*] made a comprehensive experimental study comparing the coefficient $h_{c,o}$ for a wet-surface cooling coil to that for the same coil operated without dehumidification. Myers' experimental coil was closely similar to Surface I of Table 12.4 and Fig. 12.15. Myers analyzed his results in a manner entirely consistent with procedures of this chapter. Figure 12.20 shows Myers' correlations for heat transfer to the external surface both for dry-surface operation (cooling without dehumidification) and for wet-surface operation (cooling and dehumidification). In all wet-surface calculations, Myers used a mean water film thickness of 0.004 in. Because of the presence of the water film, core velocity is higher for the wet coil than for the dry coil for a given face velocity.

Myers also prepared approximate simplified equations showing $h_{c,o}$ as a function of *standard-air face velocity* $V_{s,f}$ both for dry-surface and for wet-surface operation. Combining his equations gives the relation

$$\frac{h_{c,o,w}}{h_{c,o,d}} = 0.626 V_{s,f}^{0.101} \qquad (12.62)$$

[*]Raymond J. Myers, "The Effect of Dehumidification on the Air Side Heat Transfer Coefficient for a Finned-tube Coil" (Master's thesis, University of Minnesota, 1967).

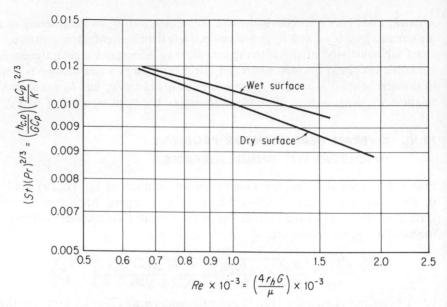

Figure 12.20 Dry-surface and wet-surface heat transfer correlations for the external surface of a finned-tube cooling coil. (Adapted by permission of the author from R. J. Myers, "The Effect of Dehumidification on the Air Side Heat Transfer Coefficient for a Finned-tube Coil," Master's thesis, University of Minnesota, 1967, p. 115.)

where $h_{c,o,w}$ applies to the wet coil and $h_{c,o,d}$ applies to the dry coil and where both coefficients are taken at the same face velocity $V_{s,f}$, ft per min. Equation (12.62) may be used for an approximate estimation of $h_{c,o}$ for wet-surface operation where values are known for the same surface for dry-surface operation.

In calculation of $U_{o,w}$ by Eq. (12.61), attention must be given to the term $b_{w,m}$. It is expressed in units of Btu/(lb$_a$)(F) and represents the slope of a curve expressing the enthalpy of saturated air as a function of temperature. In our analyses, we assumed a linear relation between h_s and t_s which is permissible over a small range of t_s such as about 10 F or less. The quantity $b_{w,m}$ should be evaluated at the mean water-film surface temperature. Where the mean water-film surface temperature change exceeds about 10 F for the entire coil, it is more accurate to separate the coil depth into two or more parts and to treat each part separately. It is particularly interesting to analyze the influence of $b_{w,m}$ in Eqs. (12.56), (12.58), and (12.61). If the quantity y_w/k_w in Eq. (12.47) is of minor importance, $b_{w,m}$ mainly affects ϕ_w since it is essentially cancelled in the term $b_{w,m}/h_{o,w}$. This circumstance is fortunate, and it is usually not necessary to make a precise evaluation of $b_{w,m}$.

Depending upon the circumstances of the problem, a cooling coil may operate with one or more rows of its initial external surface dry (no dehumidification) and with the remainder of its surface wet. If Eq. (12.58), when applied to inlet conditions, indicates a value of $t_{w,m}$ higher than the inlet-air dew-point temperature, then the initial surface will be dry. The procedures of Secs. 12.2–12.5 should be applied to this part of the coil.

For the dry section of a coil, we may show by procedures analogous to those used in obtaining Eq. (12.58) that

$$t_{F,m} = t - \phi\left(1 - \frac{U_o A_o}{h_i A_{P,i}}\right)(t - t_R) \tag{12.63}$$

When $t_{F,m}$ is equal to the inlet-air dew-point temperature $t_{d,1}$, we have

$$t = \frac{t_{d,1} - \phi(1 - U_o A_o/h_i A_{P,i})t_R}{1 - \phi(1 - U_o A_o/h_i A_{P,i})} \tag{12.64}$$

Equation (12.63) allows calculation of the mean fin temperature $t_{F,m}$ for a dry section for given values of air dry-bulb temperature t and refrigerant temperature t_R. Equation (12.64) allows calculation of air temperature t for the location where condensation just begins. It follows that the analysis for a wet coil would begin at this location.

An illustration of the fundamental calculation procedures for a fin-and-tube cooling coil is shown by the following example.

Example 12.6. A direct-expansion, fin-and-tube cooling coil using Refrigerant 12 is to be sized to process 12,000 cu ft per min of moist air from an initial condition of 82 F dry-bulb temperature and 70 F thermodynamic wet-bulb temperature, to a final condition of 54 F dry-bulb temperature and 53 F thermodynamic wet-bulb temperature. Barometric pressure is 14.696 psia. Refrigerant temperature is 40 F. Type II surface (Table 12.4) is to be used. Inlet face velocity of the air is to be 500 ft per min. Experiments indicate that h_i may be estimated as 500 Btu per (hr)(sq ft)(F). Determine (a) the required coil face area in sq ft, (b) the required total outside-surface area in sq ft, and (c) the number of rows of tubes required.

SOLUTION: (a) From the given data we may calculate the coil face area as $12,000/500 = 24$ sq ft.

(b) The outside-surface area A_o must be found by solution of Eq. (12.59). We will assume that all of the coil surface is wet. The mean air enthalpy difference is given by Eq. (12.60). By Fig. E-8, $h_1 = 33.98$ Btu/lb$_a$ and $h_2 = 22.01$ Btu/lb$_a$. By Table A.5, $h_{s,R} = 15.23$ Btu/lb$_a$. Thus

$$\Delta h_m = \frac{h_1 - h_2}{\ln\left(\dfrac{h_1 - h_{s,R}}{h_2 - h_{s,R}}\right)} = \frac{11.97}{\ln\left(\dfrac{18.75}{6.78}\right)} = 11.77 \text{ Btu/lb}_a$$

The mean air enthalpy is then $h_{s,R} + \Delta h_m = 27.00$ Btu/lb$_a$.

In order to find $U_{o,w}$ we must evaluate the individual quantities in Eq. (12.61). We will assume a mean pipe temperature $t_{P,m}$ of 50 F and a mean water-film temperature $t_{w,m}$ of 54 F as first approximations. By Eq. (12.52)

$$b'_R = \frac{20.301 - 15.230}{50 - 40} = 0.507 \text{ Btu/(lb}_a)(F)$$

By Fig. 12.18, $b_{w,m} = 0.60$ Btu/(lb$_a$)(F). An adequate estimate of $c_{p,a}$ is 0.245

Btu/(lb$_a$)(F). We will first determine a value of $h_{c,o}$ based upon dry-surface operation. We will use a mean air temperature of 68 F. By Table A.6, we obtain $\mu = 0.0439$ lb/(ft)(hr) and $c_p\mu/k = 0.709$. By Figure E-8, $W_1 = 0.0130$ lb$_w$/lb$_a$ and $v_1 = 13.94$ cu ft/lb$_a$. Thus

$$m_a = \frac{(12,000)(60)}{13.94} = 51,600 \text{ lb}_a/\text{hr}$$

By Table 12.4, the net area between the fins is 49.7 per cent of the face area. Thus

$$G_a = \frac{51,600}{(0.497)(24)} = 4330 \text{ lb}_a/(\text{hr})(\text{sq ft})$$

and

$$G = (1 + W_1)G_a = (1.013)(4330) = 4390 \text{ lb}/(\text{hr})(\text{sq ft})$$

Thus

$$\text{Re} = \frac{(0.01268)(4390)}{0.0439} = 1268$$

By Fig. 12.15 (Surface II)

$$\frac{h_{c,o}(c_p\mu/k)^{2/3}}{Gc_p} = h_{c,o}\frac{(c_p\mu/k)^{2/3}}{G_a c_{p,a}} = 0.0079$$

Thus, for dry operation

$$h_{c,o} = \frac{(0.0079)G_a c_{p,a}}{(c_p\mu/k)^{2/3}} = \frac{(0.0079)(4330)(0.245)}{(0.709)^{2/3}} = 10.5 \text{ Btu}/(\text{hr})(\text{sq ft})(\text{F})$$

The standard-air face velocity $V_{s,f}$ is given by the relation

$$V_{s,f} = \frac{m_a(1 + W_1)}{(60)(0.075)(24)} = \frac{(51,600)(1.013)}{108} = 484 \text{ ft/min}$$

By Eq. (12.62), we may estimate $h_{c,o}$ for wet-surface operation as

$$h_{c,o} = (10.5)(0.626)(484)^{0.101} = 12.3 \text{ Btu}/(\text{hr})(\text{sq ft})(\text{F})$$

By Eq. (12.47) and assuming a water-film thickness of 0.005 in., we have

$$h_{o,w} = \cfrac{1}{\cfrac{0.245}{(0.60)(12.3)} + \cfrac{0.005}{(12)(0.33)}} = 29.0 \text{ Btu}/(\text{hr})(\text{sq ft})(\text{F})$$

Thus (see Example 12.3)

$$L\sqrt{\frac{h_{o,w}}{k_F y_F}} = \frac{0.575}{12}\sqrt{\frac{(29.0)(2)(12)}{(120)(0.016)}} = 0.91$$

$$\frac{r_2}{r_1} = 2.70$$

By Fig. 12.11, $\phi_w = 0.68$. From Example 12.3, we know for the Type II surface, $A_o/A_{P,i} = 19.31$, $A_{P,o}/A_F = 0.105$. By Table 2.4, $1/h_{d,i} = 0$. By Eq. (12.61),

$$U_{o,w} = \cfrac{1}{\cfrac{(0.507)(19.31)}{500} + \cfrac{(0.60)(0.32)}{(29.0)(0.785)} + \cfrac{0.60}{29.0}} = 20.5 \text{ Btu}/(\text{hr})(\text{sq ft})(\text{Btu}/\text{lb}_a)$$

We may now check the assumed values of $t_{P,m}$ and $t_{w,m}$. By Eq. (12.57),

$$t_{P,m} = 40 + \frac{(20.5)(19.31)(11.77)}{500} = 49.3 \text{ F}$$

which is sufficiently close to our assumed value of 50 F. By Eq. (12.58),

$$h_{s,w,m} = 27.00 - \frac{(0.245)(29.0)(0.68)}{(0.60)(12.3)}\left[1 - \frac{(0.507)(20.5)(19.31)}{500}\right](27.00 - 15.23)$$

$$= 22.38 \text{ Btu/lb}_a$$

and $t_{w,m}$ is about 53.6 F which is very close to our assumed value of 54 F. For the inlet conditions of the coil, by Eq. (12.58), we find that $t_{w,m}$ is about 60 F. Since $t_{d,1} = 64.4$ F, the coil surface is wet throughout.

The required total heat-transfer rate for the coil is

$$q = m_a(h_1 - h_2) = (51,600)(33.98 - 22.01) = 618,000 \text{ Btu/hr}$$

By Eq. (12.59),

$$A_o = \frac{618,000}{(20.5)(11.77)} = 2560 \text{ sq ft}$$

(c) Table 12.4 shows that Type II surface has an outside-surface area of 22.86 sq ft/(sq ft of face area)(row). Thus

$$\text{Rows} = \frac{2560}{(22.86)(24)} = 4.67$$

Before leaving the solution of Example 12.6, we should check the mean water-film surface temperature through the coil to ascertain whether its variation exceeded about 10 F. For the outlet conditions, we find by Eq. (12.58) that $h_{s,w,m} = 18.87$ Btu/lb$_a$ and $t_{w,m} = 47$ F approximately. Thus, the variation of $t_{w,m}$ is about 11 F.

The solution of five rows of tubes for Example 12.6 is probably adequate. As a check to the solution, we could divide the coil into increments, such as two rows, and evaluate more accurate values of $b_{w,m}$ and $U_{o,w}$ for each increment. We may easily show that for Example 12.6

$$h_2 = h_{s,R} + (h_1 - h_{s,R})e^{-U_{o,w}A_o/m_a}$$

where h_1 and h_2 are the respective enthalpies of the air entering and leaving the incremental coil and A_o is the surface area of the incremental coil. In this way, we may calculate h_2 for each increment except for the last increment, where the required incremental A_o would be calculated.

12.11. CALCULATION OF COOLING COIL PERFORMANCE AT OTHER THAN DESIGN CONDITIONS

The calculation procedures given in Secs. 12.8–12.10 are adequate for calculating necessary surface area and the number of rows of tubes where operating conditions are known. Example 12.6 was such a problem where, among other data, the final air state was specified. We may, however, encounter a problem where the coil surface area is known and we need to determine the final air state for various operating conditions. The fundamental procedure previously given would provide for determination of the final air enthalpy h_2 only. In this section we will study how the air humidity ratio W also may be determined throughout a cooling coil. Unfortunately, no convenient mathematical solution is available for W as was the case for h. However, Eq. (12.38) allows us to construct the process line on the h-W psychrometric chart for moist air passing through a cooling-and-dehumidifying coil.

We will assume that the inlet air state, inlet refrigerant state, the various heat transfer coefficients, and the coil surface data are known. If the refrigerant is a liquid which does not change phase, by alteration of Eq. (12.60), we may obtain that the enthalpy of the air leaving the coil is given by

$$h_2 = \frac{h_{s,R,1}(1 - e^{-(1-c_1)c_2}) + h_1(1 - c_1)e^{-(1-c_1)c_2}}{1 - c_1 e^{-(1-c_1)c_2}} \qquad (12.65)$$

where $c_1 = m_a b_R / M C_P$ and $c_2 = U_{o,w} A_o / m_a$, and m_a is mass-flow rate (dry basis) of the air, M is the mass-flow rate of the refrigerant, C_P is the specific heat of the refrigerant, and other quantities are the same as those defined in Secs. 12.8 and 12.9. If the refrigerant temperature is constant, Eq. (12.65) reduces to

$$h_2 = h_{s,R} + (h_1 - h_{s,R})e^{-U_{o,w}A_o/m_a} \qquad (12.66)$$

The relationship between the mean water-film surface temperature $t_{w,m}$ and the air enthalpy h may be found from Eq. (12.58).

Equations (12.38), (12.58), and (12.65)–(12.66) in conjunction with the psychrometric chart provide a method for determining the change of air state through a cooling and dehumidifying coil. An illustration of the method is provided by Example 12.7.

> **Example 12.7.** A cooling coil having face dimensions of 3 ft by 6 ft is constructed of Type II surface (Table 12.4). Refrigerant 12 evaporates in the tubes at a constant temperature of 40 F. Air enters the coil at 82 F dry-bulb temperature, 70 F thermodynamic wet-bulb temperature, and 14.696 psia barometric pressure at a rate of 9000 cu ft/min. For purposes of this example, it will be sufficient to assume the same fin efficiency and heat transfer coefficients which were used for Example 12.6. Determine (a) the complete moist-air condition line on the psychrometric chart for a coil of infinite depth, (b) the final air state for a four-row coil, and (c) the final air state if the coil was eight rows deep.
>
> SOLUTION: (a) We will determine a general process line which would be applicable, regardless of the number of rows of tubes. From the given data, we obtain from the psychrometric chart, $h_1 = 33.98$ Btu/lb$_a$ and $h_{s,R} = 15.23$ Btu/lb$_a$. Also, $m_a = (9000)(60)/13.94 = 38,700$ lb$_a$/hr. With the known data from Example 12.6, Eq. (12.58) becomes
>
> $$h_{s,w,m} = 0.608\,h + 5.97$$
>
> Since we may arbitrarily choose Δh, we may write
>
> $$\Delta h_{s,w,m} = 0.608\,\Delta h$$

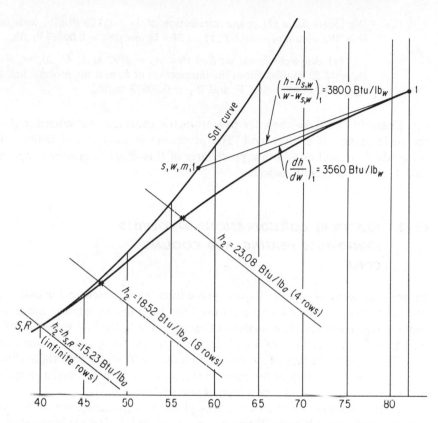

Figure 12.21 Solution of Example 12.7.

The process line may be obtained by graphical solution of Eq. (12.38). Corresponding to State (1) of the air, we have $h_{s,w,m,1} = (0.608)(33.98) + 5.97 = 26.63$ Btu/lb$_a$ and $t_{w,m,1} = 60.2$ F. Since $t_{d,1} = 64.4$ F, the coil is wet throughout. We locate States (1) and $(s, w, m, 1)$ on the psychrometric chart. With the chart protractor, we obtain $[(h - h_{s,w,m})/(W - W_{s,w,m})]_1 = 3800$ Btu/lb$_w$. At 82 F, $h_{g,t} = 1097$ Btu/lb$_w$. By Eq. (10.7), we find Le = 0.90. By Eq. (12.38),

$$\left(\frac{dh}{dW}\right)_1 = (0.90)(3800) + 1097 - (1061)(0.90) = 3560 \text{ Btu/lb}_w$$

With the aid of the chart protractor, we draw a short line segment of direction $(dh/dW)_1$ through State (1). At an arbitrary value of $h = 32.00$ Btu/lb$_a$ we locate a new State (a) on this line. We then repeat the procedure until the condition line meets with the state of saturated air at the refrigerant temperature of 40 F. Figure 12.21 shows the complete condition line.

(b) We know that the face area of the coil is 18 sq ft. By Table 12.4, the outside-surface area is 22.86 sq ft per (sq ft of face area)(row). For four rows, $A_o = 1646$ sq ft, and $U_{o,w}A_o/m_a = 0.872$. By Eq. (12.66),

$$h_2 = 15.23 + (18.75)e^{-0.872} = 23.08 \text{ Btu/lb}_a$$

We locate State (2) at the intersection of $h_2 = 23.08$ Btu/lb$_a$ with the process line. We obtain $t_2 = 56.6$ F, $t_2^\star = 54.9$ F, and $W_2 = 0.00887$ lb$_w$/lb$_a$.

(c) For eight rows, we find that $A_o = 3292$ sq ft, $U_{o,w}A_o/m_a = 1.74$, and $h_2 = 18.52$ Btu/lb$_a$. From the intersection of h_2 and the process line, we obtain $t_2 = 47.2$ F, $t_2^\star = 46.9$ F, and $W_2 = 0.00673$ lb$_w$/lb$_a$.

Example 12.7 shows how the psychrometric chart may be utilized in determining the performance of a given coil. The procedure is lengthy and tedious. However, the method, as well as all other procedures of this chapter, may be programmed for calculation by a digital computer.

12.12. PRACTICAL CONSIDERATIONS INVOLVING FINNED-TUBE HEATING AND COOLING COILS

In previous sections of this chapter, procedures were presented for calculating rates of heat transfer in finned-tube coils used for heating and cooling moist air. These fundamental methods allow analysis of many of the individual variables affecting the heat transfer performance of a coil. In the analyses, certain assumptions were necessary. In addition, the lack of information on the coefficient h_i for evaporating refrigerants, and the uncertainty of the external surface coefficient $h_{c,o}$, make experimentation necessary in the design of a new type of coil surface.

In this chapter, no discussion was given on optimizing the construction of a finned-tube heat exchanger. However, in a practical design problem, attention must be given to determining the correct fin dimensions, optimum number of fins per inch, and the optimum air mass velocity.

When a fin is made larger, its efficiency decreases, so that beyond a certain size, an increase in dimensions does not appreciably increase the rate of heat transfer. For natural-convection coils where the coefficient $h_{c,o}$ is relatively small, fins may typically be larger than for forced-convection coils.

Increasing the number of fins per inch of tube length may result in an increased rate of heat transfer provided that the boundary layer developed on one fin surface does not interfere with that of the adjacent fin surface. Thus, the length of the fins in the direction of air flow, and the air mass velocity are important factors in determining the optimum fin spacing.

A particularly important variable is the air mass velocity G. The external-surface coefficient $h_{c,o}$ is primarily a function of G. An increase of air velocity is usually beneficial to heat transfer but is simultaneously detrimental because of the increase in pressure drop of the air flowing through the coil. The air mass velocity cannot be made arbitrarily large because the fan-power costs may become excessive. As is true with any heat exchanger, the optimum design must be a compromise which results in the highest heat transfer capacity per unit of amortized costs. Costs must include operating expenditures as well as initial costs.

PROBLEMS

12.1. An air heating coil utilizes circular-plate fins of uniform thickness. The aluminum fins are 0.010 in. thick and have a diameter of 2.0 in. There are 84 fins per ft. The copper tubing has an outside diameter of 0.530 in. and an inside diameter of 0.480 in. You may assume heat transfer coefficients of 9.0 and 1000 Btu per (hr)(sq ft)(F) respectively for the outside surface and inside surface. Calculate the overall heat transfer coefficient U_o if deposit coefficients are neglected.

12.2. A steam heating coil is constructed with rectangular-plate aluminum fins having a thickness of 0.01 in. There are eight fins per inch of tube length. The copper tubes have an outer diameter of 0.50 in. and an inner diameter of 0.44 in. Tube spacing is 1.5 in. across the face and 2.0 in. between rows. Based upon laboratory tests, it is estimated that $h_i = 1200$ Btu per (hr)(sq ft)(F) and that for an inlet face velocity of 500 ft per min, $h_{c,o} = 10.0$ Btu per (hr)(sq ft)(F). Based upon saturated steam at 20 psia and saturated air entering the coil at 20 F, calculate the final air temperature for a coil one row deep, two rows deep, and four rows deep. You may assume a deposit coefficient $1/h_{d,i}$ of 0.001 on the steam side and a barometric pressure of 14.696 psia.

12.3. A heating coil is to be designed to heat 6000 cu ft per min of saturated moist air from 60 F to 110 F. Heating medium is water available at 180 F. Type II surface (Table 12.4) is to be used. Barometric pressure is 14.696 psia. The design is to be based upon an average water velocity of approximately 4 ft per sec, water temperature drop of 10 F, and an inlet face velocity of the air of 600 ft per min. The tubes are to be connected to give a counter-cross-flow arrangement. Determine (a) the face dimensions of the coil, and (b) the number of rows of tubes required.

12.4. A steam pipe passes through a space where the ambient temperature is 80 F. Temperature of the saturated steam inside the pipe is 250 F. The wrought-iron pipe (2.375 in. O.D., 2.059 in. I.D.) is covered by insulation [2-in. thickness, $k = 0.50$ Btu/(hr)(sq ft)(F/in.)]. The insulation has a highly reflective outer covering (ϵ negligibly small). Barometric pressure is 14.696 psia. The natural-convection coefficient $h_{o,o}$, Btu per (hr)(sq ft)(F), for the outside surface of the insulation is given by the equation

$$h_{c,o} = 0.27 \left(\frac{\Delta t}{D_o} \right)^{0.25}$$

where Δt is the temperature difference between the outer surface and the surrounding air, F, and D_o is the diameter, ft. Calculate the rate of heat loss from the pipe in Btu per (hr)(ft).

12.5. A single-row finned-tube coil having face dimensions of 12 in. by 36 in. was tested in the laboratory. The following test data were obtained.

t_1 F	$t_1^\star$ F	t_2 F	Inlet Air Flow cu ft/min	Steam Pressure psig
80	60	171.0	900	4.50
80	60	159.0	1200	4.50
80	60	151.0	1500	4.50
80	60	145.0	1800	4.50
80	60	140.0	2100	4.50
80	60	136.0	2400	4.50

The barometer reading for each test was 27.45 in. Hg. (a) Plot a curve expressing $U_o A_o$ in Btu per (hr)(F)(sq ft of face area) as a function of face mass velocity in lb per (hr)(sq ft). (b) Consider the performance of a similar type coil but two rows deep. Assume saturated air entering at 20 F, a face velocity of 500 ft per min, steam pressure of 10.0 psig, steam quality of 95 per cent, and sea-level pressure. Calculate the temperature of the air leaving the coil.

12.6. A cooling coil has the same basic fin-and-tube construction as in Problem 12.2 except that it has four fins per inch of tube length. You may assume $h_i = 500$ Btu per (hr)(sq ft)(F) and $h_{c,o} = 8.0$ Btu per (hr)(sq ft)(F). Air conditions are 0 F, 100 per cent relative humidity, and 14.696 psia barometric pressure. Refrigerant temperature is -12 F. Determine the individual thermal resistances, their per cent of the total resistance, and the overall heat transfer coefficient $U_{o,w}$ if (a) a uniform layer of frost [$k = 0.27$ Btu per (hr)(sq ft)(F per ft)] of 0.05 in. thickness covers all outside surfaces, and (b) if frost layer is of negligible thickness.

12.7. Derive Eq. (12.60).

12.8. Derive Eq. (12.65).

12.9. Rework Example 12.6 with chilled water as the refrigerant and a counter-cross-flow arrangement being used. Assume that water enters the coil at 40 F and leaves at 48 F. Assume a water velocity of approximately 7.0 ft per sec. Determine (a) the required face dimensions of the coil, and (b) the number of rows of tubes required.

12.10. A refrigerated storage room is maintained at an average uniform temperature of 0 F by a bare-pipe direct-expansion coil mounted at ceiling level. Relative humidity in the room may be assumed as 100 per cent. Barometric pressure is 14.696 psia. The pipes are arranged in a single horizontal bank. Total length of piping is 1000 ft. The wrought-iron pipes have an outside diameter of 2.375 in. and an inside diameter of 2.059 in. The pipes are covered with frost [$k = 0.27$ Btu per (hr)(sq ft)(F per ft)] having an average thickness of 1.0 in. Refrigerant evaporating temperature is -20 F. It is estimated that $h_i = 500$ Btu per (sq ft of inside pipe surface)(F) and that the combined (radiation and convection) coefficient $h_o = 3.0$ Btu per (hr)(sq ft of outside surface)(F). (a) Determine the refrigerating capacity of the pipe coil in Btu per hr. (b) Determine the capacity if all frost were scraped from the pipes. Assume other conditions remain the same.

12.11. A direct-expansion, fin-and-tube cooling coil is constructed of Type II surface. The coil has a face area of 20 sq ft. It is used to process 10,000 cu ft per min of moist air which enters the coil at 100 F dry-bulb temperature and 70 F thermodynamic wet-bulb temperature. Barometric pressure is 14.696 psia. Refrigerant temperature is 44 F. You may assume that $h_{c,o}$ and h_i are 10.0 and 500 Btu per (hr)(sq ft)(F) respectively. (a) Construct the complete moist-air condition line on the psychrometric chart. (b) Determine the dry-bulb temperature and thermodynamic wet-bulb temperature of the air leaving the coil if the coil depth is two rows, four rows, and eight rows.

SYMBOLS USED IN CHAPTER 12

A	Surface area, sq ft; A_F refers to fin; A_i refers to inside of pipe; A_o refers to total outside surface.
A_c	Net cross-sectional area, sq ft.
A_P	Surface area of pipe, sq ft; $A_{P,i}$ refers to inside surface; $A_{P,m}$ refers to mean surface; $A_{P,o}$ refers to outside surface.
A_x	Surface area per unit length in x direction, sq ft per ft.
a	Coefficient in Eq. (12.40), Btu per lb_a; a_R evaluated at refrigerant temperature t_R; a_w evaluated at water-film temperature t_w.
b	Coefficient in Eq. (12.40), Btu per $(lb_a)(F)$; b_R evaluated at refrigerant temperature t_R; b_w evaluated at water-film temperature t_w; $b_{w,m}$ evaluated at mean water-film temperature $t_{w,m}$.
C_P	Specific heat at constant pressure of tube-side fluid, Btu per (lb)(F).
c_1	$m_a b_R/M C_P$, dimensionless.
c_2	$U_{o,w} A_o/m_a$, dimensionless.
c_p	Specific heat at constant pressure of fluid outside of tubes (usually air), Btu per (lb)(F).
$c_{p,a}$	Specific heat at constant pressure of moist air per unit mass of dry air, Btu per $(lb_a)(F)$.
F	$\Delta t_m/\Delta t_{m,cf}$, dimensionless.
G	Mass velocity of moist air, lb per (hr)(sq ft).
G_a	Mass velocity of dry air, (lb_a) per (hr)(sq ft).
h	Enthalpy of moist air, Btu per lb_a; h_1 refers to entering air; h_2 refers to leaving air.
$h_{c,o}$	Convection heat transfer coefficient for outside surface, Btu per (hr)(sq ft)(F).
$h_{D,o}$	Mass transfer coefficient for outside surface, lb_w per (hr)(sq ft)(lb_w per lb_a).
$h_{d,i}$	Deposit coefficient for inside surface (see Table 2.7), Btu per (hr)(sq ft)(F).
h_F	Fictitious enthalpy of moist air defined by Eq. (12.43), Btu per lb_a;

$h_{F,m}$ evaluated at mean fin temperature $t_{F,m}$; $h_{F,B}$ evaluated at temperature $t_{F,B}$ of fin base; $\Delta h_F = h - h_F$, etc.

$h_{f,w}$ Enthalpy of saturated liquid water at temperature t_w, Btu per lb_w.

$h_{g,t}$ Enthalpy of saturated water vapor at air dry-bulb temperature, Btu per lb_w.

h_i Heat transfer coefficient for inside surface, Btu per (hr)(sq ft)(F).

$h_{o,w}$ Heat transfer coefficient, Btu per (hr)(sq ft)(F), defined by Eq. (12.47).

h_s Enthalpy of saturated moist air, Btu per lb_a; $h_{s,P}$ evaluated at pipe temperature t_P; $h_{s,R}$ evaluated at refrigerant temperature t_R; $h_{s,w}$ evaluated at water-film temperature t_w; $h_{s,w,m}$ evaluated at mean water-film temperature $t_{w,m}$.

Δh_m Mean enthalpy difference for moist air, Btu per lb_a.

K_1 $U_o A_o/mc_p$, dimensionless.

K_2 $(mc_p/MC_P)(1 - e^{-K_1})$, dimensionless.

k Thermal conductivity, Btu per (hr)(sq ft)(F per ft); k_F refers to fin material; k_P refers to pipe material; k_w refers to water.

L Length, ft.

Le $h_{c,o}/h_{D,o}c_{p,a}$, dimensionless.

L_x Total length in x direction, ft.

L_y Total length in y direction, ft.

M Mass rate of flow of fluid inside tubes, lb per hr.

m Mass rate of flow of fluid outside of tubes, lb per hr.

m_a Mass rate of flow of dry air, lb_a per hr.

N Number of rows of tubes, dimensionless.

P $(t_2 - t_1)/(T_1 - t_1)$, dimensionless.

P $\sqrt{h_{c,o}/ky}$ for dry fins, ft^{-1}; $\sqrt{h_{o,w}/k_F y_F}$ for wet fins, ft^{-1}.

q Rate of heat transfer, Btu per hr; q_F refers to heat conduction in fins.

R $(T_1 - T_2)/(t_2 - t_1)$, dimensionless.

R Thermal resistance term used with heat exchangers with dry fins, (hr)(sq ft)(F) per Btu; R_i refers to inside surface; R_F refers to fins; R_P refers to pipe wall; R_o refers to outside surface; R_t is total value.

R_w Thermal resistance term used with heat exchangers with wet fins, (hr)(sq ft)(Btu per lb_a) per Btu; $R_{i,w}$ refers to inside surface; $R_{F,w}$ refers to fins; $R_{o,w}$ refers to outside surface; $R_{t,w}$ refers to total value.

Re $4r_h G/\mu$ for fluid outside of tubes, dimensionless; $D_i G/\mu$ for fluid inside of tubes, dimensionless.

r_1 Outside radius of tube, ft.

r_2 Outside radius of circular fin, ft; $\sqrt{ac/\pi}$ for rectangular fin, ft (see Fig. 12.13).

r_h LA_c/A_o, ft.

T Temperature of fluid inside of tubes, F (not absolute temperature); T_1 refers to entering fluid; T_2 refers to leaving fluid.

t Temperature of fluid outside of tubes, F; t_1 refers to entering fluid; t_2 refers to leaving fluid.

$t^\star$ Thermodynamic wet-bulb temperature of moist air, F.

t_F Temperature of fin, F; $t_{F,B}$ refers to base of fin; $t_{F,m}$ refers to mean fin temperature; $\Delta t_F = t_F - t$, etc.

t_f	Temperature of fluid inside of tubes, F.
t_P	Temperature of pipe or tube, F; $t_{P,i}$ refers to inside surface; $t_{P,o}$ refers to outside surface.
t_R	Refrigerant temperature, F.
t_s	Temperature of saturated moist air, F.
t_w	Temperature of water film, F; $t_{w,m}$ refers to mean value.
Δt_m	Mean temperature difference between fluids, F; $\Delta t_{m,cf}$ refers to pure counter-flow.
U_o	Overall heat transfer coefficient, Btu per (hr)(sq ft)(F).
$U_{o,w}$	Overall heat transfer coefficient for heat exchangers with wet fins, Btu per (hr)(sq ft)(Btu per lb_a).
v	Volume of moist air, cu ft per lb_a.
W	Humidity ratio of moist air, lb_w per lb_a; $W_{s,w}$ refers to saturated air at water-film temperature t_w.
x_P	Thickness of pipe wall, ft.
y_F	One-half of fin thickness, ft.
y_w	Thickness of water film, ft.
ϵ	Emissivity, dimensionless.
μ	Fluid viscosity, lb per (ft)(hr).
ϕ	Fin efficiency for dry fins defined by Eq. (12.19), dimensionless.
ϕ_w	Fin efficiency for wet fins defined by Eq. (12.50), dimensionless.

Part IV

SOLAR RADIATION

13

SOLAR RADIATION

13.1. INTRODUCTION

We now begin a new phase in our study of thermal environmental engineering—*solar radiation*. The external thermal environment of a locality results from the combined influences of solar radiation and meteorological effects. Physical influences such as topography and ocean currents may also be of great importance to the climate of a given locality.

In the final analysis, the sun is the source of most energy on the earth and is a primary factor in determining the thermal environment of a locality. It is important for an engineer to have a working knowledge of the earth's relationship to the sun. He should be able to make estimates of solar radiation intensity and know how to make simple solar radiation measurements. He should also understand the thermal effects of solar radiation and know how to control or utilize them.

13.2. THE SOLAR SYSTEM

The sun, its family of planets, and the satellites which accompany the planets, make up the *solar system*. Each planet revolves about the sun in an approximately circular orbit.

279

TABLE 13.1

VARIOUS PHYSICAL DATA FOR THE SOLAR SYSTEM

	Sun	Mercury	Venus	Earth	Mars	Jupiter	Saturn	Uranus	Neptune	Pluto
Sidereal period (Earth years)	. . .	0.241	0.616	1.00	1.88	11.86	29.46	84.02	164.8	247.7
Mean distance from sun (millions of miles)	. . .	36.0	67.2	93.0	141.7	484	887	1,785	2,797	3,675
Orbital speed (miles per second)	. . .	29.8	21.8	18.5	15.0	8.5	6.0	4.2	3.4	3.0
Length of day (Earth days)	25	88	30	1.0	1.0	0.41	0.44	0.45	0.66	. . .
Mean diameter (miles)	865,400	3,108	7,707	7,919	4,270	86,860	71,530	31,700	31,000	7,900
Ratio of mass to Earth's mass	332,000	0.04	0.82	1.00	0.11	318.3	95.3	14.7	17.3	1.0
Density (lb per cu ft)	88.0	237.1	303.3	344.4	246.1	83.0	44.3	78.6	99.8	. . .
Ratio of surface gravity to Earth's gravity	28.0	0.27	0.86	1.00	0.37	2.64	1.17	0.92	1.44	. . .
Maximum surface temperature, F	10,000	770	200	140	86	−216	−243	−300	−330	−348
Atmosphere	H_2 He	None	CO_2	Air	H_2O	CH_4 NH_3	CH_4 NH_3	CH_4 NH_3	CH_4 NH_3	None
Albedo	. . .	0.07	0.59	0.29	0.15	0.44	0.42	0.45	0.52	0.04
Number of moons	. . .	None	None	1	2	12	9	5	2	None

*Data on planets from Cecilia Payne-Gaposchkin, *Introduction to Astronomy* (New York: Prentice-Hall, Inc., 1954) p. 180.

In order of increasing distance from the sun, the planets are Mercury, Venus, Earth, Mars, Jupiter, Saturn, Uranus, Neptune, and Pluto.

Table 13.1 shows physical data for the sun and the planets. The sun has a tremendous size and has a very high temperature as compared to the planets. The *major planets* Jupiter, Saturn, Uranus, and Neptune have large diameters and low densities compared to the other planets. Because of their great distances from the sun, the major planets and Pluto too have extremely low surface temperatures. The *terrestrial planets* Mercury, Venus, Earth, Mars, and Pluto are relatively small and dense. The orbital planes of all of the planets except Pluto are nearly the same. The plane of Pluto's orbit is inclined about 17 degrees from the common plane. The orbit of each planet is an almost circular ellipse.

13.3. THE SUN

The sun is one of the stars of the universe and, relative to the earth, is the most important heavenly body. Because of its extremely high temperature, the sun is entirely gaseous. The diameter of the sharp circular boundary of the sun is about 865,400 miles. The mass of the sun is about 332,000 times that of the earth. Observations on sunspot movements have established that the sun rotates about its axis, but not as a rigid body. The period of rotation varies from about 25 earth days at its equator to about 27 days at 40 degrees latitude.

Astrophysicists* generally divide the structure of the sun into three regions—the *solar interior,* the *photosphere,* and the *solar atmosphere.* The *solar interior* which is the main mass of the sun is believed to have a central temperature of many millions of degrees and a central gas pressure of perhaps a billion atmospheres. It is here that the sun's energy output is generated. Astrophysicists have established that the sun's main constituent is hydrogen. It is believed that the sun's energy output occurs from conversion of hydrogen into helium in the presence of carbon and nitrogen.

The *photosphere* is a thin layer of gas which forms the bright boundary of the sun. Both pressure and density are very low in this layer. The temperature of the photosphere is about 10,000 F. The greatest part of the sun's thermal radiation is emitted from the photosphere.

Beyond the photosphere, or surface of the sun, is the *solar atmosphere,* which is composed of the *chromosphere* and the *corona.* Both regions of the solar atmosphere offer little resistance to radiation from the photosphere and in addition contribute very little to the sun's radiation output. The *chromosphere* appears as the ring of red light surrounding the sun which may be seen during a total eclipse. It is a relatively thin layer of gas at extremely low pressure and low density. The *corona* is the extremity of the sun consisting of rarified gases which extend beyond the chromosphere for a million miles or more.

*W. H. McCrea, *Physics of the Sun and Stars* (London: Hutchinson's University Library, 1950) pp. 62, 78, 79, 85.

13.4. THE EARTH

The planet earth is nearly spherical with a diameter of about 7,900 miles. It makes one rotation about its axis every 24 hours and completes a revolution about the sun in a period of $365\frac{1}{4}$ days approximately. The earth's mean density is about 5.52 times that of water.

The earth's internal structure has been studied extensively by geophysicists from records of earthquake waves.* It is believed that the earth has a *central core* about 1600 miles in diameter which is more rigid than steel. Beyond the central core is the *mantle* which forms about 70 per cent of the earth's mass, and beyond this is the *outer crust* which forms about one per cent of the total mass.

The earth revolves about the sun in an approximately circular path, with the sun located slightly off center of the circle. The earth's mean distance to the sun is about 92,900,000 miles. About January 1, the earth is closest to the sun while on about July 1 it is most remote, being about 3.3 per cent farther away. Since intensity of solar radiation incident upon the top of the atmosphere varies inversely with the square of the earth-sun distance, the earth receives about seven per cent more radiation in January than in July. The earth's axis of rotation is tilted 23.5 degrees with respect to its orbit about the sun. The earth's tilted position is of profound significance, for, together with the earth's daily rotation and yearly revolution, it accounts for the

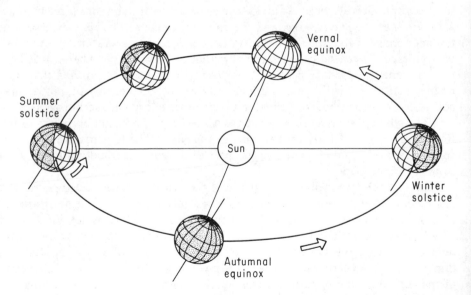

Figure 13.1 The earth's revolution about the sun.

*H. N. Russell, R. S. Dugan, and J. Q. Stewart, *Astronomy I. The Solar System* (Boston: Ginn and Company, 1945) p. 131.

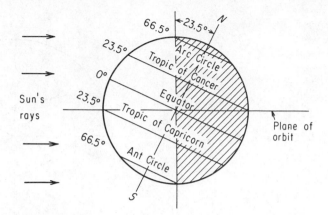

Figure 13.2 Position of earth in relation to sun's rays at the time of the winter solstice.

distribution of solar radiation over the earth's surface, the changing length of hours of daylight and darkness, and the changing of the seasons.

Figure 13.1 schematically shows the effect of the earth's tilted axis at various times of the year. Figure 13.2 shows the position of the earth relative to the sun's rays at the time of the *winter solstice*. At the winter solstice (December 22 approximately), the North Pole is inclined 23.5 degrees away from the sun. All points on the earth's surface north of 66.5 degrees north latitude are in total darkness while all regions within 23.5 degrees of the South Pole receive continuous sunlight. At the time of the *summer solstice* (June 22 approximately), the situation is reversed. At the times of the two *equinoxes* (March 22 and September 22 approximately), both poles are equidistant from the sun and all points on the earth's surface have 12 hours of daylight and 12 hours of darkness.

Because of its tilted axis, the earth's surface has been divided into five zones. The *torrid zone* includes all locations where the sun is at the zenith (vertically overhead) at least once yearly. The torrid zone extends 23.5 degrees on either side of the equator. The *temperate zones* include all locations where the sun appears above the horizon each day but never at the zenith. The temperate zones extend from latitudes 23.5 degrees to 66.5 degrees (north and south). The *frigid zones* include all locations where the sun is below the horizon (and above) for at least one full day yearly. The frigid zones extend 23.5 degrees from the poles.

13.5. BASIC EARTH-SUN ANGLES

The position of a point P on the earth's surface with respect to the sun's rays is known at any instant if the *latitude l* and *hour angle h* for the point, and the *sun's declination d* are known. These fundamental angles are shown by Fig. 13.3. Point P represents a location on the northern hemisphere.

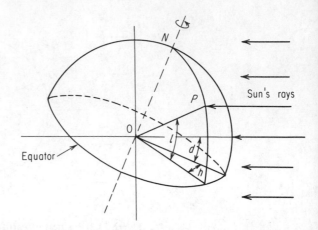

Figure 13.3 Latitude, hour angle, and sun's declination.

The latitude l is the angular distance of the point P north (or south) of the equator. It is the angle between the line $\overline{O\,P}$ and the projection of $\overline{O\,P}$ on the equatorial plane. Point O represents the center of the earth.

The hour angle h is the angle measured in the earth's equatorial plane between the projection of $\overline{O\,P}$ and the projection of a line from the center of the sun to the

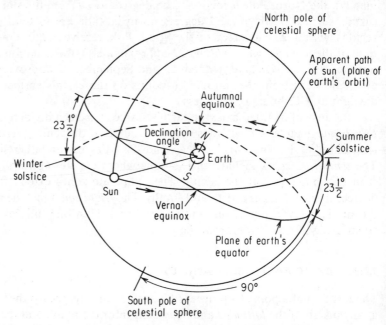

Figure 13.4 Schematic celestial sphere showing apparent path of sun and sun's declination angle.

center of the earth. At solar noon, the hour angle is zero. The hour angle expresses the time of day with respect to solar noon. One hour of time is represented by 360/24 or 15 degrees of hour angle.

The sun's declination d is the angular distance of the sun's rays north (or south) of the equator. It is the angle between a line extending from the center of the sun to the center of the earth and the projection of this line upon the earth's equatorial plane.

Figure 13.4 shows a schematic *celestial sphere* where the earth is taken as the center of the universe. The sun would appear to move in the plane of the earth's orbit. Figure 13.4 shows the sun's angle of declination. At the time of the winter solstice, the sun's rays would be 23.5 degrees south of the earth's equator; $(d = -23.5°)$. At the time of the summer solstice, the sun's rays would be 23.5 degrees north of the earth's equator $(d = 23.5°)$. At the equinoxes, the sun's declination would be zero. Figure 13.5 shows approximately the variation of the sun's declination throughout

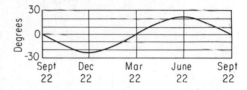

Figure 13.5 Variation of sun's declination.

the year. Because the period of the earth's complete revolution about the sun does not coincide exactly with a calendar year, the declination varies slightly on the same day from year to year. Precise values may be obtained for a particular year from an ephemeris such as that in the *American Ephemeris and Nautical Almanac*.* For ordinary calculations it is sufficiently accurate to use values for any year. Table 13.2 shows weekly values of the sun's declination. For any one day, the declination may be assumed constant.

13.6. TIME

Solar radiation calculations must be made in terms of *solar time*. In a discussion of time, numerous designations may be used. We will consider here only a brief description to allow us to convert local watch time to solar time for engineering calculations.

Time reckoned from midnight at the Greenwich meridian (zero longitude) is known as *Greenwich Civil Time* GCT or *Universal Time*. Such time is expressed on an hour scale from zero to 24. Thus, midnight is 0^h and noon is 12^h. *Local Civil Time* LCT is reckoned from the precise longitude of the observer. On any particular meridian, LCT is more advanced at the same instant than on any meridian further west

*U. S. Nautical Almanac Office, *The American Ephemeris and Nautical Almanac* (Washington, D. C., U. S. Naval Observatory, Annual).

TABLE 13.2

THE SUN'S DECLINATION AND EQUATION OF TIME

Day →	1		8		15		22	
Month	Dec. Deg: Min	Eq. of Time Min: Sec	Dec. Deg: Min	Eq. of Time Min: Sec	Dec. Deg: Min	Eq. of Time Min: Sec	Dec. Deg: Min	Eq. of Time Min: Sec
January	−(23:08)	− (3:16)	−(22:20)	− (6:26)	−(21:15)	− (9:12)	−(19:50)	−(11:27)
February	−(17:18)	−(13:34)	−(15:13)	−(14:14)	−(12:55)	−(14:15)	−(10:27)	−(13:41)
March	−(7:51)	−(12:36)	−(5:10)	−(11:04)	−(2:25)	−(9:14)	0:21	− (7:12)
April	4:16	− (4:11)	6:56	− (2:07)	9:30	− (0:15)	11:57	1:19
May	14:51	2:50	16:53	3:31	18:41	3:44	20:14	3:30
June	21:57	2:25	22:47	1:15	23:17	− (0:09)	23:27	− (1:40)
July	23:10	− (3:33)	22:34	− (4:48)	21:39	− (5:45)	20:25	− (6:19)
August	18:12	− (6:17)	16:21	− (5:40)	14:17	− (4:35)	12:02	− (3:04)
September	8:33	− (0:15)	5:58	2:03	3:19	4:29	0:36	6:58
October	− (2:54)	10:02	− (5:36)	12:11	− (8:15)	13:59	−(10:48)	15:20
November	−(14:12)	16:20	−(16:22)	16:16	−(18:18)	15:29	−(19:59)	14:02
December	−(21:41)	11:14	−(22:38)	8:26	−(23:14)	5:13	−(23:27)	1:47

and less advanced than on any meridian further east. The difference amounts to four minutes of time for each degree difference in longitude. Thus at 90 deg west longitude, LCT is less advanced than GCT by six hours.

Time as measured by the apparent diurnal motion of the sun is called *Apparent Solar Time* or *Solar Time*. Whereas a *civil day* is precisely 24 hours, a *solar day* is slightly different due to irregularities of the earth's rotation, obliquity of the earth's orbit and other factors.

The difference between Local Solar Time LST and Local Civil Time LCT is called the *Equation of Time*. Thus

$$LST = LCT + \text{Eq. of Time} \tag{13.1}$$

Table 13.2 shows weekly values of the Equation of Time. For any one day, the Equation of Time may be considered constant.

At a given locality, watch time may differ from civil time. Clocks are usually set for the same reading throughout an entire zone covering about 15 deg of longitude. The United States is divided into four time zones. The time kept in each zone is the Local Civil Time of a selected meridian near the center of the zone. Such time is called *Standard Time*. The four standard meridians in the United States are at west longitudes of 75 deg (*Eastern Standard Time* EST), 90 deg (*Central Standard Time* CST), 105 deg (*Mountain Standard Time* MST), and 120 deg (*Pacific Standard Time* PST). In many localities, clocks are advanced one hour beyond Standard Time in summer. In the United States, such time is called *Daylight Saving Time*.

> **Example 13.1.** Determine the local solar time LST corresponding to 10:00 a.m. CST on February 8 for a location in the United States at 95 deg west longitude.
>
> SOLUTION: Central Standard Time CST is Local Civil Time LCT at 90 deg west longitude. At 95 deg west longitude, Local Civil Time is (5)(4) = 20 min less advanced than Local Civil Time at 90 deg west longitude. Thus, at 95 deg west longitude
>
> $$10:00 \text{ a.m. CST} = 9:40 \text{ a.m. LCT}$$
>
> By Table 13.2, Eq. of Time = −(14 min, 14 sec), say − 14 min. By Eq. (13.1),
>
> $$LST = 9:40 − 0:14 = 9:26 \text{ a.m.}$$

13.7. DERIVED SOLAR ANGLES

Besides the three basic angles, latitude, hour angle, and sun's declination, several other angles are useful in solar radiation calculations. Such angles include the sun's *zenith angle* ψ, *altitude angle* β, and *azimuth angle* γ. For a particular surface orientation, the sun's *incidence angle* θ, and *wall-solar azimuth angle* α may be defined. All of these additional angles may be expressed in terms of the three basic angles.

To an observer on the earth, the sun appears to move across the sky following the path of a circular arc from horizon to horizon. Figure 13.6 schematically shows one apparent solar path and defines the sun's zenith, altitude, and azimuth angles. Point P represents the position of the observer, Point O is the center of the earth, and I_N is a vector representing the sun's rays. The zenith angle ψ is the angle between the sun's rays and a line perpendicular to the horizontal plane at P (extension of $\overline{OP}$). The altitude angle β is the angle in a vertical plane between the sun's rays and the projection of the sun's rays on the horizontal plane. It follows that $\beta + \psi = \pi/2$. The azimuth angle γ is the angle in the horizontal plane measured from north to the horizontal projection of the sun's rays.

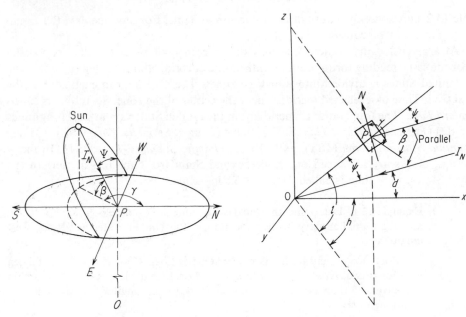

Figure 13.6 Definition of sun's zenith, altitude, and azimuth angles.

Figure 13.7 Relation of a point on the earth's surface to sun's rays.

Figure 13.7 shows a coordinate system with the z axis coincident with the earth's axis. The xy plane coincides with the earth's equatorial plane. The vector I_N representing the sun's rays lies in the xz plane (coinciding with a line drawn from the center of the sun to the center of the earth). The line $\overline{PN}$ pointing north from Point P is perpendicular to $\overline{OP}$ and lies in the plane containing $\overline{OP}$ and the z axis.

In Fig. 13.7, let a_1, b_1, and c_1 be the direction cosines of $\overline{OP}$ with respect to the x, y, and z axes. Also let a_2, b_2, and c_2 be the corresponding direction cosines of I_N. Thus,

$$a_1 = \cos l \cos h, \qquad b_1 = \cos l \sin h, \qquad c_1 = \sin l,$$
$$a_2 = \cos d, \qquad\qquad b_2 = 0, \qquad\qquad\qquad c_2 = \sin d$$

The sun's zenith angle ψ is the angle between $\overline{OP}$ and I_N. By a common equation from analytic geometry, we have

$$\cos \psi = a_1 a_2 + b_1 b_2 + c_1 c_2$$

Thus

$$\cos \psi = \cos l \cos h \cos d + \sin l \sin d \tag{13.2}$$

Since $\beta = \pi/2 - \psi$, we may write

$$\sin \beta = \cos l \cos h \cos d + \sin l \sin d \tag{13.3}$$

By similar methods, we may show that the sun's azimuth γ in Fig. 13.7 is given by the relation

$$\cos \gamma = \sec \beta (\cos l \sin d - \cos d \sin l \cos h) \tag{13.4}$$

By Eqs. (13.3) and (13.4), and with appropriate trigonometric identities, we may obtain the relation

$$\sin \gamma = \sec \beta \cos d \sin h \tag{13.5}$$

Figure 13.8 shows the lines of Fig. 13.7 for the case of solar noon. At solar noon, $h = 0$, and $\gamma = \pi$ if $l > d$, and $\gamma = 0$ if $l < d$. For the case of $l = d$, γ is undefined for $h = 0$. From Fig. 13.8, we may deduce that

$$\beta_{\text{noon}} = \frac{\pi}{2} - |(l - d)| \tag{13.6}$$

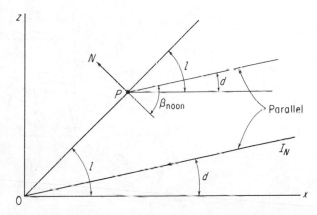

Figure 13.8 Relation of a point on the earth's surface to noon sun's rays.

where $|(l - d)|$ is the absolute value of $(l - d)$. Equation (13.6) allows rapid determination of the daily maximum altitude of the sun for a given location.

Equations (13.2)–(13.4) allow calculation of the sun's zenith, altitude, and azimuth angles if the declination, hour angle, and latitude are known. In applying these equations, attention must be given to correct signs for the latitude and declination angles. If north latitudes are considered positive and south latitudes negative, the sun's declination will be positive for the summer period between the vernal equinox and autumnal equinox (March 22 to September 22 approximately) and negative at other times. The hour angle is measured on either side of solar noon. Thus h is limited to values between zero and π. If $h < \pi/2$, cos h is positive and if $h > \pi/2$, cos h is negative.

Calculations for the sun's zenith angle by Eq. (13.2) and for the altitude angle by Eq. (13.3) are straightforward. In applying Eq. (13.4), no difficulty results if we consider that the azimuth is measured clockwise from north for hour angles before noon and counterclockwise from north for hour angles after noon. Thus, the azimuth is limited to values between zero and π. In Eq. (13.4), if cos γ is positive, γ is less than $\pi/2$, and if cos γ is negative, γ is greater than $\pi/2$.

Extensive tables showing calculated values of the sun's altitude and azimuth,* and charts giving less exact solutions† are available. Table A.7 of the Appendix shows values of the sun's altitude and azimuth for each daylight hour for the first day of each month for north latitudes of 30, 36, 42, and 48 deg.

In calculations involving other than horizontal surfaces, it may be convenient to express the sun's position relative to the surface in terms of the *incidence angle θ*. For vertical surfaces, use of the *wall-solar azimuth α* may also be helpful. Figure 13.9

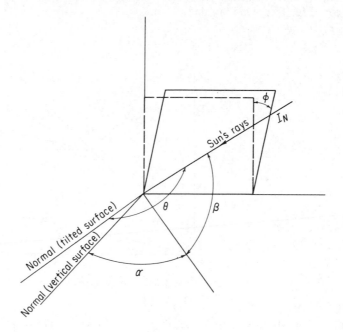

Figure 13.9 Relation of sun's rays to a tilted surface.

*Tables of Computed Altitude and Azimuth, Vols. 1–9, U. S. Hydrographic Office, Publication No. 214 (Washington, D. C.: Government Printing Office, 1940).

†W. J. Wilson and J. M. Van Swaay, "Where is the Sun?" *Heating and Ventilating*, Vol. 39, May 1942, pp. 41–46; June 1942, pp. 59–61.

shows a surface tilted by an angle ϕ from the vertical position. The sun's angle of incidence θ is the angle between the sun's rays and the normal to the tilted surface. It is associated with a definite surface position. The wall-solar azimuth α is the angle measured in a horizontal plane between the normal to the *vertical* surface and the horizontal projection of the sun's rays. Thus, α is associated with a definite vertical-wall position and may be found from the sun's azimuth γ.

In Fig. 13.9, for the tilted surface we may derive

$$\cos \theta = \cos \beta \cos \alpha \cos \phi + \sin \beta \sin \phi \qquad (13.7)$$

If the surface is vertical ($\phi = 0$), then

$$\cos \theta = \cos \beta \cos \alpha \qquad (13.8)$$

If the surface is horizontal ($\phi = \pi/2$), then

$$\cos \theta = \sin \beta = \cos \psi \qquad (13.9)$$

Thus for a horizontal surface, the incidence angle is equal to the zenith angle.

Example 13.2. Calculate the sun's altitude and azimuth angles at 7:30 a.m. solar time on August 1 for a location at 40 degrees north latitude.

SOLUTION: From the given data, $l = 40°$, $h = 67° 30'$. By Table 13.2, $d = 18° 12'$. By Eq. (13.3),

$$\sin \beta = (0.7660)(0.3827)(0.9500) + (0.6428)(0.3123) = 0.4792$$

and $\beta = 28° 38'$. By Eq. (13.4),

$$\cos \gamma = (1.1393)[(0.7660)(0.3123) - (0.9500)(0.6428)(0.3827)] = 0.0063$$

and $\gamma = 89° 38'$ east of north.

Example 13.3. Determine the solar time and azimuth angle for sunrise in Example 13.2.

SOLUTION: At sunrise (or sunset), $\beta = 0$. If we let H be the hour angle for sunrise, by Eq. (13.3) we have

$$\cos l \cos H \cos d + \sin l \sin d = 0$$

and

$$\cos H = -\tan l \tan d = -(0.8391)(0.3288) = -0.2759$$

Thus, $H = 106° 01'$, and sunrise occurs 7.07 hrs prior to solar noon or at 4:56 a.m. solar time. Sunset occurs at 7:04 p.m. solar time. By Eq. (13.4),

$$\cos \gamma = (1.00)[(0.7660)(0.3123) - (0.9500)(0.6428)(-0.2759)] = 0.4077$$

and $\gamma = 65° 56'$ east of north.

Example 13.4. Calculate the sun's incidence angle for a south-facing surface tilted back from the vertical position by 30 deg at 3:00 p.m. solar time on June 8 for a location at 36 deg north latitude.

SOLUTION: From the given data, $l = 36°$, $h = 45°$. By Table 13.2, $d = 22° 47'$. By the procedures identical to those in Example 13.2, we find $\beta = 49°02'$

and $\gamma = 96°08'$ west of north. By the given data, $\phi = 30°$. Also for a south-facing vertical surface, $\alpha = 180 - \gamma = 83°52'$. By Eq. (13.7),

$$\cos \theta = (0.6556)(0.1068)(0.8660) + (0.7551)(0.5000) = 0.4382$$

and $\theta = 64°01'$.

13.8. SOLAR RADIATION INTENSITY AT OUTER LIMIT OF ATMOSPHERE

When the earth is at its mean distance from the sun, the solar radiation intensity incident upon a surface normal to the sun's rays and at the outer limit of the atmosphere is known as the *solar constant*. Expressed in the *International Scale of 1956*, the solar constant has been determined by Johnson* to be 444.7 Btu per (hr) (sq ft) with a probable error of ± 2.0 per cent.

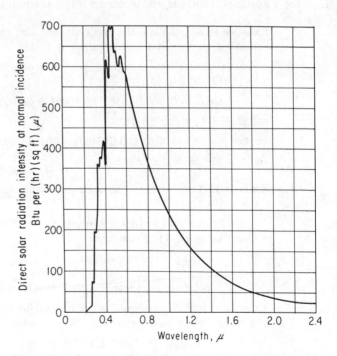

Figure 13.10 Spectral distribution of solar radiation incident upon a surface normal to the sun's rays at the outer limit of the atmosphere (corrected to mean solar distance). (Adapted by permission from F. S. Johnson, "The Solar Constant," *Journal of Meteorology*, Vol. 11, December 1954, p. 434.)

*F. S. Johnson, "The Solar Constant," *Journal of Meteorology*, Vol. 11, December 1954, pp. 431–439.

Figure 13.10 shows spectral distribution of solar radiation intensity at the outer limit of the atmosphere. Ultraviolet radiation includes the wavelength range of about 0.2 to 0.4 microns; visible radiation is contained between about 0.4 to 0.7 microns, and infrared radiation occurs at the higher wavelengths. Maximum intensity occurs within the visible range. The area under the entire curve is the solar constant.

The intensity of solar radiation $I_{N,o}$ normal to the sun's rays at the outer limit of the atmosphere varies with the earth-sun distance. Figure 13.11, calculated from data

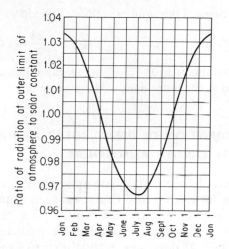

Figure 13.11 Ratio of solar radiation intensity at outer limit of atmosphere to the solar constant. (Reprinted by permission from *ASHAE Trans.*, Vol. 64, p. 58.)

in the *Smithsonian Physical Tables** shows the factor to be multiplied by the solar constant to give $I_{N,o}$.

An interesting problem is to calculate the solar radiation $Q_{H,o}$, Btu per (sq ft) (day) incident upon a horizontal surface at the outer limit of the atmosphere over a complete day. By Eqs. (13.3) and (13.9), we may write

$$dQ_{H,o} = I_{H,o}\,dt = \frac{I_{N,o}(\cos l \cos h \cos d + \sin l \sin d)\,dh}{\omega}$$

where dh is the change in hour angle in radians during time dt hr, and ω is the earth's angular velocity, $\pi/12$ radians per hr.

For one day, $I_{N,o}$ and d, as well as l, may be considered as constants. Since hour angles are symmetrical with respect to solar noon, we have

$$Q_{H,o} = \frac{24\,I_{N,o}}{\pi}\left[\cos l \cos d \int_0^H \cos h\,dh + \sin l \sin d \int_0^H dh\right]$$

Smithsonian Physical Tables, 6th ed. (Washington, D.C.: Smithsonian Institution, 1914) Table 181.

where H is the hour angle of sunrise and sunset. Integration gives

$$Q_{H,o} = \frac{24}{\pi} I_{N,o} \sin l \sin d \, (H - \tan H) \tag{13.10}$$

where H (radians) is found from Eq. (13.3) for $\beta = 0$. For a location on the equator ($l = 0$), we have

$$Q_{H,o} = \frac{24}{\pi} I_{N,o} \cos d \tag{13.11}$$

For the times of the equinoxes ($d = 0$), we have

$$Q_{H,o} = \frac{24}{\pi} I_{N,o} \cos l \tag{13.12}$$

Equations (13.10)–(13.12) allow rapid calculation of the daily total solar energy delivered at the outer limit of the atmosphere.

13.9. DEPLETION OF DIRECT SOLAR RADIATION BY EARTH'S ATMOSPHERE

The effects of the earth's atmosphere upon solar radiation have been studied by scientists for many years. This research has shown that when solar radiation passes through the atmosphere, part of it may be intercepted by constituents such as dry air molecules, water molecules, and dust particles, resulting in a scattering of radiation in practically all directions. Secondly, part of the radiation may be absorbed, particularly by ozone in the upper atmosphere and by water vapor nearer the earth's surface. The remaining portion of the original direct radiation quantity may reach the earth's surface unchanged in wavelength.

Some of the radiation intercepted by the atmosphere and turned aside from the direct beam reaches the earth's surface. This radiation of diffuse nature comes from the entire sky vault. Thus a surface on the earth receives solar energy of two forms—direct radiation and diffuse radiation.

The depletion of solar radiation by the atmosphere is large even during clear days, while with heavy cloudiness almost complete extinction may occur. Most environmental control problems at the earth's surface occur during clear days when the heating effect of the sun's rays is a maximum. Fortunately, the results of depletion of solar radiation by a cloudless atmosphere are rather well established.

To understand how a clear atmosphere depletes solar radiation, we must give attention to the composition and structure of the atmosphere and consider the spectral nature of solar radiation. Experimental observations have shown that, for monochromatic radiation, the quantity of radiation depleted by absorption and scattering increases arithmetically with the intensity of the radiation and geometrically with the quantity of material passed through by the sun's rays. Thus

$$I_{N,\lambda} = \tau_\lambda^m I_{No,\lambda} \tag{13.13}$$

where $I_{N,\lambda}$ is monochromatic intensity normal to the sun's rays after passage through m units of material, $I_{No,\lambda}$ is the initial monochromatic intensity, and τ_λ is the monochromatic transmissivity for unit depth of material.

In solar radiation calculations, unit depth of the atmosphere is taken as the depth when the sun is at the zenith. The *air mass m* is the ratio of the length of path of the sun's rays through the atmosphere to the length of path if the sun were at the zenith. Except for very low solar altitude angles, the air mass is equal to the cosecant of the altitude angle.

Moon* correlated the work of several investigators and calculated overall trans-

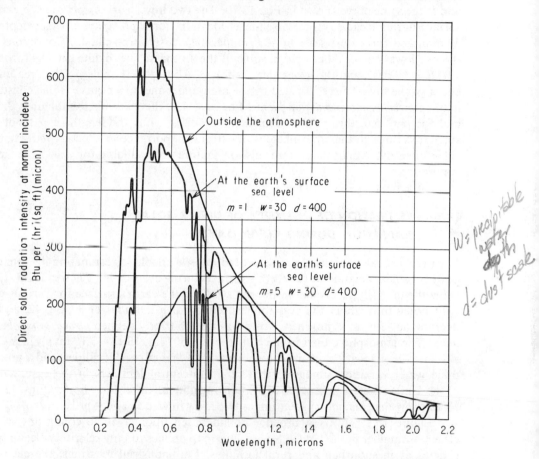

Figure 13.12 Spectral distribution of direct solar radiation intensity at normal incidence for the upper limit of the atmosphere and at the earth's surface during clear days. (Adapted by permission from *ASHAE Trans.*, Vol. 64, p. 50.)

*P. Moon, "Proposed Standard Radiation Curves for Engineering Use," *Journal of the Franklin Institute*, Vol. 230, November 1940, pp. 583–617.

missivities for monochromatic radiation including the effects of scattering by dry air molecules, water vapor molecules, and dust particles, and the effects of absorption by ozone and water vapor. At any single wavelength, the overall transmissivity multiplied by the intensity at the outer limit of the atmosphere gives the intensity at the earth's surface. Such calculations carried out for the entire solar spectrum allow energy distribution curves to be drawn for a location at the earth's surface or for any point within the atmosphere.

Figure 13.12 shows spectral distribution curves for direct solar radiation for three sets of conditions. The upper curve applies at the outer limit of the atmosphere and is the same curve shown in Fig. 13.10. The two lower curves apply to the earth's surface during clear days, for a sea-level location, for a precipitable water depth of 30 mm, and for a dust scale of 400 (moderately dusty atmosphere). Conditions for the two lower curves differ only in value of the air mass. The middle curve is for $m = 1.0$ ($\beta = 90$ deg), and the lower curve is for $m = 5.0$ ($\beta = 11.5$ deg). The area under either of the lower curves divided by the area under the upper curve (solar constant) is the *atmospheric transmission factor*. For $m = 1.0$, the transmission factor is 0.633, and for $m = 5.0$, the transmission factor is 0.276. Thus the length of path of the sun's rays through the atmosphere is of extreme importance in affecting reduction of solar intensity. Atmospheric transmission factors are available for a wide range of conditions.*

13.10. ESTIMATION OF INTENSITY OF DIRECT SOLAR RADIATION DURING CLEAR DAYS

For practical calculations, a simple procedure is available for estimating the intensity of direct solar radiation at the earth's surface during clear days.† Figures 13.13–13.16 show the intensity I_N normal to the sun's rays at the earth's surface for various daylight hours throughout the year for respective north latitudes of 30, 36, 42, and 48 degrees, and for an atmosphere which has an arbitrarily defined *clearness number of unity*. The atmosphere used in the construction of Figs. 13.13–13.16 involved a clear sky, a sea-level location, a dust content similar to that of rural localities, and a precipitable water variation similar to that of north-central latitudes in the eastern part of the United States. Except for early morning and late afternoon hours, Figs. 13.13–13.16 show that maximum solar intensity occurs in March and April.

Figure 13.17 shows a broad estimate based upon average conditions of the clearness number in the United States. This map applies to only relatively clean atmospheres as in suburban and rural localities. For industrial locations, the clearness number may be 15 per cent or more below that shown.

Estimates of direct solar radiation incident upon a surface normal to the sun's rays for average clear days in the United States may be made by multiplying values read from Figs. 13.13–13.16 by clearness numbers read from Fig. 13.17. This procedure

---

*J. L. Threlkeld and R. C. Jordan, "Direct Solar Radiation Available on Clear Days," *ASHAE Trans.*, Vol. 64, 1958, pp. 45–56.

†*ASHAE Trans.*, Vol. 64, pp. 62–67.

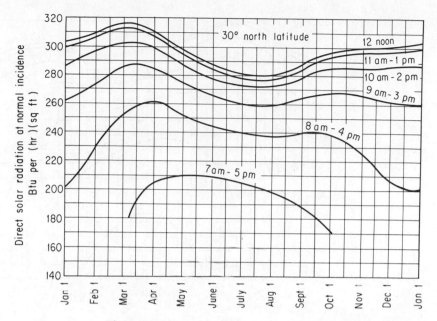

Figure 13.13 Intensity of direct solar radiation normal to the sun's rays at 30 degrees north latitude for an atmospheric clearness number of unity. (Reprinted by permission from *ASHAE Trans.*, Vol. 64, p. 64.)

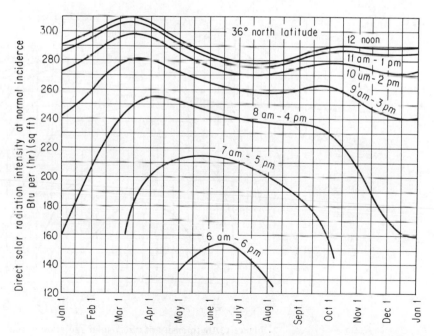

Figure 13.14 Intensity of direct solar radiation normal to the sun's rays at 36 degrees north latitude for an atmospheric clearness number of unity. (Reprinted by permission from *ASHAE Trans.*, Vol. 64, p. 64.)

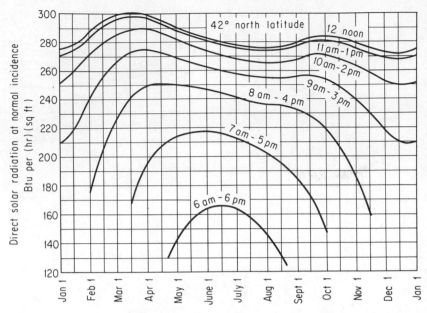

Figure 13.15 Intensity of direct solar radiation normal to the sun's rays at 42 degrees north latitude for an atmospheric clearness number of unity. (Reprinted by permission from *ASHAE Trans.*, Vol. 64, p. 64.)

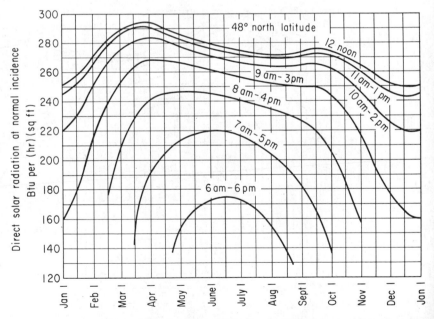

Figure 13.16 Intensity of direct solar radiation normal to the sun's rays at 48 degrees north latitude for an atmospheric clearness number of unity. (Reprinted by permission from *ASHAE Trans.*, Vol. 64, p. 64.)

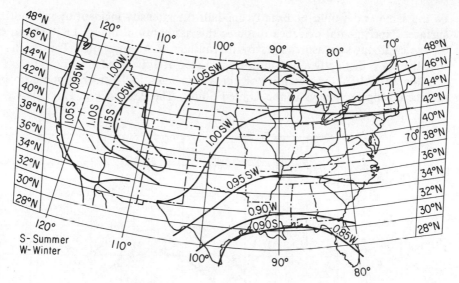

Figure 13.17 Estimated atmospheric clearness numbers in the United States for non-industrial localities. (Reprinted by permission from *ASHAE Trans.*, Vol. 64, p. 67.)

establishes an I_N value at the earth's surface suitable for many types of design problems.

This author* has experimentally studied the clearness number for clear days at Minneapolis and found good agreement between the mean value and the value of 1.02 given by Fig. 13.17. There is a need for experimental studies on the clearness number in other localities, particularly for industrial regions.

13.11. INCIDENCE OF DIFFUSE SKY RADIATION DURING CLEAR DAYS

Because of scattering of direct solar radiation by atmospheric constituents, solar radiation of a diffuse nature reaches the ground from the sky vault. This radiation is typically of rather short wavelength since short-wavelength radiation is scattered more by the atmosphere. Although diffuse solar radiation on clear days is usually small compared to direct radiation, it cannot be ignored in engineering calculations. During extremely cloudy days, only diffuse solar radiation may reach the ground.

Because of its non-directional nature, diffuse solar radiation is more difficult to analyze than direct solar radiation; consequently, less is known about it. A common belief held by many is that the sky is a uniform radiator of diffuse radiation. If so, intensity of diffuse radiation incident upon vertical surfaces of all orientations would

*J. L. Threlkeld, "Solar Irradiation of Surfaces on Clear Days," *ASHRAE Trans.*, Vol. 69, 1963, p. 27.

be the same and would be exactly one-half the intensity incident upon a horizontal surface. Experimental evidence disputes this assumption. Figure 13.18 shows results of one clear day's measurements by the author at Minneapolis. The results show that the sky is distinctly a non-uniform radiator of diffuse radiation.

Figure 13.19 shows intensity of diffuse sky radiation incident upon a horizontal surface during clear days in the United States. The data in Fig. 13.19, based upon

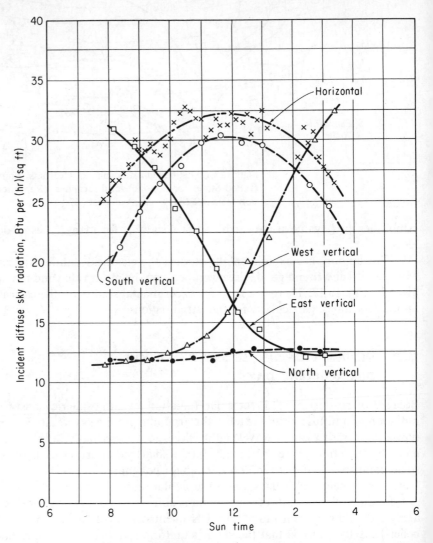

Figure 13.18 Variation of diffuse solar radiation from a clear sky incident upon various surfaces. (Reprinted by permission from *ASHAE Trans.*, Vol. 69, p. 29.)

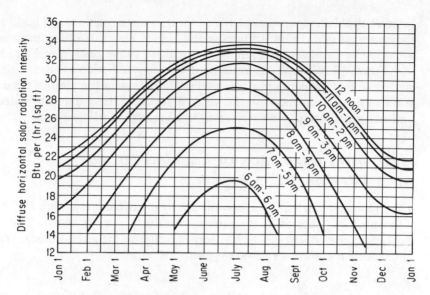

Figure 13.19 Intensity of diffuse sky radiation incident upon a horizontal surface.

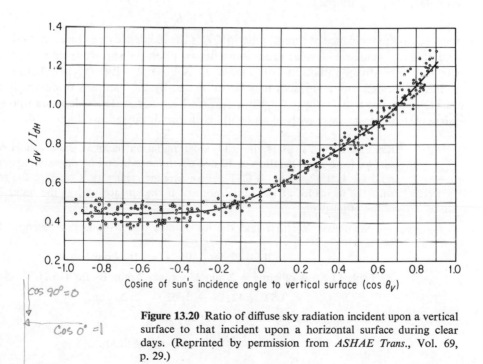

$\cos 90° = 0$

$\cos 0° = 1$

Figure 13.20 Ratio of diffuse sky radiation incident upon a vertical surface to that incident upon a horizontal surface during clear days. (Reprinted by permission from *ASHAE Trans.*, Vol. 69, p. 29.)

research by this author, were calculated for the same conditions as Fig. 13.15; however, they are applicable to all latitudes in the United States. It is also recommended that no adjustment for clearness number be made to values read from Fig. 13.19.

Figure 13.20 shows the ratio of diffuse sky radiation incident upon a vertical surface to that incident upon a horizontal surface as a function of the cosine of the sun's angle of incidence for the vertical surface. These data are based upon this author's experimental measurements for clear days at Minneapolis.

13.12. TOTAL IRRADIATION OF A SURFACE DURING CLEAR DAYS

The irradiation of a surface at any instant by direct solar radiation may be calculated if the intensity I_N normal to the sun's rays at the location is known. This requires that we find the component I_D which is perpendicular to the given surface. With the angle of incidence θ known, we have the general relation

$$I_D = I_N \cos \theta \qquad (13.14)$$

where $\cos \theta$ is given by Eq. (13.7) for a tilted surface, and by Eqs. (13.8) and (13.9) respectively for a vertical surface and for a horizontal surface.

The total solar radiation I incident upon a surface at any moment may be calculated by

$$I = I_D + I_d + I_R \qquad (13.15)$$

where I_D is the incidence of direct radiation, I_d is the incidence of diffuse sky radiation, and I_R is the incidence of solar radiation reflected upon the surface from surrounding surfaces. The direct radiation component I_D is given by Eq. (13.14). Diffuse sky radiation incident upon surfaces may be estimated by the methods discussed in Sec. 13.11. Solar radiation reflected upon a surface by other surfaces depends upon the particular circumstances involved and will be further discussed in Chapter 14.

> **Example 13.5.** Calculate the total solar radiation incident upon a horizontal surface for a rural location at 42 deg north latitude and 84 deg west longitude on June 1 at 12:00 noon CST for an average clear day. Compare the result with the incidence at the same moment if the surface was at the outer limit of the atmosphere.
>
> SOLUTION: We first find the local solar time. Similar to Example 13.1, we have
>
> $$LCT = 12:00 + 0:24 = 12:24 \text{ p.m.}$$
>
> By Table 13.2, Eq. of Time = 2 min 25 sec, say 2 min. By Eq. (13.1),
>
> $$LST = 12:24 + 0:02 = 12:26 \text{ p.m.}$$
>
> The hour angle corresponding to 12:26 p.m. is
>
> $$h = \frac{(26)(15)}{60} = 6.5 \text{ deg or } 6°30'$$

By Table 13.2, $d = 21°57'$. By Eq. (13.3),

$$\sin \beta = (0.7431)(0.9936)(0.9275) + (0.6691)(0.3738) = 0.9349$$

By Fig. 13.15, $I_N = 281$ Btu/(hr) (sq ft) for a clearness number $CN = 1$. By Fig. 13.17, $CN = 0.99$. Thus

$$I_N = (281)(0.99) = 278 \text{ Btu/(hr)(sq ft)}$$

By Eqs. (13.9) and (13.14),

$$I_{DH} = (278)(0.9349) = 260 \text{ Btu/(hr)(sq ft)}$$

By Fig 13.19, $I_{dH} = 33.0$ Btu/(hr) (sq ft). By Eq. (13.15) with $I_R = 0$,

$$I_H = 260 + 33 = 293 \text{ Btu/(hr) (sq ft)}$$

If the surface was located at the outer limit of the atmosphere, with Fig. 13.11 we would have

$$I_{N,o} = (444.7)(0.971) = 432 \text{ Btu/(hr) (sq ft)}$$

Thus, for outside the atmosphere,

$$I_{H,o} = (432)(0.9349) = 404 \text{ Btu/(hr) (sq ft)}$$

which is 37.9 per cent higher than I_H at the earth's surface.

13.13. HEAT BALANCE OF THE EARTH AND ATMOSPHERE

In preceding sections of this chapter, we studied incidence of solar radiation during clear days. In thermal environmental problems, clear days are usually of primary concern.

The availability of solar energy at the earth's surface is highly variable from day to day because of the occurrence of clouds. Example 13.5 shows that during clear days the atmosphere may significantly reduce the energy of the sun's rays. During days with heavy cloudiness, practically no solar energy may directly reach the earth's surface. However, atmospheric water vapor and clouds also form a barrier to loss of long-wavelength radiation from the earth's surface.

A topic of primary importance in meteorology is the heat balance of the earth and its atmosphere. If we consider the earth as a thermodynamic system, the external environment is outer space. Thus radiation is the only mode of heat transfer that can maintain the overall heat balance. Energy exchanges between the earth and the atmosphere are complicated and involve processes of convection as well as radiation.

We will now consider the average effects of solar radiation for all days, taking the average cloudiness for the earth as about 52 per cent. For this condition, Byers* has estimated that of the total solar energy incident at the top of the atmosphere, 33 per cent is reflected from the tops of clouds back to outer space, 9 per cent is lost to outer space by atmospheric scattering, 15 per cent is absorbed by atmospheric

*H. R. Byers, *General Meteorology* (New York: McGraw-Hill Book Company, 1944) p. 33.

constituents (principally water vapor), 27 per cent reaches the earth's surface as direct radiation, and 16 per cent reaches the earth's surface as diffuse radiation. Thus 43 per cent of the energy incident at the outer limit of the atmosphere reaches the earth.

The surface of the earth continually emits long-wave (infrared) radiation. Maximum intensity of emission occurs at a wavelength of about 10 microns. Some of this radiation escapes directly to outer space, but the majority is absorbed by atmospheric water vapor. In the band of wavelengths from about 8.5 to 11 microns, atmospheric water vapor is largely transparent to terrestrial radiation, but at other wavelengths it is practically opaque.

Figure 13.21, prepared from Byers'* data, shows the overall heat balance for the earth and its atmosphere. The left side shows the disposal of 100 energy units of incoming solar radiation, as discussed before. The right side shows the energy exchanges between the earth and its atmosphere. The earth's surface emits 131 units of long-wavelength radiation. Of these, 120 units are absorbed by atmospheric water vapor and 11 units escape directly to outer space. Evaporation of water at the earth's surface delivers 23 units to the atmosphere, where this energy is released to clouds by

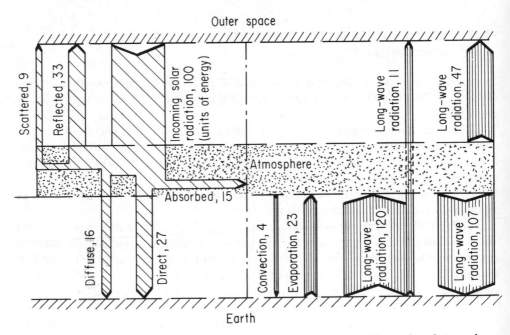

Figure 13.21 Heat balance of the earth and atmosphere.

*Byers, *General Meteorology*, p. 33.

condensation of the vapor. The atmosphere thus receives 158 total units of energy. The atmosphere disposes of this energy by radiating 47 units to outer space, radiating 107 units back to the earth, and delivering 4 units to the earth by convection.

We should fully appreciate the part played by atmospheric water vapor in the earth's heat balance. Figure 13.21 shows that only 15 per cent of the incoming solar radiation is absorbed by the atmosphere, but about 92 per cent of the earth's radiation is absorbed. Thus the atmosphere, and principally the water vapor constituent, exerts a greenhouse effect on the earth, allowing the earth's surface to be maintained at a much higher and more uniform temperature than would otherwise be possible.

Localities over the earth's surface differ widely in their zonal heat balances, primarily because of non-uniform distribution of solar radiation. At higher latitudes, the angle of incidence of the sun with respect to a horizontal surface is much greater than it is at localities closer to the equator. Byers* has concluded that over the half of the earth's surface between latitudes 30 deg N and 30 deg S the earth gains heat by radiation, while between 30 deg and the poles there is a net loss of heat by radiation. Thus, without atmospheric circulation, the tropics would become progressively hotter, while higher-latitude areas would continually become cooler. Atmospheric circulation or winds provide for transport of heat poleward from the equator.

Atmospheric circulation results from actions similar to those of a simple heat engine. Heat is applied by equatorial regions. Heat is rejected in polar regions. The work produced by the atmospheric engine appears in the kinetic energy of the winds. This work is dissipated by friction.

Besides the winds, there are other controls which influence the climate of a locality. Ocean currents and mountain ranges have far-reaching effects on many local climates.

13.14. STANDARD PYRHELIOMETERS AND SCALES
OF SOLAR RADIATION

Experimental determination of the energy transferred to a surface by solar radiation requires instruments which will measure the heating effect of direct solar radiation and diffuse solar radiation. Such instruments are called *pyrheliometers*. There are two general classes of pyrheliometers: one type allows separate measurement of direct radiation normal to the sun's rays while the other type measures total radiation. A total-radiation type of pyrheliometer may be used for measuring diffuse radiation alone by shading the sensing element from the sun's direct rays.

For almost a century, various solar scientists have attempted to perfect an instrument for measuring the thermal energy of the sun's rays. Just prior to 1900,

*Byers, *General Meteorology*, p. 215.

Angstrom* in Sweden developed the first reliable pyrheliometer. This instrument is still widely used today, particularly in Europe, practically unchanged from Angstrom's original design. The Angstrom pyrheliometer is based on the principle of electrical compensation. Solar energy is absorbed by a thin blackened metallic strip at the base of a cylindrical tube. A similar strip, shaded from the sun's rays, is heated electrically such that, at the time of a reading, both strips are at the same temperature. Response of the Angstrom pyrheliometer defines the *Angstrom Scale of Solar Radiation*.

Most development work in the United States on normal-incidence-type pyrheliometers has been done by the Smithsonian Institution. In its early work, a mercury-type pyrheliometer was used. Response of this instrument defined the *Smithsonian Original Scale of Solar Radiation*. Later, the Smithsonian Institution adopted the first version of Abbot's† water-flow-type pyrheliometer. Response of this instrument defined the *Smithsonian Scale of 1913*. In 1932, the Smithsonian Institution adopted Abbot's‡ improved version of the water-flow pyrheliometer as its standard. Response of this instrument defined the *Smithsonian Scale of 1932*.

Abbot's 1913 pyrheliometer was basically a calorimeter. Solar energy trapped by a well-insulated, cylindrical black-body chamber was absorbed by water circulating through a coil. Measurement of temperature rise of the water and its flow rate made the calculation of the incident solar radiation possible. Abbot's 1932 pyrheliometer included modifications to the 1913 type. The device embodied two identical cylindrical chambers whose walls were cooled by water. The water flow was divided equally between the two chambers. While one chamber was open to entry of solar radiation, the other was closed and its receiving surface was heated electrically. The electrical input was varied such that the water left each chamber at the same temperature. The heat gained by the exposed tube was equated to the electrical input of the closed tube.

Over the years many direct comparisons have been made between the Angstrom and Smithsonian pyrheliometers. These comparisons have shown that the *Smithsonian Scale of 1913* is about 3.5 per cent higher than the *Angstrom Scale* and about 2.4 per cent higher than the *Smithsonian Scale of 1932*. In 1956 at the International Radiation Conference held at Davos in Switzerland, a new solar radiation scale called the *International Scale of 1956*§ was recommended for worldwide adoption. No new pyrheliometer was involved. The new scale was defined as the *Angstrom Scale plus 1.5 per cent* or the *Smithsonian Scale of 1913 minus 2.0 per cent*. The *International Scale* is considered to be within ± 1.0 per cent of a true absolute scale. Figure 13.22 shows the earlier scales of solar radiation as percentages of the International Scale.

When using pyrheliometers, close attention must be given to the scale against

---

*K. Angstrom, "The Absolute Determination of the Radiation of Heat with the Electric Compensation Pyrheliometer," *Astrophysical Journal*, Vol. 9, 1899, pp. 332–340.

†C. G. Abbot, F. E. Fowle, and L. B. Aldrich, "Improvements and Tests of Solar-Constant Methods and Apparatus," *Annals of the Astrophysical Observatory*, Vol. 3, 1913, pp. 39–72.

‡L. B. Aldrich and W. B. Hoover, "Pyrheliometry," *Annals of the Astrophysical Observatory*, Vol, 7, 1954, pp. 99–104.

§A. J. Drummond and H. W. Greer, "Fundamental Pyrheliometry," *The Sun at Work*, Vol. 3, June 1958, pp. 3–5, 11.

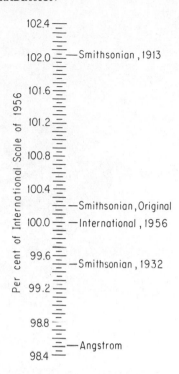

Figure 13.22 Comparison of the Smithsonian and Angstrom scales of solar radiation with the International Scale of 1956.

which the instruments were calibrated. Prior to 1957, practically all secondary-type pyrheliometers made in the United States were calibrated in terms of the Smithsonian Scale of 1913. Since July 1, 1957, most commercially available pyrheliometers have been calibrated in terms of the International Scale. Likewise, one must be careful in analyses using published radiation data. Prior to July 1, 1957, the U.S. Weather Bureau reported measurements using the Smithsonian Scale of 1913, but has subsequently used the International Scale.

13.15. SECONDARY PYRHELIOMETERS

The Angstrom pyrheliometer is the only standard type which is convenient for everyday use. The Smithsonian water-flow pyrheliometer requires auxiliary equipment and is not convenient for field measurements. For this reason, several secondary-type pyrheliometers have been developed.

The most widely accepted secondary pyrheliometer for measuring solar radiation at normal incidence is the *Smithsonian silver-disk pyrheliometer*.* This instrument,

*C. G. Abbot, "The Silver Disk Pyrheliometer," *Smithsonian Miscellaneous Collections*, Vol. 56, No. 19, 1911, pp. 1–10.

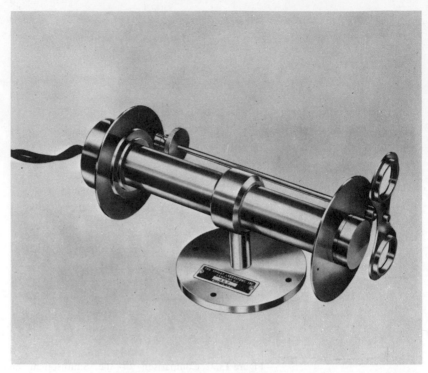

Figure 13.23 Thermoelectric type of pyrheliometer for measuring normally incident direct solar radiation. (Courtesy of The Eppley Laboratory, Inc.)

also designed by Dr. Abbot, consists of a blackened silver disk placed at the bottom of a brass tube through which the solar rays are admitted. The rate of increase of temperature of a pool of mercury under the disk is used as an indication of solar intensity. Figure 13.23 shows a thermoelectric version of the Smithsonian silver-disk pyrheliometer. The sensing element is an eight-junction, copper-constantan thermopile. This instrument is convenient to use and has received wide application.

Several pyrheliometers have been developed for measuring the combined incidence of direct and diffuse radiation. In most cases, a receiving surface is contained under a glass bulb permitting full vision of the sky. Figure 13.24 shows a thermoelectric type of pyrheliometer for measuring total radiation. The receiving surface includes a blackened inner ring and a white outer ring hermetically sealed in a soda-lime-glass bulb. A 50-junction thermopile measures the voltage induced by the temperature difference between the two rings. The instrument is widely used by the U.S. Weather Bureau. It is intended for use in a horizontal position but may also be placed in other positions, if suitably calibrated. This author* has described calibration and use of this type of pyrheliometer.

*J. L. Threlkeld, "Solar Irradiation of Surfaces on Clear Days," *ASHRAE Trans.*, Vol. 69, 1963, pp. 24–36.

Figure 13.24 Thermoelectric type of pyrheliometer for measuring total direct and diffuse solar radiation. (Courtesy of The Eppley Laboratory, Inc.)

A total-radiation type of pyrheliometer of the kind shown in Fig. 13.24 may be used for separate measurement of diffuse sky radiation by shading the receiving element from the direct rays of the sun and maintaining essentially full vision of the sky vault.

PROBLEMS

13.1. Calculate the sun's altitude and azimuth angles at 9:00 a.m. solar time on September 1 at 42 deg north latitude. Compare your results with values shown in Table A.7.

13.2. Determine the solar time and azimuth angle for sunrise at 50 deg north latitude on (a) June 15 and (b) December 15.

13.3. Determine the altitude angle of the sun at a time on July 15 at 45 deg north latitude when the horizontal projection of the sun's rays is normal to a west-facing vertical surface.

13.4. What is the maximum altitude angle of the sun for a location at (a) 45 deg north latitude, (b) 23.5 deg north latitude, and (c) the equator?

13.5. Derive Eq. (13.4).

13.6. Assuming existence of Eqs. (13.3) and (13.4), derive Eq. (13.5).

13.7. Prove that at the times of the equinoxes, the sun rises due east ($\gamma = 90$ deg) for all locations on the earth.

13.8. Derive Eq. (13.7).

13.9. For a location at 85 deg west longitude and 43 deg north latitude on July 8, determine (a) the incidence angle of the sun for a horizontal surface at 4:00 p.m. central daylight saving time and (b) the time of sunset in central daylight saving time.

13.10. Calculate the angle of incidence at 11:00 a.m. EST on July 15 for a location at 36 deg north latitude and 80 deg west longitude for (a) a horizontal surface, (b) a south-facing vertical surface, and (c) an inclined surface tilted 65 deg from the vertical and facing 30 deg south of due east.

13.11. Consider a flat plate located at the equator. The plate faces due north and is tilted toward north by 30 deg from the equatorial plane. Determine the number of hours that the *south-facing* side would be sunlit on June 22.

13.12. Assuming the earth's solar constant to be 444.7 Btu per (hr) (sq ft), calculate the equivalent surface temperature of the sun, if the sun is assumed to be a black-body radiator.

13.13. Calculate the solar constant for each of the other planets of the solar system.

13.14. Calculate the daily total solar radiation in Btu per (day) (sq ft) incident upon a horizontal surface at the outer limit of the atmosphere on March 1 at 48 deg north latitude.

13.15. A solar collector device is located at the outer limit of the atmosphere at 45 deg north latitude. The collector faces due south with its flat receiving surface placed in a vertical position. The collector is capable of absorbing 50 per cent of the incident solar radiation. Calculate the total energy absorbed by the collector per sq ft of surface during one full day on June 15.

13.16. Calculate the intensity of direct solar radiation incident upon a south-facing vertical surface at solar noon for a location having a clearness number of 1.0 at 42 deg north latitude for (a) June 22, and (b) December 22.

13.17. Based upon average clear-day conditions, calculate the incidence of total solar radiation at 12:00 noon CST on July 1 upon a flat roof of a building in (a) suburban Minneapolis, and (b) suburban Dallas.

13.18. Determine the total direct and diffuse sky radiation in Btu per (hr) (sq ft) incident upon a south-facing vertical surface on February 1 at 12 : 00 noon solar time at a location having a clearness number of 0.95 and at 42 deg north latitude.

SYMBOLS USED IN CHAPTER 13

CN	Atmospheric clearness number, dimensionless.
CST	Central Standard Time.
d	Sun's angle of declination.
EST	Eastern Standard Time.
GCT	Greenwich Civil Time.
H	Sun's hour angle at sunrise or sunset.
h	Sun's hour angle.
I	Incidence of total solar radiation upon a surface, Btu per (hr) (sq ft); I_H refers to horizontal surface; $I_{H,o}$ refers to a horizontal surface at the outer limit of the atmosphere.
I_N	Incidence of direct solar radiation upon a surface normal to sun's rays, Btu per (hr) (sq ft); $I_{N,o}$ refers to a surface at the outer limit of the atmosphere; $I_{N,\lambda}$ refers to monochromatic radiation.
I_D	Incidence of direct solar radiation upon a surface, Btu per (hr) (sq ft); I_{DH} refers to a horizontal surface.
I_d	Incidence of diffuse sky radiation upon a surface, Btu per (hr) (sq ft); I_{dH} refers to a horizontal surface; I_{dV} refers to a vertical surface.
I_R	Incidence of solar radiation upon a surface, Btu per (hr) (sq ft), which is reflected from surrounding surfaces.
LCT	Local Civil Time.
LST	Local Solar Time.
l	Latitude angle.
MST	Mountain Standard Time.
m	Air mass, ratio of length of path of sun's rays through atmosphere to length of path when sun is at the zenith position, dimensionless.
PST	Pacific Standard Time.
$Q_{H,o}$	Incidence of solar radiation upon a horizontal surface at outer limit of atmosphere, Btu per (day) (sq ft).
t	Elapsed time, hr.
α	Wall-solar azimuth angle for a vertical surface.
β	Sun's altitude angle.
γ	Sun's azimuth angle.
θ	Sun's angle of incidence for a surface.
τ_λ	Transmissivity for monochromatic radiation, dimensionless.
ϕ	Angle by which a surface is tilted from vertical position.
ψ	Sun's zenith angle.
ω	Earth's angular velocity, $\pi/12$ radians per hr.

14

HEAT TRANSMISSION IN BUILDINGS
AND SOLAR RADIATION EFFECTS
UPON STRUCTURES

14.1. INTRODUCTION

To control the thermal environment within a space, measures to counteract the heat and moisture gains (or losses) of the space are necessary. In this chapter we will study methods of heat transmission between the interior of a structure and the external thermal environment.

Heat transmission occurs through solid boundaries of a structure if a temperature difference exists between the internal and external environments. Heat and moisture transfers may also occur through exchange of air by leakage through cracks and other openings. Solar radiation may significantly affect energy transmission through opaque boundaries, but its effect is much greater when transmission is through diathermanous materials such as glass or some plastics.

Because of fluctuating changes in temperature and in incidence of solar radiation, the external thermal environment is constantly changing. Thus, steady-state heat transmission seldom occurs in a building structure. However, assumption of the steady state provides a convenient procedure for some calculations. To understand fully the nature of heat exchanges through boundaries of a building, we must consider their periodic nature.

14.2. STEADY-STATE HEAT TRANSMISSION THROUGH SOLID BOUNDARIES OF A STRUCTURE

In this section we will assume a steady-state condition for heat transmission through solid boundaries of a building. Figure 14.1 shows the steady-state problem. We assume a plane wall whose height and width are extremely large compared to its thickness x.

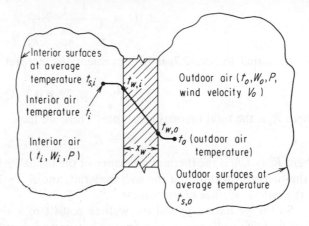

Figure 14.1 Steady-state temperature relationships between two thermal environments separated by a homogeneous wall.

Thus, heat transfer through the wall is one-dimensional. The interior thermal environment includes moist air at a dry-bulb temperature t_i, humidity ratio W_i, and barometric pressure P. The interior air is essentially still (velocity less than about 50 ft per min). The interior thermal environment also includes surfaces (interior walls, etc.) which are at a mean temperature $t_{s,i}$. Likewise, the external thermal environment includes moist air (t_o, W_o, P, wind velocity V_o) and surfaces (nearby buildings, sky, etc.), but *not solar radiation*.

At this time we will not be concerned with moisture transmission. Proceeding as in Sec. 2.7, for the rate of heat transfer q in Btu per (hr)(sq ft) we may write

$$q = h_i(t_i - t_{w,i}) = \frac{k_w}{x_w}(t_{w,i} - t_{w,o}) = h_o(t_{w,o} - t_o) \qquad (14.1)$$

where

$$h_i = h_{c,i} + h_{R,i}\left(\frac{t_{s,i} - t_{w,i}}{t_i - t_{w,i}}\right) \qquad (14.2)$$

$$h_o = h_{c,o} + h_{R,o}\left(\frac{t_{w,o} - t_{s,o}}{t_{w,o} - t_o}\right) \qquad (14.3)$$

In Eqs. (14.1)–(14.3), h_c represents a convection heat transfer coefficient, Btu per (hr)(sq ft)(F), h_R represents a radiation coefficient, Btu per (hr)(sq ft)(F), k_w is the wall thermal conductivity, Btu per (hr)(sq ft)(F per in.), and x_w is the wall thickness, in.

By definition of an overall heat transfer coefficient U, Btu per (hr)(sq ft)(F), we may write

$$q = U(t_i - t_o) \tag{14.4}$$

By Eqs. (14.1) and (14.4)

$$U = \frac{1}{\dfrac{1}{h_i} + \dfrac{x_w}{k_w} + \dfrac{1}{h_o}} \tag{14.5}$$

As stated in Sec. 2.7, it is convenient to interpret the overall coefficient U as

$$U = \frac{1}{R_t} = \frac{1}{\Sigma R} \tag{14.6}$$

where R_t is the total thermal resistance. Thus, we have

$$R_t = R_i + R_w + R_o$$

where $R_i = 1/h_i$ is the thermal resistance of the inside surface of the wall, $R_w = x_w/k_w$ is the thermal resistance of the wall material, and $R_o = 1/h_o$ is the thermal resistance of the outside surface of the wall.

So far we have assumed the wall to consist of a single homogeneous material. Most building walls are of a composite nature and may include layers of non-homogeneous materials (hollow block, etc.) and also cavities or air spaces. Figure 14.2 shows a schematic wall section of a hollow building block. We assume that one surface is at a higher temperature than the other surface. Heat transfer through the block would occur through a complex combination of conduction, convection, and radiation pro-

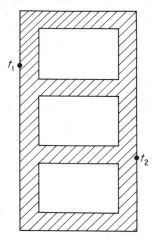

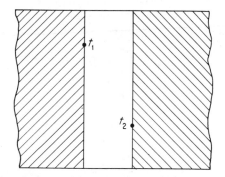

Figure 14.2 Schematic example of non-homogeneous wall construction.

Figure 14.3 Schematic illustration of a wall cavity or air space.

cesses. Practically, the problem is handled by the use of a *thermal conductance C*, Btu per (hr)(sq ft)(F), defined by the equation

$$q = C(t_1 - t_2) \tag{14.7}$$

where numerical values of C are determined experimentally for the particular material and for the necessary range of environmental temperature. It follows that the thermal resistance is equal to $1/C$.

Figure 14.3 shows two wall sections separated by a cavity or air space. Heat transfer across the cavity occurs by convection and radiation. Assuming the same convection coefficient for both surfaces, we have

$$q = \frac{h_c}{2}(t_1 - t_2) + h_R(t_1 - t_2) = \left(\frac{h_c}{2} + h_R\right)(t_1 - t_2) \tag{14.8}$$

If we approximate by assuming the surfaces of the air space as infinite parallel planes, we have

$$h_R = \frac{0.1713E}{(t_1 - t_2)}\left[\left(\frac{T_1}{100}\right)^4 - \left(\frac{T_2}{100}\right)^4\right] \tag{14.9}$$

where

$$E = \frac{1}{\frac{1}{\epsilon_1} + \frac{1}{\epsilon_2} - 1} \tag{14.10}$$

and ϵ_1 and ϵ_2 are the emissivities of the surfaces of the air space.

Practically, heat transmission across an air space is determined in the same way as for a non-homogeneous material (Eq. 14.7). Values of $C = h_c/2 + h_R$ are directly found from experiments.

The U factor of a composite wall may be calculated by the same procedure as was given for a wall of a single material. Equation (14.5) may be written in the general form

$$U = \frac{1}{\frac{1}{h_i} + \Sigma\frac{x_w}{k_w} + \Sigma\frac{1}{C} + \frac{1}{h_o}} = \frac{1}{\Sigma R} = \frac{1}{R_t} \tag{14.11}$$

For a known wall, the U factor may be calculated if the surface coefficients h_i and h_o, the required thermal conductivities (k_w values), and the required thermal conductances (C values) are available. Fortunately, a vast amount of experimental data have been published covering a wide variety of building materials. The *ASHRAE Handbook of Fundamentals** shows extensive tables for the various coefficients and also calculated U factors for many wall constructions. Table 14.1 shows experimentally determined values of the surface coefficients h_i and h_o. These values are generally applicable to most building wall problems. They apply specifically for the case where the ambient air and surrounding surfaces are at the same temperature ($h = h_c + h_R$). Table

ASHRAE Handbook of Fundamentals (New York: American Society of Heating, Refrigerating and Air Conditioning Engineers, 1967) pp. 429–453.

TABLE 14.1*†

SURFACE HEAT TRANSFER COEFFICIENTS h_i AND h_o, BTU PER (HR)(SQ FT)(F)

Description of Surface	Direction of Heat Flow	Surface Emissivity ϵ		
		0.90	0.20	0.05
Horizontal (still air)	Up	1.63	0.91	0.76
Horizontal (still air)	Down	1.08	0.37	0.22
Sloping, 45 deg (still air)	Up	1.60	0.88	0.73
Sloping, 45 deg (still air)	Down	1.32	0.60	0.45
Vertical (still air)	Horizontal	1.46	0.74	0.59
Any position (15 MPH wind)	any	6.00	. . .	. . .
Any position (7.5 MPH wind)	any	4.00	. . .	. . .

*Abstracted by permission from *ASHRAE Handbook of Fundamentals*, 1967, p. 429.

†For surfaces of stated emissivity facing virtual black-body surroundings at the same temperature as ambient air. Coefficients based on a surface temperature of 70 F and an air temperature of 60 F.

14.2 shows experimentally determined values of the thermal conductance C for a variety of plane air spaces. In both Tables 14.1 and 14.2, we should observe the significant influence of direction of heat flow and the effect of surface emissivities upon the coefficients.

Table 14.3 shows thermal conductivities and thermal conductances for a few selected building materials.

Example 14.1. Figure 14.4 shows a schematic wall consisting of 0.5 in. of cement plaster, 4 in. of hollow clay tile, a 4-in. air space, 8 in. of concrete block, and 4 in. of face brick. Assume an outside wind velocity of 15 MPH. Calculate the U factor of the wall.

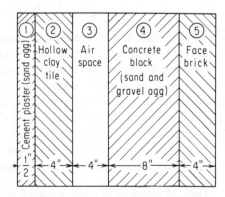

Figure 14.4 Schematic wall section for Example 14.1.

TABLE 14.2*

THERMAL CONDUCTANCE C FOR PLANE AIR SPACES

Position of Air Space	Direction of Heat Flow	Air Space Width in.	Air Space Mean Temp. F	Air Space Temp. Diff. F	C, Btu per (hr) (sq ft) (F), for Various Values of E† $E = 0.82$	$E = 0.50$	$E = 0.20$	$E = 0.05$
Horizontal	Up	0.75	50	10	1.15	0.86	0.59	0.45
		4	50	10	1.07	0.78	0.50	0.37
		0.75	90	10	1.32	0.96	0.61	0.44
		4	90	10	1.24	0.88	0.53	0.36
Horizontal	Down	0.75	50	...	0.98	0.69	0.42	0.28
		1.5	50	...	0.88	0.59	0.31	0.17
		4	50	...	0.81	0.52	0.25	0.11
		0.75	90	...	1.19	0.82	0.48	0.31
		1.5	90	...	1.07	0.71	0.36	0.19
		4	90	...	1.01	0.64	0.30	0.12
Sloping: 45 deg	Up	0.75	50	10	1.06	0.77	0.50	0.36
		4	50	10	1.04	0.74	0.47	0.33
		0.75	90	10	1.24	0.87	0.53	0.36
		4	90	10	1.21	0.85	0.51	0.33
Sloping: 45 deg	Down	0.75	50	10	0.98	0.69	0.42	0.28
		4	50	10	0.93	0.64	0.36	0.23
		0.75	90	10	1.19	0.82	0.48	0.31
		4	90	10	1.11	0.74	0.40	0.23
Vertical	Horizontal	0.75	50	10	0.99	0.70	0.42	0.29
		4	50	10	0.99	0.70	0.43	0.29
		0.75	90	10	1.19	0.82	0.48	0.31
		4	90	10	1.17	0.81	0.46	0.29

*Abstracted by permission from *ASHRAE Handbook of Fundamentals*, 1967, pp. 429–430.

†E given by Eq. (14.10).

SOLUTION: It is convenient to number the wall materials in some way, such as in Fig. 14.4. By Table 14.1,

$$h_i = 1.46 \text{ Btu/(hr)(sq ft)(F)} \qquad h_o = 6.0 \text{ Btu/(hr)(sq ft)(F)}$$

By Table 14.3,

$$k_{w,1} = 5.0 \text{ Btu/(hr)(sq ft)(F/in.)} \qquad C_2 = 0.90 \text{ Btu/(hr)(sq ft)(F)}$$
$$C_4 = 0.90 \text{ Btu/(hr)(sq ft)(F)} \qquad k_{w,5} = 9.0 \text{ Btu/(hr)(sq ft)(F/in.)}$$

TABLE 14.3*

PHYSICAL PROPERTIES OF SELECTED BUILDING MATERIALS

Material	Description	Specific Heat Btu per (lb)(F)	Density lb per cu ft	k_w Btu per (hr)(sq ft)(F per in.)	C Btu per (hr) (sq ft) (F)
Woods . . .	Plywood	......	34	0.80	......
	Maple, oak, and similar hard woods . .	0.57	45	1.10	......
	Fir, pine, and similar soft woods	0.65	32	0.80	......
Insulating materials . .	*Blanket and Batt* Cotton fiber	0.31	0.8–2.0	0.26	......
	Mineral wool, glass fiber	0.16	1.5–4.0	0.27	......
	Wood fiber	......	3.2–3.6	0.25	......
	Board and Slab Glass fiber	0.16	4.0–9.0	0.25	......
	Sheathing (impreg. or coated)	......	20.0	0.38	......
	Cellular glass	......	9.0	0.40	......
	Corkboard (without added binder) . . .	0.45	6.5–8.0	0.27	......
	Expanded polystyrene	......	1.9	0.26	......
Masonry materials . .	Concrete (sand, gravel agg.)	0.21	140	12.0	......
	Cement mortar or plaster	0.19	116	5.0	......
	Brick, common . . .	0.20	120	5.0	......
	Brick, face	0.20	130	9.0	......
	Hollow Clay Tile 1 cell deep—4 in. . .	......	......	......	0.90
	2 cells deep—8 in. . .	......	......	......	0.54
	3 cells deep—12 in. .	......	......	......	0.40
	Hollow Concrete Block (sand, gravel agg.) Three oval core, 4 in. .	......	......	......	1.40
	8 in. .	......	......	......	0.90
	12 in. .	......	......	......	0.78
Roofing materials . .	Asphalt shingles . . .	0.35	70	......	2.27
	Built-up roofing, ⅜ in.	......	70	......	3.00
	Wood shingles . . .	......	...	......	1.06
Miscellaneous	Metal lath, gypsum plaster, ¾ in. . . .	......	......	......	7.7
	Gypsum plaster, sand agg.	0.26	105	5.6	......
	Wood siding, bevel, ½ in. by 8 in..	......	......	......	1.23

*Density, thermal conductivity, and thermal conductance values abstracted by permission from *ASHRAE Handbook of Fundamentals*, 1967, pp. 431–434. Specific heat values are more approximate and were taken from various sources.

By Table 14.2, $C_3 = 0.99$ Btu/(hr)(sq ft)(F) assuming non-reflective surfaces ($E = 0.82$), a mean temperature of 50 F, and a temperature difference of 10 F. By Eq. (14.11),

$$R_t = \frac{1}{h_i} + \frac{x_1}{k_{w,1}} + \frac{1}{C_2} + \frac{1}{C_3} + \frac{1}{C_4} + \frac{x_5}{k_{w,5}} + \frac{1}{h_o}$$

$$= \frac{1}{1.46} + \frac{0.5}{5.0} + \frac{1}{0.90} + \frac{1}{0.99} + \frac{1}{0.90} + \frac{4}{9.0} + \frac{1}{6.0}$$

$$= 0.685 + 0.100 + 1.111 + 1.010 + 1.111 + 0.444 + 0.167$$

$$= 4.628 \text{ (hr)(sq ft)(F)/Btu}$$

and

$$U = \frac{1}{R_t} = \frac{1}{4.63} = 0.22 \text{ Btu/(hr)(sq ft)(F)}$$

Example 14.2. Calculate the U factor for Example 14.1 if the air space were filled with a batt of glass fiber insulation.

SOLUTION: All thermal resistances would remain the same except $1/C_3$. The total resistance for Example 14.1 less the air space is $4.63 - 1.01 = 3.62$. By Table 14.3, $k_{w,3} = 0.27$ Btu/(hr)(sq ft)(F/in.). Thus

$$R_t = 3.62 + \frac{x_3}{k_{w,3}} = 3.62 + \frac{4}{0.27} = 3.62 + 14.81$$

$$= 18.43 \text{ (hr)(sq ft)(F)/Btu}$$

and

$$U = \frac{1}{R_t} = \frac{1}{18.43} = 0.054 \text{ Btu/(hr)(sq ft)(F)}$$

which is 75 per cent less than the result for Example 14.1.

Example 14.3. Calculate the U factor for Example 14.1 if one of the surfaces of the air space were covered with aluminum foil ($\epsilon = 0.05$).

SOLUTION: One surface of the air space remains non-reflective ($\epsilon = 0.90$). By Eq. (14.10),

$$E = \frac{1}{\frac{1}{0.90} + \frac{1}{0.05} - 1} = 0.0497 \text{ or } 0.05$$

By Table 14.2 with $E = 0.05$, $C_3 = 0.29$ Btu/(hr)(sq ft)(F), if we still assume the same mean temperature and temperature difference. Thus

$$R_t = 3.62 + \frac{1}{C_3} = 3.62 + \frac{1}{0.29} = 3.62 + 3.45 = 7.07 \text{ (hr)(sq ft)(F)/Btu}$$

and

$$U = \frac{1}{R_t} = \frac{1}{7.07} = 0.14 \text{ Btu/(hr)(sq ft)(F)}$$

which is 36 per cent less than the result for Example 14.1.

Examples 14.1–14.3 show that the use of insulating materials may significantly reduce the overall coefficient U. Particularly interesting is the use of reflective foils in conjunction with an air space. Example 14.3 shows that the covering of one surface of the air space with aluminum foil gave an air-space resistance of 3.45 units which is equivalent to the resistance of a non-reflective air space plus about 0.7 in. of fibrous insulation. Table 14.2 shows that the thermal conductance of a vertical air space is approximately the same for widths varying from 0.75 to 4 in. In Example 14.1, the 4-in. air space could have been divided by thin sheets of aluminum foil into 5 spaces, each with a width of 0.75 in. or more. The insulating effect of such a combination is treated in Problem 14.3.

To be effective, aluminum foil insulations must be properly installed. Adequate separation of the foil sheets is a necessity.

Walls insulated with fibrous materials (mineral wool, glass fibers, wood fibers, etc.) which are permeable to movement of water vapor must be provided with suitable vapor barriers. Vapor barriers are necessary for refrigerated structures in all climates and for all insulated structures, including residences, in cold climates. A typical practice is to provide a single vapor barrier suitably placed on the *warm* side of the insulation. The vapor barrier should be located such that its temperature will be higher than the dew-point temperature of the air on the humid side of the wall. In cold-storage wall construction, multiple vapor barriers are often applied when the insulation is installed in multiple layers.

Often a membrane type of vapor barrier is used. Membrane barriers include aluminum foil sheets, laminates of aluminum foil and treated papers, coated felts and papers, and plastic films. Rigid sheets and coatings made of asphaltic, resinous, or polymeric materials may be used as vapor barriers.

Vapor barriers do not necessarily prevent all flow of water vapor. They should be of sufficient effectiveness so that *vapor-flow continuity* can be maintained and condensation of moisture in interior parts of walls can be minimized.

A thorough discussion of insulations, vapor-proofing of walls, and effects of moisture in building construction is given in the *ASHRAE Handbook of Fundamentals*.*

14.3. SOLAR RADIATION PROPERTIES OF SURFACES

Heat transmission through the boundaries of a building may be markedly influenced by the incidence of solar radiation upon external surfaces. As shown in Fig. 2.2 and discussed in Sec. 2.3, part of the incident radiation may be reflected, part absorbed, and the remainder transmitted through the material. In general, we have

$$\rho + \alpha + \tau = 1 \tag{14.12}$$

where ρ is the *reflectivity* of the surface, α the *absorptivity*, and τ the *transmissivity*.

ASHRAE Handbook of Fundamentals (New York: American Society of Heating, Refrigerating and Air Conditioning Engineers, 1967) pp. 241–282.

We will first concern ourselves with opaque materials ($\tau = 0$), for which

$$\rho + \alpha = 1 \tag{14.13}$$

We should fix in our minds that for a surface irradiated by the sun, the emissivity ϵ is not equal to the absorptivity α. Although Kirchhoff's law ($\epsilon = \alpha$) is true for long-wave (low-temperature) radiation exchange, it is invalid when a surface is sunlit. Table 2.2 shows absorptivities for various surfaces, both for low-temperature radiation

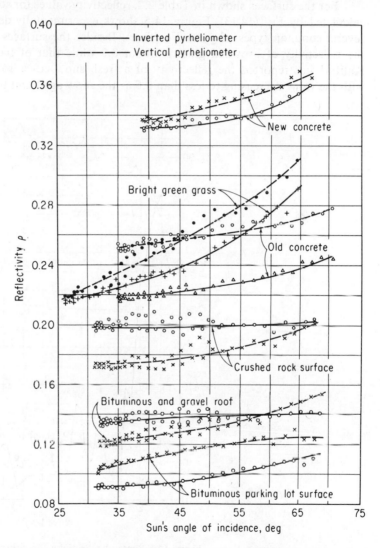

Figure 14.5 Solar reflectivity for various ground surfaces. (Reprinted by permission from *ASHRAE Trans.*, Vol. 69, p. 31.)

and for solar radiation. Whereas color of the surface is unimportant with low-temperature radiation, it is highly important with solar radiation. Table 2.2 shows that for the temperature range of 50 to 100 F, a black surface and a white surface have essentially the same absorptivity. For solar radiation, a glossy-white surface may have an absorptivity of about 0.3 while a dull-black surface may have an absorptivity as high as 0.98. Table 2.2 also shows that the absorptivity of metallic surfaces is generally much higher for solar radiation than for low-temperature radiation.

For the surfaces shown in Table 2.2, reflectivity values for solar radiation may be calculated by Eq. (14.13). Figure 14.5 shows experimentally determined values for several common types of ground surfaces. Besides the surfaces shown in Fig. 14.5, the reflectivity of browned grass is about 0.2 while that of bare soil is about 0.1. Kalitin* has reported the reflectivity of a fresh snow cover to be as high as 0.87, with the value decreasing to less than 0.5 as the snow aged and became dirty.

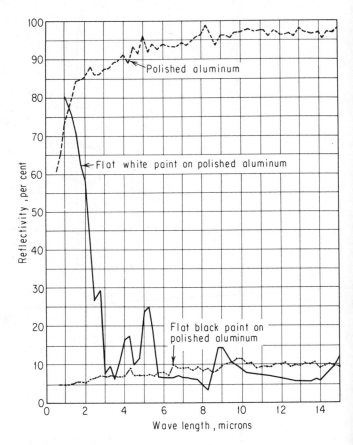

Figure 14.6 Spectral reflectivity for polished aluminum, flat white paint, and flat black paint.

*N. N. Kalitin, "The Measurements of the Albedo of a Snow Cover," *Monthly Weather Review*, Vol. 58, February 1930, p. 60.

As discussed in Sec. 2.3, reflection from a surface may be *specular* or *diffuse*. Figure 2.2 shows the case of specular reflection where the reflection angle θ'' equals the incidence angle θ and where the reflected beam, incident beam, and surface normal lie in the same plane. Specular reflection may occur from surfaces such as glass, water, and polished metals. Diffuse reflection occurs from irregular or roughened surfaces; the surfaces shown in Fig. 14.5 are essentially diffuse reflectors. Few real surfaces exhibit either pure specular or pure diffuse reflection characteristics. With the exception of glass, most building surfaces may be considered as diffuse reflectors.

Except for differentiating between low-temperature radiation and solar radiation, we have ignored the variation of ρ and α with wavelength of the incident radiation. Thus, the solar absorptivities shown in Table 2.2 and the reflectivities shown in Fig. 14.5 are average or total values applying to the entire solar spectrum. However, the reflectivity and absorptivity of some surfaces may vary with wavelength in an erratic manner. Figure 14.6 prepared from data given by Dunkle and Gier,[*] shows that the reflectivity of flat white paint and polished aluminum varies considerably in the solar spectrum ($\lambda = 0.3 - 2.3\mu$). However, the reflectivity of flat black paint is essentially constant.

14.4. SOLAR RADIATION PROPERTIES OF DIATHERMANOUS MATERIALS

A *diathermanous* material is one capable of transmitting thermal radiation. The most important example in building construction is glass. We will study in some detail the disposal of solar radiation incident upon glass, much of our discussion being based upon the treatise by Parmelee.[†]

Figure 14.7 shows the disposal of a quantity of monochromatic direct solar radiation I_λ incident upon a single sheet of glass of thickness L. Part of the incident radiation is reflected from the front surface and part is absorbed by the glass material. Because of successive internal reflections, the reflected, absorbed, and transmitted radiation quantities are given by the sums of infinite series. Let r be the fraction of each component reflected, and a be the fraction of each component available after absorption. The total monochromatic transmissivity τ_λ is given by

$$\tau_\lambda = (1 - r)^2a + r^2(1 - r)^2a^3 + r^4(1 - r)^2a^5 + \ldots$$

Since this is a convergent geometric series, we have

$$\tau_\lambda = \frac{(1 - r)^2a}{1 - r^2a^2} \tag{14.14}$$

In a similar way, we obtain the total monochromatic reflectivity ρ_λ as

$$\rho_\lambda = r + \frac{r(1 - r)^2a^2}{1 - r^2a^2} \tag{14.15}$$

[*]R. V. Dunkle and J. T. Gier, *Progress Report for the Year June* 27, 1952 *to June* 27, 1953, Contract No. DA-11-190-ENG-3. Series No. 62, Issue No. 1. (Berkeley: University of California Institute of Engineering Research, June 27, 1953) pp. 43, 45, 57.

[†]G. V. Parmelee, "Transmission of Solar Radiation through Flat Glass," *ASHVE Trans.*, Vol. 51, 1945, pp. 317–350.

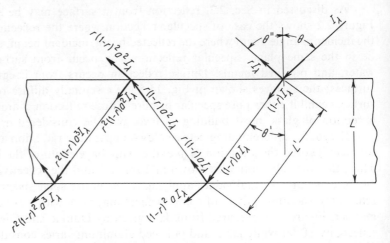

Figure 14.7 Multiple reflections of direct solar radiation by a single sheet of glass.

and since $\alpha_\lambda = 1 - \tau_\lambda - \rho_\lambda$, we have

$$\alpha_\lambda = 1 - r - \frac{(1-r)^2 a}{1-ra} \tag{14.16}$$

To evaluate Eqs. (14.14) to (14.16) we must know the quantities r and a. We will first consider the absorption coefficient a. We assume that the absorbed radiation is proportional to the intensity of incident radiation and to the length of path of the refracted beam. Thus

$$-dI_\lambda = KI_\lambda dL'$$

and

$$a = \frac{I_{\lambda,2}}{I_{\lambda,1}} = e^{-KL'} \tag{14.17}$$

where K is called the *extinction coefficient*. Table 14.4 shows values of K for several types of window glass. Since $L' = L/\cos\theta'$, by Eq. (2.9) we have

$$L' = \frac{L}{\sqrt{1 - \dfrac{\sin^2\theta}{n^2}}} \tag{14.18}$$

where L is the glass thickness, θ is the angle of incidence of the sun's rays, and n is the *index of refraction* for the glass.

The component reflectivity r in Eqs. (14.14)–(14.16) may be found from the Fresnel relations.* Natural or unpolarized light may be assumed to consist of two vibrating components—one vibrating in a plane normal to the plane of the glass and

*Joseph Valasek, *Introduction to Theoretical and Experimental Optics* (New York: John Wiley & Sons, Inc., 1949) pp. 199–201.

TABLE 14.4*

EXTINCTION COEFFICIENT K FOR VARIOUS TYPES OF GLASS

Type of Glass	K 1/in.
1. Double strength, A quality	0.194
2. Clear plate	0.174
3. Heat absorbing	3.30
4. Heat absorbing	6.89

*Abstracted by permission from *ASHVE Trans.*, Vol. 51, p. 332.

the other vibrating in a plane parallel to the plane of the glass. If the components are of equal intensity

$$r = \frac{1}{2}\left[\frac{\sin^2(\theta - \theta')}{\sin^2(\theta + \theta')} + \frac{\tan^2(\theta - \theta')}{\tan^2(\theta + \theta')}\right] \tag{14.19}$$

Figure 14.8 shows the solution of Eq. (14.19) for glass having an index of refraction 1.526.

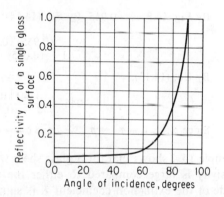

Figure 14.8 Reflection of direct solar radiation from a single surface of glass having an index of refraction of 1.526.

The use of two separated sheets of glass is common in buildings. Parmelec* has shown that for double glass

$$\tau_{\lambda 1,2} = \frac{\tau_{\lambda 1}\tau_{\lambda 2}}{1 - \rho_{\lambda 1}\rho_{\lambda 2}} \tag{14.20}$$

$$\rho_{\lambda 1,2} = \rho_{\lambda 1} + \frac{\tau_{\lambda 1}^2 \rho_{\lambda 2}}{1 - \rho_{\lambda 1}\rho_{\lambda 2}} \tag{14.21}$$

where the subscript $\lambda 1,2$ refers to the double-glass combination, $\lambda 1$ refers to the first

**ASHVE Trans.*, Vol. 51, pp. 330–331.

sheet of glass considered separately, and $\lambda2$ refers to the second sheet of glass considered separately.

In heat transmission calculations it is more useful to know the solar absorption in each glass sheet rather than the absorption for both sheets. We find

$$\alpha_{\lambda,1 \text{ of } 2} = \frac{[1 - (\rho_{\lambda1} + \tau_{\lambda1})][1 - \rho_{\lambda2}(\rho_{\lambda1} - \tau_{\lambda1})]}{1 - \rho_{\lambda1}\rho_{\lambda2}} \tag{14.22}$$

$$\alpha_{\lambda,2 \text{ of } 2} = \frac{[1 - (\tau_{\lambda2} + \rho_{\lambda2})]\tau_{\lambda1}}{1 - \rho_{\lambda1}\rho_{\lambda2}} \tag{14.23}$$

Example 14.4. Calculate the monochromatic transmissivity, reflectivity, and absorptivity for clear plate glass ($n = 1.526$) 0.25 in. thick, when the sun's angle of incidence is 30 deg.

SOLUTION: By Eq. (14.18),

$$L' = \frac{0.25}{\sqrt{1 - \left(\frac{0.5}{1.526}\right)^2}} = 0.26 \text{ in.}$$

By Table 14.4, $K = 0.174$ in.$^{-1}$. By Eq. (14.17),

$$a = e^{-(0.174)(0.26)} = 0.956$$

By Fig. 14.8, $r = 0.045$. By Eq. (14.14),

$$\tau_\lambda = \frac{(0.955)^2(0.956)}{1 - (0.045)^2(0.956)^2} = 0.874$$

By Eq. (14.15),

$$\rho_\lambda = 0.045 + \frac{(0.045)(0.955)^2(0.956)^2}{1 - (0.045)^2(0.956)^2} = 0.083$$

Since $\alpha_\lambda = 1 - \tau_\lambda - \rho_\lambda$, we have $\alpha_\lambda = 0.043$.

Example 14.4 shows that for a glass sheet with a small extinction coefficient, the transmissivity is large compared to either the reflectivity or the absorptivity. The magnitude of the extinction coefficient K is significant. If we solve Example 14.4 for heat-absorbing glass with $K = 6.89$ in.$^{-1}$, we obtain $\tau_\lambda = 0.152$, $\rho_\lambda = 0.046$, and $\alpha_\lambda = 0.802$.

Parmelee* has stated that the composition of window glasses usually lies within the following percentage ranges:

Si O$_2$	Na$_2$ O	Ca O	Mg O	Al$_2$ O$_3$	Fe$_2$ O$_3$
70–73	12–15	9–14	0–3	0–1.5	0–0.15

The percentage of ferrous oxide is of particular importance, since iron accounts for

*ASHVE Trans., Vol. 51, p. 331.

most of absorption which occurs in glass. Heat-absorbing glass may have 0.5 per cent or more Fe_2O_3. For almost all types of window glass, the index of refraction may be taken as 1.526.

Figure 14.9, given by Dietz,[*] shows variation of transmissivity with wavelength for window glass containing different percentages of Fe_2O_3. Since the solar spectrum

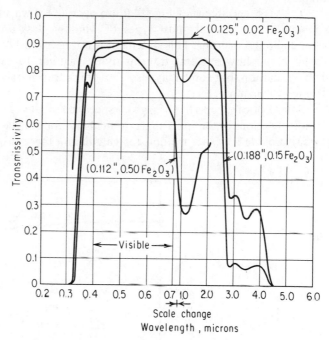

Figure 14.9 Spectral transmissivity of window glass containing various amounts of Fe_2O_3.

extends from about 0.3–2.3 μ, Fig. 14.9 shows that glass containing little Fe_2O_3 has an essentially constant transmissivity over the entire solar spectrum. However, the transmissivity of heat-absorbing glass (0.5 per cent Fe_2O_3) varies considerably with wavelength. Since the spectral distribution of solar radiation varies with atmospheric conditions, it follows that an overall transmissivity for a given angle of incidence is not a constant for the same glass sample.

Parmelee et al.[†] have made extensive experimental studies on the transmission

---

[*]A. G. H. Dietz, "Diathermanous Materials and Properties of Surfaces," in *Space Heating with Solar Energy*, Proceedings of a Course Symposium held at the Massachusetts Institute of Technology, 1950 (Cambridge: Massachusetts Institute of Technology, 1954) p. 33.

[†]G. V. Parmelee, W. W. Aubele, and R. G. Huebscher, "Measurements of Solar Heat Transmission through Flat Glass," *ASHVE Trans.*, Vol. 54, 1948, pp. 165–186, and G. V. Parmelee and W. W. Aubele, "Solar and Total Heat Gain through Double Flat Glass," *ASHVE Trans.*, Vol. 54, 1948, pp. 407–428.

of solar radiation through window glass. For single sheets of window glass, they found generally good agreement between measured and calculated transmissivities. For double glass, good agreement was obtained for sheets of glass whose spectral transmissivities were reasonably constant. For some combinations considerable differences were found, with the experimental transmissivity generally higher than the calculated value.

The glass specimens of Fig. 14.9 are opaque for wavelengths greater than about 4.5 microns. In general, most types of window glass are opaque for wavelengths beyond about 5 microns. This statement is not necessarily true for plastic films. Many plastic films show spectral transmissivities similar to ordinary window glass for visible radiation, but some may also show high transmissivities for wavelengths up to 10 microns or more.

Solar-optical properties of a given specimen of glass for monochromatic radiation are sole functions of the sun's angle of incidence for the glass sheet. Equations (14.14)–(14.23) provide an exact method for calculating these properties for a single wave-

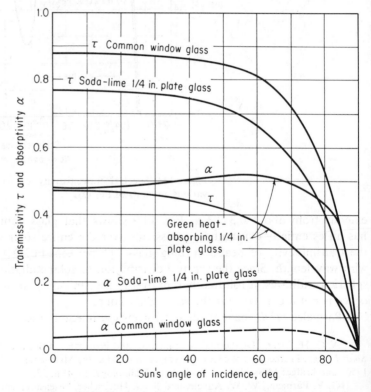

Figure 14.10 Transmissivity and absorptivity for various window glasses. (Adapted by permission from *ASHRAE Trans.*, Vol. 69, p. 423.)

length. However, in practical calculations, it is expedient to know average values of τ, ρ, and α which apply to the whole spectrum of incident solar radiation.

Figure 14.10 shows transmissivity and absorptivity for three glass specimens as experimentally determined by Yellot.* Reflectivity may be found from Eq. (14.12). Figure 14.10 shows that τ, ρ, and α are little affected by angle of incidence for angles less than about 50 degrees. The curves of Fig. 14.10 closely follow those which would be calculated from extension of the normally incident ($\theta = 0$) values by Eqs. (14.14) and (14.16).

14.5. CALCULATION OF THE IRRADIATION OF SURFACES BY SOLAR RADIATION REFLECTED FROM SURROUNDING SURFACES

In Chapter 13, we studied the irradiation of surfaces by direct solar radiation and by diffuse sky radiation during clear days. Besides these components, additional solar radiation may impinge upon building surfaces through reflection from the ground, from nearby buildings, and from other surfaces. As discussed in Sec. 14.3, *diffuse* reflection, *specular* reflection, or some combination of the two types, may occur. We will first consider the irradiation of a surface by diffusely reflected solar radiation.

Figure 14.11 shows a small surface dA_2 whose plane is inclined by an angle ϕ from

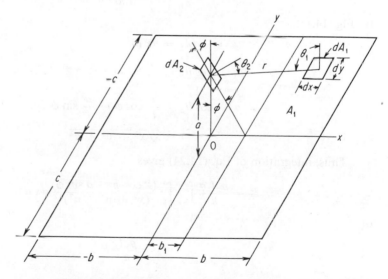

Figure 14.11 Relation of a small surface dA_2 to a diffusely reflecting large surface A_1.

*John L. Yellot, "Selective Reflectance—A New Approach to Solar Heat Control," *ASHRAE Trans.*, Vol. 69, 1963, pp. 418–438.

a perpendicular position with respect to a large surface A_1. The distance between the two surfaces is a. The large surface is rectangular and is assumed to diffusely reflect solar radiation incident upon it. The reflectivity ρ_1 of the large surface is given by

$$q = \rho_1 I_1 A_1$$

or

$$dq = \rho_1 I_1\, dA_1$$

where dq is the radiation in Btu per hr diffusely reflected by dA_1 upon a hemisphere of radius r, and I_1 is the intensity of solar radiation (sum of direct, diffuse, and reflected) in Btu per (hr)(sq ft) incident upon A_1. Since A_1 is assumed to be a pure diffuse reflector, we may show that

$$dq = \pi r^2 I_{R,N}$$

where $I_{R,N}$ is the intensity of reflected radiation along the normal to dA_1 and at a distance r away.

Let dI_R be the intensity incident upon dA_2 due to radiation diffusely reflected by dA_1. We may write

$$dI_R = I_{R,N} \cos \theta_1 \cos \theta_2$$

Thus

$$dI_R = \frac{\rho_1 I_1\, dA_1 \cos \theta_1 \cos \theta_2}{\pi r^2} \qquad (14.24)$$

By Fig. 14.11,

$$dA_1 = dx\, dy$$

$$\cos \theta_1 = \frac{a}{r}$$

$$\cos \theta_2 = \frac{x}{r} \cos \phi - \frac{a}{r} \sin \phi$$

$$r = \sqrt{x^2 + y^2 + a^2}$$

Thus, integration of Eq. (14.24) gives

$$I_R = \frac{\rho_1 I_1 a}{\pi} \int_{-c}^{c} \int_{b_1}^{b} \frac{(x \cos \phi - a \sin \phi)}{(x^2 + y^2 + a^2)^2}\, dx\, dy \qquad (14.25)$$

or

$$I_R = \frac{\rho_1 I_1}{\pi} F \qquad (14.26)$$

where

$$F = a \int_{-c}^{c} \int_{b_1}^{b} \frac{(x \cos \phi - a \sin \phi)}{(x^2 + y^2 + a^2)^2}\, dx\, dy \qquad (14.27)$$

The general solution for F in Eq. (14.27) is

$$F = \cos \phi \left[\frac{a}{\sqrt{a^2 + b_1^2}} \tan^{-1} \frac{c}{\sqrt{a^2 + b_1^2}} - \frac{a}{\sqrt{a^2 + b^2}} \tan^{-1} \frac{c}{\sqrt{a^2 + b^2}} \right]$$

$$+ \sin \phi \left[\frac{b_1}{\sqrt{a^2 + b_1^2}} \tan^{-1} \frac{c}{\sqrt{a^2 + b_1^2}} - \frac{b}{\sqrt{a^2 + b^2}} \tan^{-1} \frac{c}{\sqrt{a^2 + b^2}} \right]$$

$$+ \frac{c \sin \phi}{\sqrt{a^2 + c^2}} \left[\tan^{-1} \frac{b_1}{\sqrt{a^2 + c^2}} - \tan^{-1} \frac{b}{\sqrt{a^2 + c^2}} \right] \tag{14.28}$$

Table 14.5 shows F for several special cases. Cases B-3, B-4, C-2, and C-3 are of particular application to building problems. Equation (14.26) with Table 14.5 allow estimates of the irradiation of a small surface, such as a window, by diffusely reflected radiation from the ground or from nearby building surfaces.

> **Example 14.5.** For the same conditions as in Example 13.5, calculate the total incidence of solar radiation upon a south-facing window whose center-line is located 6 ft above the ground. In front of the window is a new concrete parking area which extends 50 ft south, 50 ft east, and 50 ft west of the window. Ignore set-back of the window.
>
> SOLUTION: For Example 13.5, $\beta = 69° 13'$, $\gamma = 162° 42'$, $I_N = 278$ Btu/(hr)(sq ft), $I_H = 293$ Btu/(hr)(sq ft), and $I_{dH} = 33$ Btu/(hr)(sq ft).
>
> For the south-facing vertical window, $\alpha = 180° - \gamma = 17° 18'$ and by Eq. (13.8),
>
> $$\cos \theta = \cos \beta \cos \alpha = (0.3548)(0.9548) = 0.339$$
>
> By Eq. (13.14),
>
> $$I_{DV} = I_N \cos \theta = (278)(0.339) = 94 \text{ Btu/(hr)(sq ft)}$$
>
> By Fig. 13.20, $I_{dV}/I_{dH} = 0.75$. Thus
>
> $$I_{dV} = (0.75)(33) = 25 \text{ Btu/(hr)(sq ft)}$$
>
> By Fig. 14.5, we may estimate the concrete reflectivity $\rho_1 = 0.33$. Case B-3 of Table 14.5 applies for the calculation of I_R. We have $a = 6$ ft, $b = 50$ ft, and $c = 50$ ft. Thus
>
> $$F = \tan^{-1} \frac{50}{6} - \frac{6}{\sqrt{(6)^2 + (50)^2}} \tan^{-1} \frac{50}{\sqrt{(6)^2 + (50)^2}} = 1.36$$
>
> By Eq. (14.26) with $I_1 = I_H$, we have
>
> $$I_{RV} = \frac{(0.33)(293)(1.36)}{\pi} = 42 \text{ Btu/(hr)(sq ft)}$$
>
> We see that I_{RV} is a significant part of the total radiation.

We will now consider *specular* reflection. Figure 14.12 shows a small surface dA_2 whose plane is inclined by an angle ϕ from a perpendicular position with respect to a large surface A_1. The distance between the two surfaces is a. The angle θ_1 is the sun's angle of incidence with respect to A_1, the angle β is the sun's altitude angle, and the

TABLE 14.5
FACTOR F IN EQ. (14.26) FOR SEVERAL SPECIAL CASES

Case	F
A. $\tan\phi \geq b/a$, $b_1 = b$	$F = 0$
B. $-b/a \leq \tan\phi \leq b/a$, $b_1 = a\tan\phi$ 1. General case	$F = \tan^{-1}\dfrac{c\cos\phi}{a} - \dfrac{(a\cos\phi + b\sin\phi)}{\sqrt{a^2+b^2}}\tan^{-1}\dfrac{c}{\sqrt{a^2+b^2}}$ $+ \dfrac{c\sin\phi}{\sqrt{a^2+c^2}}\left(\tan^{-1}\dfrac{a\tan\phi}{\sqrt{a^2+c^2}} - \tan^{-1}\dfrac{b}{\sqrt{a^2+c^2}}\right)$
2. Same as Case B-1 with b and c infinitely large and a finite	$F = \dfrac{\pi}{2}(1 - \sin\phi)$
3. $\phi = 0$, $b_1 = 0$ (Plane of dA_2 perpendicular to A_1)	$F = \tan^{-1}\dfrac{c}{a} - \dfrac{a}{\sqrt{a^2+b^2}}\tan^{-1}\dfrac{c}{\sqrt{a^2+b^2}}$
4. Same as Case B-3 but with b and c infinitely large and a finite	$F = \dfrac{\pi}{2}$
C. $\tan\phi \leq -b/a$, $b_1 = -b$ 1. General case	$F = -2\sin\phi\left(\dfrac{b}{\sqrt{a^2+b^2}}\tan^{-1}\dfrac{c}{\sqrt{a^2+b^2}}\right.$ $\left.+ \dfrac{c}{\sqrt{a^2+c^2}}\tan^{-1}\dfrac{b}{\sqrt{a^2+c^2}}\right)$
2. $\phi = -\pi/2$ (dA_2 parallel to and facing A_1)	$F = \dfrac{2b}{\sqrt{a^2+b^2}}\tan^{-1}\dfrac{c}{\sqrt{a^2+b^2}}$ $+ \dfrac{2c}{\sqrt{a^2+c^2}}\tan^{-1}\dfrac{b}{\sqrt{a^2+c^2}}$
3. Same as Case C-2 but with b and c infinitely large and a finite	$F = \pi$

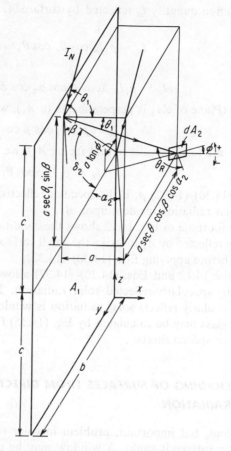

Figure 14.12 Relation of a small surface dA_2 to a specularly reflecting large surface A_1.*

angle α_2 is the wall-solar azimuth angle for dA_2 when $\phi = 0$. The large surface is assumed to reflect direct solar radiation specularly.

For Fig. 14.12, we may show that

$$\cos \theta_R = \cos \beta \cos \alpha_2 \cos \phi - \cos \theta_1 \sin \phi \qquad (14.29)$$

*Radiation reflected by A_1 will strike dA_2 if the following requirements are fulfilled:

For $0 \leq \phi \leq \pi/2$	For $-\pi/2 \leq \phi \leq 0$	For $-\pi \leq \phi \leq -\pi/2$
$\tan \phi \leq \sec \theta_1 \cos \beta \cos \alpha_2$ $\leq b/a$ $\sec \theta_1 \sin \beta \leq c/a$	$\sec \theta_1 \cos \beta \cos \alpha_2 \leq b/a$ $\sec \theta_1 \sin \beta \leq c/a$	$\tan \phi \geq \sec \theta_1 \cos \beta \cos \alpha_2$ $\leq b/a$ $\sec \theta_1 \sin \beta \leq c/a$

The radiation quantity I_R reflected by surface A_1 which is incident upon the surface dA_2 is

$$I_R = \rho_1 I_N \cos \theta_1 \cos \theta_R = \rho_1 I_{D,1} \cos \theta_R$$

or

$$I_R = \rho_1 I_{D,1}(\cos \beta \cos \alpha_2 \cos \phi - \cos \theta_1 \sin \phi) \tag{14.30}$$

If $\phi = 0$ (Plane of dA_2 is perpendicular to A_1), we have

$$I_R = \rho_1 I_{D,1} \cos \beta \cos \alpha_2 \tag{14.31}$$

If $\phi = -\pi/2$ (dA_2 parallel to and facing A_1), we have

$$I_R = \rho_1 I_{D,1} \cos \theta_1 \tag{14.32}$$

In Eqs. (14.30)–(14.32) ρ_1 is the specular reflectivity of A_1, and $I_{D,1}$ is the intensity of *direct* solar radiation incident upon A_1.

The footnote to Fig. 14.12 shows the conditions which must be fulfilled for solar radiation reflected by A_1 to strike the small surface dA_2. These requirements should be checked before applying Eqs. (14.30)–(14.32).

Figure 14.12 and Eqs. (14.30)–(14.32) allow calculation of the irradiation of a surface by specularly-reflected solar radiation. The most common building surface which specularly reflects solar radiation is window glass. The spectral reflectivity of window glass may be calculated by Eq. (14.15) for a single sheet and by Eq. (14.21) for two air-spaced sheets.

14.6. SHADING OF SURFACES FROM DIRECT SOLAR RADIATION

An obvious, but important, problem in solar radiation calculations is determining whether a surface is sunlit. A window may be partially shaded because of set-back from the external plane of a building; a flat roof may be partially shaded by parapet walls. Any external building surface may be partially or wholly shaded by nearby buildings.

A second class of problems arises when architectural projections are to be designed to prevent or control the irradiation of a surface. (Shading of windows by overhangs and awnings are examples.) Typically, the shading device is designed to exclude the sun's rays in summer but to admit them in winter.

Although each shading problem must be analyzed individually, the approach to all such problems is similar. In most cases, it is desirable to make an isometric sketch showing the relationship of the sun's rays to the surfaces involved. Figure 14.13 shows a flat roof partially shaded by two parapet walls. We may easily show that the shaded dimensions of the roof are

$$x = a \cot \beta \cos \alpha_1$$
$$y = a \cot \beta \cos \alpha_2$$

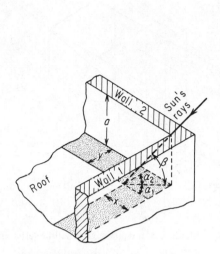
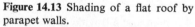

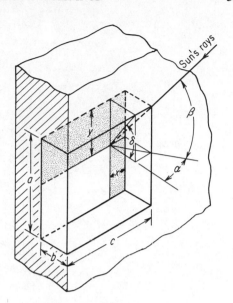

Figure 14.13 Shading of a flat roof by parapet walls.

Figure 14.14 Shading of a window set back from the plane of a building.

where β is the sun's altitude angle, α_1 is the wall-solar azimuth angle for Wall 1, and α_2 is the wall-solar azimuth angle for Wall 2.

Figure 14.14 shows a common shading problem. A window set back from the outside plane of a building will be partially shaded. We may show that

$$x = b \tan \alpha$$

$$y = b \tan \delta$$

where the *profile angle* δ is related to the sun's altitude angle β and the wall-solar azimuth angle α by

$$\tan \delta = \tan \beta / \cos \alpha \tag{14.33}$$

For Fig. 14.14, we may also derive that the sunlit fraction of the window surface F_s is given by

$$F_s = 1 - r_1 \tan \delta - r_2 \tan \alpha + r_1 r_2 \tan \delta \tan \alpha \tag{14.34}$$

where $r_1 = b/a$ and $r_2 = b/c$.

Figure 14.15 shows a window completely shaded by a solid overhang. Dimensions shown for the overhang are minimum values or those required to just shade the window for the given position of the sun's rays. By Fig. 14.15, the minimum dimensions for f and g are given by

$$f = (a + e) \cot \delta - b \tag{14.35}$$

$$g = \tan \alpha \, [(a + e) \cot \delta - b] = f \tan \alpha \tag{14.36}$$

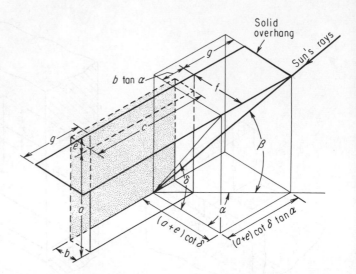

Figure 14.15 Shading of a window by a solid overhang.

Example 14.6. A south-facing window 4 ft high and 6 ft wide is set back from the plane of a building by 8 in. It is desired to install a solid overhang 1 ft above the window, of such dimensions that the window will be completely shaded each day from 9: 00 a.m. to 3: 00 p.m. solar time during the period April 15 to August 29. Location is at 42 deg north latitude. Find the minimum width f and length $(c + 2g)$ of the overhang.

SOLUTION: April 15 and August 29 are symmetrical with respect to June 22. Thus, solar angles on the two days are essentially identical for the same solar time. Calculations will be made for 9: 00 a.m. on April 15. By Table 13.2, $d = 9°\ 30'$. At 9: 00 solar time, $h = 45°$. As given data, $l = 42°$. By Eq. (13.3)

$$\sin \beta = (0.7431)(0.7071)(0.9863) + (0.6691)(0.1650) = 0.6286$$

and $\beta = 38°\ 57'$. By Eq. (13.4),

$$\cos \gamma = (1.286)[(0.7431)(0.1650) - (0.9863)(0.6691)(0.7071)] = -0.4424$$

and $\gamma = 116°\ 15'$ east of north. Thus, $\alpha = 180° - 116°\ 15' = 63°\ 45'$. By Eq. (14.33),

$$\cot \delta = \frac{0.4424}{0.8083} = 0.5471$$

and $\delta = 61°\ 19'$. By Eq. (14.35),

$$f = (48 + 12)(0.5471) - 8 = 24.8 \text{ in.}$$

By Eq. (14.36),

$$g = f \tan \alpha = (24.8)(2.028) = 50.3 \text{ in.}$$

Thus, the required length $(c + 2g)$ is 172.6 in. The window will be continuously shaded from 9: 00 a.m. to 3: 00 p.m. solar time on April 15, since the sun's profile angle δ will be larger and the wall-solar azimuth angle α will be smaller than the 9:00 a.m. (or 3:00 p.m.) values. Furthermore, the window will be shaded at least from 9:00 a.m. to 3:00 p.m. every day from April 15 to August 29 since the sun's declination is continuously larger than the value for April 15. This circumstance leads to a larger β, smaller α, and larger δ for the same solar time compared to April 15 (or August 29).

Example 14.7. Using the given latitude and window and the calculated dimensions of the overhang for Example 14.6, determine the percentage of the window surface which would be sunlit at solar noon on December 22.

SOLUTION: By Table 13.2, $d = -23° 27'$. At solar noon, $h = 0$, and for a south-facing vertical surface, $\alpha = 0$. By Eq. (13.3),

$$\sin \beta = (0.7431)(1.000)(0.9174) + (0.6691)(-0.3980) = 0.4154$$

and $\beta = 24° 33'$. Since $\alpha = 0$, $\delta = \beta$. We may write

$$e' = f \tan \beta = (24.8)(0.4568) = 11.33 \text{ in.}$$

Since e' is less than 12 in., the overhang does not shade the window. The window is shaded only by set-back. Thus

$$a' = b \tan \beta = (8)(0.4568) = 3.65 \text{ in.}$$

and 92.4 per cent of the window is sunlit.

Examples (14.6)–(14.7) show that the entry of direct solar radiation through windows may be controlled to a large extent by solid overhangs. An overhang, of course, does not need to be solid or fixed. At low solar altitude angles, an overhang is incapable of shading a window; thus, lateral-type shading devices, such as vertical fins, free-standing screens, or foliage may be utilized. The necessary dimensions and positions for such devices may be calculated by the same basic procedures given in this section.

14.7. ENERGY TRANSMISSION THROUGH DIATHERMANOUS MATERIALS

Energy transmission through window glass may result from solar radiation effects and from a temperature difference between the internal and external thermal environments.

Figure 14.16 shows the schematic problem. We assume no external shading except by set-back, and no internal shades such as Venetian blinds. Direct solar radiation, diffuse sky radiation, and reflected solar radiation may be incident upon the outer surface of the window. Part of this radiation may be directly transmitted through the glass, part may be reflected, and part may be absorbed. Energy exchange by convection may occur between the glass outer surface and the outside air. Also, the glass outer surface may transfer heat by long-wave radiation exchange with the sky and surrounding objects. Similar convection and radiation exchanges may occur between the glass inner surface and the interior of the structure. In general, we may

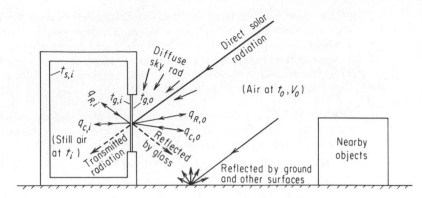

Figure 14.16 Energy exchanges through a glass window.

state that q_i, the rate of heat gain (or loss) by the interior of the structure through the glass, Btu per (hr)(sq ft of glass), is given by

$$q_i = F_s \tau_D I_D + \tau_d I_d + \tau_R I_R + h_i(t_{g,i} - t_i) \tag{14.37}$$

Since the angles of incidence for direct radiation, diffuse sky radiation, and reflected radiation may differ, separate calculations for the products τI are required. Equations (14.14) and (14.20) or charts such as Fig. 14.10 allow determination of each transmissivity. The angle of incidence θ for direct solar radiation is given by Eqs. (13.7)–(13.9). Parmelee* has given a graphical-integration method for determining the transmissivity, reflectivity, and absorptivity of window glass for diffuse sky radiation. For ordinary calculations, he states that for a vertical window, a *mean angle of incidence of 60 deg* may be used. The angle of incidence for specularly-reflected radiation incident upon the window is given by Eq. (14.29). Similarly to diffuse sky radiation, diffusely-reflected radiation may strike the window from a variety of angles. In the usual problem, it is satisfactory to estimate a mean angle of incidence. The sunlit fraction of the window surface F_s in Eq. (14.37) is given by Eq. (14.34). Shading of the window from diffuse sky radiation and reflected radiation are assumed to be negligible. The inside-surface coefficient h_i is given by Eq. (14.2).

We may also write an energy balance for the glass sheet itself. We have

$$F_s I_D \alpha_D + I_D \alpha_D + I_R \alpha_R = h_i(t_{g,i} - t_i) + h_o(t_{g,o} - t_o) \pm q_{g,s} \tag{14.38}$$

The left side of Eq. (14.38) represents gain of heat by the glass because of absorption of solar radiation. The right side represents heat dissipation where the surface coefficients h_i and h_o are given by Eqs. (14.2) and (14.3) respectively. The term $q_{g,s}$ is the rate of change of heat stored in the glass sheet.

In many problems, Eqs. (14.37) and (14.38) may be simplified. Since $t_{g,i}$ may differ only slightly from $t_{g,o}$, an average glass temperature t_g may be used for each. For glass having a low solar absorptivity, the term $q_{g,s}$ is small and may be neglected.

*ASHVE Trans., Vol. 51, pp. 322–324, 334–335.

However, for thick heat-absorbing glasses $q_{g,s}$ may be substantial. With these approximations, we may show by Eqs. (14.37) and (14.38) for a window glass with low absorptivity that

$$q_i = (F_s\tau_D I_D + \tau_d I_d + \tau_R I_R) + \frac{(F_s\alpha_D I_D + \alpha_d I_d + \alpha_R I_R)}{1 + (h_o/h_i)} + U(t_o - t_i) \quad (14.39)$$

where

$$U = \frac{1}{(1/h_i) + (1/h_o)} \quad (14.40)$$

Further comments should be made concerning the coefficients h_i and h_o in Eqs. (14.37)–(14.40). Equation (14.2) shows that an average interior-surfaces temperature $t_{s,i}$ must be known. In most cases, we may assume $t_{s,i} = t_i$. The average temperature of outdoor radiating objects $t_{s,o}$ in Eq. (14.3) is difficult to estimate. Where the window glass views principally the ground and nearby buildings, it is usually satisfactory to assume $t_{s,o} = t_o$. This approximation is generally valid also where the window glass views the sky during daylight hours. However, at night the effective sky temperature for radiation exchange may be substantially less than the outdoor air temperature. Mackey and Wright* recommend the use of -60 F as the effective sky temperature at night.

Figure 14.16 and Eqs. (14.37)–(14.39) make no allowance for either external or internal shading of the window from direct solar radiation except for window set-back. Section 14.6 covers methods for analyzing external shading devices. Internal shading devices such as Venetian blinds, roller shades, and drapes are more difficult to analyze than external shades.

Example 14.8. Calculate the instantaneous rate of heat gain through the window of Example 14.5. Assume that the window glass is double strength, A quality, 0.125 in. thick, with an extinction coefficient of 0.194 in.$^{-1}$ Assume the outside temperature $t_o = t_{s,o} = 90$ F. Assume the inside temperature $t_i = t_{s,i} = 75$ F. Use an outside wind velocity of 7.5 MPH.

SOLUTION: By Example 14.5, $I_D = 94$ Btu/(hr)(sq ft), $I_d = 25$ Btu/(hr) (sq ft), $I_R = 42$ Btu/(hr)(sq ft), and $\cos\theta_D = 0.339$. Thus, $\theta_D = 70°\ 11'$. By Fig. 14.8, $r_D = 0.17$. By Eq. (14.18),

$$L' = \frac{0.125}{\sqrt{1 - \left(\frac{0.941}{1.526}\right)^2}} = 0.159 \text{ in.}$$

By Eq. (14.17),

$$a_D = e^{-(0.194)(0.159)} = 0.97$$

By Eq. (14.14),

$$\tau_D = \tau_{\lambda,D} = \frac{(0.83)^2(0.97)}{1 - (0.17)^2(0.97)^2} = 0.69$$

*C. O. Mackey and L. T. Wright, Jr., "The Sol-Air Thermometer—A New Instrument," *ASHVE Trans.*, Vol. 52, 1946, p. 277.

By Eq. (14.16),

$$\alpha_D = \alpha_{\lambda,D} = 1 - 0.17 - \frac{(0.83)^2(0.97)}{1 - (0.17)(0.97)} = 0.03$$

We will assume that the mean angle of incidence θ_d for diffuse sky radiation is 60 deg. Inspection shows that the mean angle of incidence θ_R for reflected radiation is less than 50 deg. Since the transmissivity and absorptivity for this type of window glass are essentially constant from 0 to 50 deg angle of incidence, assume $\theta_R = 50$ deg. By procedures identical to those for direct radiation, we find $\tau_d = 0.81$, $\alpha_d = 0.03$, $\tau_R = 0.86$, $\alpha_R = 0.03$. By Table 14.1, $h_i = 1.46$ Btu/(hr)(sq ft)(F) and $h_o = 4.00$ Btu/(sq ft)(F). These values are only approximately correct but are adequate for a problem of this type. Use a sunlit fraction $F_s = 1.0$, since no set-back is assumed. By Eq. (14.40),

$$U = \frac{1}{1/1.46 + 1/4.00} = 1.07 \text{ Btu/(hr)(sq ft)(F)}$$

By Eq. (14.39),

$$q_i = (0.69)(94) + (0.81)(25) + (0.86)(42) + \frac{(0.03)(94) + (0.03)(25) + (0.03)(42)}{1 + 4.00/1.46}$$

$$+ (1.07)(15)$$
$$= 139 \text{ Btu/(hr)(sq ft)}$$

14.8. PERIODIC HEAT TRANSFER THROUGH WALLS AND ROOFS

Earlier in this chapter we studied steady-state heat transmission through building walls and the transmission of heat and solar radiation through window glass. Steady-state heat transmission may be approached only when both the internal and external thermal environments have remained constant over an extended period of time. Because of almost continuous fluctuations in the external thermal environment, steady-state heat transmission rarely occurs in outside walls of a building.

In our treatment of heat transmission through window glass in Sec. 14.7, we neglected heat storage effects in the glass substance. Thus, the analysis was applicable to changing thermal environments. However, in the case of most building boundaries made of masonry and other opaque materials, we are not permitted to neglect heat storage effects. Neither may we assume the steady state.

Generally, the two principal factors of the external thermal environment are outdoor air temperature and solar radiation intensity. Both are subject to erratic fluctuations. However, if we limit our analysis to clear days, solar radiation intensity upon any given surface follows essentially a periodic variation. Also, variation of outdoor temperature may be essentially periodic. In our analysis in this section, we will assume a constant internal thermal environment and periodic variations of outdoor air temperature and solar radiation intensity.

Figure 14.17 shows a diurnal variation of outdoor air temperature which may be

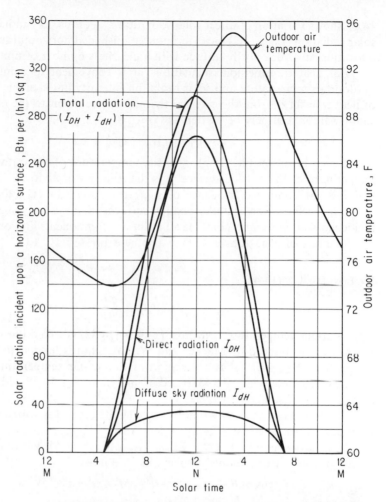

Figure 14.17 Typical variations of outdoor air temperature and intensity of solar radiation incident upon a horizontal surface for a location at 42 deg north latitude on July 1 for a clearness number of unity.

typical for a hot June–August day for most locations in the United States east of the Rocky Mountains. Minimum temperature usually occurs just before sunrise while maximum temperature usually occurs some 3–4 hr after solar noon.

Typical variations of incidence of direct solar and diffuse sky radiation during a clear day may be determined through use of Figs. 13.13–13.20. If significant, reflected radiation may be calculated by the procedures given in Sec. 14.5, and allowance for external shading of the surface may be calculated by the methods of Sec. 14.6. Figure

14.17 shows variation of direct solar radiation I_{DH}, diffuse sky radiation I_{dH}, and total solar radiation $(I_{DH} + I_{dH})$ incident upon an unshaded horizontal surface on July 1 for a location at 42 deg north latitude with a clearness number of unity.

For heat transmission calculations, it is convenient to combine the effects of outdoor air temperature and solar radiation intensity into a single quantity. The rate of heat transfer q_o from the external thermal environment to the outside surface of a sunlit wall or roof may be written as

$$q_o = h_o(t_o - t_{w,o}) + \alpha I \qquad (14.41)$$

where h_o is given by Eq. (14.3), t_o is the temperature of the outdoor air, $t_{w,o}$ is the temperature of the outside surface, α is the solar absorptivity of the outside surface, and I is the combined incidence of solar radiation (direct, diffuse, and reflected) upon the surface. For ordinary opaque surfaces, α varies only slightly with angle of incidence, so that we may combine all radiation components into one quantity I.

The rate of heat transfer q_o may also be expressed as

$$q_o = h_o(t_e - t_{w,o}) \qquad (14.42)$$

By Eqs. (14.41)–(14.42),

$$t_e = t_o + \frac{\alpha I}{h_o} \qquad (14.43)$$

The fictitious temperature t_e in Eqs. (14.42)–(14.43) is called the *sol-air temperature*. It was first introduced by Mackey and Wright,[*] and is limited to a specific surface. Curve A of Fig. 14.18 shows variation of the sol-air temperature for a horizontal surface calculated by Eq. (14.43) based upon data of Fig. 14.17 and assuming $\alpha = 0.9$, and $h_o = 4.00$ Btu per (hr)(sq ft)(F).

The variations shown by Fig. 14.17 and Curve A of Fig. 14.18 may be assumed to be repetitive for successive 24-hr cycles. Any of the curves may be mathematically expressed in terms of a Fourier series. Thus, if $t_e = f(\theta)$, where θ is the number of hours measured from midnight solar time, we have

$$t_e = t_{e,m} + M_1 \cos \omega_1\theta + N_1 \sin \omega_1\theta + M_2 \cos \omega_2\theta + N_2 \sin \omega_2\theta + \ldots \qquad (14.44)$$

where the coefficients $t_{e,m}$, M_n, and N_n are given by

$$t_{e,m} = \frac{1}{24} \int_0^{24} t_e \, d\theta$$

$$M_n = \frac{1}{12} \int_0^{24} t_e \cos \omega_n\theta \, d\theta$$

$$N_n = \frac{1}{12} \int_0^{24} t_e \sin \omega_n\theta \, d\theta$$

and $\omega_1 = \pi/12$ radians per hr or 15 deg per hr and $\omega_n = n\omega_1$. Alternatively, Eq. (14.44) may be written as

$$t_e = t_{e,m} + \sqrt{M_1^2 + N_1^2} \cos (\omega_1\theta - \psi_1) + \sqrt{M_2^2 + N_2^2} \cos (\omega_2\theta - \psi_2) + \ldots$$

[*]C. O. Mackey and L. T. Wright, Jr., "Periodic Heat Flow—Homogeneous Walls or Roof," *ASHVE Trans.*, Vol. 50, 1944, p. 293.

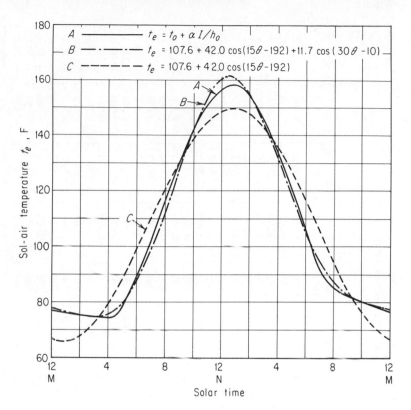

Figure 14.18 Variation of sol-air temperature for a horizontal surface based upon $\alpha = 0.9$, $h_o = 4.00$ Btu per (hr)(sq ft)(F), and t_o and $(I_{DH} + I_{dH})$ curves of Fig. 14.17.

or

$$t_e = t_{e,m} + t_{e,1} \cos(\omega_1 \theta - \psi_1) + t_{e,2} \cos(\omega_2 \theta - \psi_2) + \dots \qquad (14.45)$$

where

$$\tan \psi_n = \frac{N_n}{M_n} \qquad (14.46)$$

In Eq. (14.46), the quadrant in which ψ_n lies is determined by the requirement that $\sin \psi_n$ must have the sign of N_n and that $\cos \psi_n$ must have the sign of M_n.

As an illustration of the method, Example 14.9 shows a harmonic analysis for the sol-air temperature given by Curve A of Fig. 14.18. The procedures follow those given by Alford, Ryan, and Urban.*

Example 14.9. Make a harmonic analysis for sol-air temperature given by Curve A of Fig. 14.18. Include two harmonics.

*J. S. Alford, J. E. Ryan, and F. O. Urban, "Effect of Heat Storage and Variation in Outdoor Temperature and Solar Intensity on Heat Transfer through Walls," *ASHVE Trans.*, Vol. 45, 1939, pp. 393–395.

TABLE 14.6
HARMONIC ANALYSIS FOR SOL-AIR TEMPERATURE SHOWN BY FIG. 14.18

1	2	3	4	5	6	7	8	9	10	11	12
Solar Time	t_e F	$\omega_1\theta$ deg	$\cos\omega_1\theta$	$t_e\cos\omega_1\theta$ F	$\sin\omega_1\theta$	$t_e\sin\omega_1\theta$ F	$\omega_2\theta$ deg	$\cos\omega_2\theta$	$t_e\cos\omega_2\theta$ F	$\sin\omega_2\theta$	$t_e\sin\omega_2\theta$ F
12 M	77	0	1.000	77.0	0.000	0.0	0	1.000	77.0	0.000	0.0
1 a.m.	76	15	0.966	73.4	0.259	19.7	30	0.866	65.8	0.500	38.0
2	75.5	30	0.866	65.4	0.500	37.8	60	0.500	37.8	0.866	65.4
3	75	45	0.707	53.0	0.707	53.0	90	0.000	0.0	1.000	75.0
4	74	60	0.500	37.0	0.866	64.1	120	−0.500	−37.0	0.866	64.1
5	78	75	0.259	20.2	0.966	75.3	150	−0.866	−67.5	0.500	39.0
6	89	90	0.000	0.0	1.000	89.0	180	−1.000	−89.0	0.000	0.0
7	102	105	−0.259	−26.4	0.966	98.5	210	−0.866	−88.3	−0.500	−51.0
8	116	120	−0.500	−58.0	0.866	100.5	240	−0.500	−58.0	−0.866	−100.5
9	130	135	−0.707	−91.9	0.707	91.9	270	0.000	0.0	−1.000	−130.0
10	143	150	−0.866	−123.8	0.500	71.5	300	0.500	71.5	−0.866	−123.8
11	151	165	−0.966	−145.9	0.259	39.1	330	0.866	130.8	−0.500	−75.5
12 N	157	180	−1.000	−157.0	0.000	0.0	360	1.000	157.0	0.000	0.0
1 p.m.	159	195	−0.966	−153.6	−0.259	−41.2	30	0.866	137.7	0.500	79.5
2	154	210	−0.866	−133.4	−0.500	−77.0	60	0.500	77.0	0.866	133.4
3	145	225	−0.707	−102.5	−0.707	−102.5	90	0.000	0.0	1.000	145.0
4	134	240	−0.500	−67.0	−0.866	−116.0	120	−0.500	−67.0	0.866	116.0
5	121	255	−0.259	−31.3	−0.966	−116.9	150	−0.866	−104.8	0.500	60.5
6	106	270	0.000	0.0	−1.000	−106.0	180	−1.000	−106.0	0.000	0.0
7	92	285	0.259	23.8	−0.966	−88.9	210	−0.866	−79.7	−0.500	−46.0
8	85	300	0.500	42.5	−0.866	−73.6	240	−0.500	−42.5	−0.866	−73.6
9	83	315	0.707	58.7	−0.707	−58.7	270	0.000	0.0	−1.000	−83.0
10	81	330	0.866	70.1	−0.500	−40.5	300	0.500	40.5	−0.866	−70.1
11	79	345	0.966	76.3	−0.259	−20.5	330	0.866	68.4	−0.500	−39.5
	2582.5			−493.4		−101.4			123.7		22.9

$t_{e,m} = 107.6$

$M_1 = -41.1$ $N_1 = -8.4$

$M_2 = 10.3$ $N_2 = 1.9$

$\sqrt{M_1^2 + N_1^2} = 42.0$

$\psi_1 = \tan^{-1}\dfrac{N_1}{M_1} = 192\ \text{deg}$

$\sqrt{M_2^2 + N_2^2} = 11.7$

$\psi_2 = \tan^{-1}\dfrac{N_2}{M_2} = 10\ \text{deg}$

SOLUTION: The solution is shown in Table 14.6. Column 2 shows values of t_e read from Curve A of Fig. 14.18. Column 3 shows values of $\omega_1 \theta$ in deg where $\omega_1 = 360/24 = 15$ deg/hr. Zero time is taken as midnight. Columns 1, and 4–7 are self-explanatory. The mean value $t_{e,m}$ is obtained by summing the values of Column 2 and dividing by 24. The coefficient M_1 is obtained by adding the values of Column 5 and dividing by 12. The coefficient N_1 is obtained by adding the values of Column 7 and dividing by 12. Columns 8–12 are a repetition of the procedure for the second harmonic where $\omega_2 = 2\omega_1 = 30$ deg/hr. The expression for t_e including the second harmonic is by Eq. (14.45),

$$t_e = 107.6 + 42.0 \cos(15\theta - 192) + 11.7 \cos(30\theta - 10)$$

Curve B of Fig. 14.18 shows a plot of this equation while curve C shows a similar plot but without the second harmonic. We see that Curve B rather closely fits the given variation of sol-air temperature (Curve A).

The methods described in this section may be used to calculate sol-air temperature variations for clear days at any time of year at any location in the United States. Representative diurnal outdoor-air temperature variations may be found from local Weather Bureau data. Figures 13.13–13.20 may be used for solar radiation intensities. The entire procedure of Example 14.9 can be programmed on a digital computer for calculation through an arbitrary number of harmonics.

Example 14.9 shows that sol-air temperature variation for a horizontal surface can be adequately represented by a Fourier series with two harmonics. However, variation of t_e for vertical surfaces differs much more from pure sine-wave behavior than for a horizontal surface. In the case of north, east, and west vertical surfaces, six or more harmonics may be needed.

Generally, periodic heat transfer through a sunlit vertical wall is small compared to heat transfer through a sunlit flat roof of the same area. (Compare answers of Example 14.10 and Problem 14.21.) Thus, it is more important usually to have accurate sol-air temperature information for a horizontal surface than for vertical surfaces.

Following the same procedures described here, James C. Dunn* developed a computer program and calculated diurnal sol-air temperature variations for horizontal and vertical surfaces for clear days at Minneapolis for six different times of year. Table 14.7 shows his results for a horizontal surface. Constants in Eq. (14.45) are shown through two harmonics. Constants for the winter design day were determined for clear December-February days whose minimum temperatures were approximately −20 F. Constants for the summer design day were determined for clear June-August days whose maximum temperatures were approximately 95 F.

We will now consider the periodic transfer of heat through a wall formed by a single homogeneous material. Figure 14.19 shows the schematic problem. We will assume that (1) the wall is of infinite height and length and that heat transfer occurs only in the x direction, (2) the wall is homogeneous with constant material properties,

*James C. Dunn, "Sol-Air Temperature Data at Minneapolis, Minnesota" (Plan B Master's Degree paper, University of Minnesota, 1965).

TABLE 14.7
CONSTANTS IN EQ. (14.45) FOR A HORIZONTAL SURFACE FOR CLEAR DAYS
AT MINNEAPOLIS

Time of Year	$t_{e,m}$ F	$t_{e,1}$ F	ψ_1 deg	$t_{e,2}$ F	ψ_2 deg
Average January day	7.9	14.4	202.0	8.5	11.9
Winter design day	−8.6	13.6	204.0	8.2	11.9
Average April day	66.0	36.2	196.0	11.9	0.0
Average July day	95.7	39.2	193.0	9.3	−1.8
Summer design day	105.5	40.3	191.0	9.2	−1.8
Average October day	63.8	27.0	199.0	12.0	10.2

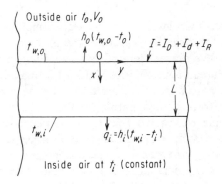

Figure 14.19 Schematic diagram for periodic heat transfer analysis.

(3) the surface coefficients h_i and h_o are constants, (4) the solar absorptivity of the outside surface is independent of angle of incidence and is constant, (5) the variation of t_o and I are periodic (identical with time on consecutive days), and (6) the internal thermal environment is constant.

At any location within the wall, we may write (see Sec. 2.2)

$$\frac{\partial t_w}{\partial \theta} = \alpha_w \frac{\partial^2 t_w}{\partial x^2} \tag{14.47}$$

where $\alpha_w = k_w/\rho_w c_w$ is the thermal diffusivity of the wall.

Equation (14.47) must be solved subject to two boundary conditions. At the inside surface, we must have

$$q_i = -k_w \left(\frac{\partial t_w}{\partial x}\right)_{x=L} = h_i(t_{w,i} - t_i) \tag{14.48}$$

At the outside surface, by Eqs. (14.41)–(14.42) we must have

$$q_o = -k_w \left(\frac{\partial t_w}{\partial x}\right)_{x=0} = h_o(t_e - t_{w,o}) \tag{14.49}$$

where t_e is given by Eq. (14.45).

The solution for Eqs. (14.47)–(14.49) has the form

$$t_w = A + Bx + \sum_1^\infty (C_n \cos p_n mx + D_n \sin p_n mx)e^{-m^2\omega_n\theta}$$

where A, B, C_n, D_n, and p_n are constants and $m = \sqrt[4]{-1}$. The coefficients A, B, C_n, and D_n may be either real or complex numbers, but the solution involves only the real parts.

The complete solution for Eqs. (14.47)–(14.49) has been given by Alford, Ryan, and Urban.* The temperature of the inside-wall surface $t_{w,i}$ may be written as

$$t_{w,i} = t_i + \frac{1}{h_i}[U(t_{e,m} - t_i) + V_1 t_{e,1} \cos(\omega_1\theta - \psi_1 - \Phi_1)$$
$$+ V_2 t_{e,2} \cos(\omega_2\theta - \psi_2 - \Phi_2) + \ldots] \quad (14.50)$$

where

$$U = \frac{1}{\dfrac{1}{h_i} + \dfrac{L}{k_w} + \dfrac{1}{h_o}} \quad (14.51)$$

$$V_n = \frac{h_o h_i}{\sigma_n k_w \sqrt{Y_n^2 + Z_n^2}} \quad (14.52)$$

$$\sigma_n = \sqrt{\frac{\omega_n}{2\alpha_w}} \quad (14.53)$$

$$Y_n = \left(\frac{h_o h_i}{2\sigma_n^2 k_w^2} + 1\right) \cos \sigma_n L \sinh \sigma_n L + \left(\frac{h_o h_i}{2\sigma_n^2 k_w^2} - 1\right) \sin \sigma_n L \cosh \sigma_n L$$
$$+ \frac{(h_o + h_i)}{\sigma_n k_w} \cos \sigma_n L \cosh \sigma_n L \quad (14.54)$$

$$Z_n = \left(\frac{h_o h_i}{2\sigma_n^2 k_w^2} + 1\right) \sin \sigma_n L \cosh \sigma_n L - \left(\frac{h_o h_i}{2\sigma_n^2 k_w^2} - 1\right) \cos \sigma_n L \sinh \sigma_n L$$
$$+ \frac{(h_o + h_i)}{\sigma_n k_w} \sin \sigma_n L \sinh \sigma_n L \quad (14.55)$$

$$\Phi_n = \tan^{-1} \frac{Z_n}{Y_n} \quad (14.56)$$

In Eq. (14.56), $\sin \Phi_n$ has the sign of Z_n and $\cos \Phi_n$ has the sign of Y_n.

The rate of heat transfer to the interior is given by

$$q_i = h_i(t_{w,i} - t_i) \quad (14.57)$$

By Eqs. (14.50) and (14.57),

$$q_i = U\{[t_{e,m} + \lambda_1 t_{e,1} \cos(\omega_1\theta - \psi_1 - \Phi_1)$$
$$+ \lambda_2 t_{e,2} \cos(\omega_2\theta - \psi_2 - \Phi_2) + \ldots] - t_i\} \quad (14.58)$$

where

$$\lambda_n = \frac{V_n}{U} \quad (14.59)$$

*ASHVE Trans., Vol. 45, pp. 387–392.

The form of Eq. (14.58) is of interest. Although q_i may be continually changing, it can be calculated by multiplying the overall coefficient U for steady-state heat transmission by an *equivalent temperature difference* which accounts for the periodic variation of the sol-air temperature t_e and for the heat storage characteristics of the wall. The quantity λ_n in Eq. (14.58) is called the *decrement factor*. The angle Φ_n is the angular displacement or lag between an harmonic of the sol-air temperature and the same harmonic of the inside-surface temperature.

Example 14.10. The flat roof of a building is sunlit throughout a clear day on July 1. Its location is 42 deg north latitude. Variation of sol-air temperature throughout the day is given by Fig. 14.18. The roof consists of 6 in. of concrete, covered by a thin layer of roofing which may be neglected in the solution. The temperature of the inside air and interior surrounding surfaces is 80 F. Determine (a) the rate of heat transmission through the roof to the room below over the full day and (b) the rate of heat transmission if heat storage effects of the concrete slab are neglected.

SOLUTION: (a) Since we are using Fig. 14.18, we have $h_o = 4.00$ Btu/(hr) (sq ft)(F). By Table 14.1, $h_i = 1.08$ Btu/(hr)(sq ft)(F). By Table 14.3,

$$\rho_w = 140 \text{ lb/cu ft}, \quad k_w = 12/12 = 1.0 \text{ Btu/(hr)(sq ft)(F/ft)}, \quad c_w = 0.21 \text{ Btu/(lb)(F)}$$

Thus, $\alpha_w = 1.0/(140)(0.21) = 0.034$ sq ft/hr. By Eq. (14.51),

$$U = \frac{1}{\dfrac{1}{1.08} + \dfrac{0.5}{1.0} + \dfrac{1}{4.00}} = 0.60 \text{ Btu/(hr)(sq ft)(F)}$$

The fundamental angular velocity $\omega_1 = 2\pi/24 = 0.2618$ radians/hr $= 15$ deg/hr. By Eq. (14.53),

$$\sigma_1 = \sqrt{\frac{0.2618}{(2)(0.034)}} = 1.96 \text{ ft}^{-1}$$

and $\sigma_2 = \sqrt{2}\,\sigma_1 = 2.77$ ft^{-1}. Thus $\sigma_1 L = 0.98$ and $\sigma_2 L = 1.385$. By Eq. (14.54),

$$
\begin{aligned}
Y_1 &= \left[\frac{(4.0)(1.08)}{2(1.96)^2(1.0)^2} + 1\right](0.5570)(1.145) \\
&\quad + \left[\frac{(4.0)(1.08)}{2(1.96)^2(1.0)^2} - 1\right](0.8305)(1.520) + \left[\frac{4.0 + 1.08}{(1.96)(1.0)}\right](0.5570)(1.520) \\
&= 2.638
\end{aligned}
$$

By Eq. (14.55),

$$
\begin{aligned}
Z_1 &= \left[\frac{(4.0)(1.08)}{2(1.96)^2(1.0)^2} + 1\right](0.8305)(1.520) \\
&\quad - \left[\frac{(4.0)(1.08)}{2(1.96)^2(1.0)^2} - 1\right](0.5570)(1.145) + \left[\frac{4.0 + 1.08}{(1.96)(1.0)}\right](0.8305)(1.145) \\
&= 4.716
\end{aligned}
$$

Likewise, we find $Y_2 = -0.335$ and $Z_2 = 5.802$. By Eq. (14.56), $\Phi_1 = \tan^{-1}$

$4.716/2.638 = 61$ deg and $\Phi_2 = \tan^{-1} 5.802/(-0.335) = 93$ deg. By Eqs. (14.52) and (14.59),

$$\lambda_1 = \frac{h_o h_i}{U \sigma_1 k \sqrt{Y_1^2 + Z_1^2}} = \frac{(4.0)(1.08)}{(0.60)(1.96)(1.0)\sqrt{(2.638)^2 + (4.716)^2}} = 0.680$$

Likewise, we find $\lambda_2 = 0.449$. By Example 14.9, we know that

$$t_e = 107.6 + 42.0 \cos (15\theta - 192) + 11.7 \cos (30\theta - 10) + \ldots$$

We may now write the expression for q_i, including the second harmonic, by Eq. (14.58). We have

$$q_i = 0.60 [27.6 + 28.6 \cos (15\theta - 253) + 5.3 \cos (30\theta - 103)]$$

Curve A of Fig. 14.20 shows the variation of q_i throughout the day.

(b) If heat storage effects are neglected, $\lambda_n = 1.00$ and $\Phi_n = 0$. The expression for q_i is in phase with the sol-air temperature t_e. Equation (14.58) reduces to

$$q_i = 0.60 [27.6 + 42.0 \cos (15\theta - 192) + 11.7 \cos (30\theta - 10)]$$

Curve B of Fig. 14.20 shows the fictitious variation of q_i where we neglect heat storage effects. Curve A lags Curve B by approximately 3.7 hr which

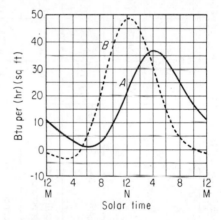

Figure 14.20 Solution for Example 14.10.

closely compares with the fundamental time lag $\Phi_1/15 = 4.1$ hr. Figure 14.20 also shows that the maximum value of q_i (Curve A) is about 75 per cent of the maximum value which would occur if heat storage were absent.

The development of this section on periodic heat transfer is valid only for a homogeneous wall. Mackey and Wright* have extended a similar analysis to cover composite walls. Stewart† has applied the procedures of Mackey and Wright to several

*C. O. Mackey and L. T. Wright, Jr., "Periodic Heat Flow—Composite Walls or Roofs," *ASHVE Trans.*, Vol. 52, 1946, pp. 283–296.

†J.P. Stewart, "Solar Heat Gain through Walls and Roofs for Cooling Load Calculations," *ASHVE Trans.*, Vol. 54, 1948, pp. 361–388.

TABLE 14.8

CONSTANTS FOR Eq. (14.58) FOR VARIOUS FLAT-ROOF CONSTRUCTIONS

No.	Description	U Btu/(hr)(sq ft)(F) Winter	Summer	λ_1 Winter	Summer	Φ_1 deg Winter	Summer	λ_2 Winter	Summer	Φ_2 deg Winter	Summer
1	A: 2 in. cellular glass insulation B: 2 in. concrete	0.137	0.128	0.506	0.447	66.0	70.0	0.280	0.241	86.6	89.2
2	A: 2 in. cellular glass insulation B: 4 in. concrete	0.134	0.125	0.279	0.241	86.2	88.5	0.142	0.121	106.5	108.0
3	A: 4 in. concrete B: 2 in. cellular glass insulation	0.134	0.125	0.566	0.519	68.5	72.2	0.320	0.286	97.3	100.2
4	Same as No. 1 except no air space and no ceiling	0.159	0.150	0.788	0.674	44.5	54.1	0.536	0.414	70.3	78.5
5	Same as No. 2 except no air space and no ceiling	0.155	0.146	0.518	0.405	70.8	78.4	0.286	0.213	97.3	102.2
6	A: 4 in. expanded polystyrene insulation B: 4 in. concrete	0.0653	0.0637	0.444	0.348	76.2	82.7	0.240	0.018	101.8	105.8
7	A: 4 in. cellular glass insulation B: 4 in. concrete	0.0773	0.0838	0.444	0.345	89.5	93.8	0.222	0.171	127.3	127.5
8	A: 8 in. corkboard B: 4 in. concrete	0.0321	0.0317	0.164	0.125	204.3	211.0	0.035	0.026	297.5	301.9

Built-up roofing
A
B
Air space
Ceiling (metal lath and plaster)

Built-up roofing
A
B
0.5 in. gypsum plaster

practical wall constructions. Several authors* have applied electrical analogue techniques to the same general problem. Stephenson and Mitalis† have described a different approach to the problem using thermal response factors.

The general form of the solution to Eqs. (14.47)–(14.49) for composite walls is identical to Eq. (14.58). However, the formulations for λ_n and Φ_n become progressively more complicated as the number of layers is increased. Bullock‡ developed solutions for various composite-wall constructions. He programmed his solutions on a digital computer and calculated U, λ_n, and Φ_n values for a large number of wall and roof constructions. Table 14.8 shows Bullock's values for these constants through two harmonics for various flat-roof constructions. Only sol-air temperature data are further required to solve Eq. (14.58) for the roof constructions of Table 14.8. Table 14.7 provides these data for clear days at Minneapolis. Values shown in Table 14.7 for a summer design day would apply approximately to most regions in the United States east of the Rocky Mountains.

Example 14.11. Calculate the instantaneous rate of heat gain, Btu per (hr) (sq ft), through a flat-roof construction similar to No. 2 of Table 14.8 at 4:00 p.m. solar time for an average clear July day at Minneapolis. Assume an interior temperature of 75 F.

SOLUTION: By Table 14.7, we have $t_{e,m} = 95.7$ F, $t_{e,1} = 39.2$ F, $\psi_1 = 193.0$ deg, $t_{e,2} = 9.3$ F, and $\psi_2 = -1.8$ deg. By Table 14.8, we have $U = 0.125$ Btu/(hr)(sq ft)(F), $\lambda_1 = 0.241$, $\Phi_1 = 88.5$ deg, $\lambda_2 = 0.121$, and $\Phi_2 = 108$ deg. By Eq. (14.58)

$$q_i = 0.125\{[95.7 + (0.241)(39.2)\cos(240 - 193.0 - 88.5)$$
$$+ (0.121)(9.3)\cos(480 + 1.8 - 108)] - 75\}$$
$$= 0.125[95.7 + 7.1 + 1.0 - 75]$$
$$= 3.6 \text{ Btu/(hr)(sq ft)}$$

PROBLEMS

14.1. An outside wall of a building consists of 0.5 in. of gypsum plaster, 8 in. of common brick, and 4 in. of face brick. Calculate the overall heat transfer coefficient U, assuming a wind velocity of 15 MPH.

14.2. A schematic ceiling construction is shown by Fig. 14.21. You may neglect the presence of the wood joists. Determine (a) the U coefficient for summer (heat transfer down), and (b) the U coefficient for winter (heat transfer up).

*See *ASHVE Trans.*, Vol. 60, 1954, pp. 59–102; *ASHAE Trans.*, Vol. 61, 1955, pp. 125–150, 339–386; *ASHAE Trans.*, Vol. 64, 1958, pp. 111–128.

†See *ASHRAE Trans.*, Vol. 73, Part I, 1967, pp. III. 1.1–III. 1.7, III. 2.1–III. 2.10.

‡Charles E. Bullock, "Periodic Heat Transfer in Walls and Roofs" (Plan B Master's Degree paper, University of Minnesota, 1961).

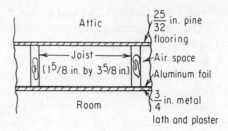

Figure 14.21 Schematic ceiling construction for Problem 14.2.

14.3. Calculate the U coefficient for the wall of Example 14.1 if the air space was divided into 5 spaces of equal width by sheets of kraft paper backed on one side with aluminum foil. Assume each space has one reflective surface.

14.4. An outside wall of a building is constructed of 0.75 in. metal lath and plaster, 8 in. hollow concrete block, and 4 in. of face brick. Outside wind velocity is 15 MPH. Barometric pressure is 14.696 psia. Interior air temperature is 75 F. Outdoor temperature is -10 F. Assuming steady-state conditions, determine the maximum relative humidity permissible for the inside air without condensation of moisture upon the inside surface of the wall.

14.5. Figure 14.22 shows schematic views of two outside-wall constructions. Assume inside air conditions of 75 F dry-bulb temperature and 60 F

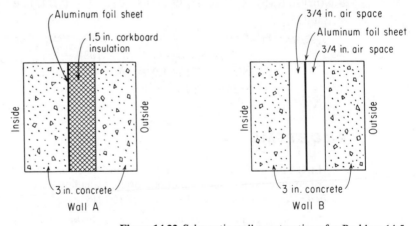

Figure 14.22 Schematic wall constructions for Problem 14.5.

thermodynamic wet-bulb temperature. Barometric pressure is 14.696 psia. Outside temperature is 10 F and outside wind velocity is 15 MPH. Which wall is thermally superior?

14.6. Figure 14.23 schematically shows a ceiling electric heating panel. The panel is made of a thin sheet of metal and is kept at a uniform temperature of 110 F. Assume steady-state conditions. Calculate the thickness of insulation required if 5 per cent of the output of the panel is lost to the attic.

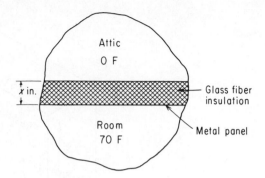

Figure 14.23 Schematic ceiling panel for Problem 14.6.

14.7. A spherical space vehicle, rotating continuously, is exposed to solar radiation. Its surface temperature is maintained at 65 F by covering part of the surface with a material (Material 1) having a solar absorptivity of 0.6 and an emissivity of 0.4 and covering the rest of the surface with a "gray" material (Material 2) having an absorptivity (and emissivity) of 0.1. Intensity of solar radiation normal to the sun's rays is 400 Btu per (hr)(sq ft). What fraction of the surface needs to be covered by each material?

14.8. Measurements on a clear day with a pyrheliometer indicate that a certain glass sample ($n = 1.526$), 0.25 in. thick, has a transmissivity of 0.82 when the sun's angle of incidence is zero. Through use of monochromatic formulae, calculate the transmissivity for an angle of incidence of 75 deg.

14.9. A window faces due east. The glass ($n = 1.526$, $K - 3.30$ in.$^{-1}$) is 0.25 in. thick. Calculate the monochromatic absorptivity of the glass on July 1 at 9:00 a.m. solar time for a location at 42 deg north latitude.

14.10. Check F for Case B-3 in Table 14.5 by performing the integration in Eq. (14.27).

14.11. Determine the magnitudes of direct, diffuse sky, and reflected solar radiation incident upon a south-facing vertical surface on March 1 at solar noon for a location at 48 deg north latitude having a clearness number of 0.95. The reflecting surface is the ground (infinite extent) covered by a fresh layer of snow having a diffuse reflectivity of 0.85.

14.12. A building is oriented as shown in Fig. 14.24. Assume that the east-facing wall has an average diffuse solar reflectivity of 0.5. Calculate the intensity of solar radiation incident upon the window in the south wall which is reflected by the east wall at 9:00 a.m. solar time on January 1. Location is at 36 deg north latitude. Assume a clearness number of 0.95.

14.13. For the same conditions as Problem 14.12, calculate the intensity of reflected radiation incident upon the window if the east-facing wall has a specular reflectivity of 0.5.

14.14. A window panel 8 ft high in a south-facing vertical wall is located

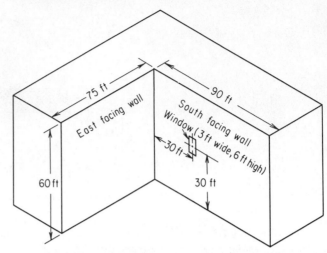

Figure 14.24 Schematic building for Problems 14.12, 14.13, and 14.16.

at 45 deg north latitude. The glass has no set-back. It is desired to design a solid overhang such that the window will just be completely shaded at solar noon on the day of the summer solstice, but further that the window will just be completely sunlit at solar noon on the day of the winter solstice. Determine the width of the overhang and the distance above the top of the window at which the overhang should be located.

14.15. Two downtown buildings are arranged as schematically shown in Fig. 14.25. Location is at 42 deg north latitude. Find the dimensions of the south

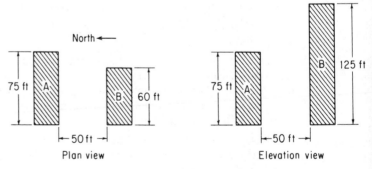

Figure 14.25 Schematic buildings for Problem 14.15.

side of Building A shaded from direct sunlight by Building B at 11 : 00 a.m. solar time on September 1.

14.16. For the building of Fig. 14.24 and for the same location as Problem 14.12, calculate the dimensions of the south-facing wall shaded from direct sunlight by the east-facing wall at 1 : 00 p.m. solar time on July 1.

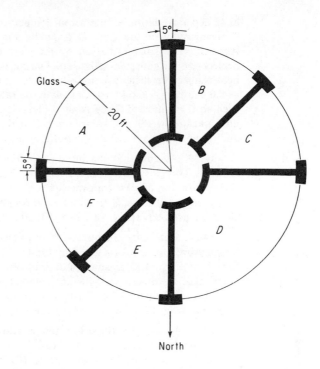

Figure 14.26 Schematic building for Problems 14.17 and 14.19.

14.17. Figure 14.26 shows a schematic plan view of one floor of a cylindrically-shaped office building. Location is at 45 deg north latitude and 93 deg west longitude. The glass panels are 9 ft high. Even with the top of each glass panel is a continuous solid overhang whose outer edge describes an arc of a circle of radius 22 ft. Find the surface area of the glass wall of Office A shaded from direct sunlight on a clear August 1 day at 11:00 a.m. Central Standard Time. Describe the approximate shape of the shaded area.

14.18. Figure 14.27 shows a schematic cross section of a greenhouse. Roof A faces due east and has an unobstructed vision of the sky. Location is

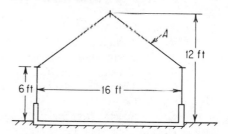

Figure 14.27 Schematic cross-section of a greenhouse.

at 42 deg north latitude. Interior air temperature is maintained constant at 60 F. Transmissivity of the glass is given by Fig. 14.10 (common window glass). Absorption of solar radiation by the glass may be neglected. You may also neglect ground-reflected radiation. During daylight hours, the outside-surface heat transfer coefficient h_o may be taken as 6.0 Btu per (hr)(sq ft)(F), at night as 8.0 Btu per (hr)(sq ft)(F). Calculate the rate of energy transmission through each sq ft of glass of Roof A on a clear January 1 day (clearness number of unity) at (a) 10:00 a.m. solar time when the outdoor air temperature is 0 F, and (b) at 10:00 p.m. when the outdoor air temperature is −10 F.

14.19. Assume that the building of Fig. 14.26 is located at 36 deg north latitude. The glass panels are 9 ft high. Figure 14.10 (soda-lime plate glass) shows variation of glass transmissivity. Calculate the rate in Btu per hr at which direct solar radiation is transmitted through the glass wall of Office A on a clear August 1 day (C.N. = 1.0) at 10:00 a.m. solar time.

14.20. Determine the numerical coefficients for the equation for the sol-air temperature t_e of the form given by Eq. (14.45) for the same conditions as those shown in Fig. 14.17 except for a south-facing vertical surface. Include two harmonics. Assume a wall-surface solar absorptivity of 0.70 and an outside-surface coefficient h_o of 4.00 Btu per (hr)(sq ft)(F). Assume that the ground surface in front of the wall is of infinite extent with a diffuse reflectivity of 0.30.

14.21. Based upon the sol-air temperature variation determined in Problem 14.20, calculate the rate of heat transmission q_i to the interior through each sq ft of a 12 in. common brick wall at (a) 10:00 a.m. solar time, and (b) at 8:00 p.m. solar time.

14.22. Calculate the instantaneous rate of heat gain, Btu per (hr)(sq ft), through a sunlit flat-roof construction similar to No. 5 of Table 14.8 at 3:00 p.m. solar time on an average clear day at Minneapolis for (a) January, (b) April, (c) July, and (d) October. Assume a constant interior temperature of 75 F.

14.23. Roofs 2 and 3 of Table 14.8 differ only in order of placement of Layers A and B. Compare the maximum rate of heat gain and corresponding solar time for these two roof constructions for a summer design day at Minneapolis. Assume an interior temperature of 75 F.

SYMBOLS USED IN CHAPTER 14

A	Constant, F.
A	Surface area, sq ft.
a	Absorption coefficient, dimensionless.
B	Constant, F per ft.
C	Thermal conductance, Btu per (hr)(sq ft)(F).
C_n	Constants, F.
c_w	Specific heat for wall material, Btu per (lb)(F).
D_n	Constants, F.
d	Sun's declination angle.

E Effective emissivity defined by Eq. (14.10), dimensionless.

F Factor defined by Eq. (14.27), dimensionless.

F_s Sunlit fraction of window, dimensionless.

h Sun's hour angle.

h_c Convection heat transfer coefficient, Btu per (hr)(sq ft)(F); $h_{c,i}$ for wall-inside surface; $h_{c,o}$ for wall-outside surface.

h_i Combined convection and radiation heat transfer coefficient for wall-inside surface as defined by Eq. (14.2), Btu per (hr)(sq ft)(F).

h_o Combined convection and radiation heat transfer coefficient for wall-outside surface as defined by Eq. (14.3), Btu per (hr)(sq ft)(F).

h_R Radiation heat transfer coefficient, Btu per (hr)(sq ft)(F); $h_{R,i}$ for wall-inside surface; $h_{R,o}$ for wall-outside surface.

I Incidence of total solar radiation upon a surface, Btu per (hr)(sq ft); I_H refers to horizontal surface; I_V refers to vertical surface.

I_D Incidence of direct solar radiation upon a surface, Btu per (hr)(sq ft); I_{DH} refers to horizontal surface; I_{DV} refers to vertical surface.

I_d Incidence of diffuse sky radiation upon a surface, Btu per (hr)(sq ft); I_{dH} refers to horizontal surface; I_{dV} refers to vertical surface.

I_N Incidence of direct solar radiation upon a surface normal to sun's rays, Btu per (hr)(sq ft).

I_R Incidence of solar radiation upon a surface, Btu per (hr)(sq ft), which is reflected by surrounding surfaces; I_{RV} refers to a vertical surface.

I_λ Incidence of monochromatic solar radiation upon a surface, Btu per (hr)(sq ft).

K Extinction coefficient for glass, in.$^{-1}$ or ft^{-1}.

k_w Thermal conductivity of wall material, Btu per (hr)(sq ft)(F per ft) or Btu per (hr)(sq ft)(F per in.).

L Thickness of material, in. or ft.

l Latitude angle.

M_n Constants in Eqs. (14.44)–(14.46), F.

N_n Constants in Eqs. (14.44)–(14.46), F.

n Index of refraction for glass as defined by Eq. (2.9), dimensionless.

n Order of harmonic in Fourier series, dimensionless, 1 for first harmonic, 2 for second harmonic, etc.

P_n Constants, ft^{-1}.

q Rate of heat transfer, Btu per (hr)(sq ft); q_i refers to rate of heat transfer from wall-inside surface to interior; q_o refers to rate of heat transfer from external thermal environment to wall-outside surface.

$q_{g,s}$ Rate of change of heat stored in glass, Btu per (hr)(sq ft).

R Thermal resistance, (hr)(sq ft)(F) per Btu; R_t refers to total resistance; R_i refers to wall-inside surface; R_o refers to wall-outside surface; R_w refers to wall material.

r Specular reflectivity for single radiation component, dimensionless.

T Absolute temperature, R; T_1 refers to warm surface, T_2 refers to cold surface.

t Temperature, F; t_1 refers to warm surface; t_2 refers to cold surface; t_i refers to inside air; t_o refers to outside air.

t_e Sol-air temperature, F; $t_{e,m}$ refers to 24 hr-mean value; $t_{e,n}$ refers to harmonic coefficient.

t_g Glass temperature, F; $t_{g,i}$ refers to inside surface; $t_{g,o}$ refers to outside surface.

t_s Temperature of surrounding surfaces, F; $t_{s,i}$ refers to interior surrounding surfaces; $t_{s,o}$ refers to exterior surrounding surfaces and/or sky.

t_w Wall temperature, F; $t_{w,i}$ refers to inside surface; $t_{w,o}$ refers to outside surface.

U Overall heat transfer coefficient, Btu per (hr)(sq ft)(F).

V_n Factor given by Eq. (14.52), Btu per (hr)(sq ft)(F).

x Thickness of wall, in.

Y_n Factor defined by Eq. (14.54), dimensionless.

Z_n Factor defined by Eq. (14.55), dimensionless.

α Absorptivity for solar radiation, dimensionless; α_λ refers to monochromatic radiation; α_D refers to direct radiation; α_d refers to diffuse sky radiation; α_R refers to reflected radiation.

α Wall-solar azimuth angle.

α_w Thermal diffusivity $k_w/\rho_w c_w$, sq ft per hr.

β Sun's altitude angle.

γ Sun's azimuth angle.

δ Sun's profile angle.

ϵ Emissivity for long-wave radiation, dimensionless.

θ Time, hr.

θ Incidence angle.

θ' Refraction angle (see Fig. 14.7).

θ'' Specular reflection angle (see Fig. 14.7).

λ_n Decrement factor given by Eq. (14.59), dimensionless.

μ Microns.

ρ Surface reflectivity, dimensionless; ρ_λ refers to monochromatic radiation.

ρ_w Wall density, lb per cu ft.

σ_n Factor given by Eq. (14.53), ft^{-1}.

τ Transmissivity, dimensionless; τ_λ refers to monochromatic radiation; τ_D refers to direct solar radiation; τ_d refers to diffuse sky radiation; τ_R refers to reflected radiation.

Φ_n Lag angle defined by Eq. (14.56).

ϕ Angle of tilt from vertical position.

ψ_n Angle defined by Eq. (14.46).

ω_n Angular velocity of sinusoidal wave, radians per hr.

Part V

APPLICATIONS

15

SOME EFFECTS
OF THERMAL ENVIRONMENT

15.1. INTRODUCTION

A topic of fundamental importance in our study of thermal environmental engineering
is the effect of the thermal environment upon people, processes, and materials.

From an engineering point of view, the human body may be likened to a heat
engine. The body functions to convert the chemical energy of its food into work
and heat. As is true with an engine, the harder we exercise or work, the greater is the
amount of heat we reject. With the human body, heat rejection occurs primarily
from the body surface. Through blood circulation, heat is transported to the skin
from which it is transferred to the environment. The body must continually reject
heat, in summer as well as in winter. Temperature and humidity of the environment
may profoundly influence the body's skin and interior temperature. Control of the
thermal environment is necessary if comfort conditions are to be maintained or if
physiological hazards are to be avoided in hot industries.

There are many requirements in industry for control of the thermal environment
apart from human considerations. Many processes in manufacturing involve hygroscopic materials or chemical reactions. Temperature and/or humidity control may be
mandatory if a satisfactory result is to be produced. We may also consider the growing

361

of plants and animals in the scope of industrial processing. The use of controlled environments offers much promise in the increase of food production.

An important use of thermal environmental control occurs in the storage of various materials. Corrosion of metallic products may be substantially reduced or eliminated by maintenance of proper atmospheric humidity. In the food industry, various products may be maintained in excellent condition for long periods of time by storage under proper environmental conditions.

15.2. THERMAL EXCHANGES OF BODY WITH ENVIRONMENT

Heat exchanges between the body and its environment may be generally expressed as

$$q_M = \pm q_c \pm q_R \pm q_E \pm q_s \qquad (15.1)$$

where q_M is the rate of body heat production, Btu per hr; q_c is the rate of convection heat exchange with surrounding air, Btu per hr; q_R is the rate of radiation heat exchange with surrounding surfaces, Btu per hr; q_E is the rate of heat loss by evapora-

TABLE 15.1*

ESTIMATES OF BODY HEAT PRODUCTION FOR VARIOUS TYPES
OF ACTIVITY

Kind of Work	Activity	q_m† Btu/hr
Light Work	Sleeping	250
	Sitting quietly	400
	Sitting, moderate arm and trunk movements (e.g., desk work, typing)	450–550
	Sitting, moderate arm and leg movements (e.g., playing organ, driving car in traffic)	550–650
	Standing, light work at machine or bench, mostly arms	550–650
Moderate Work	Sitting, heavy arm and leg movements	650–800
	Standing, light work at machine or bench, some walking about	650–750
	Standing, moderate work at machine or bench, some walking about	750–1000
	Walking about, with moderate lifting or pushing	1000–1400
Heavy Work	Intermittent heavy lifting, pushing or pulling (e.g., pick and shovel work)	1500–2000
	Hardest sustained work	2000–2400

*Reprinted by permission from *ASHRAE Handbook of Fundamentals*, 1967, p. 119.
†Values apply for a 154 lb man and do not include rest pauses.

tion, Btu per hr; and q_s is the rate of change of heat stored in the body, Btu per hr. Assuming an unclothed man, Eq. (15.1) may be written as

$$q_M = h_c A_b(t_b - t) + h_R A_b'(t_b - t_s) + h_D A_b(W_{s,b} - W)h_{fg,b} \pm q_s \qquad (15.2)$$

where h_c and h_r are respectively convection and radiation heat transfer coefficients, Btu per (hr) (sq ft) (F); h_D is a convection mass transfer coefficient, lb_w per (hr) (sq ft) (lb_w per lb_a); t_b, t, and t_s are respectively the temperature of the mean body surface, the air, and surrounding surfaces, F; A_b is the body surface area, sq ft; A_b' is the body surface area which sees surrounding surfaces, sq ft; W and $W_{s,b}$ are respectively the humidity ratio of surrounding air and of saturated air at t_b, lb_w per lb_a; and $h_{fg,b}$ is the latent heat of vaporization of water at t_b, Btu per lb_w. For an average adult male, A_b is about 20 sq ft and A_b' is about 16 sq ft.

The body heat production rate q_M is largely dependent upon muscular activity and may vary from less than 400 Btu per hr for an adult seated at rest to more than 2500 Btu per hr for heavy work. Table 15.1 shows rates of body heat production for an average adult male for various types of activity. Table 15.2 shows rates of heat gain and moisture gain from occupants of conditioned spaces.

Equation (15.2) shows that the convection rate of heat loss may be increased by lowering the air dry-bulb temperature or by increasing the convection heat transfer

TABLE 15.2*

RATES OF HEAT GAIN AND MOISTURE GAIN FROM OCCUPANTS
OF CONDITIONED SPACES[a]

Degree of Activity	Typical Application	Sensible Heat Gain Btu/hr	Latent Heat Gain Btu/hr	Total Heat Gain Btu/hr	Moisture Gain lb/hr
Seated at rest	Theater-matinee . . .	225	105	330	0.11
	Theater-evening . . .	245	105	350	0.11
Seated, very light work . . .	Offices, hotels	245	155	400	0.16
Moderately active office work, standing, light work, or walking slowly	Offices, etc., department store, drug store, bank	250	200	450	0.21
Sedentary work	Restaurant[b]	275	275	550	0.29
Light bench work	Factory	275	475	750	0.50
Moderate dancing	Dance hall	305	545	850	0.58
Moderately heavy work . .	Factory	375	625	1000	0.66
Heavy work	Factory	580	870	1450	0.92

*Adapted by permission from a similar table in *ASHRAE Handbook of Fundamentals*, 1967, p. 497.

[a]Tabulated values for 80 F dry-bulb temperature. Heat and moisture gains adjusted for normal percentage of men, women, and children present.

[b]Includes an allowance for heat and moisture gains from food.

coefficient for a given body surface temperature. The convection coefficient may be substantially increased by increase of air velocity.

The radiation loss is primarily dependent upon the mean temperature of surrounding surfaces for a given body surface temperature. Typically, the temperature t_s is approximately equal to the air dry-bulb temperature t, which allows combination of convection and radiation transfer into one expression. This combination represents the rate of sensible heat loss from the body, provided that the environmental temperature is less than the body surface temperature.

The evaporation loss is dependent upon the mass transfer coefficient and the air humidity ratio for a given body surface temperature. Similar to convection loss, the coefficient h_D is sensitive to air velocity. The rate of evaporation may be significantly increased by increase of air movement across the body.

The body is seldom in a steady-state heat-transfer relation with its environment, so that q_s in Eq. (15.2) is ordinarily not zero. This quantity may vary from one of negligibly small magnitude where body exercise and environmental conditions are relatively stable to one of sizable magnitude where conditions are unstable and where abnormal exposure occurs.

Figure 15.1 shows rates of heat production and heat loss for a clothed subject seated at rest. The quantity q_s in Eq. (15.2) may be estimated from Fig. 15.1 as the difference between heat production and total heat loss. Figure 15.1 was determined

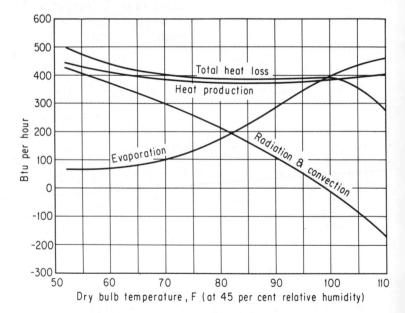

Figure 15.1 Body heat production and environmental heat exchanges for a healthy, young, clothed man seated at rest. (Adapted by permission from *ASHRAE Handbook of Fundamentals*, 1967, p. 114.)

from calorimeter tests on healthy, male subjects, 18 to 24 years of age. It shows that at lower air temperatures, body heat loss is accomplished primarily by convection and radiation; at higher temperatures, the evaporation loss becomes controlling. At elevated environmental temperatures, rates of transfer by convection and radiation become negative and evaporation must offset these heat gains as well as body heat production.

15.3. BODY REGULATORY PROCESSES AGAINST
HEAT OR COLD

A man may knowingly assist his body in maintaining its proper heat balance under exposure conditions by measures such as providing suitable clothing. However, the human body has involuntarily-initiated regulation means for adjusting itself to either heat or cold exposures. The basic purpose of these adjustments is to prevent abnormal change of interior body temperature which would impair the vital organs.

For a man exercising with some constant and moderate rate of exertion, there is a narrow range of atmospheric conditions in which his body needs to take no particular action to maintain its proper heat balance. This range of conditions may be called the *neutral zone* since the body is neutral to feelings of heat or cold. The man's degree of activity and amount of clothing affect the level of the neutral zone. For a resting nude man, the neutral zone occurs within a range of approximately 81–86 F air dry-bulb temperature.

For atmospheric conditions colder than those of the neutral zone, the body may be able to preserve a proper temperature in its deep tissues by decrease of blood flow at the body surface and by additional heat production. For mild exposure to cold there is a range of atmospheric conditions immediately below the neutral zone called the *zone of vaso-motor regulation against cold*. Within this zone, blood vessels adjacent to the skin constrict, preventing flow of blood and transport of heat to the immediate outer surface. The outer skin tissues essentially become an insulating layer. The quantity q_s in Eq. (15.2) becomes negative and the body surface temperature decreases somewhat, but cooling of deep tissues may be avoided. For still lower environmental temperatures, restriction of surface blood flow does not provide adequate protection. This range of conditions may be called the *zone of metabolic regulation against cold*. Through spontaneous increase of activity and by shivering, body heat production is increased in an effort to prevent q_s from remaining negative and to prevent the body surface temperature from further decreasing. Beyond this range of conditions the body is unable to combat cooling of its tissues and disastrous results may occur. This final range of conditions is known as the *zone of inevitable body cooling*.

On the warm side of the neutral zone there exists first a *zone of vaso-motor regulation against heat*. Here the surface blood vessels dilate and allow blood flow as close as possible to the outer surface. The skin temperature increases, providing a greater temperature difference for loss of heat by convection and radiation and allowing $W_{s,b}$ in Eq. (15.2) to become larger. For still warmer conditions there is a *zone of evaporative regulation against heat* where the body reacts in a powerful manner to prevent further rise of skin temperature. The sweat glands become highly active drenching the body surface with perspiration. If air humidity and velocity permit sufficiently rapid evaporation, further rise in body temperature may be prevented. This is the last line of defense, however, and if q_s remains positive the body enters the *zone of inevitable body heating*.

15.4. PHYSIOLOGICAL HAZARDS RESULTING FROM HEAT EXPOSURE

When an individual encounters the zone of inevitable body heating, q_s in Eq. (15.2) may be strongly positive and interior body temperature will rise. Several physiological hazards exist, the severity of which depends upon the extent and time duration of body temperature rise.

Heat exhaustion is due to failure of normal blood circulation. Symptoms of heat exhaustion include fatigue, headache, dizziness, vomiting, and abnormal mental reactions such as irritability. Severe heat exhaustion may cause fainting. Heat exhaustion usually causes no permanent injury, and recovery is usually rapid when the subject is removed to a cool place.

Heat cramps result from loss of salt due to an excessive rate of body perspiration. They are painful muscle spasms which may be largely avoided by proper use of salt tablets.

Heat stroke is the most serious hazard. When a man is exposed to excessive heat, body temperature may climb rapidly to 105 F or higher. At such elevated temperatures, sweating ceases and the subject may enter a coma, with death imminent. People experiencing a heat stroke may have permanent damage to the brain.

15.5. APPLICATION OF PHYSIOLOGICAL PRINCIPLES TO COMFORT AIR CONDITIONING PROBLEMS

Human comfort is influenced by psychological factors as well as physiological factors. There is no precise method of stating what thermal environmental condition will effect a comfort feeling in a human being. Man's thermal comfort is influenced not only by ambient air temperature and humidity, but also by rate of air movement, body activity, and amount of clothing.

Acclimatization is an important factor affecting comfort. In winter we become adjusted to somewhat cooler temperatures than in summer. People living in warm climates feel comfortable in somewhat warmer surroundings than their northern neighbors.

It is difficult to specify a single physical quantity for evaluating human comfort.

For a single individual, mean skin temperature would probably most closely suffice.

The American Society of Heating, Refrigerating and Air Conditioning Engineers (ASHRAE) has studied human reactions to environmental temperature, humidity, and air movement for many years. Its research resulted in the concept of a single empirical index called *effective temperature*, which is defined as that index which correlates the combined effects of air temperature, air humidity, and air movement upon human comfort. It was established by experiments involving the use of trained subjects who compared comfort conditions in adjoining air conditioned test rooms. The numerical value of effective temperature associated with some atmospheric condition is given by the temperature of slowly moving saturated air which indicates a like feeling of warmth or cold.

Practical application of the concept of effective temperature is presented by the ASHRAE *Comfort Chart* reproduced in Fig. 15.2. This chart is applicable to reasonably still air conditions (velocity of 15 to 25 ft per min), to situations where occupants are seated at rest or doing light work, and to spaces whose enclosing surfaces are at a mean temperature equal to the air dry-bulb temperature. If surrounding surfaces are below the air dry-bulb temperature, comfort would occur at a higher effective temperature than that indicated by Fig. 15.2. An example could be a person seated adjacent to a large window expanse on a winter day. On the other hand, with panel-heated rooms, somewhat lower effective temperatures are customary. Counter-radiation between occupants in densely populated spaces requires somewhat lower effective temperatures than in more sparsely occupied rooms.

Field studies performed for ASHRAE have indicated that, typically, women of all age groups prefer an effective temperature about one degree higher than that preferred by men, while both men and women over 40 years of age prefer an effective temperature about one degree higher than that desired by younger people.

Geographical location must also be considered. Many engineers allow one degree increase in effective temperature for each five degrees reduction in latitude. Figure 15.2 applies directly to cities at about 42 degrees north latitude.

Figure 15.3 shows how air velocity alters the normal effective temperature scale. For the example shown, atmospheric conditions of 76 F dry-bulb temperature and 62 F thermodynamic wet-bulb temperature correspond to about 70 F effective temperature with nominally still air, but correspond to less than 64 F at an air velocity of 700 ft per min.

Activity of the occupants in an air conditioned space and the duration of occupancy are additional factors. Figure 15.2 applies to people doing sedentary work. People working more actively will be comfortable at lower effective temperatures. Particularly in summer, duration of occupancy must be considered. In spaces where the normal occupant stays only a short time, a higher effective temperature should be carried than that indicated by Fig. 15.2 which considers occupancy of three hours or more.

Figure 15.2 indicates that in winter an effective temperature of 67–68 F is optimum for normally clothed people. Such a comfort condition may supposedly be

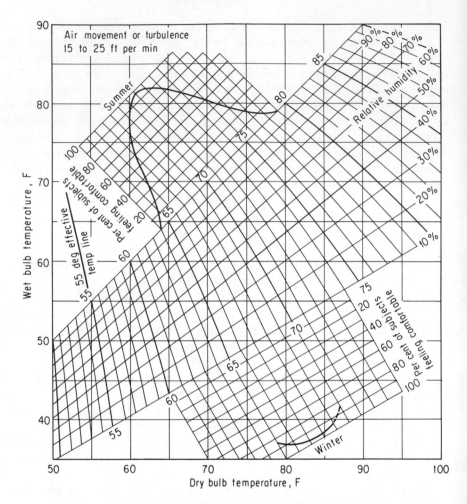

Figure 15.2 ASHRAE Comfort Chart. (Adapted by permission from *ASHRAE Handbook of Fundamentals*, 1967, p. 122.)

attained by various combinations of temperature and humidity. In northern climates, space relative humidity during cold outside weather cannot exceed much above 25–30 per cent without condensation of moisture upon inner surfaces of double-glazed windows. However, if no artificial humidification is provided during extremely low-temperature weather, space relative humidity may fall to less than 10 per cent. Such a dry condition may cause over-dehydration of skin and membrane surfaces in people and excessive drying-out of floors, veneered furniture, books, and other hygroscopic materials.

Figure 15.2 shows that in summer an effective temperature of about 71 F should be optimum. Common practice is to design comfort air conditioning systems for summer

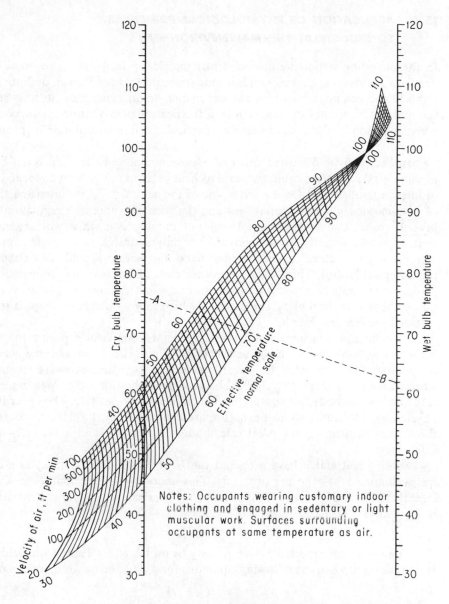

Figure 15.3 Variation of effective temperature with air velocity. (Adapted by permission from *ASHRAE Handbook of Fundamentals*, 1967, p. 117.)

to provide space conditions of about 76 F dry-bulb temperature and 50 per cent relative humidity.

15.6. APPLICATION OF PHYSIOLOGICAL PRINCIPLES TO INDUSTRIAL THERMAL ENVIRONMENTS

In the preceding section we discussed physiological principles as applied to human comfort. An ideal comfort air conditioning system would produce an optimum space-atmospheric condition in which the occupants would experience no heat stress or thermal strain. A comfort condition is in the neutral zone of atmospheric conditions where a person's body needs to take no particular action to maintain its proper heat balance.

In Sec. 15.3 we discussed zones of exposure conditions for which the body by physiological regulation could maintain its heat balance and prevent abnormal change of interior temperature. On the warm side of the neutral zone, we discussed the zone of vaso-motor regulation against heat and the zone of evaporative regulation against heat. There are many environmental situations in industry in which workers encounter heat exposure. Usually, it is impossible or impracticable to attempt comfort air conditioning in such cases. On the other hand, the worker should not be exposed to physiological hazard. Thus, there is another class of thermal environmental control problems, separate from comfort air conditioning, but still involving human reactions to the thermal environment. The material which follows has been adapted from the work of Belding and Hatch.*

In industrial environmental control problems, some heat exposure in excess of comfort conditions must usually be accepted. The worker is under some heat stress, but it should be kept below the hazardous level. Under heat exposure conditions, a working man may establish a thermal balance but only with active sweating, elevated skin and interior body temperature, and an accelerated heart rate. The upper limits of physiological tolerance to heat exposure are determined by the highest limit to which the indices of heat strain—sweat rate, body temperature, and heart rate—may be safely elevated.

Belding and Hatch have proposed the *Heat Stress Index* (H.S.I.) as a concept for evaluating heat exposure problems. This concept is based upon certain physiological criteria and upon the fundamental body heat balance equation. If we assume q_s in Eq. (15.1) is negligible, then

$$q_M \pm q_c \pm q_R = q_E \tag{15.3}$$

The body heat production rate q_M may be estimated by Table 15.1. Belding and Hatch have proposed the following equations for q_c, q_R, and q_E for a nude man:

$$q_c = 2V^{0.5}(t_b - t) \tag{15.4}$$

*H. S. Belding and T. F. Hatch, "Index for Evaluating Heat Stress in Terms of Resulting Physiological Strains," *ASHAE Trans.*, Vol. 62, 1956, pp. 213–236.

$$q_R = 22(t_b - t_s) \tag{15.5}$$

$$q_E = 517V^{0.4}(P_{w,s,b} - P_w) \tag{15.6}$$

where V is air velocity, ft per min; $P_{w,s,b}$ is the pressure of saturated water vapor, psia, at the body surface temperature t_b; P_w is the partial pressure of water vapor in the air, psia; and other terms are the same as those defined in Eq. (15.2). For sea-level pressure, Eq. (15.6) may also be expressed approximately as

$$q_E = 11,700V^{0.4}(W_{s,b} - W) \tag{15.7}$$

where $W_{s,b}$ and W are respectively the humidity ratio of saturated air at t_b and the humidity ratio of the bulk air stream, lb_w per lb_a.

Two important physiological criteria in the H.S.I. concept are (1) in order to limit rise of body temperature, average skin temperature should not exceed 95 F, and (2) in order to limit loss of body fluids, the sweat rate should not exceed one litre per hour (2400 Btu per hr). The Heat Stress Index is defined by

$$\text{H.S.I.} = (100)\frac{q_E}{q_{E,max} \le 2400} \tag{15.8}$$

where

$$q_E = q_M + 2V^{0.5}(t - 95) + 22(t_s - 95) \tag{15.9}$$

$$q_{E,max} = 517V^{0.4}(0.816 - P_w) \tag{15.10}$$

The quantity q_E in Eqs. (15.8) and (15.9) is the evaporative cooling required to balance body heat production and convection and radiation heat gains under the given environmental conditions with the average body surface temperature at 95 F. The quantity $q_{E,max}$ is the available evaporative cooling rate under the given environmental conditions and with the average temperature of the wet body surface at 95 F, providing that the cooling rate is not greater than 2400 Btu per hr. If the numerical value of the H.S.I. is less than 100, the average skin temperature is less than 95 F.

> **Example 15.1.** A man performs moderate work ($q_M = 900$ Btu per hr) in an environment where the air is at 100 F dry-bulb temperature and 70 F thermodynamic wet-bulb temperature. Surrounding surfaces are at a mean temperature of 90 F. Air velocity is 75 ft per min. Assume sea-level pressure. Calculate the H.S.I.
>
> SOLUTION: By Fig. E-8, $W = 0.00883$ lb_w/lb_a. By Eq. (8.22),
>
> $$P_w = \frac{1.608PW}{1 + 1.608W} = \frac{(1.608)(14.696)(0.00883)}{1 + (1.608)(0.00883)} = 0.206 \text{ psia}$$
>
> By Eq. (15.9)
>
> $$q_E = 900 + (2)(75^{0.5})(5) - (22)(5) = 877 \text{ Btu/hr}$$
>
> By Eq. (15.10),
>
> $$q_{E,max} = (517)(75^{0.4})(0.61) = 1774 \text{ Btu/hr}$$
>
> By Eq. (15.8),
>
> $$\text{H.S.I.} = \frac{(100)(877)}{1774} = 49$$

The physiological criteria and definition of the H.S.I. is based upon a so-called standard young man (5 ft 8 in. tall, 154 lb) in good physical condition and acclimatized to heat. Belding and Hatch postulated that such a man could safely engage in simple physical work in a certain thermal environment over an eight-hour day provided his average skin temperature did not exceed 95 F and his average sweat rate did not exceed one litre per hour. Table 15.3 shows an interpretation of the H.S.I. in regard to its physiological and hygienic implications as given by Belding and Hatch. They believe that no more than 10 to 20 per cent of the working population could be expected to work successfully under an H.S.I. of 100.

The H.S.I. concept provides a seemingly sound basis for dealing with thermal problems in industrial environmental control. One problem in its application is evaluation of the effect of clothing. Determination of endurable values of the H.S.I. for a particular situation is a medical problem, but the problem of altering or controlling the thermal environment is an engineering one.

TABLE 15.3*

EVALUATION OF HEAT STRESS INDEX

Index of Heat Stress	Physiological and Hygienic Implications of 8-hr Exposures to Various Heat Stresses
−20 −10	Mild cold strain. This condition frequently exists in areas where men recover from exposure to heat.
0	No thermal strain.
+10 20 30	Mild to moderate heat strain. Where a job involves higher intellectual functions, dexterity, or alertness, subtle to substantial decrements in performance may be expected. In performance of heavy physical work, little decrement expected unless ability of individuals to perform such work under no thermal stress is marginal.
40 50 60	Severe heat strain, involving a threat to health unless men are physically fit. Break-in period required for men not previously acclimatized. Some decrement in performance of physical work is to be expected. Medical selection of personnel desirable because these conditions are unsuitable for those with cardiovascular or respiratory impairment or with chronic dermatitis. These working conditions are also unsuitable for activities requiring sustained mental effort.
70 80 90	Very severe heat strain. Only a small percentage of the population may be expected to qualify for this work. Personnel should be selected (a) by medical examination and (b) by trial on the job (after acclimatization). Special measures are needed to assure adequate water and salt intake. Amelioration of working conditions by any feasible means is highly desirable, and may be expected to decrease the health hazard while increasing efficiency on the job. Slight "indisposition" which in most jobs would be insufficient to affect performance may render workers unfit for this exposure.
100	The maximum strain tolerated daily by fit, acclimatized young men.

*Reprinted by permission from *ASHAE Trans.*, Vol. 62, p. 226.

In studying an industrial environmental problem where a lower H.S.I. is required, it is particularly useful to analyze Eqs. (15.9) and (15.10) in order to recognize a feasible control measure. The H.S.I. may be reduced by decrease of ambient air temperature. This may be possible through general ventilation of the space with outdoor air, through spot cooling with refrigerated air or by other means. Another effective control measure may be increase of air velocity over the worker.

Thermal radiation is often the mechanism of heat exchange which causes a high H.S.I. The temperature t_s in Eq. (15.9) may be reduced through insulating of hot surfaces or by the use of radiation shields. The coefficient of 22 was determined on a basis of the worker's surface being non-reflective. Through use of aluminumized clothing this coefficient may be reduced and the H.S.I. reduced, providing that the evaporation rate is not diminished by the metallic covering.

It should be realized that throughout this chapter, heat exchange equations between the body and its environment have made no allowance for exposure to solar radiation. A high value of the H.S.I. may occur due to such exposure.

15.7. INDUSTRIAL PROCESSES

There are many industrial processes which must be carried out under controlled atmospheric conditions. Representative processes include brewing, candy making, packaging of cereals, drying of leather, manufacture of pharmaceuticals, printing, and textile manufacture. An engineer who deals with industrial air conditioning must have a thorough understanding of the process involved and the effect of atmospheric conditions upon it. A comprehensive treatment of the subject is beyond the scope of this book. More detailed information may be found in the *ASHRAE Guide and Data Book** and in references given therein.

The *ASHRAE Guide and Data Book* has classified industrial air conditioning problems into eight categories. These are the control of (1) regain, (2) rate of chemical reactions, (3) rate of biochemical reactions, (4) rate of crystallization, (5) temperature for close tolerance machining and grinding, (6) dew point for protection of highly polished surfaces, (7) humidity for static electricity elimination, and (8) conditions for material test laboratories. The *ASHRAE Guide and Data Book* also gives recommended atmospheric conditions for many industrial processes. Conditions vary widely depending upon the material and the nature of the process. In particular, low air humidity is required for many processes.

Many manufacturing processes involve *hygroscopic* materials, which may contain *bound* moisture. A hygroscopic material has sorbent properties and its moisture content will approach equilibrium with the moisture content of the surrounding air. The *equilibrium moisture content* is that value to which a material may be dried under

ASHRAE Guide and Data Book, Applications (New York: American Society of Heating, Refrigerating and Air Conditioning Engineers, 1968) pp. 89–94.

definite conditions of air temperature and humidity. The equilibrium moisture content of various materials differs widely.

The total moisture content of a material is the sum of its *free moisture content* and its *equilibrium moisture content*. The equilibrium moisture content is also known as *regain*. Regain is usually expressed as a percentage of the bone-dry weight of the material. Table 15.4 shows equilibrium moisture content for several hygroscopic materials. Values shown are for exposure to air at 75 F and various relative humidities. However, temperature exerts a minor effect as relative humidity is the principal property which influences moisture content in hygroscopic materials. Table 15.4 shows that for all the materials, equilibrium moisture content increases with increase of relative humidity, and in the case of many materials, it assumes large percentage values of the dry weight.

With materials of the type shown in Table 15.4, particularly those of organic origin, atmospheric conditions may have an important influence upon the weight, strength, and appearance of the finished product. The moisture content of the material must be controlled in order to achieve rapid production with a minimum of material damage. This requires control of the thermal environment to which the material is exposed.

Many manufacturing processes involve chemical reactions. Temperature and humidity control are important since the rate of the reaction is dependent upon temperature, while evaporation from a solution surface is affected by air humidity. Examples of such processes include manufacture of rayon, drying of varnish, and curing of fruits such as lemons and bananas.

Only two more of the many classes of processes where control of atmospheric conditions are necessary will be mentioned here. When the product involves highly-polished metallic surfaces, great care may be needed to avoid deposition of perspiration from the workers' hands upon the product. The ambient temperature and humidity must be kept sufficiently low to prevent sweating of the hands. In processes where explosive conditions may be present, consideration must be given to the prevention of sparks due to static electricity. Presence of static electricity is greatly reduced by increase of air humidity. If possible, a relative humidity greater than 55 per cent should be carried.

15.8. DEHUMIDIFIED STORAGE OF INDUSTRIAL MATERIALS

The long-term storage of some materials may require controlled atmospheric conditions or else elaborate preservation and packaging of the products. In this section we will consider storage of manufactured products and other materials, excluding foodstuffs.

The requirement in long-term storage is the prevention of deterioration of the materials. Depending upon the nature of the product, deterioration may result from many causes, which may include sunlight, atmospheric oxygen and ozone, atmospheric contaminants, acids, bases, moisture, heat and/or cold, wind and other influences. However, moisture is the greatest single deteriorating factor. The military services have found that a wide variety of military equipment and supplies may be kept for long periods through proper control of surrounding air humidity.

Moisture as water vapor in the air, as a thin film of condensed liquid, as rain, steam, or ice, may contribute to deterioration in many ways. Its effects have been extensively discussed by Greathouse and Wessel.* Moisture may cause discoloration, weakening, or even disintegration of papers and textiles. It is an essential agent in the corrosion of iron and steel. It may cause damage to lenses and other glass items since constituents in some kinds of glass are soluble in water. It is highly detrimental to electrical equipment, and influences the dimensions and strength of wood products.

Although most deterioration problems with moisture are caused by its excess, there are also situations where damage occurs because of its deficiency. Examples are brittle paper, stiff leather, and dusting of carbon materials.

Dehumidified storage was pioneered by the U.S. Navy. Tests conducted by the Philadelphia Naval Base† over several years involving various metallic materials showed that, for practical purposes, corrosion could be prevented if the ambient relative humidity did not exceed 40 per cent. In spaces subject to rapid temperature change, 30 per cent relative humidity was considered necessary. These limits were established to control deterioration of metals. Research has shown that moisture-supported life such as mildew and fungi may be controlled at higher humidity than that required to control corrosion.

Dehumidified storage is accomplished by controlling ambient humidity so that the amount of water deposited on surfaces is insufficient to support corrosion. The air may be dehumidified by sorbent or mechanical refrigeration methods. The most common method employed by the military services is the solid adsorbent system using either activated alumina or silica gel as the desiccant.

The largest application of dehumidified storage occurred in the inactivation of the Reserve Fleet following World War II. In this tremendous engineering accomplishment, which has been described by Gates,‡ the Navy "mothballed" approximately 2300 vessels. The problem of ship inactivation included four procedures: preservation of the underwater hull, preservation of surfaces exposed to the weather, preservation of the interior structure exposed to the ambient air, and preservation of working surfaces not exposed to the ambient air. Preservation of the interior structure was accomplished through use of thin-film preservatives and the circulation of dehumidified air. Space relative humidity of about 30 per cent has been maintained.

*Glenn A. Greathouse and Carl J. Wessel, *Deterioration of Materials* (New York: Reinhold Publishing Corporation, 1954), pp. 117–132.

†*Report on Laboratory Tests to Determine any Adverse Effect of High or Low Humidity on Materials and Equipment Found Aboard U.S. Naval Vessels*, Report 3014-P (Philadelphia, Industrial Test Laboratory, U.S. Naval Base).

‡Albert S. Gates, "The Mothball Fleet Pays Off," *Heating, Piping & Air Conditioning*, Vol. 24, July 1952, pp. 92–95; September 1952, pp. 94–97.

TABLE 15.4*

EQUILIBRIUM MOISTURE CONTENT FOR VARIOUS HYGROSCOPIC MATERIALS

Classification	Material	Description	Moisture Content in Per Cent of Dry Weight for Exposure to Air at 75 F and Various Relative Humidities								
			10	20	30	40	50	60	70	80	90
Natural textile fibers	Cotton	American cloth	2.6	3.7	4.4	5.2	5.9	6.8	8.1	10.0	14.3
	Cotton	Absorbent	4.8	9.0	12.5	15.7	18.5	20.8	22.8	24.3	25.8
	Wool	Australian	4.7	7.0	8.9	10.8	12.8	14.9	17.2	19.9	23.4
	Silk	Raw chevennes	3.2	5.5	6.9	8.0	8.9	10.2	11.9	14.3	18.3
	Linen	Dry spun yarn	3.6	5.4	6.5	7.3	8.1	8.9	9.8	11.2	13.8
	Hemp	Manila and sisal	2.7	4.7	6.0	7.2	8.5	9.9	11.6	13.6	15.7
Rayons	Viscose Nitrocellulose	Average skein	4.0	5.7	6.8	7.9	9.2	10.8	12.4	14.2	16.2
	Cellulose Acetate		0.8	1.1	1.4	1.9	2.4	3.0	3.6	4.3	5.3
Paper	M.F. newsprint	Wood pulp—24% ash	2.1	3.2	4.0	4.7	5.3	6.1	7.2	8.7	10.6
	White bond	Rag—1% ash	2.4	3.7	4.7	5.5	6.5	7.5	8.8	10.8	13.2
	Kraft wrapping	Coniferous	3.2	4.6	5.7	6.6	7.6	8.9	10.5	12.6	14.9
Miscellaneous organic materials	Leather	Sole oak—tanned	5.0	8.5	11.2	13.6	16.0	18.3	20.6	24.0	29.2
	Rubber	Solid tires	0.11	0.21	0.32	0.44	0.54	0.66	0.76	0.88	0.99
	Wood	Timber (average)	3.0	4.4	5.9	7.6	9.3	11.3	14.0	17.5	22.0
	Soap	White	1.9	3.8	5.7	7.6	10.0	12.9	16.1	19.8	23.8
	Tobacco	Cigarette	5.4	8.6	11.0	13.3	16.0	19.5	25.0	33.5	50.0
Foodstuffs	White bread		0.5	1.7	3.1	4.5	6.2	8.5	11.1	14.5	19.0
	Flour		2.6	4.1	5.3	6.5	8.0	9.9	12.4	15.4	19.1
	Starch		2.2	3.8	5.2	6.4	7.4	8.3	9.2	10.6	12.7
	Gelatin		0.7	1.6	2.8	3.8	4.9	6.1	7.6	9.3	11.4
Miscellaneous inorganic materials	Asbestos fiber	Finely divided	0.16	0.24	0.26	0.32	0.41	0.51	0.62	0.73	0.84
	Activated charcoal	Steam activated	7.1	14.3	22.8	26.2	28.3	29.2	30.0	31.1	32.7
	Sulfuric acid	H_2SO_4	33.0	41.0	47.5	52.5	57.0	61.5	67.0	73.5	82.5

*Abstracted by permission from *ASHRAE Guide and Data Book, Applications*, 1968, p. 93.

Other military applications of dehumidified storage have been utilized in warehouses and in limestone caves. Higgs* has stated that, following World War II, the Navy converted about 8,500,000 square feet of warehouse space into dehumidified storage space. In warehouses, the Navy has found that a relative humidity of 40 per cent is adequately low for long-term storage.

15.9. *ENVIRONMENTAL CONTROL IN AGRICULTURE*

The possibilities of environmental control in agriculture are far-reaching. Traditionally, farm animals are raised in regions close to feed supplies and/or close to the markets for the animal products. In some regions, particularly during certain parts of the year, local climatic conditions may not be conducive to efficient conversion of feed to product. This is particularly true in summer in the Corn Belt of the United States. Environmental control, in a partial sense, has been practiced for centuries in the greenhouse. However, the typical greenhouse is essentially a hot house, providing only for temperatures higher than the outside temperature. Considerable attention is now being given to more complete environmental control in plant growing. Some ASHRAE publications†,‡ give extensive information on environmental control for animals and plants.

Man and farm animals have many similarities as thermal machines. With both, body heat production increases with food consumption and activity. However, man is a sweating species while farm animals, excepting the burro, mule, and horse, are not. Cattle, swine, poultry, and sheep do not sweat, but ordinarily rely upon convection and radiation for most body heat dissipation. At high ambient temperatures, non-sweating animals dissipate excess body heat by panting.

Although many animals do not sweat, evaporation is still an important mechanism of heat loss. Respiratory evaporative cooling plus evaporation from moistened surfaces such as the snout, ears, etc., provide effective means for heat removal. Brody§ has stated that per unit area of outer surface, cows evaporate six times as much moisture from the respiratory tract as man.

Most farm animals, particularly cattle and swine, experience no difficulty in cold environments if properly fed, but are subject to considerable thermal stress at high temperatures. Body temperature rises rapidly once some critical ambient temperature is reached. Concurrently, the animal decreases its feed intake in order to reduce heat production. As a result, milk production of dairy cows and growth rate of swine

---

*George W. Higgs, Jr., "Economics of Dehumidified Storage," *Heating, Piping & Air Conditioning*, Vol. 29, April 1957, pp. 159–162.

†*ASHRAE Handbook of Fundamentals* (New York: American Society of Heating, Refrigerating and Air Conditioning Engineers, 1967) pp. 137–160.

‡*ASHRAE Guide and Data Book, Applications* (New York: American Society of Heating, Refrigerating and Air Conditioning Engineers, 1968) pp. 179–206.

§Samuel Brody, "Do Cow Barns Need Air Conditioning," *Refrigerating Engineering*, Vol. 64, April 1956, p. 41.

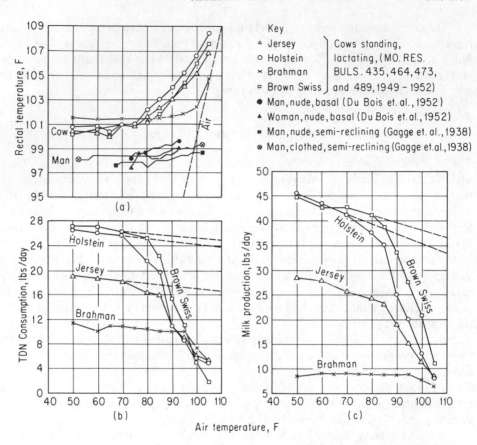

Figure 15.4 Effects of environmental temperature upon dairy cattle. (Reprinted by permission from *Refrigerating Engineering*, Vol. 64, April 1956, p. 41.)

decline. Figure 15.4 shows results of measurements on dairy cattle by Brody.* Figure 15.4(a) shows variation of rectal temperature with ambient temperature for Holstein, Brown Swiss, Jersey, and Brahman cows. Also shown, for comparative purposes, are variations of rectal temperature for nude and clothed men and women. Figure 15.4(b) shows variation of feed intake (TDN = total digestive nutrients) and Fig. 15.4(c) shows variation of milk production for the same cows. Normal rectal temperature is about 101 F. Above an ambient temperature of about 75 F, the rectal temperature of European-evolved cows rises sharply and is accompanied by a rapid decrease in feed consumption and milk production. The Brahman cow is much more heat resistant than the European breeds.

The *ASHRAE Handbook of Fundamentals*† gives data on rates of body heat

---

Refrigerating Engineering, Vol. 64, pp. 39–43.
†*ASHRAE Handbook of Fundamentals*, 1967, p. 138.

production for various farm animals under conditions of normal metabolic activity. Body rates of heat loss for mature, producing dairy cows varies approximately linearly from about 3.5 Btu per (hr) (lb of body weight) at 0 F to about 2.5 Btu per (hr) (lb of body weight) at 100 F. Rate of heat loss per unit body weight for growing hogs depends upon body size as well as upon ambient temperature. At 75 F, rates of heat loss vary from about 4.5 Btu per (hr)(lb) for 50 lb hogs to about 1.3 Btu per (hr) (lb) for 200 lb hogs.

Optimal environmental temperatures for swine lie between 60 F and 70 F. In this range of temperature, conversion of feed to pork and daily weight gain are at maximum values. The *ASHRAE Handbook of Fundamentals** also states that the level of ambient temperature at which swine are grown has an influence upon carcass quality. Formation of lean meat is highest in hogs grown at temperatures between 60 F and 70 F.

The thermal environment also has an important influence on reproduction of farm animals. Apparently, low temperatures do not affect reproduction, but sustained temperatures above about 85 F may significantly reduce breeding efficiency in cattle, swine, and sheep. Additionally, when mating and gestation occur at relatively high temperatures, the size of litter may decrease, the number of live births may decrease, and there may be an increase in the proportion of small, weak offspring.

Environmental control with plants is somewhat broader than most air conditioning problems. In addition to temperature, humidity, purity, and motion of air, we have the important factors of light intensity, spectral quality of light, light duration, and CO_2 content of the air to consider. Discussion of these factors and their influences upon plant growth are given in the *ASHRAE Handbook of Fundamentals*† and its references.

15.10. FOOD PROCESSING AND STORAGE

Outside of comfort air conditioning, the largest application for refrigerating equipment occurs in food preservation. Environmental control is required in processing of fresh foods, processing of frozen foods, storage of food products, and in the distribution of foods. Environmental control in food preservation is so large a subject that only a cursory mention of it may be made here. An excellent source of detailed information is the *ASRE Data Book, Refrigeration Applications*.‡

Many classes of foods are processed at temperatures somewhat above freezing temperature. Important industries requiring refrigeration in the processing of fresh foods include meat packing, poultry products, fish products, butter manufacture, cheese manufacture, and candy manufacture.

**ASHRAE Handbook of Fundamentals*, 1967, p. 143.

†*ASHRAE Handbook of Fundamentals*, 1967, pp. 145–150.

‡*Air Conditioning, Refrigerating Data Book, Refrigeration Applications*, Vol. 1, No. 1 (New York: The American Society of Refrigerating Engineers, 1959).

For any one type of foodstuff, many diverse processes may be involved, with each requiring different environmental conditions. In meat packing, refrigerated spaces are needed for chilling, holding, and possible aging of carcasses; making of sausage; cutting, trimming, and curing; and chilling of lard. In addition, refrigeration may be required in the processing of by-products as in hide curing.

The frozen foods industry has grown rapidly since World War II. Important foods which are frozen include fruits and vegetables, fish products, meat products, poultry products, fruit juice concentrates, and ice cream. Food products may be processed through *sharp freezing* or *quick freezing*. Sharp freezing, also called slow freezing, usually involves placing products in a room held at 5 F to −20 F. The products are frozen by natural convection cooling. Quick freezing is more rapid and is accomplished by one of the following methods: direct immersion of the product in a refrigerating medium, freezing by indirect contact with a refrigerant, or by freezing in a high velocity stream of cold air. Advantages claimed for quick freezing as compared to sharp freezing include formation of smaller ice crystals in the product causing less cellular damage, less time for diffusion of salts and separation of water in product, less decomposition during freezing from growth of bacteria and mold, and much greater product capacity.

Fresh or unfrozen products may be stored for a few days or perhaps a few weeks prior to distribution. Frozen products may be stored for many months. The *ASRE Data Book** gives recommended environmental temperatures and humidities and approximate storage life for many food commodities.

Refrigeration is the principal technique for retarding food spoilage. However, there are several supplements to refrigeration which may be effective in providing better preservation environments. These supplements include use of modified atmospheres, ultraviolet light, fumigation, antiseptic washes, and irradiation with cathode and gamma rays.

PROBLEMS AND QUESTIONS

15.1. Using fundamental heat transfer and mass transfer considerations, discuss the reasons for a person feeling more comfortable on a warm summer day if seated in front of an electric fan.

15.2. Determine whether the following conditions should be comfortable for typical people doing light work and exposed to air velocities of 15 to 25 ft/min.

a. Summer: $t = 75$ F, $t^\star = 65$ F; $t = 80$ F, $t^\star = 70$ F; $t = 95$ F, $t^\star = 75$ F.

b. Winter: $t = 70$ F, $\phi = 20\%$; $t = 75$ F, $\phi = 30\%$; $t = 80$ F, $\phi = 40\%$.

Air Conditioning, Refrigerating Data Book, Refrigeration Applications, Vol. 1, pp. 23-02 to 23-05.

15.3. Discuss why the effective temperature lines in Fig. 15.2 become more nearly vertical as the air dry-bulb temperature is reduced.

15.4. The mean body surface temperature of a certain unclothed person is 85 F. The mean temperature of surrounding surfaces is 70 F. The air within the room is at 90 F dry-bulb temperature, 90 per cent relative humidity, and 14.696 psia pressure. By what methods will this person be able to lose body heat?

15.5. In order to maintain body thermal comfort, what change (increase or decrease) should be made in ambient air temperature to compensate for (a) an increase in activity of occupants, (b) a decrease in air velocity, (c) the wearing of heavier clothing, (d) a decrease in temperature of surrounding surfaces, and (e) a decrease in air humidity ratio?

15.6. During an examination in a summer-session class, the instructor made the following measurements in the class room:

Air dry-bulb temperature = 90 F (assume $t_s = t$)
Air relative humidity = 53 per cent
Average air velocity = 40 ft per min

The instructor later gave these data to the class and asked it to quantitatively determine if the conditions were those favorable to good mental effort. He suggested that a body heat production rate of 500 Btu per hr should apply. What should have been the conclusion?

15.7. You have been asked to investigate a heat-exposure problem in a local company. The problem concerns 25 men doing light work requiring little mental effort. The men are selected by medical examination. The labor union believes that working conditions are too hot and that the men are being subjected to severe heat strain. In your survey, you observe the following conditions:

Air dry-bulb temperature = 90 F
Air wet-bulb temperature = 75 F (assume $t^* = 75$ F)
Temperature of surrounding surfaces = 90 F (estimate)
Barometric pressure = 14.15 psia
Air velocity over worker's torso = 150 ft per min
Activity of workers: standing, light work at machines,
                 mostly arm movements

What is your conclusion?

15.8. In an industrial heat-exposure problem, labor and management have agreed that the heat stress index should not exceed 30. Measurements in the working space indicate the following conditions:

Air dry-bulb temperature = 92 F
Temperature of surrounding surfaces = 90 F (estimate)
Air relative humidity = 40 per cent

It is estimated that the average body heat production rate is 700 Btu per (hr) (person). Determine the minimum allowable air velocity over each worker's body.

SYMBOLS USED IN CHAPTER 15

A_b Body surface area, sq ft.

A'_b Body surface area which sees surrounding surfaces, sq ft.

H.S.I.   Heat Stress Index, dimensionless.

h_c Convection heat transfer coefficient, Btu per (hr) (sq ft) (F).

h_D Convection mass transfer coefficient, lb_w per (hr) (sq ft) (lb_w per lb_a).

$h_{fg,b}$ Latent heat of vaporization of water at t_b, Btu per lb.

h_R Radiation heat transfer coefficient, Btu per (hr) (sq ft) (F).

P Barometric pressure, psia.

P_w Partial pressure of water vapor in moist air, psia.

$P_{w,s,b}$ Pressure of saturated water vapor at t_b psia.

q_c Rate of heat transfer by convection, Btu per hr.

q_E Rate of heat transfer by evaporation, Btu per hr.

$q_{E,max}$ Rate of heat transfer by evaporation given by Eq. (15.10), Btu per hr.

q_M Rate of body heat production, Btu per hr.

q_R Rate of heat transfer by radiation, Btu per hr.

q_s Rate of change of heat stored in body, Btu per hr.

t Air dry-bulb temperature, F.

$t^\star$ Air thermodynamic wet-bulb temperature, F.

t_b Mean surface temperature of nude body, F.

t_s Mean temperature of surrounding surfaces, F.

V Air velocity, ft per min.

W Humidity ratio of bulk-air stream, lb of water vapor per lb of dry air.

W_{sb} Humidity ratio of saturated moist air at t_b, lb of water vapor per lb of dry air.

ϕ Relative humidity, dimensionless.

16

AIR CONDITIONING
CALCULATIONS

16.1. INTRODUCTION

The design of a comfort or industrial air conditioning system involves several integrated procedures. For a given problem, the designer must (a) establish the design conditions, (b) estimate the thermal loads, (c) choose the type of system to be employed, including methods of control, (d) calculate the capacity requirements of each component of the system, (e) select equipment if commercially available and/or design special equipment, (f) design the air distribution system and miscellaneous piping systems, and (g) prepare drawings and specifications for the system. A successful design engineer must be cognizant of commercially-available equipment and current application techniques.

In this chapter we will study procedures involved with estimating thermal loads and making psychrometric calculations. Our emphasis will continue to be on basic factors. The reader should consult the *ASHRAE Handbook of Fundamentals** and the *ASHRAE Guide and Data Books*†,‡ for specific application data and for procedures on designing air distribution systems and piping systems.

ASHRAE Handbook of Fundamentals (New York: American Society of Heating, Refrigerating and Air Conditioning Engineers, Inc., 1967).

†*ASHRAE Guide and Data Book, Systems and Equipment* (New York: American Society of Heating, Refrigerating and Air Conditioning Engineers, Inc., 1969).

‡*ASHRAE Guide and Data Book, Applications* (New York: American Society of Heating, Refrigerating and Air Conditioning Engineers, Inc., 1968).

16.2. ESTIMATING THERMAL LOADS FOR ENCLOSED SPACES

One of the most important steps in the design of a thermal environmental control system is estimation of the space heat and moisture loads. The extent, construction, and use of the space must be known. The schedule of occupancy and activity of occupants must be estimated. All heat- and/or moisture-emitting equipment in the space and schedules of operation must be determined.

The space thermal environmental conditions to be maintained must be specified. In Chapter 15, the effects of the thermal environment upon people, processes, and materials were discussed. Figure 15.2 may be used in establishing desirable space conditions for comfort air conditioning. The *ASHRAE Handbook of Fundamentals** recommends a design dry-bulb temperature of 75 F for both winter and summer for comfort systems. For summer operation, a relative humidity of 50 per cent is typically specified.

The thermal environment external to the space must also be specified. Most design problems involve clear days, and the procedures of Chapter 13 may be used to establish solar radiation intensity values. Table 16.1 shows outdoor-air design conditions for several cities (airport stations) in the United States. The *ASHRAE Handbook of Fundamentals*† shows similar data for several hundred other locations in the United States; data are also shown for Canada and for most other foreign countries.

In the Winter part of Table 16.1, three temperature columns are shown. The median of annual extremes is the median of the lowest temperature recorded each year for periods of 25 or 30 years. The 99 and 97½ per cent columns show temperatures which on the average are equalled or exceeded for the respective percentages of the total hours (2160) in December, January, and February. Thus typically at Atlanta, 99 per cent of the time during these three months, the outdoor air temperature is 18 F or higher. Selection of winter design temperatures requires judgment by the heating engineer. As a guide, the *ASHRAE Handbook of Fundamentals*‡ suggests that for structures with low heat capacity and/or large glass areas, the median of annual extremes be selected; for buildings with moderate heat capacity, the 99 per cent value be chosen; and for massive institutional buildings with small glass areas, the 97½ per cent value be selected.

In the Summer portion of Table 16.1, dry-bulb and wet-bulb temperatures are shown which are equalled or exceeded on the average for the indicated percentages of the total hours (2928) during June through September. Thus typically at Atlanta, one per cent of the time (30 hours approximately) during the four summer months, the outdoor air dry-bulb temperature is 95 F or higher; 2½ per cent of the time (75 hours approximately), the temperature is 92 F or higher; and 5 per cent of the time

---

**ASHRAE Handbook of Fundamentals*, 1967, p. 371.
†*ASHRAE Handbook of Fundamentals*, 1967, pp. 373–392.
‡*ASHRAE Handbook of Fundamentals*, 1967, p. 372.

TABLE 16.1*

OUTDOOR AIR DESIGN CONDITIONS

City	Winter				Summer						
	Median of Annual Extremes F	99%	97½%	Coincident Wind Velocity†	Design Dry-bulb Temperature F			Daily Range Dry-bulb Temp. F	Design Wet-bulb Temperature F		
					1%	2½%	5%		1%	2½%	5%
Atlanta, Georgia	14	18	23	H	95	92	90	19	78	77	76
Chicago, Illinois	−9	−4	0	M	93	90	87	20	77	75	74
Dallas, Texas	14	19	24	H	101	99	97	20	79	78	77
Denver, Colorado	−9	−2	3	L	92	90	89	28	65	64	63
Kansas City, Mo.	−2	4	8	M	100	97	94	20	79	77	76
Los Angeles, Calif.	36	41	43	VL	86	83	80	15	69	68	67
Minneapolis, Minn.	−19	−14	−10	L	92	89	86	22	77	75	74
New York, N.Y.	12	17	21	H	91	87	84	16	77	76	75
Phoenix, Arizona	25	31	34	VL	108	106	104	27	77	76	75
Seattle, Washington	14	20	24	L	85	81	77	22	66	64	63
Washington, D.C.	12	16	19	M	94	92	90	18	78	77	76

*Abstracted by permission from *ASHRAE Handbook of Fundamentals*, 1967, pp. 373–385.

†VL = Very light, 70 per cent or more of cold hours ≤ 7 mph.

L = Light, 50 to 59 per cent of cold hours ≤ 7 mph.

M = Moderate, 50 to 74 per cent of cold hours > 7 mph.

H = High, 75 per cent or more of cold hours > 7 mph, and 50 per cent > 12 mph.

(150 hours approximately), the temperature is 90 F or higher. The daily range of temperature shown represents the difference between the average maximum and average minimum temperatures during the warmest month.

Figure 16.1 shows various space heat and moisture transfers which may be present. The symbol q_s represents a sensible heat-transfer rate while the symbol m_w

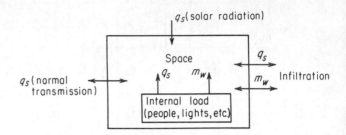

Figure 16.1 Schematic illustration of space heat and moisture transfers.

represents a moisture-transfer rate. A transfer of moisture also involves a transfer of energy given by the product $m_w h_w$, where h_w is the specific enthalpy of the added (or removed) moisture. Transfers due to solar radiation and internal loads are always gains upon the space. Heat transfer across solid boundaries due to a temperature difference (normal transmission) may represent a gain or a loss; transfers because of infiltration may also represent gains or losses.

Energy transmission through opaque boundaries and glass was discussed in Chapter 14. Equation (14.58) applies to calculation of periodic heat transfer through walls and roofs. It is of practical use if the necessary coefficients for sol-air temperature and wall properties are available; at present, only limited data are available. Equation (14.58) may also be written in the form

$$q_i = U \, \Delta t_E \qquad (16.1)$$

where

$$\Delta t_E = [t_{e,m} + \lambda_1 t_{e,1} \cos (\omega_1 \theta - \psi_1 - \Phi_1) \\ + \lambda_2 t_{e,2} \cos (\omega_2 \theta - \psi_2 - \Phi_2) + \dots] - t_i \qquad (16.2)$$

The quantity Δt_E may be considered as an equivalent temperature difference, F, which includes solar radiation and normal heat transmission effects and heat storage characteristics of the wall or roof.

Tables 16.2 and 16.3, respectively, show equivalent temperature differentials (Δt_E values) for roofs and walls. Tables 16.2 and 16.3 were calculated for August 1 at 40 deg north latitude, but for estimating purposes they may be used for all localities in the United states for the June–August period.

Heat transmission through unshaded window glass may be calculated by Eq. (14.39). Table 16.4 shows the solar transmission factor F for various sunlit window

TABLE 16.2*†

EQUIVALENT TEMPERATURE DIFFERENTIALS FOR HEAT GAIN THROUGH ROOFS

Roof Construction (Includes $\frac{3}{8}$ in. felt roofing with or without slag, or shingles)	Solar Time								
	a.m.			p.m.					
	8	10	12	2	4	6	8	10	12
Roofs Exposed to Sun									
1 in. wood	12	38	54	62	50	26	10	4	0
2 in. concrete	6	30	48	58	50	32	14	6	2
4 in. concrete	0	20	38	50	52	40	22	12	6
6 in. concrete	4	6	24	38	46	44	32	18	12
Light construction, 1 in. water	0	4	16	22	18	14	10	2	0
Heavy construction, 1 in. water	−2	−2	−4	10	14	16	14	10	6
Roofs in Shade									
Light construction	−4	0	6	12	14	12	8	2	0
Medium construction	−4	−2	2	8	12	12	10	6	2
Heavy construction	−2	−2	0	4	8	10	10	8	4

*Abstracted by permission from *ASHRAE Guide and Data Book, Fundamentals and Equipment* (New York: American Society of Heating, Refrigerating and Air Conditioning Engineers, 1965) p. 501.

†For August 1 at 40 deg north latitude. For a difference of 15 F between maximum outdoor air temperature and room temperature. Where maximum temperature difference is not 15 F, algebraically add the excess to values in table.

shades. Where F factors apply, Eq. (14.39) becomes

$$\frac{q_s}{A} = F[F_s\tau_D I_D + \tau_d I_d + \tau_R I_R] + \frac{F_s I_D \alpha_D + I_d \alpha_d + I_R \alpha_R}{1 + (h_o/h_i)} + U(t_o - t_i) \qquad (16.3)$$

The F factors shown in Table 16.4 are average values and thus are only approximately correct. Actually, they would vary to some extent with angle of incidence.

Infiltration is the leakage of outdoor air into a building through cracks and openings caused by a pressure difference across the boundary surfaces. Infiltration is always accompanied by *exfiltration*. The net exchange of air may lead to both heat and moisture gains (or losses) for the space. These transfers are given by

$$q_t = m_a(h_o - h_i) \qquad (16.4)$$

$$m_w = m_a(W_o - W_i) \qquad (16.5)$$

where q_t is the rate of energy transfer, Btu per hr; m_a is the rate of exchange of dry air, lb_a per hr; h_o and h_i are respectively the enthalpy of the outdoor and indoor air, Btu per lb_a; m_w is the rate of moisture transfer, lb_w per hr; and W_o and W_i are respectively the humidity ratio of the outdoor and indoor air, lb_w per lb_a.

Estimation of the rate of infiltration for a structure is often the most uncertain

TABLE 16.3*†

EQUIVALENT TEMPERATURE DIFFERENTIALS FOR HEAT GAIN THROUGH WALLS

Wall and Orientation	Solar Time																	
	a.m.						p.m.											
	8		10		12		2		4		6		8		10		12	
	D	L	D	L	D	L	D	L	D	L	D	L	D	L	D	L	D	L
Frame																		
E	30	14	36	18	32	16	12	12	14	14	14	14	10	10	6	6	2	2
S	−4	−4	4	0	22	12	30	20	26	20	16	14	10	10	6	6	2	2
W	−4	−4	0	0	6	6	20	12	40	28	48	34	22	22	8	8	2	2
N	−4	−4	−2	−2	4	4	10	10	14	14	12	12	8	8	4	4	0	0
8 in. brick or 12 in. tile block																		
E	8	6	8	6	14	8	18	10	18	10	14	8	14	10	14	10	12	10
S	4	2	4	2	4	2	4	2	10	6	16	10	16	12	12	10	10	8
W	8	4	6	4	6	6	8	6	10	6	14	8	20	16	24	16	24	16
N	0	0	0	0	0	0	0	0	2	2	6	6	8	8	8	8	6	6
12 in. brick																		
E	12	8	12	8	12	8	10	6	12	8	14	10	14	10	14	8	14	8
S	8	6	8	6	6	4	6	4	6	4	8	4	10	6	12	8	12	8
W	12	8	12	8	12	8	10	6	10	6	10	6	10	6	12	8	16	10
N	4	4	2	2	2	2	2	2	2	2	2	2	2	2	4	4	6	6

*Abstracted by permission from *ASHRAE Guide and Data Book*, 1965, p. 503.

†For August 1 at 40 deg north latitude. For a difference of 15 F between maximum outdoor air temperature and room temperature. Where maximum temperature difference is not 15 F, algebraically add the excess to values in table. Columns headed D refer to dark surfaces having a solar absorptivity of 0.9, while columns headed L refer to light surfaces having a solar absorptivity of 0.5.

TABLE 16.4*

SOLAR TRANSMISSION FACTORS FOR VARIOUS WINDOW SHADES

Shade	Finish on Side Exposed to Sun	Transmission Factor F
Canvas awning, sides open	Dark or medium	0.25
Inside roller shade, fully drawn	White, cream	0.41
Inside roller shade, fully drawn	Dark	0.81
Inside Venetian blind, slats at 45 deg	White, cream	0.56
Inside Venetian blind, slats at 45 deg	Aluminum	0.45

*Abstracted by permission from *Heating, Ventilating, Air Conditioning Guide*, Vol. 38, p. 205.

calculation of a thermal load estimate. The rate of air flow into (or out of) a building depends upon pressure differences existing across boundary surfaces and upon resistances to air flow offered by cracks and other openings. Pressure differences may arise because of wind pressure or because of density differences between inside and outside air (chimney effect). The *ASHRAE Handbook of Fundamentals** gives an extensive discussion of air infiltration in various kinds of structures.

In residences and low-height commercial buildings, pressure differences arise mainly from wind forces. The velocity head p_v equivalent to a given wind velocity V_w may be calculated by the relation

$$p_v = 0.000482V_w^2 \tag{16.6}$$

where p_v is in inches of water and V_w is in miles per hour. The pressure difference across a wall because of wind forces is the difference between the wind pressure p_v and the inside pressure. Wind causes an increase of pressure inside a building. To account for this effect, the pressure difference across a windward wall is commonly taken as 0.64 p_v.†

Table 16.5 shows rates of infiltration in cu ft per (hr)(ft of crack) through double-hung wood sash windows for various pressure differences.

In residences, it is common practice to estimate infiltration on an air-change method. Table 16.6 shows air changes under average conditions in residences. Table

TABLE 16.5*
INFILTRATION THROUGH DOUBLE-HUNG WOOD WINDOWS†

Type of Window	Pressure Difference in Inches Water				
	0.1	0.2	0.3	0.4	0.5
A. Wood double-hung window (locked)					
1. Non-weatherstripped, loose fit	77	122	150	194	225
2. Non-weatherstripped, average fit	27	43	57	69	80
3. Weatherstripped, loose fit	28	44	58	70	81
4. Weatherstripped, average fit	14	23	30	36	42
B. Frame-wall leakage					
1. Around frame in masonry wall, not caulked	17	26	34	41	48
2. Around frame in masonry wall, caulked	3	5	6	7	8
3. Around frame in wood frame wall	13	21	29	35	42

*Reprinted by permission from *ASHRAE Handbook of Fundamentals*, 1967, p. 409.
†Expressed in cu ft per (hr)(ft of crack).

**ASHRAE Handbook of Fundamentals*, 1967, pp. 405–418.
†*ASHRAE Handbook of Fundamentals*, 1967, p. 409.

TABLE 16.6*

AIR CHANGES OCCURRING UNDER AVERAGE CONDITIONS IN
RESIDENCES, EXCLUSIVE OF AIR PROVIDED FOR VENTILATION†

Kind of Room or Building	Number of Air Changes Occurring per Hour
Rooms with no windows or exterior doors	$\frac{1}{2}$
Rooms with windows or exterior doors on one side	1
Rooms with windows or exterior doors on two sides	$1\frac{1}{2}$
Rooms with windows or exterior doors on three sides	2
Entrance halls	2

*Reprinted by permission from *ASHRAE Handbook of Fundamentals*, 1967, p. 409.
†For rooms with weatherstripped windows or with storm sash, use $\frac{2}{3}$ of these values.

TABLE 16.7*

INFILTRATION THROUGH DOORS

Type of Door	cu ft per (person) (passage)
72 in. revolving	40–75
36 in. swinging	20–100

*Abstracted by permission from *Heating, Ventilating, Air Conditioning Guide*, Vol. 38, p. 152.

16.6 may be used to determine an infiltration allowance for each room; the total allowance for a residence is the sum of the allowances of the individual rooms.

In commercial buildings, infiltration may occur mostly from door openings. Table 16.7 shows an estimate for infiltration by door openings.

Infiltration should not be confused with the *outdoor air requirement* of a space. An air conditioned space requires positive introduction of outdoor (fresh) air (1) for dilution of tobacco smoke and odors, and (2) as make-up air for exhaust fans. Except in some residences, infiltration should never be relied upon to provide the outdoor air requirement. Table 16.8 shows recommended outdoor air requirements for several applications. As will be shown later, outdoor air positively introduced into a system is first processed before it enters the condition space. Thus, it does not represent a space load.

The internal heat and moisture gains for a space may be due to people, lights, appliances, and other sources within the space. Heat and moisture gains from people may be estimated from Table 15.2. Table 16.9 shows rates of heat gain from electric lights and rates of heat and moisture gain for a few appliances. The *ASHRAE Hand-*

TABLE 16.8*

OUTDOOR AIR REQUIREMENTS

Application	Amount of Smoking	cu ft per (min) (person)	cu ft per (min) (sq ft of floor)
Barber shop	Considerable	15	
Cocktail bar	Heavy	40	
Department store	None	7	
Drug store	Considerable	10	
Hotel room	Heavy	30	
Kitchen of restaurant	..		4.0
Meeting room	Very heavy	50	
Cafeteria	Considerable	15	
Toilet room	..		2.0

*Abstracted by permission from *ASHRAE Handbook of Fundamentals*, 1967, p. 468.

book of Fundamentals* gives rates of heat gain for many other types of equipment.

The procedure for estimating a heat and/or moisture load for a space is first dependent upon the type of system to be designed. For example, if a heating system is to furnish only sensible heat, then the load estimation should involve sensible heat transfers only. In design of heating systems, it is usual to base the design upon a fictitious steady-state condition in which all intermittent heat sources (solar effects, people, lights, etc.) are ignored. A winter outdoor-air design temperature is selected from Table 16.1 and an inside design temperature such as 75 F is chosen. The maximum capacity of the heating system is determined from steady-state calculations; generally only normal heat transmission and infiltration are considered.

In a cooling load estimation, transient and periodic heat sources add to the load, and thus must be included. All calculations should be based on the same moment of time. Where the time of the maximum rate of heat gain cannot be estimated with sureness, it may be necessary to make more than one set of calculations. In a large building having many diverse spaces, the time-occurrences of the peak heat gains among the various spaces may differ widely. Such a situation requires zoning of the building and the use of separate air-handling systems for each zone. However, a central refrigeration plant would need to satisfy only the instantaneous needs of the building as a whole. This required capacity would generally be much less than the sum of the individual peak gains. For this kind of problem, it might be necessary to calculate the heat gains for each space at several times of day.

Although the terms "cooling load" and "net instantaneous rate of heat gain" are used interchangeably here, there may be a substantial difference between them. The radiative portion of an instantaneous heat gain is absorbed by solid objects and heats them. This absorbed heat is released to the air stream at some later time. At

ASHRAE Handbook of Fundamentals, 1967, pp. 498–499.

TABLE 16.9*
HEAT AND MOISTURE GAINS FROM EQUIPMENT

Item	Manufacturer's Input Rating		Probable Maximum Hourly Input, Btu/hr	Without Hood				With Hood, All Sensible Gain, Btu/hr
				Recommended Rate of Heat Gain, Btu/hr			Rate of Moisture Gain, lb/hr	
	Watt	Btu/hr		Sensible	Total			
Electric lights, incandescent	...	...	3.41 per watt	...	...		...	...
Electric lights, fluorescent	...	...	4.09 per watt	...	...		...	...
Coffee urn, 5 gal., gas	...	15,000	7500	5250	7500		1.95	1500
Coffee urn, 5 gal., electric	3000	10,200	5100	3850	5100		1.09	1600
Toaster, 360 slices per hr, gas	...	12,000	6000	3600	6000		2.09	1200
Toaster, 360 slices per hr, elec.	2200	7,500	3700	1960	3700		1.51	1200
Steam table, gas-burning, per sq ft of top	...	2,500	1250	750	1250		0.44	250

*Adapted by permission from *ASHRAE Handbook of Fundamentals*, 1967, p. 498.

present, there is no precise way to account for such heat storage effects in practical calculations. Engineering judgment must be used. Depending upon the heat capacity of the interior structure, the radiative portion of the heat gain may be averaged over a period of two hours or more prior to the time of the calculation.

At present, there is an accelerating trend toward the use of a digital computer in thermal calculations for buildings. The thermal requirements of a building may be determined much more exactly than when manual calculations are used. In turn, better selections of equipment and more economical modes of equipment operation may be realized.

Example 16.1. Figure 16.2 shows a schematic plan of a restaurant in Minneapolis. Additional data are as follows: (a) Outside walls: 4 in. face brick, 6 in. hollow tile, plastered on inside $[U = 0.34$ Btu per (hr)(sq ft)(F)]. (b) Partition

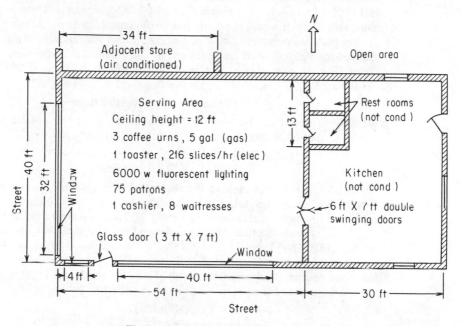

Figure 16.2 Schematic plan of restaurant for Example 16.1.

wall: 2 in. × 4 in. studs, metal lath and plaster both sides $[U = 0.39$ Btu per (hr)(sq ft)(F)]. (c) Doors to kitchen: 1 in. wood $[U = 0.45$ Btu per (hr)(sq ft)(F)]. (d) Roof and ceiling: flat roof, 2 in. concrete, built-up roofing, steel joists, hung ceiling $[U = 0.36$ Btu per (hr)(sq ft)(F)]. Neglect shading of roof by parapet walls. (e) Windows: $\frac{1}{4}$ in. plate glass $[U = 1.10$ Btu per (hr)(sq ft)(F)]. Windows are 8 ft high, set back 5 in., and equipped with white Venetian blinds. Bottom of windows is 30 in. above floor. (f) Exhaust fan requirements for the kitchen and rest rooms are 4500 cu ft per min. Make-up air comes from the serving area. It is probable that the rest rooms are maintained at essentially

the same temperature as the serving area, while the kitchen may be assumed
5 F warmer. (g) Rush hours are 7:00–8:30 a.m., 11:30 a.m.–1:30 p.m., and
5:30–7:30 p.m. Daylight saving time (1 hr advance) is in effect.

Based upon space conditions of 75 F dry-bulb temperature and 50 per cent
relative humidity, calculate the peak rate of heat gain in Btu per hr and the peak
rate of moisture gain in lb_w per hr for the serving area.

SOLUTION: We will calculate the gains for August 1. The latitude of Minneapolis is 45° north and the longitude is 93°16′ west. By the method explained in
Sec. 13.6, we find that solar time on August 1 is less advanced than daylight
saving time by approximately 80 min. Thus, rush hours occur at solar times of
5:40–7:10 a.m., 10:10 a.m.–12:10 p.m., and 4:10–6:10 p.m. We may estimate
that the largest components of the cooling load will be due to the roof, people,
and west window. Thus, the probable maximum rate of heat gain occurs at
4:10 p.m. We will make the calculations for 4:00 p.m. solar time. By Table
16.1 (1% columns), $t_o = 92$ F and $t_o^\star = 77$ F for Minneapolis. These are maximum values which would occur at about 3:00 p.m. solar time. At other times,
the dry-bulb temperature t_o may be assumed to vary as shown by Fig. 14.17.
The outside air dew-point temperature may be assumed constant over a 24 hr
period. For the roof, $\Delta t_E = 50$ F by Table 16.2 for $(t_o - t_i)_{max} = 15$ F. Since
our value of $(t_o - t_i)_{max} = 17$ F, we use $\Delta t_E = 52$ F. By Eq. (16.1),

$$q_s = (0.36)(2160)(52) = 40{,}440 \text{ Btu/hr}$$

We may assume that the heat storage characteristic of the outside walls is
similar to that of an 8 in. brick wall. For the north wall, $\Delta t_E = 2$ F by Table
16.3. We use $\Delta t_E = 4$ F. By Eq. (16.1),

$$q_s = (0.34)(240)(4) = 330 \text{ Btu/hr}$$

Heat transmission rates through the south and west walls are found similarly.
Heat transmission rates through the partition wall and doors to the kitchen are
calculated for steady-state conditions, using the kitchen temperature as 80 F.

Heat transmission through the south windows and door may be calculated
by Eq. (16.3). By Table 13.2, $d = 18°12′$. We also know that $l = 45°$ and
$h = 60°$. By Eq. (13.3), we find $\beta = 33°49′$. By Eq. (13.4), we find $\gamma = 97°57′$,
measured counter-clockwise from north. Thus, the south side is sunlit at 4:00
p.m. solar time and the wall-solar azimuth $\alpha = 82°03′$. By Eq. (13.8),

$$\cos \theta = (0.8308)(0.1383) = 0.115$$

and for the south-facing vertical surface the angle of incidence $\theta = 83°24′$.
By Eq. (14.34), the shading factor F_s due to setback for the large south window
is

$$F_s = 1 - \left(\frac{5}{96}\right)\frac{(0.6699)}{(0.1383)} - \frac{5}{480}(7.161) + \left(\frac{5}{96}\right)\left(\frac{5}{480}\right)\left(\frac{0.6699}{0.1383}\right)(7.161) = 0.692$$

By Table 14.4, we will assume an extinction coefficient $K = 0.174$ in.$^{-1}$ for the
plate glass. By Eq. (14.18), we find $L' = 0.33$ in., and by Eq. (14.17), $a = 0.944$.
By Fig. 14.8, $r = 0.54$, and by Eq. (14.14),

$$\tau_D = \frac{(0.46)^2(0.944)}{1 - (0.54)^2(0.944)^2} = 0.27$$

By Eq. (14.16),

$$\alpha_D = 1 - 0.54 - \frac{(0.46)^2(0.944)}{1 - (0.54)(0.944)} = 0.05$$

Interpolation between Figs. 13.15 and 13.16 gives $I_N = 236$ Btu/(hr)(sq ft) for a clearness number of unity. By Fig. 13.17, CN $= 1.02$ for Minneapolis. Thus, by Eq. (13.14),

$$I_D = (1.02)(236)(0.115) = 28 \text{ Btu/(hr)(sq ft)}$$

We will assume that both the south side and west side have an unobstructed view of the sky. By Fig. 13.19, $I_{dH} = 28$ Btu per (hr)(sq ft). By Fig. 13.20, $I_{dv}/I_{dH} = 0.61$. Thus

$$I_d = (0.61)(28) = 17 \text{ Btu/(hr)(sq ft)}$$

We will assume the average angle of incidence for diffuse sky radiation to be 60 deg. By methods identical to those used in finding τ_D and α_D, we obtain $\tau_d = 0.79$ and $\alpha_d = 0.05$.

We will assume that the ground surface outside the building is of large extent and consists of old concrete. By Fig. 14.5, $\rho = 0.25$. By Eq. (13.14),

$$I_{DH} = (1.02)(236)(0.557) = 134 \text{ Btu/(hr)(sq ft)}$$

Thus $I_H = 134 + 28 = 162$ Btu/(hr)(sq ft). Case B.4 of Table 14.5 applies. Thus

$$I_R = \frac{(0.25)(162)}{2} = 20 \text{ Btu/(hr)(sq ft)}$$

We will assume that $\tau_R = \tau_d$ and $\alpha_R = \alpha_d$. By Table 14.1, use $h_o = 4.0$ and $h_i = 1.46$ Btu/(hr)(sq ft)(F). By Table 16.4, F $= 0.56$. Thus, by Eq. (16.3), the rate of heat gain through the large south window is

$$q_s = (320)\Big\{0.56[(0.692)(0.27)(28) + (0.79)(17) + (0.79)(20)]$$

$$+ \frac{(0.692)(28)(0.05) + (17)(0.05) + (20)(0.05)}{1 + 4.0/1.46} + (1.10)(92 - 75)\Big\}$$

$$= 12{,}250 \text{ Btu/hr}$$

We may calculate the rate of heat gain through the small south window similarly. We will assume that the glass door is in the same plane as the windows and has no Venetian blind. Thus q_s is calculated similarly except that we use $F = 1.0$. We find for the door, $q_s = 1000$ Btu/hr.

The rate of heat gain through the west window is calculated in a similar way to that for the large south window. We find $q_s = 35{,}100$ Btu/hr.

We will assume that infiltration is due only to openings of the outside door. Assuming an average stay of 30 min for each patron, we would have a rate of 5 door openings per min. By Table 16.7, we may estimate an air exchange of 60 cu ft per passage. Thus, the volume of infiltrated air is 300 cu ft per min. The elevation at Minneapolis is approximately 900 ft. Thus, we are only approximately correct when we use the sea-level pressure psychrometric chart. By Fig. E-8, we find for the outside air ($t_o = 92$ F, $t_o^\star = 77$ F) that $W_o = 0.0166$ lb$_w$/lb$_a$, $h_o = 40.35$ Btu/lb$_a$, and $v_o = 14.28$ cu ft/lb$_a$. For the inside air ($t_i = 75$ F, $\phi_i = 50\%$), we find $W_i = 0.00928$ lb$_w$/lb$_a$ and $h_i = 28.15$ Btu/lb$_a$. Thus for infiltration, by Eqs. (16.4) and (16.5) we have

$$q_t = \frac{(300)(60)(40.35 - 28.15)}{14.28} = 15{,}380 \text{ Btu/hr}$$

$$m_w = \frac{(300)(60)(0.0166 - 0.00928)}{14.28} = 9.23 \text{ lb}_w/\text{hr}$$

TABLE 16.10

SOLUTION FOR EXAMPLE 16.1

Component	Net Area, sq ft or Quantity	U Factor, Btu/(hr) (sq ft)(F) or Unit Value	Temperature Difference, F	Sensible Heat Gain, q_s Btu/hr	Water Vapor Energy Gain, $m_w h_w$ Btu/hr	Total Energy Gain, q_t Btu/hr	Moisture Gain, m_w lb$_w$/hr
Roof	2160	0.36	52	40,440	· · ·	40,440	· · ·
North outside wall	240	0.34	4	330	· · ·	330	· · ·
West wall	224	0.34	12	910	· · ·	910	· · ·
South wall	275	0.34	12	1,120	· · ·	1,120	· · ·
Partition wall (kitchen)	282	0.39	5	550	· · ·	550	· · ·
Double doors to kitchen	42	0.45	5	95	· · ·	95	· · ·
Large south window	320	· · ·	· · ·	12,250	· · ·	12,250	· · ·
Small south window	32	· · ·	· · ·	1,150	· · ·	1,150	· · ·
South door	21	· · ·	· · ·	1,000	· · ·	1,000	· · ·
West window	256	· · ·	· · ·	35,100	· · ·	35,100	· · ·
Infiltration for door	· · ·	· · ·	· · ·	5,220	10,160	15,380	9.23
Occupants	84	· · ·	· · ·	23,070	24,580	47,650	25.91
Lights	6000	4.09	· · ·	24,540	· · ·	24,540	· · ·
Coffee urns	3	· · ·	· · ·	15,750	6,750	22,500	5.85
Totals				161,525	41,490	203,015	40.99

The quantity q_t is the total energy addition by the outside air. The energy addition by water vapor alone would be

$$m_w h_{g,o} = (9.23)(1101) = 10,160 \text{ Btu/hr}$$

and the sensible heat portion would be $15,380 - 10,160 = 5,220$ Btu/hr.

Heat and moisture gains from the occupants may be estimated from Table 15.2. For the patrons, we may use 275 Btu/(hr)(person) sensible, 550 Btu/(hr)(person) total, and 0.29 lb_w/(hr)(person). For the cashier, we may use 245 Btu/hr sensible, 400 Btu/hr total, and 0.16 lb_w/hr. For the waitresses, we may use 275 Btu/(hr)(person) sensible, 750 Btu/(hr)(person) total, and 0.50 lb_w/(hr)(person). Thus, for the occupants,

$$q_s = (75)(275) + (1)(245) + (8)(275) = 23,070 \text{ Btu/hr}$$

$$m_w h_w = (75)(275) + (1)(155) + (8)(475) = 24,580 \text{ Btu/hr}$$

$$q_t = 23,070 + 24,580 = 47,650 \text{ Btu/hr}$$

$$m_w = (75)(0.29) + (1)(0.16) + (8)(0.50) = 25.91 \text{ lb}_w/\text{hr}$$

Heat and moisture gains from the lights and coffee urns may be obtained from Table 16.9. We assume that the toaster is not in use.

Table 16.10 shows a summary of the calculations. The total rate of energy gain (summation of all sensible heat gains and energy gains due to the addition of water vapor) is 203,015 Btu/hr and the total rate of moisture gain is 40.99 lb_w/hr.

Where solar effects contribute largely to the heat gain, as in Example 16.1, it is desirable to estimate the heat and moisture gains for an extremely cloudy, humid day also. Such calculations may give a much lower value of the enthalpy-moisture ratio q' for the space and indicate the need for reheat in the air conditioning system, whereas the results of the peak-load calculations may not indicate such need.

After a load estimate is tentatively established, it is the engineer's responsibility to analyze the results and to study the economic feasibility of modifying the structure in order to reduce the thermal load. If one or two components are extremely large, they should be studied in detail. Transmission loads through walls and roofs may be reduced by adding thermal insulation. Table 16.2 shows that wetting of a sunlit roof surface may substantially reduce the heat gain. Solar loads through glass may be reduced by shading devices or by use of heat absorbing glass. Heat and moisture gains from appliances such as coffee urns may be substantially reduced by hoods with exhaust fans.

16.3. DEGREE DAYS AND ESTIMATION OF SEASONAL HEATING REQUIREMENT

The methods discussed in Sec. 16.2 allow estimation of peak momentary thermal loads. Such loads determine the capacity required for air conditioning equipment. However, in most cases peak loads occur infrequently.

TABLE 16.11*

AVERAGE MONTHLY DEGREE DAYS†

City	Jan.	Feb.	Mar.	Apr.	May	June	July	Aug.	Sept.	Oct.	Nov.	Dec.	Yearly Total
Atlanta, Ga.	636	518	428	147	25	0	0	0	18	124	417	648	2961
Chicago, Ill.	1209	1044	890	480	211	48	0	0	81	326	753	1113	6155
Dallas, Tex.	601	440	319	90	6	0	0	0	0	62	321	524	2363
Denver, Colo.	1132	938	887	558	288	66	6	9	117	428	819	1035	6283
Kansas City, Mo. . .	1032	818	682	294	109	0	0	0	39	220	612	905	4711
Los Angeles, Calif. . .	372	302	288	219	158	81	28	28	42	78	180	291	2061
Minneapolis, Minn. .	1631	1380	1166	621	288	81	22	31	189	505	1014	1454	8382
New York, N.Y. . . .	973	879	750	414	124	6	0	0	27	223	528	887	4811
Phoenix, Ariz. . . .	474	328	217	75	0	0	0	0	0	22	234	415	1765
Seattle, Wash. . . .	828	678	657	474	295	159	56	62	162	391	633	750	5145
Washington, D.C. . .	871	762	626	288	74	0	0	0	33	217	519	834	4224

*Abstracted by permission from *ASHRAE Guide and Data Book, Applications*, 1968, pp. 647–651.

†Based upon airport temperature readings.

It is advantageous to be able to predict the heating requirement or cooling requirement over a season for a structure or a space. As yet, no satisfactory method has been developed for the cooling requirement. In the case of heating, the concept of the *degree day* is useful.

The degree day may be defined by the following statement: During one day (24 hr), there are as many degree days as there are degrees Fahrenheit temperature difference between 65 F and the average outdoor air temperature for the day. Thus, if the average outdoor air temperature was 40 F, 25 degree days would accrue during a single day. Negative degree days have no meaning.

The American Gas Association and others have found that, for residential buildings, the fuel consumption for heating was approximately proportional to 65 F minus the average outdoor temperature for the period. Such findings applied to buildings kept at 68–72 F during the period. The reason that 65 F becomes the base temperature rather than 68–72 F may be explained partially by the existence of heat sources other than the heating system which help heat a building over a period of time. Such heat sources might include solar radiation, people, lights, and appliances.

Over a period of time the average hourly rate of heat loss q_{av} for a structure may be written as

$$q_{av} = q_d \frac{(65 - t_{o,av})}{(t_i - t_o)}$$

where q_d is the calculated design rate of sensible heat loss based upon temperatures t_i and t_o, and $t_{o,av}$ is the average outdoor air temperature for the period. For a period of N days, the total heat requirement Q would be

$$Q = 24 \, q_d \, N \frac{(65 - t_{o,av})}{(t_i - t_o)}$$

However, the quantity $N(65 - t_{o,av})$ is the number of degree days for the period. Thus

$$Q = \frac{24 \, q_d \, (\text{degree days})}{(t_i - t_o)} \tag{16.7}$$

Equation (16.7) provides a convenient but highly approximate method for estimating seasonal heating requirements. The method should be used with judgement. Table 16.11 shows average monthly degree days for several cities in the United States.

16.4. THE PSYCHROMETRIC CHART CONDITION LINE FOR A SPACE

Once the total energy and moisture loads for a space are known, the thermodynamic state of the supply air and the required air flow rate may be determined. Figure 16.3 shows the schematic problem. The quantity $\sum q_s$ represents the net sum of all rates of sensible heat gain in Btu per hr. The quantity $\sum m_w h_w$ represents the net sum of all rates of energy gain from added moisture in Btu per hr. The quantity $\sum m_w$ represents the net sum of all rates of moisture gain in lb_w per hr.

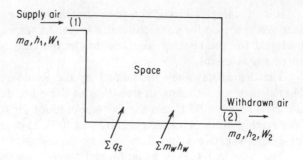

Figure 16.3 Schematic flow processes for an air conditioned space.

Assuming steady-flow conditions, we have

$$\sum q_s + \sum m_w h_w = m_a(h_2 - h_1) \qquad (16.8)$$

$$\sum m_w = m_a(W_2 - W_1) \qquad (16.9)$$

Thus

$$q' = \frac{\sum q_s + \sum m_w h_w}{\sum m_w} = \frac{h_2 - h_1}{W_2 - W_1} \qquad (16.10)$$

where q' is the *enthalpy-moisture ratio*, Btu per lb_w; h is enthalpy of the moist air, Btu per lb_a; and W is humidity ratio of the moist air, lb_w per lb_a.

Equation (16.10) reveals that for a given state of the withdrawn air, all possible psychrometric states (conditions) for the supply air must lie on the same straight line which has a direction q' and which is drawn through State (2) on the psychrometric chart. This straight line is called the *condition line*. Figure 16.4 shows three schematic condition lines. If $q' = h_{g,2}$, the condition line coincides with the dry-bulb temperature

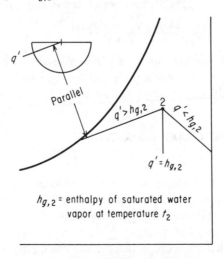

Figure 16.4 The condition line.

line t_2. Once q' is known, the condition line may be drawn through aid of the chart protractor.

According to Eq. (16.10), State (1) may lie at any location on the condition line. However, changing of State (1) requires a change in the air flow rate m_a. Practical considerations influence the location of State (1). In summer comfort air conditioning systems, $(t_2 - t_1)$ may vary from 15–25 F depending upon the method of air distribution.

Example 16.2. For the restaurant of Example 16.1, determine (a) the required dew-point temperature of the supply air, and (b) the required volume of supply air in cu ft per min. Include an allowance of an additional 10 per cent in the rate of sensible heat gain for energy input by a centrifugal fan and miscellaneous contingencies. Assume a dry-bulb temperature of 60 F for the supply air.

SOLUTION: (a) By Example 16.1, $\sum q_s + \sum m_w h_w = (1.1)(161{,}525) + 41{,}490 = 219{,}168$ Btu/hr and $\sum m_w = 40.99$ lb$_w$/hr. By Eq. (16.10),

$$q' = \frac{219{,}168}{40.99} = 5347 \text{ Btu/lb}_w$$

As mentioned in Example 16.1, the data in Fig. E-8 are only approximately correct for the elevation of Minneapolis. Following the notation of Fig. 16.3, we have $t_2 = 75$ F, $\phi_2 = 50$ per cent. By Fig. E-8, we find $W_2 = 0.00928$ lb$_w$/lb$_a$ and $h_2 = 28.15$ Btu/lb$_a$. As shown by Fig. 16.4, the condition line may be drawn with the aid of the chart protractor. At the intersection of $t_1 = 60$ F with the condition line, State (1) is established. We read $t_{d,1} = 52.3$ F.

(b) At State (1), $h_1 = 23.60$ Btu/lb$_a$, $W_1 = 0.00842$ lb$_w$/lb$_a$ and $v_1 = 13.27$ cu ft/lb$_a$. By Eq. (16.8)

$$m_a = \frac{219{,}168}{28.15 - 23.60} = 48{,}167 \text{ lb}_a/\text{hr}$$

Thus

$$\text{Supply volume} = \frac{m_a v_1}{60} = \frac{(48{,}167)(13.27)}{60} = 10{,}650 \text{ cu ft/min}$$

The procedure given by Eqs. (16.8)–(16.10) is consistent with basic thermodynamics. It is a realistic procedure, since all quantities involved have definite physical meaning. Further, the procedure is precisely compatible with the h-W type of psychrometric chart.

In the United States, it is common for air conditioning engineers to use a more approximate procedure than that given by Eqs. (16.8)–(16.10). Little use is made of the property enthalpy and no use is made of the rate of moisture gain $\sum m_w$. The concept that is followed assumes that there are two types of heat gain rates: *sensible heat gain* $\sum q_s$ and *latent heat gain* $\sum q_L$. For psychrometric calculations, it is assumed that

$$\sum q_s = m_a c_{p,a}(t_2 - t_1) \tag{16.11}$$

$$\sum q_L = m_a C(W_2 - W_1) \tag{16.12}$$

The specific heat $c_{p,a}$ is commonly taken as 0.245 Btu per (lb$_a$)(F). Depending upon the reference, values shown for the constant C vary from about 1050 to 1100 Btu per lb_w.

Instead of the enthalpy-moisture ratio q', a term called the *sensible-heat ratio* S.H.R. is used. By definition

$$\text{S.H.R.} = \frac{\Sigma q_s}{\Sigma q_s + \Sigma q_L} \tag{16.13}$$

The protractor of each of the charts, Figs. E-7 and E-8, shows a scale for the sensible-heat ratio $(\Delta H_s / \Delta H_T)$. It should be realized that this scale is only approximately correct since it is not possible to show precisely the scale on h-W coordinates.

> **Example 16.3.** Rework Example 16.2 but use the procedure given by Eqs. (16.11)–(16.13).
>
> SOLUTION: (a) By Example 16.1 (with 10% extra allowance for Σq_s), $\Sigma q_s = 177,678$ Btu/hr and $\Sigma q_L = 41,490$ Btu/hr. By Eq. (16.13)
>
> $$\text{S.H.R.} = \frac{177,678}{177,678 + 41,490} = 0.81$$

Similar to the procedure of Example 16.2, but utilizing the S.H.R. instead of q', the space condition line may be drawn on Fig. E-8 with the aid of the chart protractor. At the intersection of $t_1 = 60$ F with the condition line, State (1) is established. We read $t_{d,1} = 52.3$ F.

(b) At State (1), $v_1 = 13.27$ cu ft/lb$_a$. By Eq. (16.11) and using $c_{p,a}$ as 0.245 Btu/(lb$_a$)(F), we have

$$m_a = \frac{\Sigma q_s}{c_{p,a}(t_2 - t_1)} = \frac{177,678}{(0.245)(15)} = 48,346 \text{ lb}_a/\text{hr}$$

Thus

$$\text{Supply volume} = \frac{m_a v_1}{60} = \frac{(48,346)(13.27)}{60} = 10,690 \text{ cu ft/min}$$

We may observe that the answers to Example 16.3 are closely similar to those for Example 16.2. In later work of this chapter, the procedure given by Eqs. (16.8)–(16.10) will be emphasized.

Examples 16.2 and 16.3 show the actual supply volume required. Commercial air conditioning equipment such as centrifugal fans, heating coils, etc., are usually rated in terms of volume flow of *standard air*. Standard air may be defined as moist air having a specific density of 0.075 lb per cu ft. In general, the specific density ρ in lb per cu ft is given by

$$\rho = \frac{1 + W}{v} \tag{16.14}$$

where W is the humidity ratio, lb$_w$ per lb$_a$; and v is the volume, cu ft per lb$_a$. Equation (16.14) may be used to convert actual volume flow rates to standard air flow rates, or vice-versa.

> **Example 16.4.** Convert the volume flow rate of supply air in Example 16.2 to cu ft per min of standard air.

SOLUTION: By Eq. (16.14),

$$\rho_1 = \frac{1 + W_1}{v_1} = \frac{1 + 0.00842}{13.27} = 0.0760 \text{ lb/cu ft}$$

The mass flow rate of moist air must be the same whether actual density or standard air density is employed. Thus, in terms of standard air

$$\text{Supply volume} = \frac{\rho_1 V_1}{0.075} = \frac{(0.0760)(10,650)}{0.075} = 10,790 \text{ cu ft/min}$$

16.5. PSYCHROMETRIC SYSTEMS

After the heat and moisture loads of a space are calculated and the space condition line is drawn on the psychrometric chart, a psychrometric system may be chosen to process the moist air. The equipment requirements are greatly influenced by the space conditions desired and by the magnitude of the enthalpy-moisture ratio q'. The system should be capable of maintaining acceptable space conditions at all times. Since the thermal loads of the space may be highly variable, automatic control of the processing equipment becomes an important consideration. Haines* has extensively discussed automatic control of heating and air conditioning systems.

In this section we will discuss several types of psychrometric systems. Our purpose will be to illustrate general principles rather than to attempt to cover all combinations. Figure 16.5(a) shows a schematic system which may produce controlled temperature and humidity conditions within a space which has both heat and moisture *losses*; Fig. 16.5(b) schematically shows the psychrometric processes. We will first make a few

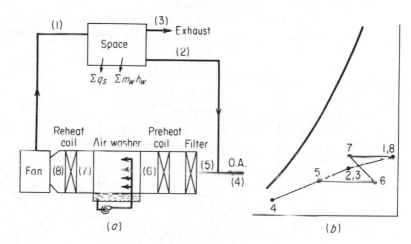

Figure 16.5 Schematic winter air conditioning system.

*J. E. Haines, *Automatic Control of Heating and Conditioning* (New York: McGraw-Hill Book Company, 1953).

general comments about Fig. 16.5 which will apply also to all other systems discussed in this section. Energy added to air by a centrifugal fan causes some rise in temperature without change in humidity ratio. This energy input may be calculated only after the ductwork is designed and processing devices are selected. Usually, the energy input is small. In our discussions we will assume that an allowance for energy input by the fan has been included in the space load calculation. Thus, in Fig. 16.5, States (1) and (8) are assumed to be the same. For each of our systems, we will schematically include filters. Although filters (pads of viscous coated fibers, electrostatic type, etc.) do not change the moist air state, they are required for control of particulate matter carried by the air. In each system, we will assume that some outdoor air is positively introduced. Such air may be required for ventilation purposes or as make-up air for exhaust fans. Introduction of outdoor air requires that an equal mass of dry air be removed from the system. We will show this removal as occurring directly from the space (Location 3 in Fig. 16.5a). Figure 16.5(a) shows only one fan. In large systems, it is common practice to use a recirculating-air fan as well as a supply-air fan. However, this circumstance does not affect the discussions to be given here.

Processing equipment in Fig. 16.5 includes a preheat coil, air washer, and reheat coil. The heating coils would be of the finned-tube type. Analysis of heat transfer in finned-tube heating coils was given in Chapter 12, and analysis of the air washer was given in Chapter 11. The preheat coil serves two functions: it prevents water in the air washer from possible freezing and, by regulation of its heat supply, the amount of water evaporated in the air washer may be controlled. The air washer serves as a humidifying device to offset the moisture losses of the space, and it also accomplishes a certain amount of air cleaning. Regulation of the heat supply by the reheat coil allows control of the space dry-bulb temperature.

The schematic process lines of Fig. 16.5(b) should be obvious. The line $\overline{1\,2}$ is the space condition line, while line $\overline{4\,5\,2}$ is the mixing line for the mixing of recirculated

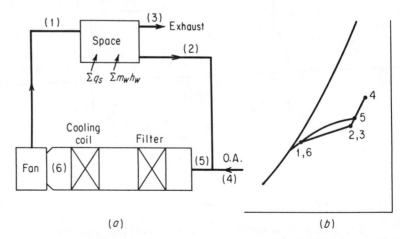

Figure 16.6 Schematic elementary summer air conditioning system.

and outdoor air. Mixing processes were discussed in Chapter 9. The process lines for both heating coils are ones of constant humidity ratio.

Figure 16.6 shows the most elementary form of a summer comfort air conditioning system; a cooling coil is the only processing device. This coil would be of the finned-tube type with the cooling medium being either chilled water or an evaporating refrigerant. Analysis of cooling coils was given in Chapter 12.

Refrigeration theory was discussed in Chapters 3–7. The mechanical vapor compression system is the most important for use with psychrometric systems. Figure 16.7

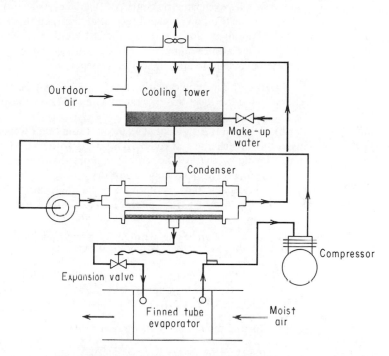

Figure 16.7 Schematic direct expansion refrigeration system for cooling and dehumidifying moist air.

shows a schematic single-stage system where the evaporating refrigerant is directly used to refrigerate moist air. Such an arrangement is called a *direct expansion* system. We assume the system to have a small to moderately large capacity. Vapor is withdrawn from the evaporator by a reciprocating compressor and discharged into a water-cooled, shell-and-tube condenser; cooling water for the condenser is reprocessed in a cooling tower. (Analysis of a cooling tower was covered in Chapter 11.) Liquid refrigerant leaves the bottom (receiver) of the condenser and is expanded into the evaporator by a thermostatic expansion valve.

The system shown in Fig. 16.7 would generally use Refrigerant 12. In large systems, used for comfort air conditioning and having several cooling coils, a central refrigeration plant might process chilled water with the water circulated to the various cooling coils. In this case, we would have an *indirect* refrigeration system. A centrifugal compressor using Refrigerant 11 would be typically used for chilling water. In industrial systems requiring moderately low temperatures, ammonia is commonly used for chilling brine, with the brine circulated to the various cooling coils. As discussed in Chapter 3, multi-stage systems may be required in low temperature applications.

Example 16.5. Assume that the restaurant of Example 16.1 is to be provided with the simple psychrometric system of Fig. 16.6. As a continuation of Example 16.2, determine (a) the volume of recirculated air, cu ft per min of standard air, (b) the thermodynamic state of the moist air entering the cooling coil, (c) the tons of refrigeration required, and (d) an energy balance in Btu per hr and a moisture balance in lb_w per hr for the system.

SOLUTION: The nomenclature of Fig. 16.6 will be used. From Example 16.2, we have the locations of States (1) and (2) on the psychrometric chart. We also have $m_{a,1} = 48,167$ lb_a/hr.

(a) Example 16.1 stated that 4,500 cu ft/min of air were required for exhaust fans in the kitchen and rest rooms. This air is withdrawn from the serving area. Table 16.8 shows that 4500 cu ft/min is more than sufficient for ventilation of the serving area. The mass flow rate of dry air $m_{a,3}$ withdrawn from the serving area for the exhaust fans would have to be identical to the mass flow rate of dry air $m_{a,4}$ introduced from outdoors. Assuming 4,500 cu ft/min of standard air, we have

$$m_{a,3} = m_{a,4} = \frac{(0.075)(4,500)(60)}{1 + W_2} = \frac{(0.075)(4,500)(60)}{1.00928} = 20,064 \text{ } lb_a/hr$$

Thus, $m_{a,2} = m_{a,1} - m_{a,3} = 28,103$ lb_a/hr, and in terms of standard air

$$\text{Recirculated volume} = \frac{m_{a,2}(1 + W_2)}{(0.075)(60)} = \frac{(28,103)(1.00928)}{(0.075)(60)} = 6303 \text{ cu ft/min}$$

(b) For the mixing process, we may write

$$W_5 = W_2 + \frac{m_{a,4}}{m_{a,5}}(W_4 - W_2)$$

$$= 0.00928 + \frac{(20,064)(0.0166 - 0.00928)}{48,167} = 0.01233 \text{ } lb_w/lb_a$$

State (5) is located at the intersection of W_5 and the line $\overline{2\,4}$. We find $t_5 = 82.2$ F, $t_5^* = 69.2$ F, and $h_5 = 33.25$ Btu/lb_a.

(c) Per lb of dry air passing through the cooling coil, we have

$$h_5 = h_6 + {}_5Q_6 + (W_5 - W_6)h_f$$

The temperature of the liquid water removed would be somewhat lower than t_6. Since the energy removed with the liquid water is extremely small, we will use $h_f = h_{f,6}$. Thus

$$_5Q_6 = (h_5 - h_6) - (W_5 - W_6)h_{f,6}$$
$$= (33.25 - 23.60) - (0.00391)(28.08) = 9.54 \text{ Btu}/lb_a$$

and

$$\text{Tons of refrigeration} = \frac{m_{a,5}(_5Q_6)}{12,000} = 38.3$$

(d) Energy balance calculations are shown in Table 16.12. We should observe the significantly large system heat and moisture gains from the outdoor air.

TABLE 16.12

SOLUTION FOR PART (d) OF EXAMPLE 16.5

Component and Calculation	Energy Btu/hr		Moisture lb$_w$/hr	
	Gain	Loss	Gain	Loss
Serving area: see examples 16.1 and 16.2	219,168		40.99	
Cooling Coil				
System energy loss $= m_{a,5}(h_5 - h_6)$		464,812		
System moisture loss $= m_{a,5}(W_5 - W_6)$ . .				188.33
Outdoor Air				
System energy gain $= m_{a,4}(h_4 - h_3)$ . . .	245,383			
System moisture gain $= m_{a,4}(W_4 - W_3)$ . .			146.86	
Totals	464,551	464,812	187.85	188.33

The simple system of Fig. 16.6 has limitations. Since the cooling coil is the only processing device, only one property of the moist air can be controlled. In comfort air conditioning systems, this would always be the space dry-bulb temperature.

Figure 16.8 shows a modification to the system of Fig. 16.6 which may be useful during partial load operation. We assume that face dampers (not shown) on the cooling coil and the by-pass damper are controlled by a motor which positions them so as to maintain a constant space dry-bulb temperature. As the sensible heat gain of the space decreases, more recirculated air is bypassed. However, the air which does pass across the coil may be more thoroughly dehumidified than when the full air quantity is handled. Thus, satisfactory space humidity conditions may be maintained during some partial load situations without the need for reheat.

Figure 16.9 shows a summer air conditioning system with a reheat coil. Reheat is required when the space condition line is so steep that a satisfactory intersection of the cooling coil process line with the saturation curve cannot be obtained. When a reheat coil is included, both temperature and humidity of a space may be controlled. However, the use of reheat results in more expensive system operation. All heat added to the air as reheat must be removed in the cooling coil.

Various combinations of psychrometric processes may be used for industrial air conditioning systems. Figure 16.10 shows an elementary system where dehumidifica-

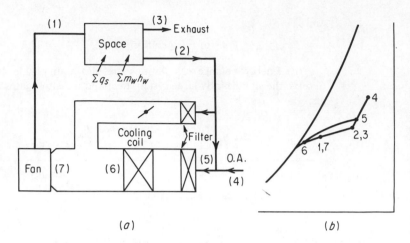

Figure 16.8 Schematic elementary summer air conditioning system with by-pass of recirculated air.

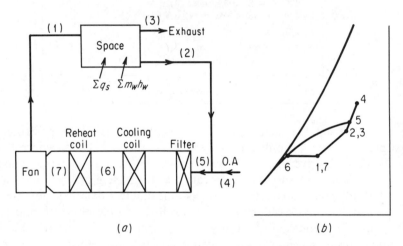

Figure 16.9 Schematic summer air conditioning system with reheat.

tion occurs in an adiabatic absorbent bed. Typical solid adsorbents used for dehumidification of moist air are silica gel and activated alumina. When moist air is passed across an adsorbent, water vapor may be separated from the air, condensed, and retained by the adsorbent as extremely small particles of liquid water. This liquid water is held under reduced vapor pressure because of very strong surface forces. Thus, a vapor pressure difference may exist between the moist air and the adsorbent, providing a potential difference for water vapor transfer. Adsorption is a physical phenomenon only, and no chemical changes occur in the adsorbent material.

Thermal processes occurring during adsorption are complex. A mathematical

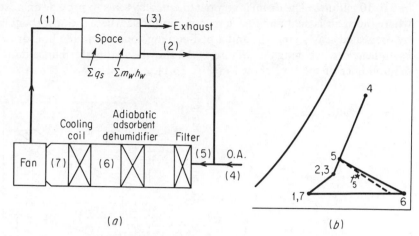

(a)   (b)

Figure 16.10 Schematic air conditioning system with an adsorbent dehumidifier.

and experimental study on adsorption has been given by Bullock and Threlkeld.*
In an adiabatically operated bed, heat of condensation of vapor plus additional heat,
called *heat of wetting*, are transferred to the air stream as sensible heat. Figure 16.10
shows in a schematic way how the moist air state may be altered during adiabatic
adsorption. Humidity ratio is decreased, dry-bulb temperature is increased, and there
is typically a small increase in thermodynamic wet-bulb temperature.

Figure 16.11 shows a schematic arrangement of an adiabatic adsorbent bed which
provides for continuous operation. A thin layer of adsorbent, such as silica gel, forms
the periphery of a rotating drum. The drum rotates slowly, completing a revolution

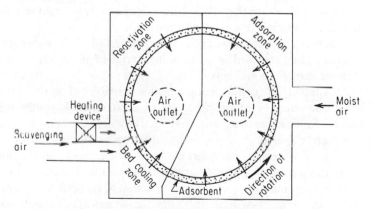

Figure 16.11 Schematic rotary drum adsorbent dehumidifier.

*C. E. Bullock and J. L. Threlkeld, "Dehumidification of Moist Air by Adiabatic Adsorption,"
ASHRAE Trans., Vol. 72, Part I, 1966, pp. 301–313.

in 5 to 10 minutes. The drum is compartmented by seals to provide an adsorption zone where moist air is dehumidified, a reactivation zone where the adsorbent is dehydrated by preheated scavenging air, and a bed-cooling zone where the hot, dry bed is cooled by unheated scavenging air. Figure 16.12 shows some performance data for a rotary dehumidifier of the type shown in Fig. 16.11.

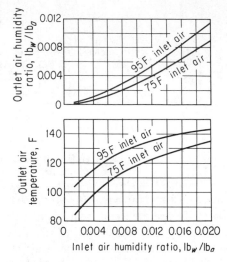

Figure 16.12 Outlet conditions for moist air from a typical rotary drum silica gel dehumidifier. (Adapted by permission from *Heating, Ventilating, Air Conditioning Guide*, Vol. 36, p. 964.)

Besides solid adsorbents, certain liquid absorbent solutions may be used for dehumidifying moist air. Examples are solutions in water of lithium chloride and of triethylene glycol. In this arrangement, moist air is brought into direct contact with the absorbent solution in a spray chamber.

Sorbent systems are applicable where relatively low dew-point temperatures are needed. They are used in various industrial applications. Comfort air conditioning systems rarely use sorbents.

Comfort air conditioning systems capable of maintaining optimum thermal conditions may be expensive to own and operate. Partially effective systems which involve much lesser costs may be attractive where finances preclude the installation of a completely effective system. In hot dry regions, *evaporative cooling systems* may be capable of providing considerable relief in enclosed spaces. Such systems may also have application in hot industrial environments. Figure 16.13 shows an elementary evaporative cooling system. We assume that State (2) is an acceptable space condition, although not necessarily an optimum one. State (3) of the outdoor air is assumed to be at a much higher temperature but lower humidity ratio than State (2). Through use of an air washer as the only processing device, an acceptable air conditioning system

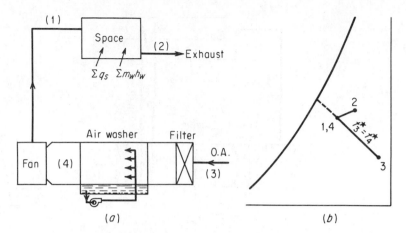

Figure 16.13 Schematic elementary evaporative cooling system.

may result. Generally, a much higher flow rate of air is used with an evaporative cooling system than with conventional systems. A high rate of air movement past a person allows the same degree of comfort but with higher effective temperatures, as compared to situations where air movement is slight.

Modifications may be made among the psychrometric systems previously discussed in order to achieve specific purposes. For example, the addition of a cooling coil between the air washer and reheater of Fig. 16.5 would provide a year-round air conditioning system. The systems of Figs. 16.5 and 16.9 might be applied as central systems to serve more than one space by removal of the reheat coil on the suction side of the fan and by use of individual reheat coils in the separate supply ducts to each space. Other modifications may also be recognized for special problems.

In designing a psychrometric system, every effort should be made to utilize heat transfers internal to the system where economically feasible so as to reduce the needs for external heat transfers. Figure 16.14 shows one case. We assume winter operation and a system which requires 100 per cent outdoor air. The outdoor air is preheated by waste heat in the exhaust air. The same recovery scheme of Fig. 16.14 might be applied for an opposite purpose in Fig. 16.13. Here, the outdoor air might be precooled by the exhaust air, allowing State (2) to be at a lower effective temperature for the same air flow rate. Other possibilities for using internal heat transfers might include the use of refrigerant-condenser heat for reheating of moist air. When outdoor air temperatures are high, reheating may be accomplished internally by a heat-exchanger system involving the reheat coil and a coil in the outdoor-air duct.

Example 16.5 showed that about 53 per cent of the required refrigerating capacity was chargeable to the outdoor air taken into the system. In such cases, and in winter systems where heating is required, the use of outdoor air should be kept to a minimum. However, for spaces which may require cooling during the entire year, provision

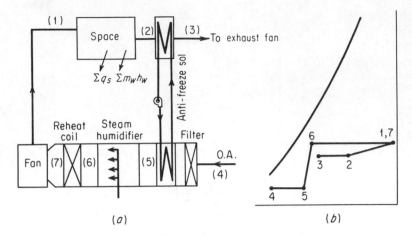

Figure 16.14 Schematic winter air conditioning system using 100 per cent outdoor air with preheating by waste heat from the exhaust air.

should be made for flexible use of outdoor air so that refrigerating equipment may be shut down during cold weather.

PROBLEMS

16.1. Calculate the design rate of sensible heat loss for winter heating for the serving area of the restaurant of Example 16.1. Assume an inside design temperature of 70 F. You may use the same U factors as given in Example 16.1.

16.2. A dairy store located in Dallas, Texas, consists of a one-story room 30 ft wide and 60 ft long. Ceiling height is 10 ft. The short side facing north has a glass-panel swinging door 4 ft by 7.5 ft, and a $\frac{1}{4}$ in. plate-glass window 6 ft high and 20 ft long. The long side facing west has a plate-glass window 6 ft high and 30 ft long. Both windows are set back by 6 in. and are provided with inside white Venetian blinds. The south wall is a partition wall to an adjacent air-conditioned space. The exposed east wall has no openings. The flat roof consists of 2 in. wood plank with built-up roofing on 2 in. by 10 in. wood joists with metal lath and plaster underneath. The outside walls consist of 4 in. of face brick, 8 in. of common brick, with an inside coat of $\frac{1}{2}$ in. of cement plaster. You may ignore ground-reflected radiation. Peak occupancy is 50 patrons and 6 clerks. Appliance energy (sensible plus water vapor) and moisture gains are respectively 15,000 Btu per hr and 5.0 lb$_w$ per hr. There are 4,000 watts of fluorescent lighting. Peak occupancy may occur from 11:00 a.m.–8:00 p.m. CST. Based upon space conditions of 80 F dry-bulb temperature and 50 per cent relative humidity, calculate the peak rates of heat and moisture gains for the space.

16.3. Estimate the cost of heating your study room at home for the October 1 to April 30 season. Base your calculations upon the type of installed heating system and current local fuel costs.

16.4. Calculate the rates of heat and moisture gain for the restaurant of Example 16.1 for an extremely cloudy day, when the outdoor air is at 82 F dry-bulb temperature and 74 F thermodynamic wet-bulb temperature. Determine whether the air conditioning system would require reheat in order to maintain space conditions of 75 F dry-bulb temperature and 50 per cent relative humidity.

16.5. A space to be maintained at 75 F dry-bulb temperature and 50 per cent relative humidity has a rate of sensible heat gain of 82,000 Btu per hr and a rate of moisture gain (average $h_w = 1100$ Btu per lb_w) of 12.0 lb_w per hr. Barometric pressure is 14.696 psia. Moist air is supplied to the room at 58 F dry-bulb temperature. Determine (a) the dew-point temperature and thermodynamic wet-bulb temperature of the supply air, and (b) the volume of supply air required in cu ft per min of standard air.

16.6. An evaporative cooling system similar to that shown in Fig. 16.13 is used for improving thermal conditions in an industrial space. The rate of sensible heat addition $\sum q_s$ is 100,000 Btu per hr and the rate of moisture addition (saturated water vapor at 90 F) is 10.0 lb per hr. The following air temperatures are known: $t_2 = 85$ F, $t_2^* = 70$ F, $t_3 = 105$ F, and $t_3^* = 67$ F. Air velocity over each worker's torso is 100 ft per min. Average temperature of surfaces surrounding the workers is 85 F. The average body-heat production rate is 1000 Btu per hr. Assume sea-level pressure. (a) Determine the required fan capacity in cu ft per min of standard air. (b) Management and labor agree that the heat stress index should not exceed 25. Determine if space (withdrawn air) thermal conditions are acceptable.

16.7. Figure 16.15 shows a schematic winter-type air conditioning system. The rate of sensible heat loss $\sum q_s$ from the space is 133,500 Btu per hr and the rate of moisture loss (average $h_w = 1100$ Btu per lb_w) is 15.0 lb per hr. Assume

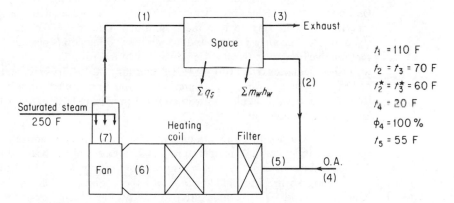

Figure 16.15 Schematic air conditioning system for Problem 16.7.

sea-level pressure. (a) Locate all statepoints on the psychrometric chart and read values of $t_1^\star$, $t_5^\star$, t_7, and $t_7^\star$. (b) Calculate the required volume of air supplied to the space in cu ft per min. (c) Find the lb per hr of steam required for the humidifier.

16.8. A space is air conditioned in winter by the schematic system of Fig. 16.5. The space has a rate of sensible heat loss $\sum q_s$ of 180,000 Btu per hr and a rate of moisture loss (average $h_w = 1092$ Btu per lb_w) of 18.5 lb_w per hr. Moist air is withdrawn from the space at 70 F dry-bulb temperature and 55 F thermodynamic wet-bulb temperature. Barometric pressure is 14.696 psia. Moist air is supplied to the space at 100 F dry-bulb temperature. Outdoor air is saturated at 35 F. The dry-air flow rate of outdoor air admitted to the system is 50 per cent of the dry-air flow rate of the air supplied to the space. You may assume that the air washer has an efficiency of 50 per cent. Determine (a) the volume of air supplied to the space in cu ft per min of standard air, (b) the spray water temperature, (c) the lb per hr of make-up water required for the air washer, and (d) the rate of heat added to the air by each heating coil in Btu per hr.

16.9. A space to be maintained at 76 F dry-bulb temperature and 65 F thermodynamic wet-bulb temperature has a rate of sensible heat gain $\sum q_s$ of 84,000 Btu per hr and a rate of moisture gain (average $h_w = 1120$ Btu per lb_w) of 20.0 lb_w per hr. Moist air enters the space at a dry-bulb temperature of 60 F. Outdoor air at 95 F dry-bulb temperature and 78 F thermodynamic wet-bulb temperature is supplied for ventilation purposes at a rate of 825 cu ft per min of standard air. The space is to be conditioned by the schematic system of Fig. 16.6. Barometric pressure is 14.696 psia. Determine (a) the dry-bulb temperature and thermodynamic wet-bulb temperature of the air entering the cooling coil, and (b) the tons of refrigeration required.

16.10. A space to be maintained at 80 F dry-bulb temperature and 50 per cent relative humidity has a rate of sensible heat gain $\sum q_s$ of 73,500 Btu per hr and a rate of moisture gain (average $h_w = 1100$ Btu per lb_w) of 15.0 lb_w per hr. The volume of air supplied to the space is 7,200 cu ft per min of standard air. Outdoor air at 95 F dry-bulb temperature and 75 F thermodynamic wet-bulb temperature is introduced into the system at a rate of 1,800 cu ft per min of standard air. The air which passes through the cooling coil is brought to 90 per cent relative humidity. Barometric pressure is 14.696 psia. The space is provided with the schematic system of Fig. 16.8. Determine (a) the dry-bulb temperature and thermodynamic wet-bulb temperature of the air supplied to the space, and (b) the volume of recirculated air in cu ft per min (standard air) which should by-pass the cooling coil.

16.11. A space to be maintained at 75 F dry-bulb temperature and 65 F thermodynamic wet-bulb temperature has a rate of sensible heat gain $\sum q_s$ of 377,000 Btu per hr and a rate of moisture gain (average $h_w = 1100$ Btu per lb_w) of 330.0 lb_w per hr. The chilled air leaves the cooling coil at 54 F dry-bulb temperature and 90 per cent relative humidity. Barometric pressure is 14.696 psia. The space is conditioned by the system shown in Fig. 16.9. Determine

(a) the dry-bulb temperature and thermodynamic wet-bulb temperature of the air supplied to the space, and (b) the rate of heat addition by the reheat coil in Btu per hr.

16.12. An interior space of a building is to be maintained at 72 F during winter. The space has a rate of sensible heat gain of 60,000 Btu per hr during working hours; moisture gain is negligible. The space is to be cooled by mixing outdoor air with recirculated air. Air is to be supplied to the space at a dry-bulb temperature of 55 F. On a particular day, the outdoor air is saturated at -10 F. Barometric pressure is 14.696 psia. (a) Determine the volume of supply air required in cu ft per min of standard air. (b) Find the percentage of the supply air (dry-air mass basis) which is outdoor air. (c) Assuming one brake horse-power per ton of refrigeration as the compressor power requirement for a mechanical refrigeration system, estimate the daily operational savings of the above system. Assume 8 hr operation and electricity costs of 2.0 cents per kwhr.

16.13. Figure 16.16 shows a schematic industrial air conditioning system. Barometric pressure is 14.696 psia. Assume that all dehumidification occurs in

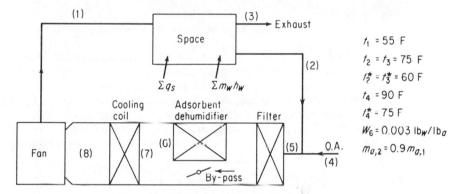

Figure 16.16 Schematic air conditioning system for Problem 16.13.

the adsorbent dehumidifier which is of the ideal adiabatic type ($t_5^* = t_6^*$). The rate of sensible heat addition $\sum q_s$ is 100,000 Btu per hr and the rate of moisture addition (saturated water vapor at 90 F) is 30.0 lb per hr. (a) Locate all statepoints on the psychrometric chart and read values of $t_1^*, t_5, t_5^*, t_6, t_6^*,$ and t_7. (b) Determine the volume rate of air flow in cu ft per min which by-passes the adsorbent dehumidifier.

16.14. A space is air conditioned in winter by a system similar to that shown by Fig. 16.14. The rate of sensible heat loss $\sum q_s$ from the space is 180,-000 Btu per hr and the rate of moisture loss (average $h_w = 1092$ Btu per lb_w) is 18.0 lb per hr. Additional known data are as follows: $t_1 = 100$ F, $t_2 = 70$ F, $t_2^* = 55$ F, $t_3 = 45$ F, $t_4 = 20$ F, $\phi_4 = 100$ per cent, and $\phi_6 = 100$ per cent. Barometric pressure is 14.696 psia. (a) Locate all statepoints on the psychrometric chart and read values of $t_1^*, t_3^*, t_5, t_5^*,$ and t_6. (b) Determine the required specific enthalpy for the steam admitted to the air by the humidifier.

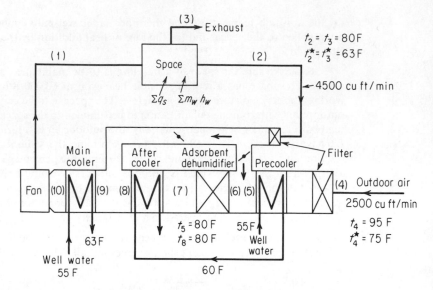

Figure 16.17 Schematic air conditioning system for Problem 16.15.

16.15. Figure 16.17 shows a schematic industrial air conditioning system. The space has a rate of sensible heat gain $\sum q_s$ of 116,000 Btu per hr and a rate of moisture gain (average $h_w = 1100$ Btu per lb_w) of 90.0 lb_w per hr. All dehumidification occurs in the adsorbent dehumidifier whose performance is given by Fig. 16.12. Barometric pressure is 14.696 psia. Determine (a) values of t, $t^\star$, h, and W for each psychrometric state, (b) the volume of recirculated air in cu ft per min which by-passes the adsorbent dehumidifier, and (c) the lb per hr of water required for each cooling coil.

SYMBOLS USED IN CHAPTER 16

A	Surface area, sq ft.
C	Constant in Eq. (16.12), Btu per lb water vapor.
$c_{p,a}$	Specific heat of moist air, Btu per (lb dry air)(F).
d	Sun's declination angle.
F	Shading factor in Eq. (16.3), dimensionless.
F_s	Shading factor due to set-back, dimensionless.
h	Hour angle of sun.
h	Enthalpy of moist air, Btu per lb of dry air; h_i for inside air, h_o for outside air.
h_g	Enthalpy of saturated water vapor, Btu per lb.
h_i	Combined convection and radiation heat transfer coefficient for wall-inside surface as defined by Eq. (14.2), Btu per (hr)(sq ft)(F).
h_o	Combined convection and radiation heat transfer coefficient for wall-outside surface as defined by Eq. (14.3), Btu per (hr)(sq ft)(F).

h_w	Specific enthalpy of moisture, Btu per lb.
I_D	Incidence of direct solar radiation upon a surface, Btu per (hr)(sq ft); I_{DH} for horizontal surface.
I_d	Incidence of diffuse sky radiation upon a surface, Btu per (hr)(sq ft); I_{dH} for horizontal surface; I_{dV} for vertical surface.
I_N	Incidence of direct solar radiation upon a surface normal to sun's rays, Btu per (hr)(sq ft).
I_R	Incidence of solar radiation upon a surface which is reflected by surrounding surfaces, Btu per (hr)(sq ft).
kwhr	Kilowatt hr.
L	Length, ft.
l	Latitude angle.
m_a	Dry-air flow rate, lb per hr.
m_w	Rate of addition of water, lb per hr.
P	Barometric pressure, psia.
P_v	Pressure due to wind velocity, inches of water.
q	Rate of heat transfer, Btu per hr.
q_{av}	Average rate of heat loss, Btu per hr.
q_d	Design rate of heat loss, Btu per hr.
q_L	Latent heat transfer (see Eq. 16.12), Btu per hr.
q_s	Rate of sensible heat transfer, Btu per hr.
q_t	Rate of energy transfer (sensible heat plus energy of water vapor), Btu per hr.
q'	$(h_2 - h_1)/(W_2 - W_1)$, Btu per lb water.
S.H.R.	Sensible heat ratio as defined by Eq. (16.13), dimensionless.
t	Dry-bulb temperature of moist air, F; t_i for inside air; t_o for outdoor air; $t_{o,av}$ for average outdoor air temperature.
t_d	Dew-point temperature of moist air, F.
$t^\star$	Thermodynamic wet-bulb temperature of moist air, F.
U	Overall heat transfer coefficient, Btu per (hr)(sq ft)(F).
V_w	Wind velocity, miles per hr.
v	Volume of moist air, cu ft per lb dry air; v_o for outdoor air.
W	Humidity ratio of moist air, lb water vapor per lb dry air; W_i for inside air; W_o for outdoor air.
α	Wall-solar azimuth angle.
α	Absorptivity for solar radiation, dimensionless; α_D for direct radiation; α_d for diffuse sky radiation; α_R for reflected radiation.
β	Sun's altitude angle.
γ	Sun's azimuth angle.
θ	Sun's angle of incidence.
ρ	Density, lb per cu ft.
τ	Transmissivity for solar radiation, dimensionless; τ_D refers to direct radiation; τ_d refers to diffuse sky radiation; τ_R refers to reflected radiation.

17

SPECIAL TOPICS

17.1. INTRODUCTION

In this final chapter we will study three diverse topics associated with thermal environmental engineering—the heat pump, solar energy collectors, and drying of materials.

More than a century ago, Lord Kelvin introduced the principle of the heat pump. It has only been since World War II that the heat pump has received widespread application in the United States.

Because of depletion of fossil-fuel reserves, increasing attention has been given to other sources of energy during recent years. Solar energy offers a vast and inexhaustible source of energy. Although its exploitation is uneconomical now except in a few isolated parts of the world, solar energy may be a significant energy source in future years.

Drying of materials is a topic usually associated with chemical engineering rather than mechanical engineering. However, convection drying of materials is closely related to psychrometric problems treated earlier in this text.

17.2. THE HEAT PUMP

Any refrigeration system acts as a heat pump. Heat is withdrawn at a low temperature level and is rejected at a higher temperature level. Where the useful purpose of the system is to produce a refrigerating effect only, the rejected heat is usually wasted. We speak of the system as a refrigeration system.

By a *heat pump*, we mean a refrigerating system where the heat rejected in the condenser is utilized for a useful purpose, at least during part of the year. The heat pump is an especially attractive consideration in year-round air conditioning systems. Refrigeration equipment may be required for cooling in summer. A logical step would be to use the same equipment for heating in winter. Typically, a heat pump would not be considered for winter heating only.

We will confine our remarks to heat pumps involving mechanical vapor-compression equipment. Analyses for mechanical vapor-compression systems were given in Chapter 3. For the theoretical single-stage cycle of Fig. 3.3, the coefficient of performance of the system *when acting as a heat pump* would be

$$\text{C.O.P.} = \frac{h_4 - h_1}{h_4 - h_3} = \frac{h_3 - h_2}{h_4 - h_3} + 1 \tag{17.1}$$

Thus, for the theoretical single-stage cycle, the C.O.P. as a heat pump is equal to the C.O.P. as a refrigeration cycle plus one. In this section, we will analyze heat pump systems using the more practical vapor cycle of Sec. 3.5 and at times using typical manufacturer's data. Such procedures are necessary if we are to make cost comparisons with other methods of heating.

A basic factor of great importance in successful application of heat pumps is the availability of a cheap, dependable source of heat for the evaporator—preferably one at a relatively high temperature. Many types of heat sources may be mentioned, including almost all forms of waste heat. Generally, three principal sources are considered—outdoor air, water (well water, surface water, city water, etc.), and the subsurface ground. Of these, outdoor air and well water are the more practicable. We will confine our discussions to these two sources.

Outdoor air is generally the most practicable heat source for a heat pump. It is universally available and in many geographical areas it exists at favorable temperatures much of the winter season. However, outdoor air temperatures are highly variable, and when the heating requirement is large, outdoor air temperature is relatively low.

Well water provides a higher-temperature heat source than outdoor air and allows attainment of higher C.O.P. values. However, many localities in the United States have inadequate supplies of ground water. In some cases where well water is available, disposal of the water after use may be difficult and costly. In addition, well water may contain dissolved solids which cause corrosion and scaling of heat-transfer equipment.

We will now develop a more practical expression for the C.O.P. of a single-stage heat pump than that given by Eq. (17.1). In general, we may write

$$\text{C.O.P.} = \frac{\text{Heat Rejected, Btu/min}}{\text{Power Supplied to Drive Compressor, Btu/min}}$$

For a system with an electric-motor-driven compressor, we may show that

$$\text{C.O.P.} = \frac{4.72\eta_e R}{\text{Hp/ton}} \tag{17.2}$$

where η_e is the efficiency of the electric motor, dimensionless; R is the ratio of heat rejected in the condenser to heat absorbed in the evaporator, dimensionless; and Hp/ton is the horsepower input to the compressor required per ton of refrigeration produced in the evaporator. It is preferable to use manufacturer's data for Hp/ton if such data are available. Values of R and also Hp/ton may be estimated by assuming a practical-type vapor cycle as discussed in Sec. 3.5.

The cost of energy for a heat pump driven by an internal combustion engine may be substantially less than that of an electric-motor-driven heat pump if most of the waste heat from the engine is recovered. Figure 17.1 schematically shows a space

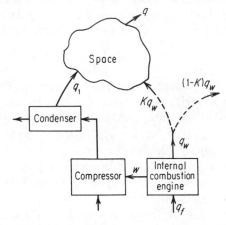

Figure 17.1 Schematic heating of a space by an engine-driven heat pump.

heated by a heat pump whose compressor is driven by an internal combustion engine. Only the compressor and condenser of the heat pump are shown. The engine is supplied fuel energy at a rate q_f. The engine delivers work to the compressor at a rate w and rejects heat at a rate q_w. The waste heat q_w is essentially equal to that absorbed by the jacket cooling water and that escaping with the exhaust gases. Through a system of heat exchangers, we assume that Kq_w is recovered for heating of the space. The condenser of the heat pump delivers heat to the space at a rate q_1. The space has a rate of heat loss q. We may write

$$q = q_1 + Kq_w = q_1 + K(q_f - w) \tag{17.3}$$

but

$$w = \eta_t q_f \tag{17.4}$$

where η_t is the thermal efficiency of the engine. Also

$$q_1 = (\text{C.O.P.})w \tag{17.5}$$

where C.O.P. is given by Eq. (17.2) with $\eta_e = 1$.

By Eqs. (17.3)–(17.5),

$$q_f = \frac{q}{\eta_t(\text{C.O.P.}) + K(1 - \eta_t)} \tag{17.6}$$

Example 17.1. A St. Paul, Minnesota residence has a calculated design rate of sensible heat loss of 75,000 Btu per hr based upon design temperatures of 75 F and −19 F. Fuel costs for an average January are to be estimated for (a) a single-stage heat pump with a well-water heat source and with electric-motor drive, (b) a natural-gas furnace, and (c) a single-stage heat pump with a well-water heat source and driven by an internal combustion engine using natural gas as fuel.

In calculating electricity costs, assume that the first 250 kwhr used are charged to other electrical appliances. Residential electric rates for St. Paul are as follows: First 50 kwhr per month, 4.74 cents per kwhr; next 150 kwhr per month, 2.63 cents per kwhr; next 500 kwhr per month, 2.11 cents per kwhr; excess kwhr per month, 1.58 cents per kwhr.

In calculating heat pump costs, assume a condensing temperature of 120 F and an evaporating temperature of 30 F. Refrigerant 12 is used. The following additional assumptions may be made about the single-stage cycle: Refrigerant liquid leaves condenser at 90 F; pressure drop in compressor suction line of 2 psi; pressure drop of 2 psi in compressor suction valves and in discharge valves; refrigerant leaves evaporator as saturated vapor and enters compressor at 45 F; vapor is further superheated to 65 F on intake stroke of compressor; compression path is isentropic; compressor clearance is 4 per cent; and compressor mechanical efficiency is 80 per cent. Otherwise, assume the theoretical cycle. Use an electric motor efficiency of 80 per cent. For the engine-driven heat pump, assume a brake-thermal efficiency of 30 per cent and assume that 75 per cent of the engine waste heat is recovered.

In calculating natural gas costs, assume a heating value of 1000 Btu per cu ft and assume that the first 3500 cu ft used are charged to other gas appliances. Use a furnace efficiency of 75 per cent. Natural gas rates for St. Paul are as follows: First 400 cu ft or less per month, $1.00; next 2600 cu ft per month, $1.50 per thousand cu ft; next 22,000 cu ft per month, 84¢ per thousand cu ft; and next 375,000 cu ft per month, 82¢ per thousand cu ft.

SOLUTION: We will first estimate the heating requirement of the residence during January. St. Paul is adjacent to Minneapolis. Table 16.11 shows that there are 1631 degree days in Minneapolis during an average January. By Eq. (16.7)

$$Q = \frac{(24)(75,000)(1631)}{94} = 31,230,000 \text{ Btu}$$

(a) Figure 17.2 shows a schematic $P\text{-}h$ diagram for the heat pump cycle. We have

$$R = \frac{h_5 - h_1}{h_3 - h_2} = \frac{98.3 - 28.7}{80.4 - 28.7} = 1.346$$

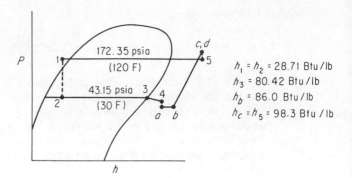

$h_1 = h_2 = 28.71$ Btu/lb
$h_3 = 80.42$ Btu/lb
$h_b = 86.0$ Btu/lb
$h_c = h_5 = 98.3$ Btu/lb

Figure 17.2 Schematic heat pump cycle for Example 17.1.

Since the compression path is isentropic we have

$$\text{Hp/ton} = \frac{4.72(h_c - h_b)}{\eta_m(h_3 - h_2)} = \frac{(4.72)(98.3 - 86.0)}{(0.80)(80.4 - 28.7)} = 1.404$$

By Eq. (17.2),

$$\text{C.O.P.} = \frac{(4.72)(0.80)(1.346)}{1.404} = 3.62$$

and

$$\text{kwhr} = \frac{Q}{(3413)(\text{C.O.P.})} = \frac{31,230,000}{(3413)(3.62)} = 2528$$

The first 450 kwhr would cost \$0.0211 per kwhr, while the remainder would cost \$0.0158 per kwhr. Thus

$$\text{Cost} = (0.0211)(450) + (0.0158)(2078) = \$42.33$$

(b) The quantity of natural gas required for the furnace is

$$\text{Volume} = \frac{Q}{(1,000)(0.75)} = \frac{31,230,000}{750} = 41,640 \text{ cu ft}$$

The first 21,500 cu ft would cost \$0.84 per thousand cu ft while the remainder would cost 82¢ per thousand cu ft. Thus

$$\text{Cost} = (21.5)(0.84) + (20.14)(0.82) = \$34.57$$

(c) For the engine-driven heat pump, the C.O.P. would be given by Eq. (17.2) with $\eta_e = 1.0$. Thus

$$\text{C.O.P.} = \frac{(4.72)(1.346)}{1.404} = 4.53$$

By Eq. (17.6),

$$Q_f = \frac{31,230,000}{(0.30)(4.53) + (0.75)(0.70)} = 16,576,000 \text{ Btu}$$

and 16,576 cu ft of gas would be required. Thus

$$\text{Cost} = (16.576)(0.84) = \$13.92$$

Example 17.1 shows that a heat pump driven by an internal combustion engine may be substantially cheaper to operate than either a gas-fired furnace or an electric-motor-driven heat pump. However, an engine-driven heat pump has serious disadvantages for residential air conditioning. Initial costs may be high, there may be a major noise problem, and considerable maintenance may be required. In commercial installations, these problems may be more easily controlled.

The most economical method of heating depends greatly upon local fuel and electricity costs. Figure 17.3 allows a rapid comparison of fuel costs for oil, gas,

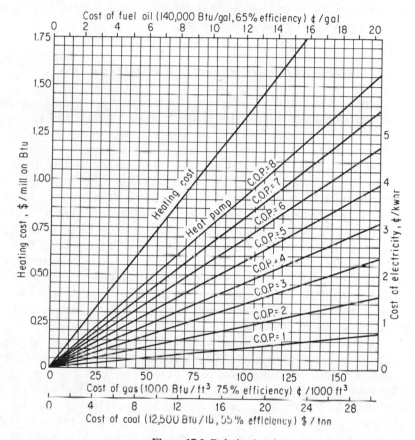

Figure 17.3 Relative heating costs for different fuels.

and coal-fired furnaces with electricity costs for an electric-motor-driven heat pump and for electric-resistance heating (C.O.P. = 1). Any vertical line across the chart is a line of equal costs. Thus, fuel costs for natural gas available at $1.00 per 1000 cu ft should be approximately equal to costs with fuel oil at 12¢ per gal, with coal at $18 per ton, with electric-resistance heating and electricity cost of 0.45¢ per kwhr, and with an electric-motor-driven heat pump (C.O.P. = 4.0) and electricity cost of 1.82¢ per kwhr. For all of these cases, the cost per million Btu supplied would be approximately $1.33.

In commercial and industrial applications of heat pumps, a wide variety of equipment arrangements are possible, depending upon whether the heat source is outdoor air or water, and upon certain special circumstances. Compressor requirements are generally determined by the cooling load. In every case, a detailed study should be made on maximum utilization of heat transfers internal to the system. Many articles have been published on commercial applications of heat pumps. Harnish* has discussed systems using outdoor air as the heat source. Kroeker et al.† have discussed systems using well water as the heat source.

17.3. THE FLAT-PLATE SOLAR ENERGY COLLECTOR

In Chapter 14, we studied influences of solar radiation upon heat transmission in structures. In most of these analyses, solar radiation represented an unwanted energy contribution. However, where the addition of thermal energy is required, as in space heating, we may wish to capture and use the energy of the sun's rays.

For many years, the world has relied upon fossil fuels as its principal energy source. Fossil fuels are expendable, whereas the sun is a continuing source of energy. Utilization of solar energy requires a solar collector. There are two general types—the *flat-plate collector* and the *concentrating collector*. A flat-plate collector basically consists of a flat surface with a high absorptivity for solar radiation. Typically, a metal plate with a blackened surface is used. Heat is transported from the absorber plate to a point of use by circulation of air or some other fluid across the solar-heated surface. Heat loss from the collector plate to the ambient environment is lessened by insulating the back side, and by covering the side facing the sun with one or more transparent covers. Glass is generally used for the transparent covers but certain plastic films may be satisfactory.

Figure 17.4 shows a schematic flat-plate collector where an air stream is heated by the back side of the collector plate. Fins attached to the plate increase the contact surface. The back side of the collector is heavily insulated with mineral wool or some other material. In front of the collector plate is a single sheet of glass, though more than one glass cover may be used. We assume that direct, diffuse sky, and reflected solar radiation may be incident upon the face of the collector.

In our analysis, we will make several simplifying assumptions. We will assume the glass cover to be thin, with a negligible solar absorptivity. We will also assume that the solar absorptivity of the collector plate is close to unity and independent

*J. R. Harnish, "Design and Use of Heat Pumps," *Air Conditioning, Heating and Ventilating*, Vol. 56, August 1959, pp. 69–80.

†See *ASHVE Trans.*, Vol. 54, 1948, pp. 221–238; Vol. 55, 1949, pp. 345–362; Vol. 60, 1954, pp. 157–176.

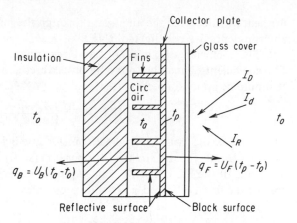

Figure 17.4 Schematic flat-plate solar collector.

of angle of incidence, and that all components of the collector have negligible heat capacity. We will assume that the fins and back side of the collector plate have highly reflective surfaces and that radiation heat transfer from these surfaces to the inside surface of the insulation is negligible.

For a unit surface element of the collector plate, we may write

$$F_s I_D \tau_D \alpha_p + I_d \tau_d \alpha_p + I_R \tau_R \alpha_p = \frac{U_o A_o}{A_p}(t_p - t_a) + U_F(t_p - t_o)$$

or

$$(F_s I_D \tau_D + I_d \tau_d + I_R \tau_R)\alpha_p = \left(U_o \frac{A_o}{A_p} + U_F\right)(t_p - t_a) + U_F(t_a - t_o) \qquad (17.7)$$

where F_s is the dimensionless sunlit fraction of the collector plate given by Eq. (14.34); I_D, I_d, and I_R are respectively the rates of direct, diffuse sky, and reflected solar radiation incident upon the front of the glass cover, Btu per (hr) (sq ft); τ_D, τ_d, and τ_R are respectively the dimensionless transmissivity of the glass cover for direct, diffuse sky, and reflected solar radiation; α_p is the dimensionless solar absorptivity of the front surface of the collector plate; U_o is the overall heat transfer coefficient for heat transfer from the collector plate to the air stream, Btu per (hr) (sq ft) (F); A_o is the total exposed surface area of the fins and back side of collector plate, sq ft; A_p is the surface area of the front side of the collector plate, sq ft; U_F is the overall heat transfer coefficient for heat transfer from the front side of the collector plate to the outside thermal environment, Btu per (hr) (sq ft) (F); and t_p, t_a, and t_o are respectively the temperature of the collector plate, air stream, and outside thermal environment, F.

For the element, we may also write

$$m_a c_{p,a} dt_a = U_o \frac{A_o}{A_p} dA_p (t_p - t_a) - U_B dA_p (t_a - t_o) \qquad (17.8)$$

where m_a is the mass flow rate of dry air, lb$_a$ per hr; $c_{p,a}$ is the specific heat at constant pressure for moist air, Btu per (lb$_a$) (F); and U_B is the overall heat transfer coefficient for heat transfer from the air stream through the insulated back to the outside thermal environment, Btu per (hr) (sq ft) (F).

The coefficients U_F and U_B may be estimated by methods discussed in Sec. 14.2. The coefficient U_o may be found by procedures similar to those discussed in Sec. 12.4. It should be approximately correct that

$$U_o = \frac{h_{c,o}(A'_p/A_F + \phi)}{1 + A'_p/A_F} \qquad (17.9)$$

where $h_{c,o}$ is the convection heat transfer coefficient for heat transfer between the surface A_o and the air stream, Btu per (hr) (sq ft) (F); A'_p is the surface area of the back side of the flat collector plate exposed to the air stream, sq ft; A_F is the surface area of the fins, sq ft; and ϕ is the dimensionless fin efficiency.

By Eqs. (17.7) and (17.8),

$$\int_{t_{a,1}}^{t_{a,2}} \frac{-K_2 \, dt_a}{K_1 - K_2(t_a - t_o)} = -K_3 \qquad (17.10)$$

where

$$K_1 = \frac{(F_s I_D \tau_D + I_d \tau_d + I_R \tau_R)\alpha_p}{1 + (U_F A_p/U_o A_o)} \qquad (17.11)$$

$$K_2 = \frac{U_F}{1 + (U_F A_p/U_o A_o)} + U_B \qquad (17.12)$$

$$K_3 = \frac{A_p}{m_a c_{p,a}} \left[\frac{U_F}{1 + (U_F A_p/U_o A_o)} + U_B \right] \qquad (17.13)$$

Integration of Eq. (17.10) gives

$$t_{a,2} = t_{a,1} + [(K_1/K_2) - (t_{a,1} - t_o)](1 - e^{-K_3}) \qquad (17.14)$$

and thus

$$q_c = m_a c_{p,a}[(K_1/K_2) - (t_{a,1} - t_o)](1 - e^{-K_3}) \qquad (17.15)$$

where q_c is the rate of heat collection, Btu per hr.

The collection efficiency η_c is given by

$$\eta_c = \frac{q_c/A_p}{I_D + I_d + I_R} \qquad (17.16)$$

Figure 17.4 and Eqs. (17.7)–(17.16) are for a collector in which air circulates behind the collector plate. Heat may also be absorbed by water or some other fluid circulating through tubes attached to the collector plate. Such an arrangement may be analyzed by assuming the tube and plate combination as tubes with bar fins.

For a theoretical ideal case, we may suppose that a flat-plate collector should be movable so that the collector plate may be kept perpendicular to the sun's rays. However, such an arrangement is not practical and the collector must usually be installed in a fixed position. An advantage of the flat-plate collector compared to the concentrating collector is that reasonably high rates of collection may be maintained over an extended period of time with the collector plate in a fixed position, whereas a concentrating collector must continuously track the sun.

Determination of the optimum orientation for a fixed, flat-plate collector is not straightforward. As shown by Eq. (17.15), the rate of heat collection is a function of several variables. Considering only direct radiation, the optimum position would vary with time, latitude, and construction of the collector; however, diffuse-sky and ground-reflected radiation must also be considered. In general, the collector plate should face due south (north in the southern hemisphere) and be tilted from the horizontal position by an angle dependent upon that portion of the year when maximum energy is needed. Hottel and Whillier* have stated that for year-round performance where the winter energy requirement is somewhat greater than that of summer, the optimum angle of tilt from the horizontal position would be from 10 to 20 degrees greater than the latitude angle.

Although the sun delivers an enormous quantity of energy to the earth's surface, utilization of solar energy for thermal purposes has distinct limitations. Primary disadvantages are the intermittency and low intensity of available energy. Solar energy is available during daylight hours only, and may be drastically reduced by clouds. Many localities may have periods of many days with almost no incident radiation. If solar energy is to be relied upon as a major heat source in a system, an extensive thermal storage arrangement is necessary.

During recent years, much research interest has been shown in the utilization of solar energy. The Massachusetts Institute of Technology has made many contributions to the field of solar energy utilization. In 1942, Hottel and Woertz† presented a pioneer paper on the performance of flat-plate collectors. The proceedings of a symposium held at Massachusetts Institute of Technology in 1950‡ provides an excellent source of information on space heating with solar energy. Papers read at the world-wide conference held in 1955 at Tucson, Arizona§ and at Phoenix, Arizona‖ covered many aspects of solar energy utilization.

17.4. THE CONCENTRATING SOLAR ENERGY COLLECTOR

The flat-plate solar collector discussed in Sec. 17.3 may provide an efficient means of capturing solar energy. However, the temperature level of the collected energy must be

---

*H. C. Hottel and A. Whillier, "Evaluation of Flat-Plate Solar-Collector Performance," in *Transactions of the Conference on the Use of Solar Energy The Scientific Basis* (Tucson: University of Arizona Press, 1958) Vol. 2, Part 1, Sec. A, p. 75.

†H. C. Hottel and B. B. Woertz, "The Performance of Flat-Plate Solar-Heat Collectors," *ASME Trans.*, Vol. 64, 1042, pp. 91–104.

‡*Space Heating with Solar Energy*, Proceedings of a Course Symposium held at the Massachusetts Institute of Technology, 1950 (Cambridge: Massachusetts Institute of Technology, 1954).

§*Transactions of the Conference on the Use of Solar Energy—The Scientific Basis*, 5 Vols. (Tucson: University of Arizona Press, 1958).

‖*Proceedings of the World Symposium on Applied Solar Energy* (Menlo Park, California: Stanford Research Institute, 1956).

relatively low. Maximum plate temperature occurs when the collection efficiency is zero, and

$$(U_F + U_B)(t_p - t_o) = (F_s I_D \tau_D + I_d \tau_d + I_R \tau_R)\alpha_p \qquad (17.17)$$

If the right-hand side of Eq. (17.17) was equal to 250 Btu per (hr) (sq ft) and $(U_F + U_B)$ was equal to 1.25 Btu (hr) (sq ft) (F), we would have $t_p - t_o = 200$ F. With heat collection, this temperature difference would be much less.

High temperatures may be attained with a solar collector by focusing or concentrating the sun's direct rays; a system of lenses or a parabolic reflector may be used. We will limit our discussion to parabolic-type reflectors. Figure 17.5 shows a schematic

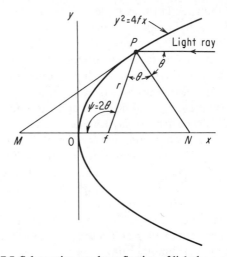

Figure 17.5 Schematic specular reflection of light by a parabola.

parabola whose vertex is at the origin and whose axis is the x axis. The equation of the parabola is $y^2 = 4fx$ where f is the *focal length*. The *focal radius* r is given by

$$r = \frac{2f}{1 + \cos \psi} \qquad (17.18)$$

The line MP is the tangent to the curve at P while NP is the normal to the curve at P. Properties of the parabola require that

$$r = f + x = Mf = fN$$

Now consider a ray of light parallel to the axis of the parabola which strikes the parabola at P. The angle of incidence of the ray is θ. If the ray is reflected specularly, the angle between the reflected ray and the incident ray must be 2θ. Since P is a point on the parabola, the path of the reflected ray would coincide with the focal radius r. Thus, all rays of light which are parallel to the parabola axis and which are specularly reflected from the parabola surface would converge at the focal point of the parabola.

In our earlier discussions of solar radiation in Chapters 13 and 14 and in this

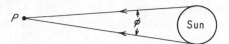

Figure 17.6 Apparent angular diameter of sun's disc as viewed from a point P on the earth's surface.

chapter, we tacitly assumed the sun as a point source of light. This is not true in fact. As shown by Fig. 17.6, rays from the sun are not parallel but converge upon a point at the earth's surface in the form of a cone. The mean value of the apparent angular diameter of the sun's disc ϕ is approximately 0.00931 radians or 32 minutes. Because of the varying earth-solar distance, ϕ varies from about 0.00948 radians in January to about 0.00917 radians in July. Although relatively unimportant in most other types of solar calculations, the sun's angular diameter is of great importance in concentrating-collector problems.

Figure 17.7 shows a parabola cut by a plane through a *paraboloid of revolution*. We assume the parabolic surface to be a perfect specular reflector. When the vertex of

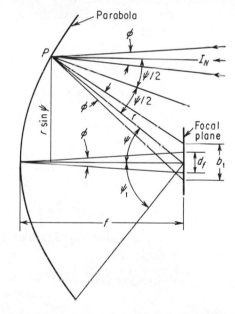

Figure 17.7 Schematic section of a paraboloidal collector.

the paraboloid is pointed directly toward the center of the sun, a circular image of the sun is formed at the focal plane. This image consists of a *focal spot* surrounded by a fringe area. Maximum intensity occurs at the focal spot while intensity decreases radially in the surrounding fringe. The circumference of the focal spot is determined by reflection from the vertex of the paraboloid. The focal spot diameter is

$$d_f = 2f \tan(\phi/2) \tag{17.19}$$

Reflections from a point such as P in Fig. 17.7 form a circular cone having an apex angle ϕ. The focal plane image of the reflected radiation is an ellipse with a major axis

$$b_1 = 2r \sec \psi \tan (\phi/2) \qquad (17.20)$$

and a minor axis

$$b_2 = 2r \tan (\phi/2) \qquad (17.21)$$

We will next consider a short segment ds tangent to the curve at P. We may show that

$$ds = r \, d\psi \sec (\psi/2)$$

We may form a circular ring of width ds and radius $r \sin \psi$. The area of the ring is

$$dA = 2\pi r \sin \psi \, ds = 2\pi r^2 \sin \psi \, (\sec \psi/2) \, d\psi$$

The rate of solar radiation incident upon the ring is

$$dq = I_N(\cos \psi/2) \, dA = 2\pi I_N r^2 \sin \psi \, d\psi \qquad (17.22)$$

If we assume a perfectly reflecting paraboloidal surface and a uniform intensity of solar radiation across the sun's disc, the rate of energy transfer to the focal spot dq_f by reflection from dA is

$$dq_f = \frac{A_f}{A_e} dq \qquad (17.23)$$

where A_f is the area of the focal spot and A_e is the area of the reflected image upon the focal plane. We have $A_f = \pi d_f^2/4$ and $A_e = (\pi/4)b_1 b_2$. By Eqs. (17.19)–(17.21), we obtain

$$\frac{A_f}{A_e} = \frac{f^2 \cos \psi}{r^2} \qquad (17.24)$$

By Eqs. (17.22)–(17.24),

$$q_f = 2\pi I_N f^2 \int_0^{\psi_1} \sin \psi \cos \psi \, d\psi = \pi I_N f^2 \sin^2 \psi_1 \qquad (17.25)$$

where q_f is the total rate of reflected radiation incident upon the focal spot and ψ_1 is the *rim angle* of the paraboloid.

The flux density at the focal spot is, by Eqs. (17.19) and (17.25),

$$q_f/A_f = I_N \sin^2 \psi_1 \cot^2 (\phi/2)$$

Since $\phi/2$ is extremely small, we may write $\phi/2 = \tan \phi/2$. Thus

$$\frac{q_f}{A_f} = \frac{4I_N \sin^2 \psi_1}{\phi^2} \qquad (17.26)$$

The *concentration ratio* C is defined by

$$C = \frac{q_f/A_f}{I_N} \qquad (17.27)$$

By Eqs. (17.26) and (17.27),

$$C = \frac{4 \sin^2 \psi_1}{\phi^2} \tag{17.28}$$

Equations (17.26) and (17.28) are based upon a uniform sun. Actually, the sun's radiation output is not uniform across the solar disc. Jose* and Simon† have made analyses of the paraboloidal collector, taking into account the sun's non-uniform radiation output. Their results show that the concentration ratio is somewhat higher than that given by Eq. (17.28).

Besides the paraboloid of revolution, the *cylindrical parabola* may be used as a concentrating device. Figure 17.8 shows a schematic cylindrical parabola collector. A *rectangular focal strip* is formed instead of the circular focal spot of the paraboloid of revolution. For a given plane through the collector, Fig. 17.7 is applicable. Analysis of the cylindrical parabola is very similar to that of the paraboloid of revolution. Assuming a uniform sun, and neglecting end effects, we may show that

$$\frac{q_f}{A_f} = \frac{2I_N \sin \psi_1}{\phi} \tag{17.29}$$

and

$$C = \frac{2 \sin \psi_1}{\phi} \tag{17.30}$$

Equation (17.30) shows that for the same rim angle, the concentration ratio of the cylindrical parabola is the square root of that for the paraboloid of revolution. Thus, much higher temperatures may be attained with the paraboloid of revolution. However, the cylindrical parabola has certain advantages. Where extremely high temperatures are not required, the focal strip offers a convenient region for utilization of solar energy.

A paraboloidal collector must be aimed directly and continuously at the sun. When the paraboloid is of large size, continuous tracking of the sun may be difficult, and working conditions at the focal plane may be awkward. Figure 17.9 shows a schematic arrangement which allows the paraboloid to be stationary. Auxiliary plane mirrors, called the *heliostat*, reflect the sun's rays upon the concentrator. The heliostat is the moving part of the system.

An important consideration with concentrating collectors is the need for high quality reflecting surfaces. Wight‡ has extensively discussed materials and surfaces for solar furnaces. Two basic reflecting surfaces are usually employed—silver and aluminum. Of the two, silver is the better reflector, but because of rapid corrosion, a

*P. D. Jose, "The Flux through the Focal Spot of a Solar Furnace," *Journal of Solar Energy Science and Engineering*, Vol. I, October 1957, pp. 19–22.

†A. W. Simon, "Calculation of the Concentration of the Solar Radiation through the Focal Spot of a Parabolic Mirror," *Journal of Solar Energy Science and Engineering*, Vol. II, April 1958, pp. 25–33.

‡R. H. Wight, "Materials and Surfaces for Solar Furnaces," *Journal of Solar Energy Science and Engineering*, Vol. I, April–July, 1957, pp. 84–93.

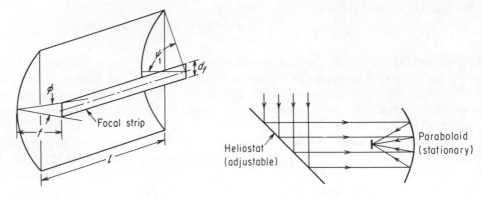

Figure 17.8 Schematic cylindrical parab-   **Figure 17.9** Schematic   paraboloidal
ola collector.   collector with a heliostat.

silver surface cannot be directly exposed to the atmosphere. In practice, a *second-surface mirror* is used whereby the silver film is protected by a layer of glass; high quality glass with a low absorption coefficient is required. Water-white polished plate glass is superior to most other types. Because of economy, aluminum surfaces have been used in some large collectors. The aluminum surface must also be protected from deterioration through use of a transparent overcoating.

Our preceding analysis of paraboloidal collectors assumed perfect surfaces with no energy losses. In an actual collector, several losses may occur which may cause the concentration ratio to be much less than the ideal value. Bliss,[*] Simon,[†] and Hukuo and Mii[‡] have analyzed some of these losses. Energy losses may occur from (1) glass absorption, (2) imperfect reflectivity of silvered surfaces, (3) defects in mirrors caused by imperfect machining and finishing, (4) shading effects caused by individual mirror frames and by other structural members, (5) improper alignment of mirrors, (6) dirty or dusty mirror surfaces, and (7) improper positioning of the specimen at focal plane. A transmission factor or efficiency term may be associated with each circumstance causing a loss. Although no single transmission factor may differ greatly from unity, the combined effects may be significant. Bliss[§] recommends that they be combined into a single *furnace factor F*, and that the actual concentration ratio C_a be calculated by

$$C_a = FC \tag{17.31}$$

[*] R. W. Bliss, "Notes on the Performance Design of Parabolic Solar Furnaces," *Journal of Solar Energy Science and Engineering*, Vol. I, January 1957, pp. 24–26.

[†] A. W. Simon, "The Loss of Energy by Absorption and Reflection in the Heliostat and Parabolic Condenser of a Solar Furnace," *Journal of Solar Energy Science and Engineering*, Vol. II, April 1958, pp. 30–33.

[‡] N. Hukuo and H. Mii, "Design Problems of a Solar Furnace," *Journal of Solar Energy Science and Engineering*, Vol. I, April–July, 1957, pp. 110–114.

[§] *Journal of Solar Energy Science and Engineering*, Vol. I, pp. 24–25.

where C is the concentration ratio for the ideal concentrator. Bliss further states that for a heliostat-paraboloid system, F may vary from about 0.35 to 0.70.

A problem of considerable importance to the designer of a concentrating collector is the equilibrium temperature to which a specimen may be heated at the focal spot of a collector. In general, the solution to this problem is given by an energy balance and appropriate heat-transfer relations. A particular problem of interest is the maximum temperature produced when one face of a specimen coincides with the focal spot and all other surfaces of the specimen are perfectly insulated from heat transfer. Since we expect a relatively high equilibrium temperature, we would have approximately

$$\epsilon \sigma T^4 = \alpha C_a I_N \tag{17.32}$$

where ϵ is the dimensionless emissivity of the surface; $\sigma = (0.1713)(10^{-8})$ Btu per (hr) (sq ft) (R^4); T is the absolute temperature of the specimen, R; α is the dimensionless solar absorptivity of the surface; C_a is the dimensionless concentration ratio given by Eq. (17.31); and I_N is the normally incident direct solar radiation, Btu per (hr) (sq ft).

If T in Eq. (17.32) is in the same order of magnitude as the sun's temperature, ϵ would be approximately equal to α, and

$$T = \left(\frac{C_a I_N}{\sigma}\right)^{1/4} \tag{17.33}$$

For a *surface outside the earth's atmosphere*, we may show that

$$I_N = \sigma T_s^4 (\phi^2/4) \tag{17.34}$$

where T_s is the average absolute black body surface temperature of the sun (about 10,000 F), and ϕ is the apparent angular diameter of the solar disc. By Eqs. (17.28), (17.31), (17.33), and (17.34),

$$T = T_s(F \sin^2 \psi_1)^{1/4} \tag{17.35}$$

Equation (17.35) shows that a perfect paraboloid of revolution ($F = 1$) with a rim angle $\psi_1 = \pi/2$ and located at the outer limit of the atmosphere could duplicate the sun's temperature at its focal spot.

Example 17.2. A paraboloidal collector is located at 42 deg north latitude at a site where the clearness number is unity. Rim angle of the paraboloid is 60 deg. Calculate the maximum temperature which may be produced at the focal spot of the collector at 12:00 noon solar time on April 1 if the furnace factor F is (a) 0.30, and (b) 0.60.

SOLUTION: By Fig. 13.15, $I_N = 300$ Btu/(hr) (sq ft). By Eq. (17.28),

$$C = \frac{(4)(0.866)^2}{(0.00931)^2} = 34{,}607$$

(a) By Eqs. (17.31) and (17.33) with $F = 0.30$,

$$T = \left[\frac{(0.30)(34{,}607)(300)}{(0.1713)(10^{-8})}\right]^{1/4} = 6530 \text{ R}$$

or
$$t = 6070 \text{ F}$$

(b) By Eqs. (17.31) and (17.33) with $F = 0.60$, we calculate $T = 7,750$ R and $t = 7,290$ F.

Example 17.2 shows that extremely high temperatures may be produced by a paraboloidal collector.

17.5. DRYING OF MATERIALS

An important topic related to thermal environmental engineering is the drying of materials. The term drying implies, in general, the removal of moisture. In our discussions, we will consider the drying of wet solids only.

Materials of many kinds must be dried. In manufacturing operations, a material must often be dried during one or more stages of fabrication. Products requiring drying during manufacture include soap, glue, gelatin, lumber, textiles, and paper. In recent years, drying of farm crops is of rapidly increasing importance.

A solid may contain moisture in one of two forms. A *hygroscopic* material may retain a certain quantity of water whose vapor pressure is less than that of free water at the same temperature. Such water is called *bound moisture. Unbound moisture* is water in excess of that which a material may retain as bound moisture. Unbound moisture exerts a vapor pressure equal to that of free water at the same temperature.

For any hygroscopic material, there is a definite relationship between the vapor pressure of bound moisture and the moisture content and temperature of the solid. Practically, the bound moisture content is closely a function of the relative saturation (relative humidity). Figure 17.10 shows moisture content of tanned sole leather plotted from Table 15.4. Such leather may retain bound moisture up to about 40 per cent of its dry weight; moisture in excess of this amount would be unbound. If the leather was exposed to air at the same temperature and at 50 per cent relative humidity, its *equilibrium moisture content* would be about 0.16 lb water per lb dry leather. Moisture retained in the leather in excess of this amount would be the *free moisture content*—the amount which could be removed if the leather was indefinitely exposed to the same air state.

If the relative saturation of a wet solid is greater than the relative humidity of moist air at the same temperature to which it is exposed, a vapor pressure difference would exist to cause vapor flow from the solid to the air. This potential difference may also be expressed in terms of a difference in humidity ratio.

A wet material may be dried in various ways. We will limit our discussion to *convection drying* where moisture is transferred to hot and/or dry air passing over the material. The physical processes occurring in the drying of a solid may be exceedingly complex. The evaporation of unbound moisture may be analyzed rather well, but the removal of bound moisture may present a formidable problem. Removal of bound moisture from a wet solid is similar in many respects to the desorption of a solid adsorbent.

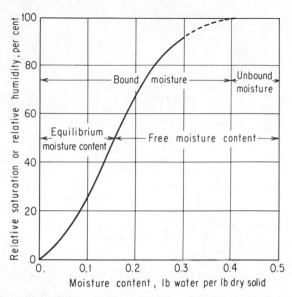

Figure 17.10 Moisture content of tanned sole leather.

Extensive research has been done to determine heat transfer and mass transfer coefficients in the drying of various materials. Treybal* has presented several correlations and has given an extensive bibliography. Generally, we must rely upon experiments to determine the rate of drying for a specific material. Fortunately, rather simple experiments may yield a considerable amount of useful information.

We will now consider the drying of a wet granular solid in a fixed tray. We will assume that (1) the sample contains both bound and unbound moisture and that the top surface of the material is completely wetted with water, (2) only the top surface of the sample is exposed to the drying medium (moist air), (3) no external, separate source of heat is supplied to the sample, and (4) the drying medium has constant conditions of temperature, humidity, and velocity. We could conduct a simple experiment where we observed the weight of the sample with respect to time. In general, we could expect the results shown by Fig. 17.11.

It is useful to convert the experimental drying curve shown by Fig. 17.11 to the relationship shown by Fig. 17.12. Here, the drying rate is shown as a function of the moisture content of the solid. We have the general relation

$$R = -\frac{1}{A_D}\frac{dx}{d\theta} \qquad (17.36)$$

where R is the drying rate, lb_w per (hr) (sq ft); A_D is the surface area of the sample, sq ft per lb of dry solid (lb_D); x is the moisture content of the wet solid, lb_w per lb_D; and θ is time, hr.

*R. E. Treybal, *Mass Transfer Operations* (New York: McGraw-Hill Book Company, 1955) pp. 524–583.

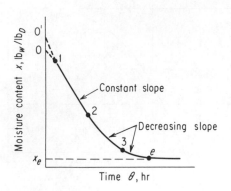

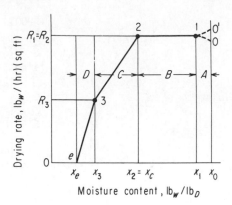

Figure 17.11 Schematic experimental drying curve for the drying of a wet solid in a fixed tray with constant air conditions.

Figure 17.12 Schematic drying rate for the drying of a wet solid in a fixed tray with constant air conditions.

Figure 17.12 shows that there may be four drying periods. Period A is a zone of unsteady operation during which the slab is adjusting itself to the steady operation of Period B.

Period B is the zone of *constant drying rate*. During this period, the wet surface of the solid behaves as a free water surface. The period continues as long as water is supplied to the surface as rapidly as evaporation takes place. The constant rate period ends when the *critical moisture content* x_c is reached. The critical moisture content is a function of the material and its thickness, the mechanisms by which water migrates to the surface, and the velocity and condition of the drying medium.

Period C is called the *first falling rate period*. This zone is characterized by a decreasing drying rate due to spot-wise recession of the wet surface into the first layers of the sample. As drying proceeds, the fraction of wet surface decreases to zero. At this point, Period C ends.

Period D is called the *second falling rate period*. It is characterized by subsurface evaporation throughout. Under prolonged operation, the period continues until the equilibrium moisture content x_e is reached.

Figure 17.12 may be used to establish relations between the drying rates for Periods B, C, and D. Thus

$$R_C = R_3 + \frac{(x - x_3)}{(x_2 - x_3)}(R_B - R_3) \tag{17.37}$$

$$R_D = \frac{(x - x_e)}{(x_3 - x_e)} R_3 \tag{17.38}$$

Example 17.3. It is desired to find the time required to dry a wet granular solid from an initial moisture content (dry-weight basis) of 35 per cent to a final moisture content of 1.0 per cent. The material has a density of 86.0 lb of dry material per cu ft. It is to be dried in layers 2.5 in. thick, in fixed trays. A drying rate diagram similar to Fig. 17.12 is available for the same average air conditions proposed for the dryer. This diagram shows that Period A is negligible, and that

$$R_1 = R_2 = 0.82 \text{ lb}_w \text{ per (hr) (sq ft)}, \qquad R_3 = 0.26 \text{ lb}_w \text{ per (hr) (sq ft)},$$

$$x_2 = 0.214 \text{ lb}_w \text{ per lb}_D, \qquad x_3 = 0.048 \text{ lb}_w \text{ per lb}_D, \qquad x_e = 0$$

SOLUTION: Equation (17.36) is the general equation. We have

$$\frac{1}{A_D} = \frac{(86)(2.5)}{12} = 17.92 \text{ lb}_D/\text{sq ft}$$

Thus, by Eq. (17.36),

$$\theta = -17.92 \int \frac{dx}{R}$$

For the constant-rate period (Period B),

$$\theta_B = \frac{(17.92)(0.136)}{0.82} = 2.97 \text{ hr}$$

By Eq. (17.37),

$$R_C = 0.26 + \frac{(x - 0.048)}{0.166}(0.56) = 0.098 + 3.37x$$

and

$$\theta_C = 17.92 \int_{x_3}^{x_2} \frac{dx}{0.098 + 3.37x} = 5.32 \ln \left[\frac{0.098 + 3.37x_2}{0.098 + 3.37x_3} \right] = 6.10 \text{ hr}$$

By Eq. (17.38),

$$R_D = \frac{(x)(0.26)}{0.048} = 5.42x$$

and

$$\theta_D = 3.31 \int_{x_4}^{x_3} \frac{dx}{x} = 3.31 \ln \frac{x_3}{x_4} = 5.19 \text{ hr}$$

Thus, $\theta = \theta_B + \theta_C + \theta_D = 14.26$ hr.

We will now examine in more detail the constant-rate drying period for a material dried in a fixed position with constant conditions of the drying medium. Figure 17.13 shows the schematic problem where t_s is the mean temperature of surfaces surround-

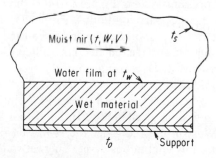

Figure 17.13 Schematic section of a fixed-position dryer.

ing the wetted surface and t_o is the environmental temperature under the supporting base. Per unit area of wetted surface, we may write

$$h_D(W_{s,w} - W)h_{fg,w} = h_c(t - t_w) + h_R(t_s - t_w) + U(t_o - t_w) \qquad (17.39)$$

where h_D is a mass transfer coefficient, lb_w per $(hr)(sq\ ft)(lb_w$ per $lb_a)$; $W_{s,w}$ is the humidity ratio of saturated air at the temperature of the water surface t_w, lb_w per lb_a; W is the humidity ratio of the air stream, lb_w per lb_a; $h_{fg,w}$ is the latent heat of vaporization of water at t_w, Btu per lb_w; h_c is a convection heat transfer coefficient, Btu per (hr) $(sq\ ft)$ (F); t is the dry-bulb temperature of the air stream, F; h_R is a radiation heat transfer coefficient, Btu per (hr) $(sq\ ft)$ (F); and U is an overall heat transfer coefficient, Btu per (hr) $(sq\ ft)$ (F).

If the last term of Eq. (17.39) is negligible and if $t_s = t$, we have

$$(W_{s,w} - W)h_{fg,w} = Le\ c_{p,a}\left(1 + \frac{h_R}{h_c}\right)(t - t_w) \qquad (17.40)$$

where $c_{p,a}$ is the moist air specific heat at constant pressure, Btu per $(lb_a)(F)$, and Le is the dimensionless Lewis number. If Le $(1 + h_R/h_c) = 1$, we may deduce from our discussion in Sec. 10.5 that t_w equals the thermodynamic wet-bulb temperature $t^\star$ of the air stream.

In most problems involving the constant rate period, we may assume $t_w = t^\star$. Thus, we may write for the drying rate that

$$R = h_D(W_s^\star - W) \qquad (17.41)$$

where $W_s^\star$ is the humidity ratio of saturated moist air at temperature $t^\star$.

In most convection dryers, it is approximately true that

$$h_D = aG^{0.8} \qquad (17.42)$$

where a is a constant, and G is the mass velocity of the air stream. Equations (17.41) and (17.42) allow knowledge of the mass transfer coefficient from a simple fixed tray experiment. Such data are useful for predicting the drying rate of the same material under other air conditions for both batch dryers and continuous dryers.

When a material is dried in a fixed position, a *batch dryer* is used. The dryer consists of a compartment or room where the trays of wet material are stacked and an air-processing system. Figure 17.14(a) shows a schematic drying compartment. Moist air in the necessary condition is supplied to the space and circulates over the individual trays. Part of the withdrawn air is exhausted and the rest is recirculated. The mixture of make-up air and recirculated air is filtered and heated before being readmitted to the compartment. Depending upon the wet material and the state of the make-up air, dehumidification equipment may also be required. Figure 17.14(b) shows the schematic psychrometric chart processes. Solution of the psychrometric problem follows the discussions given in Secs. 16.2 and 16.4–16.5.

We will now consider the *continuous dryer*. Figure 17.15(a) shows a schematic counterflow continuous dryer. The wet solid is carried through the dryer by a belt or other moving means. Moist air in contact with the solid moves in counterflow. An

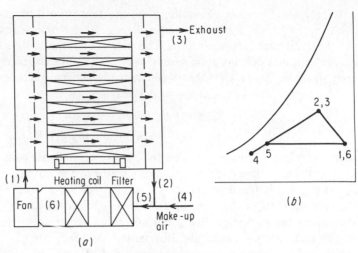

Figure 17.14 Schematic compartment dryer.

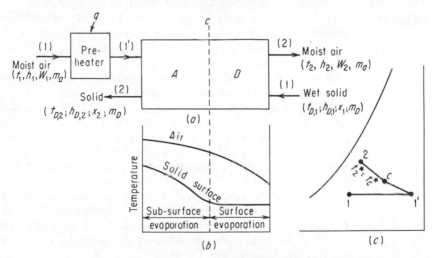

Figure 17.15 Schematic diagrams for counterflow continuous dryer.

arbitrary section c denotes the location where the critical moisture content is reached. A separate air preheater is shown. In some cases, heat may be added within the drying tunnel also.

We will first consider Section B where only surface evaporation occurs. If adiabatic conditions prevail, Eq. (17.40) must hold at any location. Thus, the wetted-surface temperature should be closely equal to the air thermodynamic wet-bulb temperature. Furthermore, Eq. (17.40) would require that both temperatures remain constant as long as only surface evaporation occurs. The drying rate would be given by Eq. (17.41). However, R would not be constant since W would change continually.

Figure 17.15(b) shows schematic changes of the moist-air dry-bulb temperature and the solid-surface temperature. The increase of the solid-surface temperature in Section A would depend upon several variables and might be relatively large or relatively small. Figure 17.15(c) shows schematic psychrometric chart processes.

An overall analysis of the system gives

$$m_w = m_a(W_2 - W_1) = m_D(x_1 - x_2) \tag{17.43}$$

$$q = m_a(h_2 - h_1) + m_D(h_{D,2} - h_{D,1}) \tag{17.44}$$

where m_w is rate of water transfer, lb_w per hr; m_a is the dry-air flow rate, lb_a per hr; W_2 and W_1 are respectively the humidity ratios of the leaving and entering moist air, lb_w per lb_a; m_D is the dry-solid flow rate, lb_D per hr; x_1 and x_2 are respectively the moisture contents of the entering and leaving solid, lb_w per lb_D; q is the rate of heat addition by the preheater, Btu per hr; h_2 and h_1 are respectively the enthalpy of the leaving and entering moist air, Btu per lb_a; and $h_{D,2}$ and $h_{D,1}$ are respectively the enthalpy of the leaving and entering wet solid, Btu per lb_D.

In order to determine the required size of a continuous dryer, knowledge of the drying rate throughout the dryer is necessary. In some cases, experiments which simulate the counterflow action must be performed. In other cases where internal resistance to moisture flow is not large, simple fixed-tray experiments may suffice.

Example 17.4. A counterflow adiabatic dryer is to be designed to dry 3,000 lb per hr of a wet granular solid from an initial moisture content of 1.0 lb_w per lb_D to a final moisture content of 0.1 lb_w per lb_D. The equilibrium moisture content of the solid is zero. Moist air enters the dryer at 200 F dry-bulb temperature and 95 F thermodynamic wet-bulb temperature at a rate of 20,000 cu ft per min. Barometric pressure is 14.696 psia. The wet solid enters and leaves at 95 F. The effective drying area is 0.3 sq ft per lb_D. The dryer holds 90 lb of dry solid per ft of length.

A sample of the wet material was dried in a fixed tray having the same effective drying area of 0.3 sq ft per lb_D. In this experiment, the moist air was at the same state as that of the entering air for the continuous dryer. The same air mass velocity was used as that proposed for the tunnel. Results of the fixed-tray experiment are shown by Fig. 17.16(a).

Estimate the required length of the dryer.

SOLUTION: We will first consider the fixed-tray experiment whose results are shown by Fig. 17.16(a). For the moist air, $t = 200$ F, $t^\star = 95$ F, and $P = 14.696$ psia. By the high temperature psychrometric chart (Fig. E-9), $W = 0.0117\ lb_w/lb_a$. By Table A.5, $W_s^\star = 0.0367\ lb_w/lb_a$. For $x \geq 0.5\ lb_w/lb_D$, by Eq. (17.41) we have

$$h_D = \frac{R}{W_s^\star - W} = \frac{1.0}{0.0367 - 0.0117} = 40.0\ lb_w/(hr)(sq\ ft)(lb_w/lb_a)$$

Thus, for $x \geq 0.5$,

$$R = 40.0(W_s^\star - W)$$

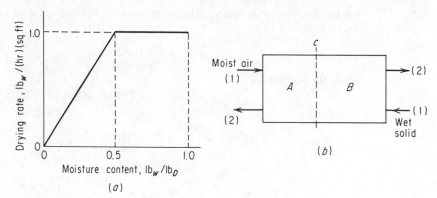

Figure 17.16 Schematic diagram for Example 17.4.

For $x \leq 0.5$, we have

$$R = 80.0(W_s^\star - W)x$$

Since these drying rates were obtained under conditions similar to those for the continuous dryer, we will assume that they are applicable to the continuous dryer. We will also make the approximation that the critical moisture content in the continuous dryer is 0.5 lb_w/lb_D (same as for the fixed-tray experiment).

We will now consider the continuous dryer. Figure 17.16(b) shows a schematic diagram. For the entering air we have $t_1 = 200$ F, $t_1^\star = 95$ F, $W_1 = 0.0117$ lb_w/lb_a, and $v_1 = 16.94$ cu ft/lb$_a$. Thus

$$m_a = \frac{(20,000)(60)}{16.94} = 70,800 \text{ lb}_a/\text{hr}$$

We also know that $m_D = 1500$ lb$_D$/hr. By Eq. (17.43),

$$W_2 = W_1 + \frac{m_D}{m_a}(x_1 - x_2) = 0.0117 + \frac{(1500)(0.9)}{70,800} = 0.0308 \text{ lb}_w/\text{lb}_a$$

Likewise, we have

$$W_c = W_1 + \frac{m_D}{m_a}(x_c - x_2) = 0.0117 + \frac{(1500)(0.4)}{70,800} = 0.0202 \text{ lb}_w/\text{lb}_a$$

At any location in the dryer, we may write

$$m_a \, dW = R A_D M_D \, dL$$

where A_D is drying area in sq ft/lb$_D$, M_D is weight of dry solid, lb$_D$/ft and L is length, ft. Thus

$$dL = \frac{m_a}{A_D M_D} \frac{dW}{R} = \frac{70,800}{(0.3)(90)} \frac{dW}{R} = 2620 \frac{dW}{R}$$

For Section B of the dryer, $R = 40.0 (W_s^\star - W)$ and $W_s^\star$ is a constant. Thus

$$L_B = 65.5 \int_{W_c}^{W_2} \frac{dW}{W_s^\star - W} = 65.5 \ln\left(\frac{W_s^\star - W_c}{W_s^\star - W_2}\right) = 67.4 \text{ ft}$$

For Section A of the dryer, $R = 80.0(W_s^\star - W)x$ and $W_s^\star$ is constant since the temperature of the wet solid remains equal to t^*. For any location in Section A, we have

$$x = x_2 + \frac{m_a}{m_D}(W - W_1)$$

Thus, for Section A,

$$R = 80.0(a_1 + a_2 W + a_3 W^2)$$

where

$$a_1 = \left(x_2 - \frac{m_a}{m_D} W_1\right) W_s^{\star}$$

$$a_2 = \frac{m_a}{m_D}(W_s^{\star} + W_1) - x_2$$

$$a_3 = -\frac{m_a}{m_D}$$

Thus

$$L_A = 32.75 \int_{W_1}^{W_c} \frac{dW}{a_1 + a_2 W + a_3 W^2} = \frac{32.75}{\sqrt{a_2^2 - 4a_1 a_3}}$$
$$\left\{\ln\left[\frac{2a_3 W_c + a_2 - \sqrt{a_2^2 - 4a_1 a_3}}{2a_3 W_c + a_2 + \sqrt{a_2^2 - 4a_1 a_3}}\right] - \ln\left[\frac{2a_3 W_1 + a_2 - \sqrt{a_2^2 - 4a_1 a_3}}{2a_3 W_1 + a_2 + \sqrt{a_2^2 - 4a_1 a_3}}\right]\right\}$$

Substitution of numerical values gives $L_A = 51.8$ ft. Thus

$$L = L_A + L_B = 119 \text{ ft}$$

Because of the uncertainties in the problem, the solution should be regarded as a rough estimate.

Efficient operation of a dryer requires that the heat input be kept to a minimum. In many cases, air discharged from a dryer has a much higher enthalpy than available make-up air. Through recirculation of part of the air or by use of heat exchangers, the required heat input may be substantially reduced.

PROBLEMS

17.1. Work Example 17.1 for your locality. Use current local gas and electricity costs. Assume the same heat loss rate from the residence per F temperature difference. You may assume a refrigerant evaporating temperature 15 F less than the temperature of available well water. Other data which are arbitrary may be taken from Example 17.1.

17.2. A flat-plate collector of the type shown in Fig. 17.4 is to be used for preheating outdoor air which serves as the heat source for a heat pump. The collector is to be designed to raise the temperature of 1000 lb_a per min of outdoor air by 20 F on a clear January 15 day at solar noon. The plane of the collector plate will be tilted by 50 deg from the horizontal position. Location is at 36 deg north latitude. Assume a clearness number of unity. You may neglect shading of the collector plate by set-back and by the glass frames. The glass has a negligible absorptivity for solar radiation. You may neglect ground-reflected radiation. The solar absorptivity of the collector plate is 0.95. It is estimated that U_F, U_o, and U_B are respectively 0.85, 6.0, and 0.10 Btu per (hr) (sq ft) (F) and that A_o/A_p is 10. Determine the required collector plate area.

17.3. Derive Eq. (17.30).

17.4. A paraboloidal collector is to be designed to have a focal spot diameter of 1.0 in. and an actual concentration ratio of 20,000. Estimated furnace factor is 0.60. You may assume a uniform sun. Determine (a) the rim angle of the paraboloid, (b) the focal length in ft, (c) the aperture or rim diameter of the paraboloid in ft, (d) the surface area of the paraboloid in sq ft, and (e) the probable maximum temperature which could be produced at the focal spot on a clear April 1 day at solar noon at Phoenix, Arizona.

17.5. Figure 17.17 shows the drying rate for a wet material when dried in a fixed tray under constant drying conditions. There are 10 lb of dry solid per

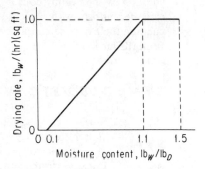

Figure 17.17 Schematic drying rate for Problem 17.5.

sq ft of drying surface. Calculate the moisture content after 10 hr in the dryer if the original moisture content was 1.50 lb_w per lb_D.

17.6. A parallel-flow adiabatic continuous dryer is to process 2,000 lb per hr of a wet material from an initial moisture content of 1.50 lb_w per lb_D to a final moisture content of 0.35 lb_w per lb_D. The equilibrium moisture content is zero. Moist air enters the dryer at a dry-bulb temperature of 200 F and a thermodynamic wet-bulb temperature of 95 F. The dry-air flow rate is 51,000 lb_a per hr. Barometric pressure is 14.696 psia. The material enters and leaves the dryer at 95 F. The effective drying area is 0.3 sq ft per lb_D and the dryer holds 20 lb of dry material per ft of length.

A sample of the wet material when dried in a fixed tray under constant air conditions identical to the entering air conditions for the tunnel showed a critical moisture content of 0.25 lb_w per lb_D. Above the critical moisture content, the drying rate was 2.0 lb_w per (hr) (sq ft). Below the critical moisture content, the drying rate fell to zero along a straight line.

Estimate the required length of the dryer.

17.7. A counterflow continuous dryer is to be designed to remove moisture from wet material which moves through the dryer at a rate of 8.0 lb of dry material per min. Moist air enters the dryer at a dry-bulb temperature of 140 F and a thermodynamic wet-bulb temperature of 86 F. The dry-air flow rate is 430 lb_a per min. Barometric pressure is 14.696 psia. Heating coils within the dryer maintain an essentially constant air dry-bulb temperature throughout. The entering wet material has a temperature of 86 F and a moisture content of 1.52 lb_w per lb_D and is to be dried to a final moisture content of 0.15 lb_w per lb_D. The dryer holds 10.0 lb of dry material per ft of length.

In the drying of this material all evaporation occurs below the critical moisture content. Laboratory experiments on the rate of drying were correlated by

$$R' = -\frac{dx}{d\theta} = 1.60(x - 0.05)(W_s^\star - W)$$

where R' is the rate of drying, lb_w per (lb_D) (min) and other quantities are the same as those defined in Sec. 17.5.

(a) Calculate the length of the dryer. (b) Calculate the total rate of heat addition by the heating coils in Btu per hr. (c) Should the output of the heating coils be uniform with respect to dryer length? Illustrate your method by calculating the Btu per hr capacity needed for each of the first three 10-ft sections from the air inlet end.

SYMBOLS USED IN CHAPTER 17

A	Surface area, sq ft; A_e for ellipse, A_F for fins, A_f for focal spot, A_o for total area of fins and exposed back side of collector plate, A_p for collector plate.
A_D	Effective drying surface, sq ft per lb of dry solid.
A_p'	$A_o - A_F$, sq ft.
b_1	Major axis of ellipse, ft.
b_2	Minor axis of ellipse, ft.
C	Ideal concentration ratio, dimensionless.
C_a	Actual concentration ratio, dimensionless.
C.O.P.	Coefficient of performance, dimensionless.
$c_{p,a}$	Specific heat at constant pressure for moist air, Btu per (lb of dry air) (F).
d_f	Diameter of focal spot, ft.
F	C_a/C, dimensionless.
f	Focal length of parabola, ft.
G	Mass velocity of air, lb per (sq ft)(hr).
h	Enthalpy of refrigerant, Btu per lb.
h_c	Convection heat transfer coefficient, Btu per (hr) (sq ft) (F).
h_D	Convection mass transfer coefficient, lb water per (hr) (sq ft) (lb water vapor per lb dry air).
h_D	Enthalpy of wet solid, Btu per lb dry solid.
$h_{fg,w}$	Latent heat of vaporization of water at t_w, Btu per lb.

h_R	Radiation heat transfer coefficient, Btu per (hr) (sq ft) (F).
I_D	Incidence of direct solar radiation upon a surface, Btu per (hr) (sq ft).
I_d	Incidence of diffuse sky radiation upon a surface, Btu per (hr) (sq ft).
I_N	Incidence of direct solar radiation upon a surface normal to sun's rays, Btu per (hr) (sq ft).
I_R	Incidence of solar radiation upon a surface which is reflected by surrounding surfaces, Btu per (hr) (sq ft).
K	Fraction of engine waste heat which is utilized, dimensionless.
K_1	Coefficient defined by Eq. (17.11), Btu per (hr) (sq ft).
K_2	Coefficient defined by Eq. (17.12), Btu per (hr) (sq ft) (F).
K_3	Coefficient defined by Eq. (17.13), dimensionless.
kwhr	Kilowatt hr.
L	Length, ft.
Le	$h_c/h_D c_{p,a}$, dimensionless.
M_D	Weight of dry solid, lb per ft of dryer length.
m_a	Dry-air flow rate, lb per hr.
m_D	Dry-solid flow rate, lb per hr.
m_w	Rate of addition of water, lb per hr.
Q	Heat transfer, Btu; Q_f refers to fuel.
q	Rate of heat transfer, Btu per hr; q_f refers to focal spot.
q_{av}	Average rate of heat loss, Btu per hr.
q_c	Rate of heat collection, Btu per hr.
q_d	Design rate of heat loss, Btu per hr.
q_f	Rate of energy input by fuel, Btu per hr.
q_w	Rate of waste heat transfer, Btu per hr.
R	Ratio of heat rejected in condenser to heat absorbed in evaporator, dimensionless.
R	Rate of drying, lb water per (hr) (sq ft).
r	Focal radius of parabola, ft.
T	Absolute temperature of surface, R; T_s refers to sun.
t	Dry-bulb temperature of moist air, F; t_a refers to air passing through collector, t_1 for inside air, t_o for outdoor air, $t_{o,av}$ for average outdoor air temperature.
t_p	Plate temperature, F.
t_s	Temperature of surrounding surfaces, F.
t_w	Temperature of water film, F.
$t\star$	Thermodynamic wet-bulb temperature of moist air, F.
U	Overall heat transfer coefficient, Btu per (hr) (sq ft) (F).
U_B	Overall coefficient for heat transfer from air stream through back side of collector, Btu per (hr) (sq ft of plate area) (F).
U_F	Overall coefficient for heat transfer from collector plate through glass covers, Btu per (hr) (sq ft) (F).
U_o	Overall coefficient for heat transfer from collector plate to air stream, Btu per (hr) (sq ft of back-side surface) (F).
W	Humidity ratio of moist air, lb water vapor per lb dry air.
W_s	Humidity ratio of saturated moist air, lb water vapor per lb dry air; $W_{s,w}$ evaluated at t_w, $W_s^\star$ evaluated at $t\star$.

w	Rate of work transfer, Btu per hr.
x	Moisture content, lb water per lb dry solid; x_c refers to critical value, x_e refers to equilibrium value.
α	Absorptivity for solar radiation, dimensionless; α_D for direct radiation, α_d for diffuse sky radiation, α_R for reflected radiation.
α_p	Absorptivity of collector plate for solar radiation, dimensionless.
ϵ	Surface emissivity, dimensionless.
η_c	Collection efficiency as defined by Eq. (17.16), dimensionless.
η_e	Efficiency of electric motor, dimensionless.
η_t	Thermal efficiency as defined by Eq. (17.4), dimensionless.
θ	Time, hr.
ρ	Density, lb per cu ft.
σ	$(0.1713)(10^{-8})$ Btu per (hr) (sq ft) (R^4).
τ	Transmissivity for solar radiation, dimensionless; τ_D refers to direct radiation, τ_d refers to diffuse sky radiation, τ_R refers to reflected radiation.
ϕ	Fin efficiency, dimensionless.
ϕ	Angular diameter of sun's disc.
ψ	Angle defined by Fig. 17.7.
ψ_1	Rim angle of paraboloid (see Fig. 17.7).

APPENDIX

TABLE A.1*

THERMODYNAMIC PROPERTIES OF WATER AT SATURATION

Fahr. Temp. $t(F)$	Absolute Pressure $p_{w,s} \times 10^5$		Specific Volume, cu ft per lb			Enthalpy, Btu per lb			Entropy, Btu per (lb)(R)		
	lb/sq in.	in. Hg	Sat. Solid v_i	Evap. $v_{ig} \times 10^{-5}$	Sat. Vapor $v_g \times 10^{-5}$	Sat. Solid h_i	Evap. h_{ig}	Sat. Vapor h_g	Sat. Solid s_i	Evap. s_{ig}	Sat. Vapor s_g
-160	0.004949	0.01008	0.01722	36070	36070	-222.05	1212.43	990.38	-0.4907	4.0456	3.5549
-155	0.009040	0.01840	0.01723	20080	20080	-220.44	1213.02	992.58	-0.4854	3.9812	3.4958
-150	0.01620	0.03298	0.01723	11390	11390	-218.82	1213.62	994.80	-0.4801	3.9188	3.4387
-145	0.02850	0.05803	0.01724	6577	6577	-217.17	1214.17	997.00	-0.4748	3.8583	3.3835
-140	0.04928	0.1003	0.01724	3864	3864	-215.49	1214.70	999.21	-0.4695	3.7996	3.3301
-135	0.08380	0.1706	0.01725	2308	2308	-213.80	1215.22	1001.42	-0.4643	3.7428	3.2785
-130	0.1403	0.2856	0.01725	1400	1400	-212.08	1215.71	1003.63	-0.4590	3.6874	3.2284
-125	0.2312	0.4708	0.01726	862.2	862.2	-210.34	1216.18	1005.84	-0.4538	3.6338	3.1800
-120	0.3757	0.7649	0.01726	538.6	538.6	-208.58	1216.63	1008.05	-0.4485	3.5815	3.1330
-115	0.6019	1.226	0.01727	341.1	341.1	-206.79	1217.05	1010.26	-0.4433	3.5308	3.0875
-110	0.9517	1.938	0.01728	218.9	218.9	-204.98	1217.45	1012.47	-0.4381	3.4815	3.0434
-105	1.486	3.025	0.01728	142.2	142.2	-203.14	1217.82	1014.68	-0.4329	3.4335	3.0006
-100	2.291	4.664	0.01729	93.52	93.52	-201.28	1218.17	1016.89	-0.4277	3.3868	2.9591
-95	3.491	7.108	0.01729	62.23	62.23	-199.40	1218.50	1019.10	-0.4225	3.3412	2.9187
-90	5.260	10.71	0.01730	41.86	41.86	-197.49	1218.80	1021.31	-0.4173	3.2969	2.8796
-85	7.841	15.96	0.01730	28.46	28.46	-195.56	1219.09	1023.52	-0.4121	3.2536	2.8415
-80	11.57	23.55	0.01731	19.55	19.55	-193.60	1219.33	1025.73	-0.4069	3.2114	2.8045
-75	16.89	34.39	0.01732	13.56	13.56	-191.62	1219.56	1027.94	-0.4017	3.1702	2.7685
-70	24.43	49.74	0.01732	9.501	9.501	-189.61	1219.76	1030.15	-0.3965	3.1301	2.7336
-65	35.01	71.28	0.01733	6.715	6.715	-187.58	1219.94	1032.36	-0.3914	3.0190	2.6996
-60	49.72	101.2	0.01734	4.788	4.788	-185.52	1220.10	1034.58	-0.3862	3.0526	2.6664
-55	70.01	142.6	0.01734	3.443	3.443	-183.44	1220.23	1036.79	-0.3810	3.0152	2.6342
-50	97.76	199.0	0.01735	2.496	2.496	-181.34	1220.34	1039.00	-0.3758	2.9786	2.6028
-48	111.5	227.0	0.01736	2.200	2.200	-180.49	1220.37	1039.88	-0.3738	2.9643	2.5905
-46	127.0	258.5	0.01736	1.941	1.941	-179.64	1220.40	1040.76	-0.3717	2.9501	2.5784
-44	144.4	294.0	0.01736	1.715	1.715	-178.78	1220.43	1041.65	-0.3696	2.9359	2.5663
-42	164.1	334.0	0.01736	1.516	1.516	-177.92	1220.45	1042.53	-0.3676	2.9219	2.5543
-40	186.1	379.0	0.01737	1.343	1.343	-177.06	1220.48	1043.42	-0.3655	2.9080	2.5425
-38	211.0	429.5	0.01737	1.191	1.191	-176.19	1220.49	1044.30	-0.3634	2.8942	2.5308
-36	238.8	486.2	0.01737	1.057	1.057	-175.32	1220.51	1045.19	-0.3614	2.8807	2.5193
-34	270.0	549.7	0.01737	0.9391	0.9391	-174.45	1220.52	1046.07	-0.3593	2.8671	2.5078
-32	304.9	620.8	0.01738	0.8355	0.8355	-173.57	1220.52	1046.95	-0.3573	2.8538	2.4965

Fahr. Temp. $t(F)$	Absolute Pressure $p_{w,s} \times 10^2$		Specific Volume, cu ft per lb			Enthalpy, Btu per lb			Entropy, Btu per (lb)(R)		
	lb/sq in.	in. Hg	Sat. Solid v_i	Evap. $v_{ig} \times 10^{-4}$	Sat. Vapor $v_g \times 10^{-4}$	Sat. Solid h_i	Evap. h_{ig}	Sat. Vapor h_g	Sat. Solid s_i	Evap. s_{ig}	Sat. Vapor s_g
−30	0.3440	0.7003	0.01738	7.441	7.441	−172.58	1220.52	1047.84	−0.3552	2.8405	2.4853
−28	0.3876	0.7891	0.01738	6.634	6.634	−171.80	1220.52	1048.72	−0.3532	2.8274	2.4742
−26	0.4363	0.8882	0.01738	5.921	5.921	−170.91	1220.51	1049.60	−0.3511	2.8143	2.4632
−24	0.4905	0.9987	0.01739	5.290	5.290	−170.01	1220.50	1050.49	−0.3490	2.8013	2.4523
−22	0.5509	1.122	0.01739	4.732	4.732	−169.12	1220.49	1051.37	−0.3470	2.7885	2.4415
−20	0.6181	1.259	0.01739	4.227	4.227	−168.21	1220.47	1052.26	−0.3449	2.7757	2.4308
−19	0.6545	1.333	0.01739	4.011	4.011	−167.76	1220.46	1052.70	−0.3439	2.7695	2.4256
−18	0.6928	1.410	0.01740	3.797	3.797	−167.31	1220.45	1053.14	−0.3429	2.7632	2.4203
−17	0.7332	1.493	0.01740	3.596	3.596	−166.85	1220.43	1053.58	−0.3418	2.7568	2.4150
−16	0.7757	1.579	0.01740	3.407	3.407	−166.40	1220.42	1054.02	−0.3408	2.7506	2.4098
−15	0.8204	1.670	0.01740	3.228	3.228	−165.94	1220.41	1054.47	−0.3398	2.7444	2.4046
−14	0.8676	1.766	0.01740	3.060	3.060	−165.48	1220.39	1054.91	−0.3388	2.7383	2.3995
−13	0.9172	1.867	0.01740	2.901	2.901	−165.03	1220.38	1055.35	−0.3377	2.7320	2.3943
−12	0.9694	1.974	0.01740	2.750	2.750	−164.57	1220.36	1055.79	−0.3367	2.7259	2.3892
−11	1.024	2.086	0.01740	2.609	2.609	−164.11	1220.34	1056.23	−0.3357	2.7198	2.3841
−10	1.082	2.203	0.01741	2.475	2.475	−163.65	1220.32	1056.67	−0.3347	2.7138	2.3791
−9	1.143	2.327	0.01741	2.349	2.349	−163.18	1220.30	1057.12	−0.3336	2.7076	2.3740
−8	1.207	2.457	0.01741	2.229	2.229	−162.72	1220.28	1057.56	−0.3326	2.7016	2.3690
−7	1.274	2.594	0.01741	2.116	2.116	−162.26	1220.26	1058.00	−0.3316	2.6956	2.3640
−6	1.344	2.757	0.01741	2.010	2.010	−161.79	1220.23	1058.44	−0.3306	2.6896	2.3590
−5	1.419	2.838	0.01741	1.909	1.909	−161.33	1220.21	1058.88	−0.3295	2.6836	2.3541
−4	1.496	3.047	0.01742	1.814	1.814	−160.86	1220.18	1059.32	−0.3285	2.6777	2.3492
−3	1.578	3.213	0.01742	1.723	1.723	−160.39	1220.15	1059.76	−0.3275	2.6718	2.3443
−2	1.664	3.388	0.01742	1.638	1.638	−159.92	1220.13	1060.21	−0.3264	2.6658	2.3394
−1	1.754	3.572	0.01742	1.557	1.557	−159.45	1220.10	1060.65	−0.3254	2.6600	2.3346
0	1.849	3.764	0.01742	1.481	1.481	−158.98	1220.07	1061.09	−0.3244	2.6541	2.3297
1	1.948	3.966	0.01742	1.408	1.408	−158.51	1220.04	1061.53	−0.3234	2.6483	2.3249
2	2.052	4.178	0.01742	1.340	1.340	−158.04	1220.01	1061.97	−0.3224	2.6425	2.3201
3	2.161	4.400	0.01743	1.275	1.275	−157.56	1219.97	1062.41	−0.3213	2.6367	2.3154
4	2.276	4.633	0.01743	1.214	1.214	−157.09	1219.94	1062.85	−0.3203	2.6309	2.3106
5	2.396	4.878	0.01743	1.155	1.155	−156.61	1219.90	1063.29	−0.3193	2.6252	2.3059
6	2.521	5.134	0.01743	1.100	1.100	−156.14	1219.88	1063.74	−0.3182	2.6194	2.3012

* Reprinted by permission from *ASHRAE Handbook of Fundamentals* (New York: American Society of Heating, Refrigerating and Air Conditioning Engineers, 1967) pp. 365–368.

TABLE A.1 (Continued)

THERMODYNAMIC PROPERTIES OF WATER AT SATURATION

Fahr. Temp. t(°F)	Absolute Pressure $p_{w,s}$		Specific Volume, cu ft per lb			Enthalpy, Btu per lb			Entropy, Btu per (lb)(R)		
	lb/sq in.	in. Hg	Sat. Solid v_i	Evap. v_{ig}	Sat. Vapor v_g	Sat. Solid h_i	Evap. h_{ig}	Sat. Vapor h_g	Sat. Solid s_i	Evap. s_{ig}	Sat. Vapor s_g
7	0.02653	0.05402	0.01743	10480	10480	−155.66	1219.84	1064.18	−0.3172	2.6138	2.2966
8	0.02791	0.05683	0.01743	9979	9979	−155.18	1219.80	1064.62	−0.3162	2.6081	2.2919
9	0.02936	0.05977	0.01744	9507	9507	−154.70	1219.76	1065.06	−0.3152	2.6025	2.2873
10	0.03087	0.06286	0.01744	9060	9060	−154.22	1219.72	1065.50	−0.3142	2.5969	2.2827
11	0.03246	0.06608	0.01744	8636	8636	−153.74	1219.68	1065.94	−0.3131	2.5912	2.2781
12	0.03412	0.06946	0.01744	8234	8234	−153.26	1219.64	1066.38	−0.3121	2.5857	2.2736
13	0.03585	0.07300	0.01744	7851	7851	−152.77	1219.59	1066.82	−0.3111	2.5801	2.2690
14	0.03767	0.07669	0.01744	7489	7489	−152.29	1219.55	1067.26	−0.3101	2.5746	2.2645
15	0.03957	0.08056	0.01744	7144	7144	−151.80	1219.50	1067.70	−0.3090	2.5690	2.2600
16	0.04156	0.08461	0.01745	6817	6817	−151.32	1219.46	1068.14	−0.3080	2.5635	2.2555
17	0.04363	0.08884	0.01745	6505	6505	−150.83	1219.41	1068.58	−0.3070	2.5581	2.2511
18	0.04581	0.09326	0.01745	6210	6210	−150.34	1219.36	1069.02	−0.3060	2.5526	2.2466
19	0.04808	0.09789	0.01745	5929	5929	−149.85	1219.31	1069.46	−0.3049	2.5471	2.2422
20	0.05045	0.1027	0.01745	5662	5662	−149.36	1219.26	1069.90	−0.3039	2.5417	2.2378
21	0.05293	0.1078	0.01746	5408	5408	−148.87	1219.21	1070.34	−0.3029	2.5364	2.2335
22	0.05552	0.1130	0.01746	5166	5166	−148.38	1219.16	1070.78	−0.3019	2.5310	2.2291
23	0.05823	0.1186	0.01746	4936	4936	−147.88	1219.10	1071.22	−0.3008	2.5256	2.2248
24	0.06105	0.1243	0.01746	4717	4717	−147.39	1219.05	1071.66	−0.2998	2.5203	2.2205
25	0.06400	0.1303	0.01746	4509	4509	−146.89	1218.98	1072.09	−0.2988	2.5150	2.2162
26	0.06708	0.1366	0.01746	4311	4311	−146.40	1218.93	1072.53	−0.2978	2.5097	2.2119
27	0.07030	0.1431	0.01746	4122	4122	−145.90	1218.87	1072.97	−0.2968	2.5045	2.2077
28	0.07365	0.1500	0.01746	3943	3943	−145.40	1218.81	1073.41	−0.2957	2.4991	2.2034
29	0.07715	0.1571	0.01747	3771	3771	−144.90	1218.75	1073.85	−0.2947	2.4939	2.1992
30	0.08080	0.1645	0.01747	3608	3608	−144.40	1218.69	1074.29	−0.2937	2.4887	2.1950
31	0.08461	0.1723	0.01747	3453	3453	−143.90	1218.63	1074.73	−0.2927	2.4835	2.1908
32	0.08858	0.1803	0.01747	3305	3305	−143.40	1218.56	1075.16	−0.2916	2.4783	2.1867
*32	0.088586	0.18036	0.01602	3304.6	3304.6	0.00	1075.16	1075.16	0.00000	2.1867	2.1867

Fahr. Temp. t(F)	Absolute Pressure $p_{w,s}$		Specific Volume, cu ft per lb			Enthalpy, Btu per lb			Entropy, Btu per (lb)(R)		
	lb/sq in.	in. Hg	Sat. Liquid v_f	Evap. v_{fg}	Sat. Vapor v_g	Sat. Liquid h_f	Evap. h_{fg}	Sat. Vapor h_g	Sat. Liquid s_f	Evap. s_{fg}	Sat. Vapor s_g
33	0.092227	0.18776	0.01502	3180.5	3180.5	1.01	1074.59	1075.60	0.00205	2.1811	2.1831
34	0.095999	0.19546	0.01502	3061.7	3061.7	2.01	1074.03	1076.04	0.00409	2.1755	2.1796
35	0.099908	0.20342	0.01602	2947.8	2947.8	3.02	1073.46	1076.48	0.00612	2.1700	2.1761
36	0.10396	0.21165	0.01602	2838.7	2838.7	4.03	1072.90	1076.92	0.00815	2.1644	2.1726
37	0.10815	0.22020	0.01602	2734.1	2734.1	5.03	1072.33	1077.36	0.01018	2.1589	2.1691
38	0.11249	0.22904	0.01602	2633.8	2633.8	6.03	1071.77	1077.80	0.01220	2.1535	2.1657
39	0.11699	0.23819	0.01602	2537.6	2537.6	7.04	1071.20	1078.24	0.01422	2.1480	2.1622
40	0.12164	0.24767	0.01602	2445.4	2445.4	8.04	1070.64	1078.68	0.01623	2.1426	2.1588
41	0.12646	0.25748	0.01602	2356.9	2356.9	9.05	1070.06	1079.11	0.01824	2.1372	2.1554
42	0.13145	0.26753	0.01602	2272.0	2272.0	10.05	1069.50	1079.55	0.02024	2.1318	2.1520
43	0.13660	0.27813	0.01602	2190.5	2190.5	11.05	1068.94	1079.99	0.02224	2.1265	2.1487
44	0.14194	0.28899	0.01602	2112.3	2112.3	12.06	1068.37	1080.43	0.02423	2.1211	2.1453
45	0.14746	0.30023	0.01602	2037.3	2037.3	13.06	1067.81	1080.87	0.02622	2.1158	2.1420
46	0.15317	0.31185	0.01602	1965.2	1965.2	14.06	1067.24	1081.30	0.02820	2.1105	2.1387
47	0.15907	0.32387	0.01602	1896.0	1896.0	15.06	1066.68	1081.74	0.03018	2.1052	2.1354
48	0.16517	0.33529	0.01602	1829.5	1829.5	16.07	1066.11	1082.18	0.03216	2.0999	2.1321
49	0.17148	0.34913	0.01602	1765.7	1765.7	17.07	1065.55	1082.62	0.03413	2.0947	2.1288
50	0.17799	0.35240	0.01602	1704.3	1704.3	18.07	1064.99	1083.06	0.03610	2.0895	2.1256
51	0.18473	0.37611	0.01602	1645.4	1645.4	19.07	1064.42	1083.49	0.03806	2.0842	2.1223
52	0.19169	0.39028	0.01602	1588.7	1588.7	20.07	1063.86	1083.93	0.04002	2.0791	2.1191
53	0.19888	0.40402	0.01603	1534.3	1534.3	21.07	1063.30	1084.37	0.04197	2.0739	2.1159
54	0.20630	0.42003	0.01603	1481.9	1481.9	22.08	1062.72	1084.80	0.04392	2.0688	2.1127
55	0.21397	0.43554	0.01603	1431.5	1431.5	23.08	1062.16	1085.24	0.04587	2.0637	2.1096
56	0.22188	0.45176	0.01603	1383.1	1383.1	24.08	1061.60	1085.68	0.04781	2.0586	2.1064
57	0.23006	0.46840	0.01603	1336.5	1336.5	25.08	1061.04	1086.12	0.04975	2.0535	2.1033
58	0.23849	0.48558	0.01603	1291.7	1291.7	26.08	1060.47	1086.55	0.05168	2.0485	2.1002
59	0.24720	0.50330	0.01503	1248.6	1248.6	27.08	1059.91	1086.99	0.05361	2.0434	2.0970
60	0.25618	0.52160	0.01603	1207.1	1207.1	28.08	1059.34	1087.42	0.05553	2.0385	2.0940
61	0.26545	0.54047	0.01604	1167.2	1167.2	29.08	1058.78	1087.86	0.05746	2.0334	2.0909
62	0.27502	0.55994	0.01604	1128.7	1128.7	30.08	1058.22	1088.30	0.05937	2.0284	2.0878
63	0.28488	0.58002	0.01604	1091.7	1091.7	31.08	1057.65	1088.73	0.06129	2.0235	2.0848
64	0.29505	0.60073	0.01604	1056.1	1055.1	32.08	1057.09	1089.17	0.06320	2.0186	2.0818
65	0.30554	0.62209	0.01604	1021.7	1021.7	33.08	1056.52	1089.60	0.06510	2.0136	2.0787
66	0.31635	0.64411	0.01604	988.65	988.63	34.07	1055.97	1090.04	0.06700	2.0087	2.0757
67	0.32750	0.66681	0.01605	956.78	956.76	35.07	1055.40	1090.47	0.06890	2.0039	2.0728
68	0.33900	0.69021	0.01605	926.08	926.06	36.07	1054.84	1090.91	0.07080	1.9990	2.0698
69	0.35084	0.71432	0.01605	896.49	896.47	37.07	1054.27	1091.34	0.07269	1.9941	2.0668
70	0.36304	0.73916	0.01605	867.97	867.95	38.07	1053.71	1091.78	0.07458	1.9893	2.0639
71	0.37561	0.76476	0.01605	840.47	840.45	39.07	1053.14	1092.21	0.07646	1.9845	2.0610

* Extrapolated to represent metastable equilibrium with undercooled liquid.

TABLE A.1 (Continued)

THERMODYNAMIC PROPERTIES OF WATER AT SATURATION

Fahr. Temp. t(F)	Absolute Pressure $p_{w,s}$ lb/sq in.	in. Hg	Specific Volume, cu ft per lb Sat. Liquid v_f	Evap. v_{fg}	Sat. Vapor v_g	Enthalpy, Btu per lb Sat. Liquid h_f	Evap. h_{fg}	Sat. Vapor h_g	Entropy, Btu per (lb)(R) Sat. Liquid s_f	Evap. s_{fg}	Sat. Vapor s_g
72	0.38856	0.79113	0.01606	813.95	813.97	40.07	1052.58	1092.65	0.07834	1.9797	2.0580
73	0.40190	0.81829	0.01606	788.38	788.40	41.07	1052.01	1093.08	0.08022	1.9749	2.0551
74	0.41564	0.84626	0.01606	763.73	763.75	42.06	1051.46	1093.52	0.08209	1.9701	2.0522
75	0.42979	0.87506	0.01606	739.95	739.97	43.06	1050.89	1093.95	0.08396	1.9654	2.0494
76	0.44435	0.90472	0.01606	717.01	717.03	44.06	1050.32	1094.38	0.08582	1.9607	2.0465
77	0.45935	0.93524	0.01607	694.88	694.90	45.06	1049.76	1094.82	0.08769	1.9560	2.0437
78	0.47478	0.96666	0.01607	673.52	673.54	46.06	1049.19	1095.25	0.08954	1.9513	2.0408
79	0.49066	0.99900	0.01607	652.91	652.93	47.06	1048.62	1095.68	0.09140	1.9466	2.0380
80	0.50701	1.0323	0.01607	633.01	633.03	48.05	1048.07	1096.12	0.09325	1.9419	2.0352
81	0.52382	1.0665	0.01608	613.80	613.82	49.05	1047.50	1096.55	0.09510	1.9373	2.0324
82	0.54112	1.1017	0.01608	595.25	595.27	50.05	1046.93	1096.98	0.09694	1.9328	2.0297
83	0.55892	1.1380	0.01608	577.34	577.36	51.05	1046.37	1097.42	0.09878	1.9281	2.0269
84	0.57722	1.1752	0.01608	560.04	560.06	52.05	1045.80	1097.85	0.10062	1.9236	2.0242
85	0.59604	1.2136	0.01609	543.33	543.35	53.05	1045.23	1098.28	0.10246	1.9189	2.0214
86	0.61540	1.2530	0.01609	527.19	527.21	54.04	1044.67	1098.71	0.10429	1.9144	2.0187
87	0.63530	1.2935	0.01609	511.60	511.62	55.04	1044.10	1099.14	0.10611	1.9099	2.0160
88	0.65575	1.3351	0.01610	496.52	496.54	56.04	1043.54	1099.58	0.10794	1.9054	2.0133
89	0.67678	1.3779	0.01610	481.96	481.98	57.04	1042.97	1100.01	0.10976	1.9008	2.0106
90	0.69838	1.4219	0.01610	467.88	467.90	58.04	1042.40	1100.44	0.11158	1.8963	2.0079
91	0.72059	1.4671	0.01610	454.26	454.28	59.03	1041.84	1100.87	0.11339	1.8919	2.0053
92	0.74340	1.5136	0.01611	441.10	441.12	60.03	1041.27	1101.30	0.11520	1.8874	2.0026
93	0.76684	1.5613	0.01611	428.38	428.40	61.03	1040.70	1101.73	0.11701	1.8830	2.0000
94	0.79091	1.6103	0.01611	416.09	416.09	62.03	1040.13	1102.16	0.11881	1.8786	1.9974
95	0.81564	1.6607	0.01612	404.17	404.19	63.03	1039.56	1102.59	0.12061	1.8741	1.9947
96	0.84103	1.7124	0.01612	392.65	392.67	64.02	1039.00	1103.02	0.12241	1.8698	1.9922
97	0.86711	1.7655	0.01612	381.51	381.53	65.02	1038.43	1103.45	0.12420	1.8654	1.9896
98	0.89388	1.8200	0.01612	370.73	370.75	66.02	1037.86	1103.88	0.12600	1.8610	1.9870
99	0.92137	1.8759	0.01613	360.30	360.32	67.02	1037.29	1104.31	0.12778	1.8566	1.9844
100	0.94959	1.9334	0.01613	350.20	350.22	68.02	1036.72	1104.74	0.12957	1.8523	1.9819
101	0.97854	1.9923	0.01614	340.42	340.44	69.01	1036.16	1105.17	0.13135	1.8480	1.9793
102	1.0083	2.0529	0.01614	330.96	330.98	70.01	1035.58	1105.59	0.13313	1.8437	1.9768
103	1.0388	2.1149	0.01614	321.80	321.82	71.01	1035.01	1106.02	0.13490	1.8394	1.9743
104	1.0700	2.1786	0.01614	312.93	312.95	72.01	1034.44	1106.45	0.13667	1.8351	1.9718
105	1.1021	2.2440	0.01615	304.34	304.36	73.01	1033.87	1106.88	0.13844	1.8309	1.9693
106	1.1351	2.3110	0.01615	296.02	296.04	74.01	1033.29	1107.30	0.14021	1.8266	1.9668
107	1.1688	2.3798	0.01616	287.96	287.98	75.00	1032.73	1107.73	0.14197	1.8224	1.9644

Fahr. Temp. t(F)	Absolute Pressure $p_{w,s}$		Specific Volume, cu ft per lb			Enthalpy, Btu per lb			Entropy, Btu per (lb)(R)		
	lb/sq in.	in. Hg	Sat. Liquid v_f	Evap. v_{fg}	Sat. Vapor v_g	Sat. Liquid h_f	Evap. h_{fg}	Sat. Vapor h_g	Sat. Liquid s_f	Evap. s_{fg}	Sat. Vapor s_g
108	1.2035	2.4503	0.01616	280.14	280.16	76.00	1032.16	1108.16	0.14373	1.8182	1.9619
109	1.2390	2.5226	0.01616	272.58	272.60	77.00	1031.58	1108.58	0.14549	1.8140	1.9595
110	1.2754	2.5963	0.01617	265.24	265.26	78.00	1031.01	1109.01	0.14724	1.8098	1.9570
111	1.3128	2.6728	0.01617	258.14	258.16	79.00	1030.44	1109.44	0.14899	1.8056	1.9546
112	1.3510	2.7517	0.01617	251.25	251.27	80.00	1029.86	1109.86	0.15074	1.8015	1.9522
113	1.3902	2.8326	0.01618	244.57	244.59	80.99	1029.30	1110.29	0.15248	1.7973	1.9498
114	1.4305	2.9125	0.01618	238.10	238.12	81.99	1028.72	1110.71	0.15423	1.7932	1.9474
115	1.4717	2.9963	0.01618	231.82	231.84	82.99	1028.15	1111.14	0.15596	1.7890	1.9450
116	1.5139	3.0823	0.01619	225.73	225.75	83.99	1027.57	1111.56	0.15770	1.7849	1.9426
117	1.5571	3.1703	0.01619	219.83	219.85	84.99	1026.99	1111.98	0.15943	1.7809	1.9403
118	1.6014	3.2606	0.01620	214.10	214.12	85.99	1026.42	1112.41	0.16116	1.7767	1.9379
119	1.6468	3.3530	0.01620	208.54	208.56	86.98	1025.85	1112.83	0.16289	1.7727	1.9356
120	1.6933	3.4477	0.01620	203.16	203.18	87.98	1025.28	1113.26	0.16461	1.7687	1.9333
121	1.7409	3.5446	0.01621	197.93	197.95	88.98	1024.70	1113.68	0.16634	1.7647	1.9310
122	1.7897	3.6439	0.01621	192.85	192.87	89.98	1024.12	1114.10	0.16805	1.7606	1.9286
123	1.8396	3.7455	0.01622	187.93	187.95	90.98	1023.54	1114.52	0.16977	1.7566	1.9264
124	1.8907	3.8496	0.01622	183.15	183.17	91.98	1022.96	1114.94	0.17148	1.7526	1.9241
125	1.9430	3.9561	0.01622	178.51	178.53	92.98	1022.39	1115.37	0.17319	1.7486	1.9218
126	1.9966	4.0651	0.01623	174.00	174.02	93.98	1021.81	1115.79	0.17490	1.7446	1.9195
127	2.0514	4.1768	0.01623	169.63	169.65	94.97	1021.24	1116.21	0.17660	1.7407	1.9173
128	2.1075	4.2910	0.01624	165.38	165.40	95.97	1020.66	1116.63	0.17830	1.7367	1.9150
129	2.1649	4.4078	0.01624	161.26	161.28	96.97	1020.08	1117.05	0.18000	1.7328	1.9128
130	2.2237	4.5274	0.01625	157.25	157.27	97.97	1019.50	1117.47	0.18170	1.7289	1.9106
131	2.2838	4.6498	0.01625	153.36	153.38	98.97	1018.92	1117.89	0.18339	1.7250	1.9084
132	2.3452	4.7750	0.01626	149.58	149.60	99.97	1018.34	1118.31	0.18508	1.7211	1.9062
133	2.4081	4.9030	0.01626	145.91	145.93	100.97	1017.76	1118.73	0.18676	1.7172	1.9040
134	2.4725	5.0340	0.01626	142.34	142.36	101.97	1017.18	1119.15	0.18845	1.7134	1.9018
135	2.5382	5.1679	0.01627	138.87	138.89	102.97	1016.59	1119.56	0.19013	1.7095	1.8996
136	2.6055	5.3049	0.01627	135.50	135.52	103.97	1016.01	1119.98	0.19181	1.7056	1.8974
137	2.6743	5.4450	0.01628	132.22	132.24	104.97	1015.43	1120.40	0.19348	1.7018	1.8953
138	2.7446	5.5881	0.01628	129.04	129.06	105.97	1014.85	1120.82	0.19516	1.6979	1.8931
139	2.8165	5.7345	0.01629	125.94	125.96	106.97	1014.26	1121.23	0.19683	1.6942	1.8910
140	2.8900	5.8842	0.01629	122.94	122.96	107.96	1013.69	1121.65	0.19850	1.6903	1.8888
141	2.9651	6.0371	0.01630	120.01	120.03	108.96	1013.11	1122.07	0.20016	1.6865	1.8867
142	3.0419	6.1934	0.01630	117.16	117.18	109.96	1012.52	1122.48	0.20182	1.6828	1.8846
143	3.1204	6.3532	0.01631	114.40	114.42	110.96	1011.94	1122.90	0.20348	1.6790	1.8825

TABLE A.1 *(Concluded)*

THERMODYNAMIC PROPERTIES OF WATER AT SATURATION

Fahr. Temp. t(F)	Absolute Pressure $p_{w,s}$		Specific Volume, cu ft per lb			Enthalpy, Btu per lb			Entropy, Btu per (lb)(R)		
	lb/sq in.	in. Hg	Sat. Liquid v_f	Evap. v_{fg}	Sat. Vapor v_g	Sat. Liquid h_f	Evap. h_{fg}	Sat. Vapor h_g	Sat. Liquid s_f	Evap. s_{fg}	Sat. Vapor s_g
144	3.2006	6.5164	0.01631	111.70	111.72	111.96	1011.35	1123.31	0.20514	1.6753	1.8804
145	3.2825	6.6832	0.01632	109.09	109.11	112.96	1010.77	1123.73	0.20679	1.6715	1.8783
146	3.3662	6.8536	0.01632	106.54	106.56	113.96	1010.18	1124.14	0.20845	1.6678	1.8763
147	3.4517	7.0277	0.01633	104.06	104.08	114.96	1009.59	1124.55	0.21010	1.6641	1.8742
148	3.5390	7.2056	0.01633	101.65	101.67	115.96	1009.01	1124.97	0.21174	1.6604	1.8721
149	3.6282	7.3872	0.01634	99.306	99.322	116.96	1008.42	1125.38	0.21339	1.6567	1.8701
150	3.7194	7.7627	0.01634	97.022	97.038	117.96	1007.83	1125.79	0.21503	1.6530	1.8680
151	3.8124	7.5722	0.01635	94.799	94.815	118.96	1007.24	1126.20	0.21667	1.6493	1.8660
152	3.9074	7.9556	0.01635	92.635	92.651	119.96	1006.66	1126.62	0.21830	1.6457	1.8640
153	4.0044	8.1532	0.01636	90.528	90.544	120.97	1006.06	1127.03	0.21994	1.6421	1.8620
154	4.1035	8.3548	0.01636	88.477	88.493	121.97	1005.47	1127.44	0.22157	1.6384	1.8600
155	4.2046	8.5607	0.01637	86.480	86.496	122.97	1004.88	1127.85	0.22320	1.6348	1.8580
156	4.3078	8.7708	0.01637	84.536	84.552	123.97	1004.29	1128.26	0.22482	1.6312	1.8560
157	4.4132	8.9853	0.01638	82.642	82.658	124.97	1003.70	1128.67	0.22645	1.6276	1.8540
158	4.5207	9.2042	0.01638	80.798	80.814	125.97	1003.11	1129.08	0.22807	1.6239	1.8520
159	4.6304	9.4276	0.01639	79.001	79.017	126.97	1002.51	1129.48	0.22969	1.6204	1.8501
160	4.7424	9.6556	0.01639	77.251	77.267	127.97	1001.92	1129.89	0.23130	1.6168	1.8481
161	4.8566	9.8882	0.01640	75.546	75.562	128.97	1001.33	1130.30	0.23292	1.6133	1.8462
162	4.9732	10.126	0.01640	73.885	73.901	129.97	1000.74	1130.71	0.23453	1.6097	1.8442
163	5.0921	10.368	0.01641	72.267	72.283	130.98	1000.13	1131.11	0.23614	1.6062	1.8423
164	5.2134	10.615	0.01642	70.690	70.706	131.98	999.54	1131.52	0.23774	1.6027	1.8404
165	5.3372	10.867	0.01642	69.153	69.169	132.98	998.94	1131.92	0.23935	1.5990	1.8384
166	5.4634	11.124	0.01643	67.654	67.670	133.98	998.35	1132.33	0.24095	1.5956	1.8365
167	5.5921	11.386	0.01643	66.194	66.210	134.98	997.75	1132.73	0.24255	1.5920	1.8346
168	5.7233	11.653	0.01644	64.770	64.786	135.98	997.16	1133.14	0.24414	1.5887	1.8328
169	5.8572	11.925	0.01644	63.382	63.398	136.99	996.55	1133.54	0.24574	1.5852	1.8309
170	5.9936	12.203	0.01645	62.029	62.045	137.99	995.95	1133.94	0.24733	1.5817	1.8290
171	6.1328	12.487	0.01645	60.710	60.726	138.99	995.36	1134.35	0.24892	1.5782	1.8271
172	6.2746	12.775	0.01646	59.423	59.439	139.99	994.76	1134.75	0.25051	1.5748	1.8253
173	6.4192	13.070	0.01647	58.168	58.184	141.00	994.15	1135.15	0.25209	1.5713	1.8234
174	6.5666	13.370	0.01647	56.944	56.960	142.00	993.55	1135.55	0.25367	1.5679	1.8216
175	6.7168	13.676	0.01648	55.750	55.766	143.00	992.95	1135.95	0.25525	1.5644	1.8197
176	6.8699	13.987	0.01648	54.586	54.602	144.00	992.35	1136.35	0.25683	1.5611	1.8179
177	7.0259	14.305	0.01649	53.450	53.466	145.00	991.75	1136.75	0.25841	1.5577	1.8161

Fahr. Temp. t(F)	Absolute Pressure $p_{w,s}$		Specific Volume, cu ft per lb			Enthalpy, Btu per lb			Entropy, Btu per (lb)(R)		
	lb/sq in.	in. Hg	Sat. Liquid v_f	Evap. v_{fg}	Sat. Vapor v_g	Sat. Liquid h_f	Evap. h_{fg}	Sat. Vapor h_g	Sat. Liquid s_f	Evap. s_{fg}	Sat. Vapor s_g
178	7.1849	14.629	0.01650	52.341	52.357	146.01	991.14	1137.15	0.25998	1.5543	1.8143
179	7.3469	14.959	0.01650	51.260	51.276	147.01	990.54	1137.55	0.26155	1.5508	1.8124
180	7.5119	15.295	0.01651	50.203	50.220	148.01	989.93	1137.94	0.26312	1.5475	1.8106
181	7.6801	15.637	0.01651	49.173	49.190	149.02	989.32	1138.34	0.26468	1.5442	1.8089
182	7.8514	15.986	0.01652	48.168	48.185	150.02	988.72	1138.74	0.26625	1.5408	1.8071
183	8.0258	16.342	0.01652	47.187	47.204	151.02	988.12	1139.14	0.26781	1.5375	1.8053
184	8.2035	16.703	0.01653	46.229	46.246	152.03	987.50	1139.53	0.26937	1.5341	1.8035
185	8.3845	17.071	0.01654	45.294	45.311	153.03	986.89	1139.92	0.27093	1.5308	1.8017
186	8.5688	17.445	0.01654	44.381	44.398	154.04	986.28	1140.32	0.27248	1.5275	1.8000
187	8.7565	17.829	0.01655	43.489	43.506	155.04	985.67	1140.71	0.27404	1.5242	1.7982
188	8.9476	18.218	0.01656	42.619	42.636	156.04	985.07	1141.11	0.27559	1.5209	1.7965
189	9.1422	18.614	0.01656	41.769	41.786	157.05	984.45	1141.50	0.27713	1.5176	1.7947
190	9.3403	19.017	0.01657	40.939	40.956	158.05	983.84	1141.89	0.27868	1.5143	1.7930
191	9.5420	19.428	0.01658	40.128	40.145	159.06	983.22	1142.28	0.28022	1.5111	1.7913
192	9.7473	19.846	0.01658	39.337	39.354	160.06	982.61	1142.67	0.28176	1.5078	1.7896
193	9.9563	20.271	0.01659	38.563	38.580	161.06	982.00	1143.06	0.28330	1.5045	1.7878
194	10.169	20.704	0.01659	37.807	37.824	162.07	981.38	1143.45	0.28484	1.5013	1.7861
195	10.386	21.145	0.01660	37.069	37.086	163.08	980.76	1143.84	0.28638	1.4980	1.7844
196	10.606	21.594	0.01661	36.348	36.365	164.08	980.15	1144.23	0.28791	1.4949	1.7828
197	10.830	22.050	0.01661	35.643	35.660	165.08	979.54	1144.62	0.28944	1.4917	1.7811
198	11.058	22.515	0.01662	34.954	34.971	166.09	978.91	1145.00	0.29097	1.4884	1.7794
199	11.290	22.987	0.01663	34.281	34.298	167.10	978.29	1145.39	0.29250	1.4852	1.7777
200	11.526	23.468	0.01663	33.623	33.640	168.10	977.68	1145.78	0.29402	1.4820	1.7760
201	11.767	23.957	0.01664	32.980	32.997	169.11	977.05	1146.16	0.29554	1.4789	1.7744
202	12.011	24.455	0.01665	32.351	32.368	170.11	976.43	1146.54	0.29706	1.4756	1.7727
203	12.260	24.961	0.01665	31.737	31.754	171.12	975.81	1146.93	0.29858	1.4725	1.7711
204	12.513	25.476	0.01666	31.136	31.153	172.12	975.19	1147.31	0.30010	1.4693	1.7694
205	12.770	26.000	0.01667	30.549	30.566	173.13	974.56	1147.69	0.30161	1.4662	1.7678
206	13.031	26.532	0.01667	29.974	29.991	174.14	973.94	1148.08	0.30312	1.4631	1.7662
207	13.297	27.074	0.01668	29.413	29.430	175.14	973.32	1148.46	0.30463	1.4600	1.7646
208	13.568	27.625	0.01669	28.863	28.880	176.15	972.69	1148.84	0.30614	1.4568	1.7629
209	13.843	28.185	0.01669	28.326	28.343	177.16	972.06	1149.22	0.30765	1.4536	1.7613
210	14.123	28.754	0.01670	27.801	27.818	178.17	971.43	1149.60	0.30915	1.4506	1.7597
211	14.407	29.333	0.01671	27.287	27.304	179.17	970.81	1149.98	0.31065	1.4474	1.7581
212	14.696	29.921	0.01671	26.784	26.801	180.18	970.17	1150.35	0.31215	1.4444	1.7565

TABLE A.2*

THERMODYNAMIC PROPERTIES OF WATER AT SATURATION (HIGH TEMPERATURES)

Temp. F t	Abs. Press. lb/sq in. p	Specific Volume, cu ft per lb		Enthalpy, Btu per lb			Entropy, Btu per (lb)(R)		
		Sat. Liquid v_f	Sat. Vapor v_g	Sat. Liquid h_f	Evap. h_{fg}	Sat. Vapor h_g	Sat. Liquid s_f	Evap. s_{fg}	Sat. Vapor s_g
212	14.696	0.01672	26.80	180.07	970.3	1150.4	0.3120	1.4446	1.7566
220	17.186	0.01677	23.15	188.13	965.2	1153.4	0.3239	1.4201	1.7440
230	20.780	0.01684	19.382	198.23	958.8	1157.0	0.3387	1.3901	1.7288
240	24.969	0.01692	16.323	208.34	952.2	1160.5	0.3531	1.3609	1.7140
250	29.825	0.01700	13.821	218.48	945.5	1164.0	0.3675	1.3323	1.6998
260	35.429	0.01709	11.763	228.64	938.7	1167.3	0.3817	1.3043	1.6860
270	41.858	0.01717	10.061	238.84	931.8	1170.6	0.3958	1.2769	1.6727
280	49.203	0.01726	8.645	249.06	924.7	1173.8	0.4096	1.2501	1.6597
290	57.556	0.01735	7.461	259.31	917.5	1176.8	0.4234	1.2238	1.6472
300	67.013	0.01745	6.466	269.59	910.1	1179.7	0.4369	1.1980	1.6350
310	77.68	0.01755	5.626	279.92	902.6	1182.5	0.4504	1.1727	1.6231
320	89.66	0.01765	4.914	290.28	894.9	1185.2	0.4637	1.1478	1.6115
330	103.06	0.01776	4.307	300.68	887.0	1187.7	0.4769	1.1233	1.6002
340	118.01	0.01787	3.788	311.13	879.0	1190.1	0.4900	1.0992	1.5891
350	134.63	0.01799	3.342	321.63	870.7	1192.3	0.5029	1.0754	1.5783
360	153.04	0.01811	2.957	332.18	862.2	1194.4	0.5158	1.0519	1.5677
370	173.37	0.01823	2.625	342.79	853.5	1196.3	0.5286	1.0287	1.5573
380	195.77	0.01836	2.335	353.45	844.6	1198.1	0.5413	1.0059	1.5471
390	220.37	0.01850	2.0836	364.17	835.4	1199.6	0.5539	0.9832	1.5371
400	247.31	0.01864	1.8633	374.97	826.0	1201.0	0.5664	0.9608	1.5272
410	276.75	0.01878	1.6700	385.83	816.3	1202.1	0.5788	0.9386	1.5174
420	308.83	0.01894	1.5000	396.77	806.3	1203.1	0.5912	0.9166	1.5078
430	343.72	0.01910	1.3499	407.79	796.0	1203.8	0.6035	0.8947	1.4982
440	381.59	0.01926	1.2171	418.90	785.4	1204.3	0.6158	0.8730	1.4887
450	422.6	0.0194	1.0993	430.1	774.5	1204.6	0.6280	0.8513	1.4793
460	466.9	0.0196	0.9944	441.4	763.2	1204.6	0.6402	0.8298	1.4700
470	514.7	0.0198	0.9009	452.8	751.5	1204.3	0.6523	0.8083	1.4606
480	566.1	0.0200	0.8172	464.4	739.4	1203.7	0.6645	0.7868	1.4513
490	621.4	0.0202	0.7423	476.0	726.8	1202.8	0.6766	0.7653	1.4419
500	680.8	0.0204	0.6749	487.8	713.9	1201.7	0.6887	0.7438	1.4325

* Abstracted by permission from J. H. Keenan and F. G. Keyes, *Thermodynamic Properties of Steam* (New York: John Wiley & Sons, Inc., 1936), pp. 31–33.

TABLE A.3†

THERMODYNAMIC PROPERTIES OF AMMONIA AT SATURATION

Temp F t	Pressure		Liquid, density	Vapor, sp vol	Enthalpy, *datum* −40 F, Btu per lb		Entropy, *datum* −40 F, Btu per lb R	
	psia	psig	lb/cu ft $1/v_f$	cu ft/lb v_g	Liquid h_f	Vapor h_g	Liquid s_f	Vapor s_g
−100	1.24	*27.4	45.51	182.90	−63.3	572.5	−0.1626	1.6055
− 95	1.52	*26.8	45.32	150.30	−58.0	574.7	−0.1481	1.5874
− 90	1.86	*26.1	45.12	124.28	−52.8	576.9	−0.1338	1.5699
− 85	2.27	*25.3	44.92	103.63	−47.5	579.1	−0.1197	1.5531
− 80	2.74	*24.3	44.73	86.54	−42.2	581.2	−0.1057	1.5368
− 75	3.30	*23.2	44.52	72.80	−37.0	583.3	−0.0920	1.5211
− 70	3.94	*21.9	44.32	61.65	−31.7	585.5	−0.0784	1.5059
− 65	4.69	*20.4	44.11	52.34	−26.5	587.5	−0.0650	1.4911
− 60	5.55	*18.6	43.91	44.73	−21.2	589.6	−0.0517	1.4769
− 58	5.93	*17.8	43.83	42.05	−19.1	590.4	−0.0464	1.4713
− 56	6.33	*17.0	43.74	39.56	−17.0	591.2	−0.0412	1.4658
− 54	6.75	*16.2	43.66	37.24	−14.8	592.1	−0.0360	1.4604
− 52	7.20	*15.3	43.58	35.09	−12.7	592.9	0.0307	1.4551
− 50	7.67	*14.3	43.49	33.08	−10.5	593.2	−0.0254	1.4487
− 48	8.16	*13.3	43.41	31.20	− 8.4	594.2	−0.0203	1.4438
− 46	8.68	*12.2	43.33	29.45	− 6.3	595.0	−0.0153	1.4388
− 44	9.23	*11.1	43.24	27.82	− 4.3	595.9	−0.0102	1.4338
− 42	9.81	*10.0	43.16	26.29	− 2.1	596.8	−0.0051	1.4290
− 40	10.41	*8.7	43.08	24.86	0.0	597.6	0.0000	1.4242
− 38	11.04	*7.4	42.99	23.53	2.1	598.3	0.0051	1.4193
− 36	11.71	*6.1	42.90	22.27	4.3	599.1	0.0101	1.4144
− 34	12.41	*4.7	42.82	21.10	6.4	599.9	0.0151	1.4096
− 32	13.14	*3.2	42.73	20.00	8.5	600.6	0.0201	1.4048
− 30	13.90	*1.6	42.65	18.97	10.7	601.4	0.0250	1.4001
− 29	14.30	*0.8	42.61	18.48	11.7	601.7	0.0275	1.3978
− 28	14.71	0.0	42.57	18.00	12.8	602.1	0.0300	1.3955
− 27	15.12	0.4	42.54	17.54	13.9	602.5	0.0325	1.3932
− 26	15.55	0.8	42.48	17.09	14.9	602.8	0.0350	1.3909
− 25	15.98	1.3	42.44	16.66	16.0	603.2	0.0374	1.3886
− 24	16.42	1.7	42.40	16.24	17.1	603.6	0.0399	1.3863
− 23	16.88	2.2	42.35	15.83	18.1	603.9	0.0423	1.3840
− 22	17.34	2.6	42.31	15.43	19.2	604.3	0.0448	1.3818
− 21	17.81	3.1	42.26	15.05	20.3	604.6	0.0472	1.3796
− 20	18.30	3.6	42.22	14.68	21.4	605.0	0.0497	1.3774
− 19	18.79	4.1	42.18	14.32	22.4	605.3	0.0521	1.3752
− 18	19.30	4.6	42.13	13.97	23.5	605.7	0.0545	1.3729
− 17	19.81	5.1	42.09	13.62	24.6	606.1	0.0570	1.3708
− 16	20.34	5.6	42.04	13.29	25.6	606.4	0.0594	1.3686
− 15	20.88	6.2	42.00	12.97	26.7	606.7	0.0618	1.3664
− 14	21.43	6.7	41.96	12.66	27.8	607.1	0.0642	1.3643
− 13	21.99	7.8	41.91	12.36	28.9	607.5	0.0666	1.3621
− 12	22.56	7.9	41.87	12.06	30.0	607.8	0.0690	1.3600
− 11	23.15	8.5	41.82	11.78	31.0	608.1	0.0714	1.3579
− 10	23.74	9.0	41.78	11.50	32.1	608.5	0.0738	1.3558
− 9	24.35	9.7	41.74	11.23	33.2	608.8	0.0762	1.3537
− 8	24.97	10.3	41.69	10.97	34.3	609.2	0.0768	1.3516
− 7	25.61	10.9	41.65	10.71	35.4	609.5	0.0809	1.3495
− 6	26.26	11.6	41.60	10.47	36.4	609.8	0.0833	1.3474
− 5	26.92	12.2	41.56	10.23	37.5	610.1	0.0857	1.3454
− 4	27.59	12.9	41.52	9.991	38.6	610.5	0.0880	1.3433
− 3	28.28	13.6	41.47	9.763	39.7	610.8	0.0904	1.3413
− 2	28.98	14.3	41.43	9.541	40.7	611.1	0.0928	1.3393
− 1	29.69	15.0	41.38	9.326	41.8	611.4	0.0951	1.3372
0	30.42	15.7	41.34	9.116	42.9	611.8	0.0975	1.3352
1	31.16	16.5	41.29	8.912	44.0	612.1	0.0998	1.3332
2	31.92	17.2	41.25	8.714	45.1	612.4	0.1022	1.3312
3	32.69	18.0	41.20	8.521	46.2	612.7	0.1045	1.3292
4	33.47	18.8	41.16	8.333	47.2	613.0	0.1069	1.3273
5	34.27	19.6	41.11	8.150	48.3	613.3	0.1092	1.3253
6	35.09	20.4	41.07	7.971	49.4	613.6	0.1115	1.3234
7	35.92	21.2	41.01	7.798	50.5	613.9	0.1138	1.3214
8	36.77	22.1	40.98	7.629	51.6	614.3	0.1162	1.3195
9	37.63	22.9	40.93	7.464	52.7	614.6	0.1185	1.3176
10	38.51	23.8	40.89	7.304	53.8	614.9	0.1208	1.3157

* Inches of mercury below one standard atmosphere (29.92 in.).
† Abstracted from National Bureau of Standards Circulars 142 and 472.

APPENDIX

TABLE A.3 (*Continued*)

THERMODYNAMIC PROPERTIES OF AMMONIA AT SATURATION

Temp F t	Pressure		Liquid, density	Vapor, sp vol	Enthalpy, *datum* −40 F, Btu per lb		Entropy, *datum* −40 F, Btu per lb R	
	psia	psig	lb/cu ft $1/v_f$	cu ft/lb v_g	Liquid h_f	Vapor h_g	Liquid s_f	Vapor s_g
11	39.40	24.7	40.84	7.148	54.9	615.2	0.1231	1.3137
12	40.31	25.6	40.80	6.996	56.0	615.5	0.1254	1.3118
13	41.24	26.5	40.75	6.847	57.1	615.8	0.1277	1.3099
14	42.18	27.5	40.71	6.703	58.2	616.1	0.1300	1.3081
15	43.14	28.4	40.66	6.562	59.2	616.3	0.1323	1.3062
16	44.12	29.4	40.61	6.425	60.3	616.6	0.1346	1.3043
17	45.12	30.4	40.57	6.291	61.4	616.9	0.1369	1.3025
18	46.13	31.4	40.52	6.161	62.5	617.2	0.1392	1.3006
19	47.16	32.5	40.48	6.034	63.6	617.5	0.1415	1.2988
20	48.21	33.5	40.43	5.910	64.7	617.8	0.1437	1.2969
21	49.28	34.6	40.38	5.789	65.8	618.0	0.1460	1.2951
22	50.36	35.7	40.34	5.671	66.9	618.3	0.1483	1.2933
23	51.47	36.8	40.29	5.556	68.0	618.6	0.1505	1.2951
24	52.59	37.9	40.25	5.443	69.1	618.9	0.1528	1.2897
25	53.78	39.0	40.20	5.334	70.2	619.1	0.1551	1.2879
26	54.90	40.2	40.15	5.227	71.3	619.4	0.1573	1.2861
27	56.08	41.4	40.10	5.123	72.4	619.7	0.1596	1.2843
28	57.28	42.6	40.06	5.021	73.5	619.9	0.1618	1.2825
29	58.50	43.8	40.01	4.922	74.6	620.2	0.1641	1.2808
30	59.74	45.0	39.96	4.825	75.7	620.5	0.1663	1.2790
31	61.00	46.3	39.91	4.730	76.8	620.7	0.1686	1.2773
32	62.29	47.6	39.86	4.637	77.9	621.0	0.1708	1.2755
33	63.59	48.9	39.82	4.547	79.0	621.2	0.1730	1.2738
34	64.91	50.2	39.77	4.459	80.1	621.5	0.1753	1.2731
35	66.26	52.6	39.72	4.373	81.2	621.7	0.1775	1.2704
36	67.63	52.9	39.67	4.289	82.3	622.0	0.1797	1.2686
37	69.02	54.3	39.63	4.207	83.4	622.2	0.1819	1.2669
38	70.43	55.7	39.58	4.126	84.6	622.5	0.1841	1.2652
39	71.87	57.2	39.54	4.048	85.7	622.7	0.1863	1.2635
40	73.32	58.6	39.49	3.971	86.8	623.0	0.1885	1.2618
41	74.80	60.1	39.44	3.897	87.9	623.2	0.1908	1.2602
42	76.31	61.6	39.39	3.823	89.0	623.4	0.1930	1.2585
43	77.83	63.1	39.34	3.752	90.1	623.7	0.1952	1.2568
44	79.38	64.7	39.29	3.682	91.2	623.9	0.1974	1.2552
45	80.96	66.3	39.24	3.614	92.3	624.1	0.1996	1.2535
46	82.55	67.9	39.19	3.547	93.5	624.4	0.2018	1.2519
47	84.18	69.5	39.14	3.481	94.6	624.6	0.2040	1.2502
48	85.82	71.1	39.10	3.418	95.7	624.8	0.2062	1.2486
49	87.49	72.8	39.05	3.355	96.8	625.0	0.2083	1.2469
50	89.19	74.5	39.00	3.294	97.9	625.2	0.2105	1.2453
51	90.91	76.2	38.95	3.234	99.1	625.5	0.2127	1.2437
52	92.66	78.0	38.90	3.176	100.2	625.7	0.2149	1.2421
53	94.43	79.7	38.85	3.119	101.3	625.9	0.2171	1.2405
54	96.23	81.5	38.80	3.063	102.4	626.1	0.2192	1.2389
55	98.06	83.4	38.75	3.008	103.5	626.3	0.2214	1.2373
56	99.91	85.2	38.70	2.954	104.7	626.5	0.2236	1.2357
57	101.8	87.1	38.65	2.902	105.8	626.7	0.2257	1.2341
58	103.7	89.0	38.60	2.851	106.9	626.9	0.2279	1.2325
59	105.6	90.9	38.55	2.800	108.1	627.1	0.2301	1.2310
60	107.6	92.9	38.50	2.751	109.2	627.3	0.2322	1.2294
61	109.6	94.9	38.45	2.703	110.3	627.5	0.2344	1.2278
62	111.6	96.9	38.40	2.656	111.5	627.7	0.2365	1.2262
63	113.6	98.9	38.35	2.610	112.6	627.9	0.2387	1.2247
64	115.7	101.0	38.30	2.565	113.7	628.0	0.2408	1.2231
65	117.8	103.1	38.25	2.520	114.8	628.2	0.2430	1.2216
66	120.0	105.3	38.20	2.477	116.0	628.4	0.2451	1.2201
67	122.1	107.4	38.15	2.435	117.1	628.6	0.2473	1.2186
68	124.3	109.6	38.10	2.393	118.3	628.8	0.2494	1.2170
69	126.5	111.8	38.05	2.352	119.4	628.9	0.2515	1.2155
70	128.8	114.1	38.00	2.312	120.5	629.1	0.2537	1.2140

TABLE A.3 (*Concluded*)

THERMODYNAMIC PROPERTIES OF AMMONIA AT SATURATION

Temp F t	Pressure		Liquid, density	Vapor, sp vol	Enthalpy, *datum* − 40 F, Btu per lb		Entropy, *datum* − 40 F, Btu per lb R	
	psia	psig	lb/cu ft $1/v_f$	cu ft/lb v_g	Liquid h_f	Vapor h_g	Liquid s_f	Vapor s_g
71	131.1	116.4	37.95	2.273	121.7	629.3	0.2558	1.2125
72	133.4	118.7	37.90	2.235	122.8	629.4	0.2579	1.2110
73	135.7	121.0	37.84	2.197	124.0	629.6	0.2601	1.2095
74	138.1	123.4	37.79	2.161	125.1	629.8	0.2622	1.2080
75	140.5	125.8	37.74	2.125	126.2	629.9	0.2643	1.2065
76	143.0	128.3	37.69	2.089	127.4	630.1	0.2664	1.2050
77	145.4	130.7	37.64	2.055	128.5	630.2	0.2685	1.2035
78	147.9	133.2	37.58	2.021	129.7	630.4	0.2706	1.2020
79	150.5	135.8	37.53	1.988	130.8	630.5	0.2728	1.2006
80	153.0	138.3	37.48	1.955	132.0	630.7	0.2749	1.1991
81	155.6	140.9	37.43	1.923	133.1	630.8	0.2769	1.1976
82	158.3	143.6	37.37	1.892	134.3	631.0	0.2791	1.1962
83	161.0	146.3	37.32	1.861	135.4	631.1	0.2812	1.1947
84	163.6	149.0	37.26	1.831	136.6	631.3	0.2833	1.1933
85	166.4	151.7	37.21	1.801	137.8	631.4	0.2854	1.1918
86	169.2	154.5	37.16	1.772	138.9	631.5	0.2875	1.1904
87	172.0	157.3	37.11	1.744	140.1	631.7	0.2895	1.1889
88	174.8	160.1	37.05	1.716	141.2	631.8	0.2917	1.1875
89	177.7	163.0	37.00	1.688	142.4	631.9	0.2937	1.1860
90	180.6	165.9	36.95	1.661	143.5	632.0	0.2958	1.1846
91	183.6	168.9	36.89	1.635	144.7	632.1	0.2979	1.1832
92	186.6	171.9	36.84	1.609	145.8	632.3	0.3000	1.1818
93	189.6	174.9	36.78	1.584	147.0	632.3	0.3021	1.1804
94	192.7	178.0	36.73	1.559	148.2	632.5	0.3041	1.1789
95	195.8	181.1	36.67	1.534	149.4	632.6	0.3062	1.1775
96	198.9	184.2	36.62	1.510	150.5	632.6	0.3083	1.1761
97	202.1	187.4	36.56	1.487	151.7	632.8	0.3104	1.1747
98	205.3	190.6	36.51	1.464	152.9	632.9	0.3125	1.1733
99	208.6	193.9	36.45	1.441	154.0	632.9	0.3145	1.1719
100	211.9	197.2	36.40	1.419	155.2	633.0	0.3166	1.1705
101	215.2	200.5	36.34	1.397	156.4	633.1	0.3187	1.1691
102	218.6	203.9	36.29	1.375	157.6	633.2	0.3207	1.1677
103	222.0	207.3	36.23	1.354	158.7	633.3	0.3228	1.1663
104	224.4	210.7	36.18	1.334	159.9	633.4	0.3248	1.1649
105	228.9	214.2	36.12	1.313	161.1	633.4	0.3269	1.1635
106	232.5	217.8	36.06	1.293	162.3	633.5	0.3289	1.1621
107	236.0	221.3	36.01	1.274	163.5	633.6	0.3310	1.1607
108	239.7	225.0	35.95	1.254	164.6	633.6	0.3330	1.1593
109	243.3	228.6	35.90	1.235	165.8	633.7	0.3351	1.1580
110	247.0	232.3	35.84	1.217	167.0	633.7	0.3372	1.1566
111	250.8	236.1	35.78	1.198	168.2	633.8	0.3392	1.1552
112	254.5	239.8	35.72	1.180	169.4	633.8	0.3413	1.1538
113	258.4	243.7	35.67	1.163	170.6	633.9	0.3433	1.1524
114	262.2	247.5	35.61	1.145	171.8	633.9	0.3453	1.1510
115	266.2	251.5	35.55	1.128	173.0	633.9	0.3474	1.1497
116	270.1	255.4	35.49	1.112	174.2	634.0	0.3495	1.1483
117	274.1	259.4	35.43	1.095	175.4	634.0	0.3515	1.1469
118	278.2	263.5	35.38	1.079	176.6	634.0	0.3535	1.1455
119	282.3	267.6	35.32	1.063	177.8	634.0	0.3556	1.1441
120	286.4	271.7	35.26	1.047	179.0	634.0	0.3576	1.1427
121	290.6	275.9	35.20	1.032	180.2	634.0	0.3597	1.1414
122	294.8	280.1	35.14	1.017	181.4	634.0	0.3618	1.1400
123	299.1	284.4	35.08	1.002	182.6	634.0	0.3638	1.1386
124	303.4	288.7	35.02	0.987	183.9	634.0	0.3659	1.1372
125	307.8	293.1	34.96	0.973	185.1	634.0	0.3679	1.1358

APPENDIX

TABLE A.4†

THERMODYNAMIC PROPERTIES OF REFRIGERANT 12 AT SATURATION

Temp. F	Pressure		Volume cu ft/lb		Density lb/cu ft		Enthalpy Btu/lb			Entropy Btu/(lb)(R)	
	psia	psig	Liquid v_f	Vapor v_g	Liquid $1/v_f$	Vapor $1/v_g$	Liquid h_f	Latent h_{fg}	Vapor h_g	Liquid s_f	Vapor s_g
−100	1.4280	27.0138*	0.009985	22.164	100.15	0.045119	−12.466	78.714	66.248	−0.032005	0.18683
−95	1.7163	26.4268*	0.010029	18.674	99.715	0.053550	−11.438	78.239	66.801	−0.029169	0.18536
−90	2.0509	25.7456*	0.010073	15.821	99.274	0.063207	−10.409	77.764	67.355	−0.026367	0.18398
−85	2.4371	24.9593*	0.010118	13.474	98.830	0.074216	− 9.3782	77.289	67.911	−0.023599	0.18267
−80	2.8807	24.0560*	0.010164	11.533	98.382	0.086708	− 8.3451	76.812	68.467	−0.020862	0.18143
−75	3.3879	23.0234*	0.010211	9.9184	97.930	0.10082	− 7.3101	76.333	69.023	−0.018156	0.18027
−70	3.9651	21.8482*	0.010259	8.5687	97.475	0.11670	− 6.2730	75.853	69.580	−0.015481	0.17916
−65	4.6193	20.5164*	0.010308	7.4347	97.016	0.13451	− 5.2336	75.371	70.137	−0.012834	0.17812
−60	5.3575	19.0133*	0.010357	6.4774	96.553	0.15438	− 4.1919	74.885	70.693	−0.010214	0.17714
−58	5.6780	18.3607*	0.010377	6.1367	96.367	0.16295	− 3.7745	74.691	70.916	−0.009174	0.17676
−56	6.0137	17.6773*	0.010397	5.8176	96.180	0.17189	− 3.3567	74.495	71.138	−0.008139	0.17639
−54	6.3650	16.9619*	0.010417	5.5184	95.993	0.18121	− 2.9386	74.299	71.360	−0.007107	0.17603
−52	6.7326	16.2136*	0.010438	5.2377	95.804	0.19092	− 2.5200	74.103	71.583	−0.006080	0.17568
−50	7.1168	15.4313*	0.010459	4.9742	95.616	0.20104	− 2.1011	73.906	71.805	−0.005056	0.17533
−48	7.5183	14.6139*	0.010479	4.7267	95.426	0.21157	− 1.6817	73.709	72.027	−0.004037	0.17500
−46	7.9375	13.7603*	0.010500	4.4940	95.236	0.22252	− 1.2619	73.511	72.249	−0.003022	0.17467
−44	8.3751	12.8693*	0.010521	4.2751	95.045	0.23391	− 0.8417	73.312	72.470	−0.002011	0.17435
−42	8.8316	11.9399*	0.010543	4.0691	94.854	0.24576	− 0.4211	73.112	72.691	−0.001003	0.17403
−40	9.3076	10.9709*	0.010564	3.8750	94.661	0.25806	0	72.913	72.913	0	0.17373
−38	9.8035	9.9611*	0.010586	3.6922	94.469	0.27084	0.4215	72.712	73.134	0.001000	0.17343
−36	10.320	8.909*	0.010607	3.5198	94.275	0.28411	0.8434	72.511	73.354	0.001995	0.17313
−34	10.858	7.814*	0.010629	3.3571	94.081	0.29788	1.2659	72.309	73.575	0.002988	0.17285
−32	11.417	6.675*	0.010651	3.2035	93.886	0.31216	1.6887	72.106	73.795	0.003976	0.17257
−30	11.999	5.490*	0.010674	3.0585	93.690	0.32696	2.1120	71.903	74.015	0.004961	0.17229
−28	12.604	4.259*	0.010696	2.9214	93.493	0.34231	2.5358	71.698	74.234	0.005942	0.17203
−26	13.233	2.979*	0.010719	2.7917	93.296	0.35820	2.9601	71.494	74.454	0.006919	0.17177
−24	13.886	1.649*	0.010741	2.6691	93.098	0.37466	3.3848	71.288	74.673	0.007894	0.17151
−22	14.564	0.270*	0.010764	2.5529	92.899	0.39171	3.8100	71.081	74.891	0.008864	0.17126
−20	15.267	0.571	0.010788	2.4429	92.699	0.40934	4.2357	70.874	75.110	0.009831	0.17102
−19	15.628	0.932	0.010799	2.3901	92.599	0.41839	4.4487	70.770	75.219	0.010314	0.17090
−18	15.996	1.300	0.010811	2.3387	92.499	0.42758	4.6618	70.666	75.328	0.010795	0.17078
−17	16.371	1.675	0.010823	2.2886	92.399	0.43694	4.8751	70.561	75.436	0.011276	0.17066
−16	16.753	2.057	0.010834	2.2399	92.298	0.44645	5.0885	70.456	75.545	0.011755	0.17055
−15	17.141	2.445	0.010846	2.1924	92.197	0.45612	5.3020	70.352	75.654	0.012234	0.17043
−14	17.536	2.840	0.010858	2.1461	92.096	0.46595	5.5157	70.246	75.762	0.012712	0.17032
−13	17.939	3.243	0.010870	2.1011	91.995	0.47595	5.7295	70.141	75.871	0.013190	0.17021
−12	18.348	3.652	0.010882	2.0572	91.893	0.48611	5.9434	70.036	75.979	0.013666	0.17010
−11	18.765	4.069	0.010894	2.0144	91.791	0.49643	6.1574	69.930	76.087	0.014142	0.16999
−10	19.189	4.493	0.010906	1.9727	91.689	0.50693	6.3716	69.824	76.196	0.014617	0.16989
−9	19.621	4.925	0.010919	1.9320	91.587	0.51759	6.5859	69.718	76.304	0.015091	0.16978
−8	20.059	5.363	0.010931	1.8924	91.485	0.52843	6.8003	69.611	76.411	0.015564	0.16967
−7	20.506	5.810	0.010943	1.8538	91.382	0.53944	7.0149	69.505	76.520	0.016037	0.16957
−6	20.960	6.264	0.010955	1.8161	91.280	0.55063	7.2296	69.397	76.627	0.016508	0.16947
−5	21.422	6.726	0.010968	1.7794	91.177	0.56199	7.4444	69.291	76.735	0.016979	0.16937
−4	21.891	7.195	0.010980	1.7436	91.074	0.57354	7.6594	69.183	76.842	0.017449	0.16927
−3	22.369	7.673	0.010993	1.7086	90.970	0.58526	7.8745	69.075	76.950	0.017919	0.16917
−2	22.854	8.158	0.011005	1.6745	90.867	0.59718	8.0898	68.967	77.057	0.018388	0.16907
−1	23.348	8.652	0.011018	1.6413	90.763	0.60927	8.3052	68.859	77.164	0.018855	0.16897
0	23.849	9.153	0.011030	1.6089	90.659	0.62156	8.5207	68.750	77.271	0.019323	0.16888
1	24.359	9.663	0.011043	1.5772	90.554	0.63404	8.7364	68.642	77.378	0.019789	0.16878
2	24.878	10.182	0.011056	1.5463	90.450	0.64670	8.9522	68.533	77.485	0.020255	0.16869
3	25.404	10.708	0.011069	1.5161	90.345	0.65957	9.1682	68.424	77.592	0.020719	0.16860
4	25.939	11.243	0.011082	1.4867	90.240	0.67263	9.3843	68.314	77.698	0.021184	0.16851
5	26.483	11.787	0.011094	1.4580	90.135	0.68588	9.6005	68.204	77.805	0.021647	0.16842
6	27.036	12.340	0.011107	1.4299	90.030	0.69934	9.8169	68.094	77.911	0.022110	0.16833
7	27.597	12.901	0.011121	1.4025	89.924	0.71300	10.033	67.984	78.017	0.022572	0.16824
8	28.167	13.471	0.011134	1.3758	89.818	0.72687	10.250	67.873	78.123	0.023033	0.16815
9	28.747	14.051	0.011147	1.3496	89.712	0.74094	10.467	67.762	78.229	0.023494	0.16807

* Inches of mercury below one atmosphere.
† Copyright 1956 by E. I. du Pont de Nemours & Company, reprinted by permission.

TABLE A.4 (Continued)

THERMODYNAMIC PROPERTIES OF REFRIGERANT 12 AT SATURATION

Temp. F	Pressure		Volume cu ft/lb		Density lb/cu ft		Enthalpy Btu/lb			Entropy Btu/(lb)(R)	
	psia	psig	Liquid v_f	Vapor v_g	Liquid $1/v_f$	Vapor $1/v_g$	Liquid h_f	Latent h_{fg}	Vapor h_g	Liquid s_f	Vapor s_g
10	29.335	14.639	0.011160	1.3241	89.606	0.75523	10.684	67.651	78.335	0.023954	0.16798
11	29.932	15.236	0.011173	1.2992	89.499	0.76972	10.901	67.539	78.440	0.024413	0.16790
12	30.539	15.843	0.011187	1.2748	89.392	0.78443	11.118	67.428	78.546	0.024871	0.16782
13	31.155	16.459	0.011200	1.2510	89.285	0.79935	11.336	67.315	78.651	0.025329	0.16774
14	31.780	17.084	0.011214	1.2278	89.178	0.81449	11.554	67.203	78.757	0.025786	0.16765
15	32.415	17.719	0.011227	1.2050	89.070	0.82986	11.771	67.090	78.861	0.026243	0.16758
16	33.060	18.364	0.011241	1.1828	88.962	0.84544	11.989	66.977	78.966	0.026699	0.16750
17	33.714	19.018	0.011254	1.1611	88.854	0.86125	12.207	66.864	79.071	0.027154	0.16742
18	34.378	19.682	0.011268	1.1399	88.746	0.87729	12.426	66.750	79.176	0.027608	0.16734
19	35.052	20.356	0.011282	1.1191	88.637	0.89356	12.644	66.636	79.280	0.028062	0.16727
20	35.736	21.040	0.011296	1.0988	88.529	0.91006	12.863	66.522	79.385	0.028515	0.16719
21	36.430	21.734	0.011310	1.0790	88.419	0.92679	13.081	66.407	79.488	0.028968	0.16712
22	37.135	22.439	0.011324	1.0596	88.310	0.94377	13.300	66.293	79.593	0.029420	0.16704
23	37.849	23.153	0.011338	1.0406	88.201	0.96098	13.520	66.177	79.697	0.029871	0.16697
24	38.574	23.878	0.011352	1.0220	88.091	0.97843	13.739	66.061	79.800	0.030322	0.16690
25	39.310	24.614	0.011366	1.0039	87.981	0.99613	13.958	65.946	79.904	0.030772	0.16683
26	40.056	25.360	0.011380	0.98612	87.870	1.0141	14.178	65.829	80.007	0.031221	0.16676
27	40.813	26.117	0.011395	0.96874	87.760	1.0323	14.398	65.713	80.111	0.031670	0.16669
28	41.580	26.884	0.011409	0.95173	87.649	1.0507	14.618	65.596	80.214	0.032118	0.16662
29	42.359	27.663	0.011424	0.93509	87.537	1.0694	14.838	65.478	80.316	0.032566	0.16655
30	43.148	28.452	0.011438	0.91880	87.426	1.0884	15.058	65.361	80.419	0.033013	0.16648
31	43.948	29.252	0.011453	0.90286	87.314	1.1076	15.279	65.243	80.522	0.033460	0.16642
32	44.760	30.064	0.011468	0.88725	87.202	1.1271	15.500	65.124	80.624	0.033905	0.16635
33	45.583	30.887	0.011482	0.87197	87.090	1.1468	15.720	65.006	80.726	0.034351	0.16629
34	46.417	31.721	0.011497	0.85702	86.977	1.1668	15.942	64.886	80.828	0.034796	0.16622
35	47.263	32.567	0.011512	0.84237	86.865	1.1871	16.163	64.767	80.930	0.035240	0.16616
36	48.120	33.424	0.011527	0.82803	86.751	1.2077	16.384	64.647	81.031	0.035683	0.16610
37	48.989	34.293	0.011542	0.81399	86.638	1.2285	16.606	64.527	81.133	0.036126	0.16604
38	49.870	35.174	0.011557	0.80023	86.524	1.2496	16.828	64.406	81.234	0.036569	0.16598
39	50.763	36.067	0.011573	0.78676	86.410	1.2710	17.050	64.285	81.335	0.037011	0.16592
40	51.667	36.971	0.011588	0.77357	86.296	1.2927	17.273	64.163	81.436	0.037453	0.16586
41	52.584	37.888	0.011603	0.76064	86.181	1.3147	17.495	64.042	81.537	0.037893	0.16580
42	53.513	38.817	0.011619	0.74798	86.066	1.3369	17.718	63.919	81.637	0.038334	0.16574
43	54.454	39.758	0.011635	0.73557	85.951	1.3595	17.941	63.796	81.737	0.038774	0.16568
44	55.407	40.711	0.011650	0.72341	85.836	1.3823	18.164	63.673	81.837	0.039213	0.16562
45	56.373	41.677	0.011666	0.71149	85.720	1.4055	18.387	63.550	81.937	0.039652	0.16557
46	57.352	42.656	0.011682	0.69982	85.604	1.4289	18.611	63.426	82.037	0.040091	0.16551
47	58.343	43.647	0.011698	0.68837	85.487	1.4527	18.835	63.301	82.136	0.040529	0.16546
48	59.347	44.651	0.011714	0.67715	85.371	1.4768	19.059	63.177	82.236	0.040966	0.16540
49	60.364	45.668	0.011730	0.66616	85.254	1.5012	19.283	63.051	82.334	0.041403	0.16535
50	61.394	46.698	0.011746	0.65537	85.136	1.5258	19.507	62.926	82.433	0.041839	0.16530
51	62.437	47.741	0.011762	0.64480	85.018	1.5509	19.732	62.800	82.532	0.042276	0.16524
52	63.494	48.798	0.011779	0.63444	84.900	1.5762	19.957	62.673	82.630	0.042711	0.16519
53	64.563	49.867	0.011795	0.62428	84.782	1.6019	20.182	62.546	82.728	0.043146	0.16514
54	65.646	50.950	0.011811	0.61431	84.663	1.6278	20.408	62.418	82.826	0.043581	0.16509
55	66.743	52.047	0.011828	0.60453	84.544	1.6542	20.634	62.290	82.924	0.044015	0.16504
56	67.853	53.157	0.011845	0.59495	84.425	1.6808	20.859	62.162	83.021	0.044449	0.16499
57	68.977	54.281	0.011862	0.58554	84.305	1.7078	21.086	62.033	83.119	0.044883	0.16494
58	70.115	55.419	0.011879	0.57632	84.185	1.7352	21.312	61.903	83.215	0.045316	0.16489
59	71.267	56.571	0.011896	0.56727	84.065	1.7628	21.539	61.773	83.312	0.045748	0.16484
60	72.433	57.737	0.011913	0.55839	83.944	1.7909	21.766	61.643	83.409	0.046180	0.16479
61	73.613	58.917	0.011930	0.54967	83.823	1.8193	21.993	61.512	83.505	0.046612	0.16474
62	74.807	60.111	0.011947	0.54112	83.701	1.8480	22.221	61.380	83.601	0.047044	0.16470
63	76.016	61.320	0.011965	0.53273	83.580	1.8771	22.448	61.248	83.696	0.047475	0.16465
64	77.239	62.543	0.011982	0.52450	83.457	1.9066	22.676	61.116	83.792	0.047905	0.16460
65	78.477	63.781	0.012000	0.51642	83.335	1.9364	22.905	60.982	83.887	0.048336	0.16456
66	79.729	65.033	0.012017	0.50848	83.212	1.9666	23.133	60.849	83.982	0.048765	0.16451
67	80.996	66.300	0.012035	0.50070	83.089	1.9972	23.362	60.715	84.077	0.049195	0.16447
68	82.279	67.583	0.012053	0.49305	82.965	2.0282	23.591	60.580	84.171	0.049624	0.16442
69	83.576	68.880	0.012071	0.48555	82.841	2.0595	23.821	60.445	84.266	0.050053	0.16438

TABLE A.4 (Continued)

THERMODYNAMIC PROPERTIES OF REFRIGERANT 12 AT SATURATION

Temp. F	Pressure		Volume cu ft/lb		Density lb/cu ft		Enthalpy Btu/lb			Entropy Btu/(lb)(R)	
	psia	psig	Liquid v_f	Vapor v_g	Liquid $1/v_f$	Vapor $1/v_g$	Liquid h_f	Latent h_{fg}	Vapor h_g	Liquid s_f	Vapor s_g
70	84.888	70.192	0.012089	0.47818	82.717	2.0913	24.050	60.309	84.359	0.050482	0.16434
71	86.216	71.520	0.012108	0.47094	82.592	2.1234	24.281	60.172	84.453	0.050910	0.16429
72	87.559	72.863	0.012126	0.46383	82.467	2.1559	24.511	60.035	84.546	0.051338	0.16425
73	88.918	74.222	0.012145	0.45686	82.341	2.1889	24.741	59.898	84.639	0.051766	0.16421
74	90.292	75.596	0.012163	0.45000	82.215	2.2222	24.973	59.759	84.732	0.052193	0.16417
75	91.682	76.986	0.012182	0.44327	82.089	2.2560	25.204	59.621	84.825	0.052620	0.16412
76	93.087	78.391	0.012201	0.43666	81.962	2.2901	25.435	59.481	84.916	0.053047	0.16408
77	94.509	79.813	0.012220	0.43016	81.835	2.3247	25.667	59.341	85.008	0.053473	0.16404
78	95.946	81.250	0.012239	0.42378	81.707	2.3597	25.899	59.201	85.100	0.053900	0.16400
79	97.400	82.704	0.012258	0.41751	81.579	2.3951	26.132	59.059	85.191	0.054326	0.16396
80	98.870	84.174	0.012277	0.41135	81.450	2.4310	26.365	58.917	85.282	0.054751	0.16392
81	100.36	85.66	0.012297	0.40530	81.322	2.4673	26.598	58.775	85.373	0.055177	0.16388
82	101.86	87.16	0.012316	0.39935	81.192	2.5041	26.832	58.631	85.463	0.055602	0.16384
83	103.38	88.68	0.012336	0.39351	81.063	2.5413	27.065	58.488	85.553	0.056027	0.16380
84	104.92	90.22	0.012356	0.38776	80.932	2.5789	27.300	58.343	85.643	0.056452	0.16376
85	106.47	91.77	0.012376	0.38212	80.802	2.6170	27.534	58.198	85.732	0.056877	0.16372
86	108.04	93.34	0.012396	0.37657	80.671	2.6556	27.769	58.052	85.821	0.057301	0.16368
87	109.63	94.93	0.012416	0.37111	80.539	2.6946	28.005	57.905	85.910	0.057725	0.16364
88	111.23	96.53	0.012437	0.36575	80.407	2.7341	28.241	57.757	85.998	0.058149	0.16360
89	112.85	98.15	0.012457	0.36047	80.275	2.7741	28.477	57.609	86.086	0.058573	0.16357
90	114.49	99.79	0.012478	0.35529	80.142	2.8146	28.713	57.461	86.174	0.058997	0.16353
91	116.15	101.45	0.012499	0.35019	80.008	2.8556	28.950	57.311	86.261	0.059420	0.16349
92	117.82	103.12	0.012520	0.34518	79.874	2.8970	29.187	57.161	86.348	0.059844	0.16345
93	119.51	104.81	0.012541	0.34025	79.740	2.9390	29.425	57.009	86.434	0.060267	0.16341
94	121.22	106.52	0.012562	0.33540	79.605	2.9815	29.663	56.858	86.521	0.060690	0.16338
95	122.95	108.25	0.012583	0.33063	79.470	3.0245	29.901	56.705	86.606	0.061113	0.16334
96	124.70	110.00	0.012605	0.32594	79.334	3.0680	30.140	56.551	86.691	0.061536	0.16330
97	126.46	111.76	0.012627	0.32133	79.198	3.1120	30.380	56.397	86.777	0.061959	0.16326
98	128.24	113.54	0.012649	0.31679	79.062	3.1566	30.619	56.242	86.861	0.062381	0.16323
99	130.04	115.34	0.012671	0.31233	78.923	3.2017	30.859	56.086	86.945	0.062804	0.16319
100	131.86	117.16	0.012693	0.30794	78.785	3.2474	31.100	55.929	87.029	0.063227	0.16315
101	133.70	119.00	0.012715	0.30362	78.647	3.2936	31.341	55.772	87.113	0.063649	0.16312
102	135.56	120.86	0.012738	0.29937	78.508	3.3404	31.583	55.613	87.196	0.064072	0.16308
103	137.44	122.74	0.012760	0.29518	78.368	3.3877	31.824	55.454	87.278	0.064494	0.16304
104	139.33	124.63	0.012783	0.29106	78.228	3.4357	32.067	55.293	87.360	0.064916	0.16301
105	141.25	126.55	0.012806	0.28701	78.088	3.4842	32.310	55.132	87.442	0.065339	0.16297
106	143.18	128.48	0.012829	0.28303	77.946	3.5333	32.553	54.970	87.523	0.065761	0.16293
107	145.13	130.43	0.012853	0.27910	77.804	3.5829	32.797	54.807	87.604	0.066184	0.16290
108	147.11	132.41	0.012876	0.27524	77.662	3.6332	33.041	54.643	87.684	0.066606	0.16286
109	149.10	134.40	0.012900	0.27143	77.519	3.6841	33.286	54.478	87.764	0.067028	0.16282
110	151.11	136.41	0.012924	0.26769	77.376	3.7357	33.531	54.313	87.844	0.067451	0.16279
111	153.14	138.44	0.012948	0.26400	77.231	3.7878	33.777	54.146	87.923	0.067873	0.16275
112	155.19	140.49	0.012972	0.26037	77.087	3.8406	34.023	53.978	88.001	0.068296	0.16271
113	157.27	142.57	0.012997	0.25680	76.941	3.8941	34.270	53.809	88.079	0.068719	0.16268
114	159.36	144.66	0.013022	0.25328	76.795	3.9482	34.517	53.639	88.156	0.069141	0.16264
115	161.47	146.77	0.013047	0.24982	76.649	4.0029	34.765	53.468	88.233	0.069564	0.16260
116	163.61	148.91	0.013072	0.24641	76.501	4.0584	35.014	53.296	88.310	0.069987	0.16256
117	165.76	151.06	0.013097	0.24304	76.353	4.1145	35.263	53.123	88.386	0.070410	0.16253
118	167.94	153.24	0.013123	0.23974	76.205	4.1713	35.512	52.949	88.461	0.070833	0.16249
119	170.13	155.43	0.013148	0.23647	76.056	4.2288	35.762	52.774	88.536	0.071257	0.16245
120	172.35	157.65	0.013174	0.23326	75.906	4.2870	36.013	52.597	88.610	0.071680	0.16241
121	174.59	159.89	0.013200	0.23010	75.755	4.3459	36.264	52.420	88.684	0.072104	0.16237
122	176.85	162.15	0.013227	0.22698	75.604	4.4056	36.516	52.241	88.757	0.072528	0.16234
123	179.13	164.43	0.013254	0.22391	75.452	4.4660	36.768	52.062	88.830	0.072952	0.16230
124	181.43	166.73	0.013280	0.22089	75.299	4.5272	37.021	51.881	88.902	0.073376	0.16226
125	183.76	169.06	0.013308	0.21791	75.145	4.5891	37.275	51.698	88.973	0.073800	0.16222
126	186.10	171.40	0.013335	0.21497	74.991	4.6518	37.529	51.515	89.044	0.074225	0.16218
127	188.47	173.77	0.013363	0.21207	74.836	4.7153	37.785	51.330	89.115	0.074650	0.16214
128	190.86	176.16	0.013390	0.20922	74.680	4.7796	38.040	51.144	89.184	0.075075	0.16210
129	193.27	178.57	0.013419	0.20641	74.524	4.8448	38.296	50.957	89.253	0.075501	0.16206

TABLE A.4 (*Concluded*)

THERMODYNAMIC PROPERTIES OF REFRIGERANT 12 AT SATURATION

Temp. F	Pressure		Volume cu ft/lb		Density lb/cu ft		Enthalpy Btu/lb			Entropy Btu/(lb)(R)	
	psia	psig	Liquid v_f	Vapor v_g	Liquid $1/v_f$	Vapor $1/v_g$	Liquid h_f	Latent h_{fg}	Vapor h_g	Liquid s_f	Vapor s_g
130	195.71	181.01	0.013447	0.20364	74.367	4.9107	38.553	50.768	89.321	0.075927	0.16202
132	200.64	185.94	0.013504	0.19821	74.050	5.0451	39.069	50.387	89.456	0.076779	0.16194
134	205.67	190.97	0.013563	0.19294	73.729	5.1829	39.588	50.000	89.588	0.077633	0.16185
136	210.79	196.09	0.013623	0.18782	73.406	5.3244	40.110	49.608	89.718	0.078489	0.16177
138	216.01	201.31	0.013684	0.18283	73.079	5.4695	40.634	49.210	89.844	0.079346	0.16168
140	221.32	206.62	0.013746	0.17799	72.748	5.6184	41.162	48.805	89.967	0.080205	0.16159
142	226.72	212.02	0.013810	0.17327	72.413	5.7713	41.693	48.394	90.087	0.081065	0.16150
144	232.22	217.52	0.013874	0.16868	72.075	5.9283	42.227	47.977	90.204	0.081928	0.16140
146	237.82	223.12	0.013941	0.16422	71.732	6.0895	42.765	47.553	90.318	0.082794	0.16130
148	243.51	228.81	0.014008	0.15987	71.386	6.2551	43.306	47.122	90.428	0.083661	0.16120
150	249.31	234.61	0.014078	0.15564	71.035	6.4252	43.850	46.684	90.534	0.084531	0.16110
155	264.24	249.54	0.014258	0.14552	70.137	6.8717	45.229	45.554	90.783	0.086719	0.16083
160	279.82	265.12	0.014449	0.13604	69.209	7.3509	46.633	44.373	91.006	0.088927	0.16053
165	296.07	281.37	0.014653	0.12712	68.245	7.8665	48.065	43.134	91.199	0.091159	0.16021
170	313.00	298.30	0.014871	0.11873	67.244	8.4228	49.529	41.830	91.359	0.093418	0.15985
175	330.64	315.94	0.015106	0.11080	66.198	9.0252	51.026	40.455	91.481	0.095709	0.15945
180	349.00	334.30	0.015360	0.10330	65.102	9.6802	52.562	38.999	91.561	0.098039	0.15900

TABLE A.5*

THERMODYNAMIC PROPERTIES OF MOIST AIR (14.696 PSIA)

Fahr. Temp. t(F)	Humidity Ratio $W_s \times 10^5$	Volume cu ft/lb dry air			Enthalpy Btu/lb dry air			Entropy Btu per (R)(lb dry air)		
		v_a	v_{as}	v_s	h_a	h_{as}	h_s	s_a	s_{as}	s_s
−160	0.0002120	7.520	0.000	7.520	−38.504	0.000	−38.504	−0.10300	0.00000	−0.10300
−155	0.0003869	7.647	0.000	7.647	−37.296	0.000	−37.296	−0.09901	0.00000	−0.09901
−150	0.0006932	7.775	0.000	7.775	−36.088	0.000	−36.088	−0.09508	0.00000	−0.09508
−145	0.001219	7.902	0.000	7.902	−34.881	0.000	−34.881	−0.09121	0.00000	−0.09121
−140	0.002109	8.029	0.000	8.029	−33.674	0.000	−33.674	−0.08740	0.00000	−0.08740
−135	0.003586	8.156	0.000	8.156	−32.468	0.000	−32.468	−0.08365	0.00000	−0.08365
−130	0.006000	8.283	0.000	8.283	−31.262	0.000	−31.262	−0.07997	0.00000	−0.07997
−125	0.009887	8.411	0.000	8.411	−30.057	0.000	−30.057	−0.07634	0.00000	−0.07634
−120	0.01606	8.537	0.000	8.537	−28.852	0.000	−28.852	−0.07277	0.00000	−0.07277
−115	0.02571	8.664	0.000	8.664	−27.648	0.000	−27.648	−0.06924	0.00000	−0.06924
−110	0.04063	8.792	0.000	8.792	−26.444	0.000	−26.444	−0.06577	0.00000	−0.06577
−105	0.06340	8.919	0.000	8.919	−25.240	0.001	−25.239	−0.06234	0.00000	−0.06234
−100	0.09772	9.046	0.000	9.046	−24.037	0.001	−24.036	−0.05897	0.00000	−0.05897
−95	0.1489	9.173	0.000	9.173	−22.835	0.002	−22.833	−0.05565	0.00000	−0.05565
−90	0.2242	9.300	0.000	9.300	−21.631	0.002	−21.629	−0.05237	0.00001	−0.05236
−85	0.3342	9.426	0.000	9.426	−20.428	0.003	−20.425	−0.04913	0.00001	−0.04912
−80	0.4930	9.553	0.000	9.553	−19.225	0.005	−19.220	−0.04595	0.00001	−0.04594
−75	0.7196	9.680	0.000	9.680	−18.022	0.007	−18.015	−0.04280	0.00002	−0.04278
−70	1.040	9.806	0.000	9.806	−16.820	0.011	−16.809	−0.03969	0.00003	−0.03966
−65	1.491	9.932	0.000	9.932	−15.617	0.015	−15.602	−0.03663	0.00005	−0.03658
−60	2.118	10.059	0.000	10.059	−14.416	0.022	−14.394	−0.03360	0.00006	−0.03354
−55	2.982	10.186	0.000	10.186	−13.214	0.031	−13.183	−0.03061	0.00009	−0.03052
−50	4.163	10.313	0.001	10.314	−12.012	0.043	−11.969	−0.02766	0.00012	−0.02754
−48	4.747	10.364	0.001	10.365	−11.532	0.049	−11.483	−0.02649	0.00013	−0.02636
−46	5.406	10.414	0.001	10.415	−11.051	0.056	−10.995	−0.02532	0.00014	−0.02518
−44	6.149	10.465	0.001	10.466	−10.571	0.064	−10.507	−0.02416	0.00016	−0.02400
−42	6.985	10.516	0.001	10.517	−10.090	0.073	−10.017	−0.02301	0.00019	−0.02282
−40	7.925	10.566	0.001	10.567	−9.609	0.083	−9.526	−0.02186	0.00021	−0.02165
−38	8.980	10.617	0.002	10.619	−9.129	0.094	−9.035	−0.02072	0.00024	−0.02048
−36	10.16	10.668	0.002	10.670	−8.648	0.106	−8.542	−0.01958	0.00026	−0.01932
−34	11.49	10.718	0.002	10.720	−8.168	0.121	−8.047	−0.01845	0.00030	−0.01815
−32	12.98	10.769	0.002	10.771	−7.687	0.136	−7.551	−0.01733	0.00034	−0.01699

Fahr. Temp. t(F)	Humidity Ratio $W_s \times 10^4$	Volume cu ft/lb dry air			Enthalpy Btu/lb dry air			Entropy Btu per (R)(lb dry air)		
		v_a	v_{as}	v_s	h_a	h_{as}	h_s	s_a	s_{as}	s_s
−30	1.464	10.820	0.002	10.822	−7.207	0.154	−7.053	−0.01621	0.00038	−0.01583
−28	1.649	10.870	0.003	10.873	−6.726	0.173	−6.553	−0.01509	0.00043	−0.01466
−26	1.856	10.921	0.003	10.924	−6.246	0.196	−6.050	−0.01398	0.00048	−0.01350
−24	2.087	10.972	0.004	10.976	−5.765	0.219	−5.546	−0.01287	0.00054	−0.01233
−22	2.344	11.022	0.004	11.026	−5.285	0.246	−5.039	−0.01177	0.00061	−0.01116
−20	2.630	11.073	0.005	11.078	−4.804	0.277	−4.527	−0.01067	0.00068	−0.00999
−19	2.785	11.098	0.005	11.103	−4.564	0.293	−4.271	−0.01012	0.00072	−0.00940
−18	2.948	11.124	0.005	11.129	−4.324	0.310	−4.014	−0.00958	0.00076	−0.00882
−17	3.120	11.149	0.006	11.155	−4.083	0.328	−3.755	−0.00904	0.00080	−0.00824
−16	3.301	11.174	0.006	11.180	−3.843	0.348	−3.495	−0.00850	0.00084	−0.00766
−15	3.491	11.200	0.006	11.206	−3.603	0.368	−3.235	−0.00796	0.00089	−0.00707
−14	3.692	11.225	0.007	11.232	−3.363	0.389	−2.974	−0.00743	0.00094	−0.00649
−13	3.903	11.250	0.007	11.257	−3.123	0.412	−2.711	−0.00689	0.00099	−0.00590
−12	4.125	11.275	0.008	11.283	−2.882	0.436	−2.446	−0.00636	0.00104	−0.00532
−11	4.359	11.301	0.008	11.309	−2.642	0.461	−2.181	−0.00582	0.00109	−0.00473
−10	4.606	11.326	0.008	11.334	−2.402	0.487	−1.915	−0.00529	0.00115	−0.00414
−9	4.865	11.351	0.008	11.359	−2.162	0.514	−1.648	−0.00475	0.00121	−0.00354
−8	5.137	11.376	0.009	11.385	−1.922	0.543	−1.379	−0.00422	0.00128	−0.00294
−7	5.423	11.401	0.010	11.411	−1.681	0.574	−1.107	−0.00369	0.00135	−0.00234
−6	5.724	11.427	0.010	11.437	−1.441	0.606	−0.835	−0.00316	0.00142	−0.00174
−5	6.040	11.452	0.011	11.463	−1.201	0.639	−0.562	−0.00263	0.00149	−0.00114
−4	6.371	11.477	0.012	11.489	−0.961	0.675	−0.286	−0.00210	0.00157	−0.00053
−3	6.720	11.502	0.013	11.515	−0.721	0.712	−0.009	−0.00157	0.00165	0.00008
−2	7.085	11.528	0.013	11.541	−0.480	0.751	0.271	−0.00105	0.00174	0.00069
−1	7.469	11.553	0.014	11.567	−0.240	0.792	0.552	−0.00052	0.00183	0.00131
0	7.872	11.578	0.015	11.593	0.000	0.835	0.835	0.00000	0.00192	0.00192
1	8.295	11.604	0.015	11.619	0.240	0.880	1.120	0.00052	0.00202	0.00254
2	8.739	11.629	0.016	11.645	0.480	0.928	1.408	0.00104	0.00212	0.00316
3	9.204	11.654	0.017	11.671	0.721	0.977	1.698	0.00156	0.00223	0.00379
4	9.692	11.679	0.018	11.697	0.961	1.030	1.991	0.00208	0.00234	0.00442
5	10.20	11.705	0.019	11.724	1.201	1.085	2.286	0.00260	0.00246	0.00506
6	10.74	11.730	0.020	11.750	1.441	1.142	2.583	0.00312	0.00258	0.00570

*Reprinted by permission from *ASHRAE Handbook of Fundamentals* (New York: American Society of Heating, Refrigerating and Air Conditioning Engineers, 1967) pp. 361–364.

TABLE A.5 *(Continued)*

THERMODYNAMIC PROPERTIES OF MOIST AIR (14.696 PSIA)

Fahr. Temp. $t(F)$	Humidity Ratio $W_s \times 10^3$	Volume cu ft/lb dry air			Enthalpy Btu/lb dry air			Entropy Btu per (R)(lb dry air)		
		v_a	v_{as}	v_s	h_a	h_{as}	h_s	s_a	s_{as}	s_s
7	1.130	11.756	0.021	11.777	1.681	1.202	2.883	0.00364	0.00271	0.00635
8	1.189	11.781	0.022	11.803	1.922	1.266	3.188	0.00415	0.00285	0.00700
9	1.251	11.806	0.024	11.830	2.162	1.332	3.494	0.00467	0.00299	0.00766
10	1.315	11.831	0.025	11.856	2.402	1.401	3.803	0.00518	0.00314	0.00832
11	1.383	11.857	0.026	11.883	2.642	1.474	4.116	0.00569	0.00330	0.00899
12	1.454	11.882	0.028	11.910	2.882	1.550	4.432	0.00620	0.00346	0.00966
13	1.528	11.907	0.029	11.936	3.123	1.630	4.753	0.00671	0.00363	0.01034
14	1.606	11.933	0.030	11.963	3.363	1.713	5.076	0.00721	0.00380	0.01101
15	1.687	11.958	0.032	11.990	3.603	1.800	5.403	0.00772	0.00399	0.01171
16	1.772	11.983	0.034	12.017	3.843	1.892	5.735	0.00822	0.00418	0.01240
17	1.861	12.009	0.035	12.044	4.083	1.988	6.071	0.00873	0.00438	0.01311
18	1.953	12.034	0.038	12.072	4.324	2.088	6.412	0.00923	0.00459	0.01382
19	2.051	12.059	0.040	12.099	4.564	2.192	6.756	0.00973	0.00481	0.01454
20	2.152	12.084	0.042	12.126	4.804	2.302	7.106	0.01023	0.00504	0.01527
21	2.258	12.110	0.044	12.154	5.044	2.416	7.460	0.01073	0.00528	0.01601
22	2.369	12.135	0.046	12.181	5.284	2.536	7.820	0.01123	0.00553	0.01676
23	2.485	12.160	0.049	12.209	5.525	2.661	8.186	0.01173	0.00579	0.01752
24	2.606	12.186	0.051	12.237	5.765	2.792	8.557	0.01223	0.00607	0.01830
25	2.733	12.211	0.054	12.265	6.005	2.929	8.934	0.01273	0.00635	0.01908
26	2.865	12.236	0.057	12.293	6.245	3.072	9.317	0.01322	0.00665	0.01987
27	3.003	12.262	0.059	12.321	6.485	3.221	9.706	0.01372	0.00696	0.02068
28	3.147	12.287	0.062	12.349	6.726	3.377	10.103	0.01421	0.00728	0.02149
29	3.297	12.312	0.065	12.377	6.966	3.540	10.506	0.01470	0.00761	0.02231
30	3.454	12.338	0.068	12.406	7.206	3.709	10.915	0.01519	0.00796	0.02315
31	3.617	12.363	0.071	12.434	7.446	3.887	11.333	0.01568	0.00832	0.02400
32	3.788	12.388	0.075	12.463	7.686	4.072	11.758	0.01617	0.00870	0.02487
32*	3.788	12.388	0.075	12.463	7.686	4.072	11.758	0.01617	0.00870	0.02487
33	3.944	12.413	0.079	12.492	7.927	4.242	12.169	0.01666	0.00904	0.02570
34	4.107	12.438	0.082	12.520	8.167	4.418	12.585	0.01715	0.00940	0.02655
35	4.275	12.464	0.085	12.549	8.407	4.601	13.008	0.01764	0.00977	0.02741
36	4.450	12.489	0.089	12.578	8.647	4.791	13.438	0.01812	0.01016	0.02828
37	4.631	12.514	0.093	12.607	8.887	4.987	13.874	0.01861	0.01056	0.02917

Fahr. Temp. $t(F)$	Humidity Ratio $W_s \times 10^3$	Volume cu ft/lb dry air			Enthalpy Btu/lb dry air			Entropy Btu per (R)(lb dry air)		
		v_a	v_{as}	v_s	h_a	h_{as}	h_e	s_a	s_{as}	s_s
38	4.818	12.540	0.097	12.637	9.128	5.191	14.319	0.01909	0.01097	0.03006
39	5.012	12.565	0.101	12.666	9.368	5.403	14.771	0.01957	0.01139	0.03096
40	5.213	12.590	0.105	12.695	9.608	5.622	15.230	0.02005	0.01183	0.03188
41	5.421	12.616	0.109	12.725	9.848	5.849	15.697	0.02053	0.01228	0.03281
42	5.638	12.641	0.114	12.755	10.088	6.084	16.172	0.02101	0.01275	0.03376
43	5.860	12.666	0.119	12.785	10.329	6.328	16.657	0.02149	0.01323	0.03472
44	6.091	12.691	0.124	12.815	10.569	6.580	17.149	0.02197	0.01373	0.03570
45	6.331	12.717	0.129	12.846	10.809	6.841	17.650	0.02245	0.01425	0.03670
46	6.578	12.742	0.134	12.876	11.049	7.112	18.161	0.02293	0.01478	0.03771
47	6.835	12.767	0.140	12.907	11.289	7.391	18.680	0.02340	0.01534	0.03874
48	7.100	12.792	0.146	12.938	11.530	7.681	19.211	0.02387	0.01591	0.03978
49	7.374	12.818	0.151	12.969	11.770	7.981	19.751	0.02434	0.01650	0.04084
50	7.658	12.843	0.158	13.001	12.010	8.291	20.301	0.02481	0.01711	0.04192
51	7.952	12.868	0.164	13.032	12.250	8.612	20.862	0.02528	0.01774	0.04302
52	8.256	12.894	0.170	13.064	12.491	8.945	21.436	0.02575	0.01839	0.04414
53	8.569	12.919	0.178	13.097	12.731	9.289	22.020	0.02622	0.01906	0.04528
54	8.894	12.944	0.185	13.129	12.971	9.644	22.615	0.02669	0.01976	0.04645
55	9.229	12.970	0.192	13.162	13.211	10.01	23.22	0.02716	0.02047	0.04763
56	9.575	12.995	0.200	13.195	13.452	10.39	23.84	0.02762	0.02121	0.04883
57	9.934	13.020	0.208	13.223	13.692	10.79	24.48	0.02809	0.02197	0.05006
58	10.30	13.045	0.216	13.261	13.932	11.19	25.12	0.02855	0.02276	0.05131
59	10.69	13.071	0.224	13.295	14.172	11.61	25.78	0.02902	0.02357	0.05259
60	11.08	13.096	0.233	13.329	14.413	12.05	26.46	0.02948	0.02441	0.05389
61	11.49	13.121	0.242	13.363	14.653	12.50	27.15	0.02994	0.02527	0.05521
62	11.91	13.147	0.251	13.398	14.893	12.96	27.85	0.03040	0.02616	0.05656
63	12.35	13.172	0.261	13.433	15.134	13.44	28.57	0.03086	0.02708	0.05794
64	12.80	13.197	0.271	13.468	15.374	13.94	29.31	0.03132	0.02803	0.05935
65	13.26	13.222	0.282	13.504	15.614	14.45	30.06	0.03177	0.02901	0.06078
66	13.74	13.247	0.292	13.539	15.855	14.98	30.83	0.03223	0.03002	0.06225
67	14.24	13.273	0.303	13.576	16.095	15.53	31.62	0.03269	0.03106	0.06375
68	14.75	13.298	0.315	13.613	16.335	16.09	32.42	0.03314	0.03213	0.06527
69	15.28	13.323	0.327	13.650	16.576	16.67	33.25	0.03360	0.03323	0.06683

TABLE A.5 (Continued)

THERMODYNAMIC PROPERTIES OF MOIST AIR (14.696 PSIA)

Fahr. Temp. $t(F)$	Humidity Ratio $W_s \times 10^2$	Volume cu ft/lb dry air			Enthalpy Btu/lb dry air			Entropy Btu per (R)(lb dry air)		
		v_a	v_{as}	v_s	h_a	h_{as}	h_s	s_a	s_{as}	s_s
70	1.582	13.348	0.339	13.687	16.816	17.27	34.09	0.03405	0.03437	0.06842
71	1.639	13.373	0.351	13.724	17.056	17.89	34.95	0.03450	0.03554	0.07004
72	1.697	13.398	0.364	13.762	17.297	18.53	35.83	0.03495	0.03675	0.07170
73	1.757	13.424	0.377	13.801	17.537	19.20	36.74	0.03540	0.03800	0.07340
74	1.819	13.449	0.392	13.841	17.778	19.88	37.66	0.03585	0.03928	0.07513
75	1.882	13.474	0.407	13.881	18.018	20.59	38.61	0.03630	0.04060	0.07690
76	1.948	13.499	0.422	13.921	18.259	21.31	39.57	0.03675	0.04197	0.07872
77	2.016	13.525	0.437	13.962	18.499	22.07	40.57	0.03720	0.04337	0.08057
78	2.086	13.550	0.453	14.003	18.740	22.84	41.58	0.03765	0.04482	0.08247
79	2.158	13.575	0.470	14.045	18.980	23.64	42.62	0.03810	0.04631	0.08441
80	2.233	13.601	0.486	14.087	19.221	24.47	43.69	0.03854	0.04784	0.08638
81	2.310	13.626	0.504	14.130	19.461	25.32	44.78	0.03899	0.04942	0.08841
82	2.389	13.651	0.523	14.174	19.702	26.20	45.90	0.03943	0.05105	0.09048
83	2.471	13.676	0.542	14.218	19.942	27.10	47.04	0.03987	0.05273	0.09260
84	2.555	13.702	0.560	14.262	20.183	28.04	48.22	0.04031	0.05446	0.09477
85	2.642	13.727	0.581	14.308	20.423	29.01	49.43	0.04075	0.05624	0.09699
86	2.731	13.752	0.602	14.354	20.663	30.00	50.66	0.04119	0.05807	0.09926
87	2.824	13.777	0.624	14.401	20.904	31.03	51.93	0.04163	0.05995	0.10158
88	2.919	13.803	0.645	14.448	21.144	32.09	53.23	0.04207	0.06189	0.10396
89	3.017	13.828	0.668	14.496	21.385	33.18	54.56	0.04251	0.06389	0.10640
90	3.118	13.853	0.692	14.545	21.625	34.31	55.93	0.04295	0.06596	0.10890
91	3.223	13.879	0.716	14.595	21.865	35.47	57.33	0.04339	0.06807	0.11146
92	3.330	13.904	0.741	14.645	22.106	36.67	58.78	0.04382	0.07025	0.11407
93	3.441	13.929	0.768	14.697	22.346	37.90	60.25	0.04426	0.07249	0.11675
94	3.556	13.954	0.795	14.749	22.587	39.18	61.77	0.04469	0.07480	0.11949
95	3.673	13.980	0.822	14.802	22.827	40.49	63.32	0.04513	0.07718	0.12231
96	3.795	14.005	0.851	14.856	23.068	41.85	64.92	0.04556	0.07963	0.12519
97	3.920	14.030	0.881	14.911	23.308	43.24	66.55	0.04600	0.08215	0.12815
98	4.049	14.056	0.911	14.967	23.548	44.68	68.23	0.04643	0.08474	0.13117
99	4.182	14.081	0.942	15.023	23.789	46.17	69.96	0.04686	0.08741	0.13427
100	4.319	14.106	0.975	15.081	24.029	47.70	71.73	0.04729	0.09016	0.13745
101	4.460	14.131	1.009	15.140	24.270	49.28	73.55	0.04772	0.09299	0.14071
102	4.606	14.157	1.043	15.200	24.510	50.91	75.42	0.04815	0.09591	0.14406
103	4.756	14.182	1.079	15.261	24.751	52.59	77.34	0.04858	0.09891	0.14749
104	4.911	14.207	1.117	15.324	24.991	54.32	79.31	0.04900	0.1020	0.1510

Fahr. Temp. t(F)	Humidity Ratio $W_s \times 10$	Volume cu ft/lb dry air			Enthalpy Btu/lb dry air			Entropy Btu per (R)(lb dry air)		
		v_a	v_{as}	v_s	h_a	h_{as}	h_s	s_a	s_{as}	s_s
105	0.5070	14.232	1.155	15.387	25.232	56.11	81.34	0.04943	0.1052	0.1546
106	0.5234	14.258	1.194	15.452	25.472	57.95	83.42	0.04985	0.1085	0.1584
107	0.5404	14.283	1.235	15.518	25.713	59.85	85.56	0.05028	0.1118	0.1621
108	0.5578	14.308	1.278	15.586	25.953	61.80	87.76	0.05070	0.1153	0.1660
109	0.5758	14.333	1.321	15.654	26.194	63.82	90.03	0.05113	0.1189	0.1700
110	0.5944	14.359	1.365	15.724	26.434	65.91	92.34	0.05155	0.1226	0.1742
111	0.6135	14.384	1.412	15.796	26.675	68.05	94.72	0.05197	0.1264	0.1784
112	0.6333	14.409	1.460	15.869	26.915	70.27	97.18	0.05239	0.1302	0.1826
113	0.6536	14.435	1.509	15.944	27.156	72.55	99.71	0.05281	0.1342	0.1870
114	0.6746	14.450	1.560	16.020	27.397	74.91	102.31	0.05323	0.1384	0.1916
115	0.6962	14.485	1.613	16.098	27.637	77.34	104.98	0.05365	0.1426	0.1963
116	0.7185	14.510	1.668	16.178	27.878	79.85	107.73	0.05407	0.1470	0.2011
117	0.7415	14.536	1.723	16.259	28.119	82.43	110.55	0.05449	0.1515	0.2060
118	0.7652	14.561	1.782	16.343	28.359	85.10	113.46	0.05490	0.1562	0.2111
119	0.7897	14.586	1.842	16.428	28.600	87.86	116.46	0.05532	0.1610	0.2163
120	0.8149	14.611	1.905	16.516	28.841	90.70	119.54	0.05573	0.1659	0.2216
121	0.8410	14.637	1.968	16.605	29.082	93.64	122.72	0.05615	0.1710	0.2272
122	0.8678	14.662	2.034	16.696	29.322	96.66	125.98	0.05656	0.1763	0.2329
123	0.8955	14.687	2.103	16.790	29.563	99.79	129.35	0.05698	0.1817	0.2387
124	0.9242	14.712	2.174	16.886	29.804	103.0	132.8	0.05739	0.1872	0.2446
125	0.9537	14.738	2.247	16.985	30.044	106.4	136.4	0.05780	0.1930	0.2508
126	0.9841	14.763	2.323	17.086	30.285	109.8	140.1	0.05821	0.1989	0.2571
127	1.016	14.788	2.401	17.189	30.526	113.4	143.9	0.05862	0.2050	0.2636
128	1.048	14.813	2.482	17.295	30.766	117.0	147.8	0.05903	0.2113	0.2703
129	1.082	14.839	2.565	17.404	31.007	120.8	151.8	0.05944	0.2178	0.2772
130	1.116	14.864	2.652	17.516	31.248	124.7	155.9	0.05985	0.2245	0.2844
131	1.152	14.889	2.742	17.631	31.489	128.8	160.3	0.06026	0.2314	0.2917
132	1.189	14.915	2.834	17.749	31.729	133.0	164.7	0.06067	0.2386	0.2993
133	1.227	14.940	2.930	17.870	31.970	137.3	169.3	0.06108	0.2459	0.3070
134	1.267	14.965	3.029	17.994	32.211	141.8	174.0	0.06148	0.2536	0.3151

TABLE A.5 *(Concluded)*

THERMODYNAMIC PROPERTIES OF MOIST AIR (14.696 PSIA)

Fahr. Temp. $t(F)$	Humidity Ratio W_s	Volume cu ft/lb dry air			Enthalpy Btu/lb dry air			Entropy Btu per (R)(lb dry air)		
		v_a	v_{as}	v_s	h_a	h_{as}	h_s	s_a	s_{as}	s_s
135	0.1308	14.990	3.132	18.122	32.452	146.4	178.9	0.06189	0.2614	0.3233
136	0.1350	15.016	3.237	18.253	32.692	151.2	183.9	0.06229	0.2695	0.3318
137	0.1393	15.041	3.348	18.389	32.933	156.1	189.0	0.06270	0.2778	0.3405
138	0.1439	15.066	3.462	18.528	33.174	161.2	194.4	0.06310	0.2865	0.3496
139	0.1485	15.091	3.580	18.671	33.414	166.5	199.9	0.06350	0.2954	0.3589
140	0.1534	15.117	3.702	18.819	33.655	172.0	205.7	0.06390	0.3047	0.3686
141	0.1584	15.142	3.829	18.971	33.896	177.7	211.6	0.06430	0.3142	0.3785
142	0.1636	15.167	3.961	19.128	34.136	183.6	217.7	0.06470	0.3241	0.3888
143	0.1689	15.192	4.098	19.290	34.377	189.7	224.1	0.06510	0.3343	0.3994
144	0.1745	15.218	4.239	19.457	34.618	196.0	230.6	0.06549	0.3449	0.4104
145	0.1803	15.243	4.386	19.629	34.859	202.5	237.4	0.06589	0.3559	0.4218
146	0.1862	15.268	4.539	19.807	35.099	209.3	244.4	0.06629	0.3672	0.4335
147	0.1924	15.293	4.698	19.991	35.340	216.4	251.7	0.06669	0.3790	0.4457
148	0.1989	15.319	4.862	20.181	35.581	223.7	259.3	0.06708	0.3912	0.4583
149	0.2055	15.344	5.033	20.377	35.822	231.3	267.1	0.06748	0.4038	0.4713
150	0.2125	15.369	5.211	20.580	36.063	239.2	275.3	0.06787	0.4169	0.4848
151	0.2197	15.394	5.396	20.790	36.304	247.3	283.6	0.06827	0.4304	0.4987
152	0.2271	15.420	5.587	21.007	36.545	255.9	292.4	0.06866	0.4445	0.5132
153	0.2349	15.445	5.788	21.233	36.785	264.7	301.5	0.06906	0.4591	0.5282
154	0.2430	15.470	5.996	21.466	37.026	273.9	310.9	0.06945	0.4743	0.5438
155	0.2514	15.496	6.213	21.709	37.267	283.5	320.8	0.06984	0.4901	0.5599
156	0.2602	15.521	6.439	21.960	37.508	293.5	331.0	0.07023	0.5066	0.5768
157	0.2693	15.546	6.675	22.221	37.749	303.9	341.7	0.07062	0.5237	0.5943
158	0.2788	15.571	6.922	22.493	37.990	314.7	352.7	0.07101	0.5415	0.6125
159	0.2887	15.597	7.178	22.775	38.231	326.0	364.2	0.07140	0.5600	0.6314
160	0.2990	15.622	7.446	23.068	38.472	337.8	376.3	0.07179	0.5793	0.6511
161	0.3098	15.647	7.727	23.374	38.713	350.1	388.8	0.07218	0.5994	0.6716
162	0.3211	15.672	8.020	23.692	38.954	363.0	402.0	0.07257	0.6204	0.6930
163	0.3329	15.698	8.326	24.024	39.195	376.5	415.7	0.07296	0.6423	0.7153
164	0.3452	15.723	8.648	24.371	39.436	390.5	429.9	0.07334	0.6652	0.7385
165	0.3581	15.748	8.985	24.733	39.677	405.3	445.0	0.07373	0.6892	0.7629
166	0.3716	15.773	9.339	25.112	39.918	420.8	460.7	0.07411	0.7142	0.7883
167	0.3858	15.799	9.708	25.507	40.159	437.0	477.2	0.07450	0.7405	0.8150
168	0.4007	15.824	10.098	25.922	40.400	454.0	494.4	0.07488	0.7680	0.8429
169	0.4163	15.849	10.508	26.357	40.641	471.8	512.4	0.07527	0.7969	0.8722

Fahr. Temp. t(F)	Humidity Ratio W_s	Volume cu ft/lb dry air			Enthalpy Btu/lb dry air			Entropy Btu per (R)(lb dry air)		
		v_a	v_{as}	v_s	h_a	h_{as}	h_s	s_a	s_{as}	s_s
170	0.4327	15.874	10.938	26.812	40.882	490.6	531.5	0.07565	0.8273	0.9030
171	0.4500	15.900	11.391	27.291	41.123	510.4	551.5	0.07603	0.8592	0.9352
172	0.4682	15.925	11.870	27.795	41.364	531.3	572.7	0.07641	0.8927	0.9691
173	0.4875	15.950	12.376	28.326	41.605	553.3	594.9	0.07680	0.9281	1.0049
174	0.5078	15.975	12.911	28.886	41.846	576.5	618.3	0.07718	0.9654	1.0426
175	0.5292	16.001	13.475	29.476	42.087	601.1	643.2	0.07756	1.005	1.083
176	0.5519	16.026	14.074	30.100	42.328	627.1	669.4	0.07794	1.047	1.125
177	0.5760	16.051	14.710	30.761	42.569	654.7	697.3	0.07832	1.091	1.169
178	0.6016	16.076	15.386	31.462	42.810	684.1	726.9	0.07870	1.137	1.216
179	0.6288	16.102	16.104	32.206	43.051	715.2	758.3	0.07908	1.187	1.266
180	0.6578	16.127	16.870	32.997	43.292	748.5	791.8	0.07946	1.240	1.319
181	0.6887	16.152	17.689	33.841	43.534	783.9	827.4	0.07984	1.296	1.376
182	0.7218	16.177	18.565	34.742	43.775	821.9	865.7	0.08021	1.357	1.437
183	0.7572	16.203	19.504	35.707	44.016	862.5	906.5	0.08059	1.421	1.502
184	0.7953	16.228	20.513	36.741	44.257	906.2	950.5	0.08096	1.490	1.571
185	0.8363	16.253	21.601	37.854	44.498	953.2	997.7	0.08134	1.565	1.646
186	0.8805	16.278	22.775	39.053	44.740	1004	1049	0.08171	1.645	1.727
187	0.9283	16.304	24.047	40.351	44.981	1059	1104	0.08208	1.731	1.813
188	0.9802	16.329	25.427	41.756	45.222	1119	1164	0.08245	1.825	1.907
189	1.037	16.354	26.934	43.288	45.463	1184	1229	0.08283	1.928	2.011
190	1.099	16.379	28.580	44.959	45.704	1255	1301	0.08320	2.039	2.122
191	1.166	16.405	30.385	46.790	45.946	1332	1378	0.08357	2.161	2.245
192	1.241	16.430	32.375	48.805	46.187	1418	1464	0.08394	2.296	2.380
193	1.324	16.455	34.581	51.036	46.428	1513	1559	0.08431	2.444	2.528
194	1.416	16.480	37.036	53.516	46.670	1619	1666	0.08468	2.609	2.694
195	1.519	16.506	39.785	56.291	46.911	1737	1784	0.08505	2.794	2.879
196	1.635	16.531	42.885	59.416	47.153	1871	1918	0.08542	3.002	3.087
197	1.767	16.556	46.402	62.958	47.394	2022	2069	0.08579	3.238	3.324
198	1.917	16.581	50.426	67.007	47.636	2195	2243	0.08616	3.507	3.593
199	2.091	16.607	55.074	71.681	47.877	2395	2443	0.08653	3.817	3.904
200	2.295	16.632	60.510	77.142	48.119	2629	2677	0.08689	4.179	4.266

TABLE A.6*

PHYSICAL PROPERTIES OF DRY AIR (14.696 PSIA)

t F	μ lb/(ft)(hr)	k Btu/(hr)(ft)(F)	$c_p\mu/k$
-100	0.03214	0.01045	0.739
-82	0.03350	0.01095	0.736
-64	0.03483	0.01143	0.732
-46	0.03614	0.01192	0.729
-28	0.03742	0.01239	0.725
-10	0.03868	0.01287	0.722
8	0.03992	0.01334	0.719
26	0.04113	0.01380	0.716
44	0.04234	0.01426	0.713
62	0.04350	0.01472	0.710
80	0.04466	0.01516	0.708
98	0.04579	0.01561	0.706
116	0.04691	0.01606	0.703
134	0.04803	0.01649	0.701
152	0.04911	0.01692	0.699
170	0.05019	0.01735	0.697
188	0.05122	0.01779	0.695
206	0.05226	0.01820	0.693

* From *Tables of Thermal Properties of Gases*, U.S. Department of Commerce, National Bureau of Standards Circular 564 (Washington, D.C.: Government Printing Office, 1955), pp. 69–71.

TABLE A.7

SOLAR ALTITUDE (β) AND AZIMUTH (γ) ANGLES FOR VARIOUS NORTH LATITUDES

30 Degrees North Latitude

Hour Angle	Solar Time		Jan. 1		Feb. 1		March 1		April 1		May 1		June 1	
Deg	a.m.	p.m.	β Deg-Min	γ Deg	β Deg-Min	γ Deg	β Deg-Min	γ Deg	β Deg-Min	γ Deg	β Deg-Min	γ Deg	β Deg-Min	γ Deg
90	6	6									7-26	76.9	10-48	70.7
75	7	5	0-39	117.2	3-29	112.7	8-46	104.6	15-14	93.6	20-14	83.9	23-16	77.1
60	8	4	11-44	125.5	15-14	121.1	21-03	113.2	28-06	101.9	33-13	91.2	36-04	83.4
45	9	3	21-37	135.6	25-42	131.5	32-28	123.9	40-31	112.0	46-08	99.8	49-02	90.3
30	10	2	29-40	148.0	34-24	144.7	42-19	138.0	51-54	126.1	58-38	111.9	61-58	99.4
15	11	1	35-04	163.1	40-27	161.1	49-22	156.8	60-50	148.1	69-38	134.1	74-21	117.2
0	12 Noon		37-0	180.0	42-30	180.0	52-0	180.0	64-30	180.0	75-0	180.0	82-0	180.0

TABLE A.7 (Continued)

SOLAR ALTITUDE (β) AND AZIMUTH (γ) ANGLES FOR VARIOUS NORTH LATITUDES

30 Degrees North Latitude

Hour Angle Deg	Solar Time a.m.	Solar Time p.m.	July 1 β Deg-Min	July 1 γ Deg	Aug. 1 β Deg-Min	Aug. 1 γ Deg	Sept. 1 β Deg-Min	Sept. 1 γ Deg	Oct. 1 β Deg-Min	Oct. 1 γ Deg	Nov. 1 β Deg-Min	Nov. 1 γ Deg	Dec. 1 β Deg-Min	Dec. 1 γ Deg
90	6	6	11-16	69.8	8-53	74.3								
75	7	5	23-41	76.1	21-34	81.1	17-12	90.0	11-24	100.3	5-32	109.7	1-34	116.0
60	8	4	36-26	82.3	34-30	87.9	30-09	97.9	23-58	108.8	17-25	118.3	12-41	124.3
45	9	3	49-23	88.9	47-28	95.8	42-48	107.6	35-50	119.4	28-15	128.8	22-44	134.5
30	10	2	62-24	97.5	60-12	106.9	54-39	121.3	46-17	133.7	37-21	142.4	30-58	147.1
15	11	1	74-53	114.0	71-49	127.9	64-20	143.8	54-01	153.9	43-41	159.7	36-31	162.6
0	12 Noon		83-0	180.0	78-0	180.0	68-30	180.0	51-0	180.0	46-0	180.0	38-30	180.0

TABLE A.7 (Continued)

SOLAR ALTITUDE (β) AND AZIMUTH (γ) ANGLES FOR VARIOUS NORTH LATITUDES

36 Degrees North Latitude

Hour Angle	Solar Time		Jan. 1		Feb. 1		March 1		April 1		May 1		June 1	
Deg	a.m.	p.m.	β Deg-Min	γ Deg	β Deg-Min	γ Deg	β Deg-Min	γ Deg	β Deg-Min	γ Deg	β Deg-Min	γ Deg	β Deg-Min	γ Deg
90	6	6									8-45	77.8	12-43	71.9
75	7	5			5-0	115.9	7-13	105.4	14-46	95.2	20-45	86.2	24-29	79.7
60	8	4	8-12	126.3	12-04	122.4	18-35	115.2	26-42	104.9	32-53	95.1	36-32	87.8
45	9	3	17-16	137.0	21-39	133.5	29-0	126.8	38-03	116.5	44-48	105.7	48-38	97.2
30	10	2	24-32	149.6	29-26	146.8	37-44	141.2	48-07	131.7	55-59	120.3	60-26	110.0
15	11	1	29-19	164.1	34-39	162.5	43-47	159.2	55-36	152.8	65-05	143.6	70-52	132.9
0	12 Noon		31-0	180.0	36-30	180.0	46-0	180.0	58-30	180.0	69-0	180.0	76-0	180.0

TABLE A.7 (Continued)

SOLAR ALTITUDE (β) AND AZIMUTH (γ) ANGLES FOR VARIOUS NORTH LATITUDES

36 Degrees North Latitude

Hour Angle	Solar Time		July 1		Aug. 1		Sept. 1		Oct. 1		Nov. 1		Dec. 1	
			β	γ	β	γ	β	γ	β	γ	β	γ	β	γ
Deg	a.m.	p.m.	Deg-Min	Deg	Deg-Min	Deg	Deg-Min	Deg	Deg-Min	Deg	Deg-Min	Deg	Deg-Min	Deg
90	6	6	13-17	71.0	10-28	75.3								
75	7	5	24-59	78.8	22-23	83.5	17-06	91.9	10-16	101.4				
60	8	4	37-01	86.7	34-30	92.1	29-08	101.3	21-55	111.2	14-30	119.3	9-16	125.3
45	9	3	49-08	95.8	46-32	102.2	40-44	112.6	32-43	122.9	24-23	131.1	18-28	136.1
30	10	2	61-0	108.3	58-0	116.2	51-15	127.8	41-59	137.8	32-31	144.9	25-53	148.9
15	11	1	71-37	130.9	67-39	139.7	59-17	149.9	48-33	157.0	38-02	161.4	30-46	163.7
0	12 Noon		77-0	180.0	72-0	180.0	62-30	180.0	51-0	180.0	40-0	180.0	32-30	180.0

TABLE A.7 *(Continued)*

SOLAR ALTITUDE (β) AND AZIMUTH (γ) ANGLES FOR VARIOUS NORTH LATITUDES

42 Degrees North Latitude

Hour Angle	Solar Time		Jan. 1		Feb. 1		March 1		April 1		May 1		June 1	
Deg	a.m.	p.m.	β Deg-Min	γ Deg	β Deg-Min	γ Deg	β Deg-Min	γ Deg	β Deg-Min	γ Deg	β Deg-Min	γ Deg	β Deg-Min	γ Deg
90	6	6									9-58	78.7	14-31	73.3
75	7	5					5-35	106.0	14-08	96.8	21-02	88.5	25-24	82.5
60	8	4			8-49	123.3	15-57	116.9	25-01	107.7	32-09	98.9	36-31	92.3
45	9	3	12-51	138.1	17-27	135.0	25-18	129.2	35-12	120.4	42-54	111.2	47-33	109.7
30	10	2	19-20	150.8	24-22	148.4	32-58	143.8	43-57	136.2	52-38	127.3	57-56	119.2
15	11	1	23-32	164.9	28-54	163.6	38-09	161.0	50-11	156.2	60-04	149.9	66-23	143.2
0	12 Noon		25-0	180.0	30-30	180.0	40-0	180.0	52-30	180.0	63-0	180.0	70-0	180.0

TABLE A.7 (Continued)

SOLAR ALTITUDE (β) AND AZIMUTH (γ) ANGLES FOR VARIOUS NORTH LATITUDES

42 Degrees North Latitude

Hour Angle Deg	Solar Time a.m.	Solar Time p.m.	July 1 β Deg-Min	July 1 γ Deg	Aug. 1 β Deg-Min	Aug. 1 γ Deg	Sept. 1 β Deg-Min	Sept. 1 γ Deg	Oct. 1 β Deg-Min	Oct. 1 γ Deg	Nov. 1 β Deg-Min	Nov. 1 γ Deg	Dec. 1 β Deg-Min	Dec. 1 γ Deg
90	6	6	15-08	72.5	11-56	76.4	5-41	83.7						
75	7	5	26-0	81.6	22-56	86.0	16-48	93.7	9-02	102.4				
60	8	4	37-07	91.3	34-04	96.1	27-48	104.5	19-38	113.3	11-27	121.0	5-46	125.9
45	9	3	48-10	102.6	44-57	108.1	38-13	117.1	29-19	125.9	20-22	133.0	14-06	137.3
30	10	2	58-38	117.8	54-58	124.1	47-20	133.1	37-25	141.0	27-33	146.8	20-42	150.2
15	11	1	67-15	142.0	62-48	147.4	53-59	154.2	42-59	159.3	32-19	162.7	25-0	164.6
0	12 Noon		71-0	180.0	66-0	180.0	56-30	180.0	45-0	180.0	34-0	180.0	26-30	180.0

TABLE A.7 (Continued)

SOLAR ALTITUDE (β) AND AZIMUTH (γ) ANGLES FOR VARIOUS NORTH LATITUDES

48 Degrees North Latitude

Hour Angle Deg	Solar Time p.m.	Solar Time a.m.	Jan. 1 β Deg-Min	Jan. 1 γ Deg	Feb. 1 β Deg-Min	Feb. 1 γ Deg	March 1 β Deg-Min	March 1 γ Deg	April 1 β Deg-Min	April 1 γ Deg	May 1 β Deg-Min	May 1 γ Deg	June 1 β Deg-Min	June 1 γ Deg
105	7	5											6-46	64.4
90	6	6									11-05	79.8	16-10	74.9
75	5	7							13-21	98.2	21-05	90.8	26-02	85.4
60	4	8			5-29	123.9	13-10	118.3	23-04	110.2	31-02	102.5	36-03	96.7
45	3	9	8-21	138.9	13-10	136.2	21-30	131.2	32-0	123.8	40-30	116.1	45-49	109.8
30	2	10	14-04	151.7	19-13	149.7	28-04	145.9	39-30	139.8	48-46	132.9	54-39	126.7
15	1	11	17-44	165.5	23-08	164.4	32-27	162.3	44-38	158.7	54-45	154.3	61-22	150.0
0	12 Noon		19-0	180.0	24-30	180.0	34-0	180.0	46-30	180.0	57-0	180.0	64-0	180.0

TABLE A.7 (*Concluded*)

SOLAR ALTITUDE (β) AND AZIMUTH (γ) ANGLES FOR VARIOUS NORTH LATITUDES

48 Degrees North Latitude

| Hour Angle | Solar Time | | July 1 | | Aug. 1 | | Sept. 1 | | Oct. 1 | | Nov. 1 | | Dec. 1 | |
Deg	a.m.	p.m.	β Deg-Min	γ Deg	β Deg-Min	γ Deg	β Deg-Min	γ Deg	β Deg-Min	γ Deg	β Deg-Min	γ Deg	β Deg-Min	γ Deg
105	5	7	7-31	63.7										
90	6	6	16-53	74.1	13-17	77.7	6-18	84.3						
75	7	5	26-44	84.6	23-14	88.5	16-20	95.5	7-42	103.2				
60	8	4	36-45	95.8	33-13	100.1	26-09	107.4	17-10	115.2	8-20	121.9		
45	9	3	46-33	108.9	42-49	113.5	35-18	121.0	25-42	128.4	16-13	134.4	9-40	138.1
30	10	2	55-28	125.7	51-20	130.4	43-04	137.4	32-40	143.6	22-29	148.3	15-28	151.1
15	11	1	62-17	149.2	57-32	152.7	48-31	157.3	37-20	161.0	26-34	163.7	19-12	165.2
0	12 Noon		65-0	180.0	60-0	180.0	50-30	180.0	39-0	180.0	28-0	180.0	20-30	180.0

ANSWERS TO PROBLEMS

Chapter 1

1.1. (a) 14.12 psia
    (b) 391.0 in. water
    (c) 40.7 ft
1.2. (a) 0.27 psi
    (b) 0.0724 lb per cu ft
    (c) 53.74 ft-lb per (lb)(R)
    (d) 0.245 Btu per (lb)(F)
1.3.   1.03 per cent
1.4.   0.997
1.5.   5.24 lb per hr

Chapter 3

3.1. (a) 4.14
    (b) 0.81
    (c) 1.14 hp per ton
    (d) 61.8 lb per min
    (e) 17.1 hp
    (f) 99.4 cu ft per min

3.2. (a) 2.47
    (b) 168 cu ft per min
3.3.   1.50 hp per ton
3.4. (a) 93.5 Btu per lb
    (b) 477.7 Btu per lb
    (c) 9.2 Btu per lb
    (d) 9.5 Btu per lb
    (e) 9.5 Btu per lb
    (f) 0.817
3.5. (b) 19.7 per cent
3.6. (a) 16.2 tons
    (b) −76 F
3.7.   21.9 lb per min
3.8.   11.8 tons
3.9. (a) 4.30; 4.17 (isentropic)
    (b) 14.0 per cent
    (c) 10.7 per cent
3.10. (b) 14.0 per cent
    (c) 16.4 per cent
3.11. (a) Yes; 522 RPM

(b) 26.9 hp

3.13.   295 cu ft per min

3.14.   2.43

3.15.   0.803

3.16. (a) 2.16

(b) 361 cu ft per min for B

231 cu ft per min for A

(c) 7.95 hp for B

24.9 hp for A

Chapter 5

5.1. (a) 0.50 lb Li Br per lb sol

(b) 200 Btu per lb

5.2. (a) 92 F

(b) 81.6 per cent liquid

18.4 per cent vapor

5.4. (a) 1116 lb per min

(b) 587 lb per min

5.5. (a) 0.453 lb NH_3 per lb sol

(b) 74,570 Btu per min

(c) 248 tons

5.6.   3237 lb per hr

5.7.   \$0.0063 per (hr)(ton) for steam

\$0.0187 per (hr)(ton) for electricity

Chapter 6

6.3. (a) 29

(b) 503 Btu per hr

(c) 0.257

(d) 117 watts

6.4. (a) 354

(b) 0.99 cents per hr

6.5. (a) 45 for first stage

121 for second stage

(b) 816 watts

6.6.   17.9 tons

6.8.   2.26

6.9.   Yes

6.10. (a) 28.5 hp

(b) 1.12 tons

Chapter 7

7.1.   307 Btu per lb

7.2. (a) 0.0734

(b) 2563 Btu per lb

7.4. (a) 0.659

(b) 1556 Btu per lb

7.5. (a) 0.1782

(b) 949 Btu per lb

(c) 1056 Btu per lb

7.6. (a) 2675 cu ft per hr

(b) 0.629

(c) −266 F

7.7. (a) 998,000 Btu per hr

(b) 17.0 per cent

7.8. (c) 4390 Btu per mole

(d) 10,830 Btu per mole

(e) 0.962

7.10.   4200 hp

Chapter 8

8.1. (a) 68.47 F

(b) 0.68

(c) 13.928 cu ft per lb_a

(d) 35.665 Btu per lb_a

8.3.   Yes

8.4.   2.11 lb_w

8.5.   33.07 Btu per lb_a

8.6. (a) 0.01388 lb_w per lb_a

(b) 14.61 cu ft per lb_a

8.7.   0.536, 75.5 F

8.8.   33.64 Btu per lb_a

8.9.   134.4 F

Chapter 9

9.1. (a) 57.7 F

(b) 67.4 F

(c) 0.0103 lb_w per lb_a

(d) 31.70 Btu per lb_a

(e) 13.95 cu ft per lb_a

9.2.   −0.937

9.5.   14.0 tons

9.6. $t_2 = 60.0$ F, $t_2^* = 54.4$ F

9.7. (a) 11.4 lb_w per min

(b) 57 F

9.8. $t_3 = 76.5$ F, $t_3^* = 64.7$ F

9.9. (a) 31 lb_a per min

(b) 59,040 Btu per hr

9.10. (a) 55.1 F

(b) 80,060 Btu per hr

Chapter 10

10.1. 0.399

10.2. 0.01024 lb_w per lb_a

10.3. $t = 86.73$ F, $t^\star = 58.8$ F

$W = 0.00432$ lb_w per lb_a

10.4. 86.93 F

10.5. 99.2 F

10.6. (a) 100 ft per min

(b) 600 ft per min

Chapter 11

11.1. $t_2 = 73.7$ F, $t^\star_2 = 65.0$ F

11.2. (a) $t_2 = 96.3$ F , $t^\star_2 = 61.2$ F

(b) 778,200 Btu per hr

(c) 0.58 GPM

(d) 167 cu ft

11.3. (a) 299,000 cu ft per min

(b) 55.2 GPM

11.4. 17,930 cu ft

11.5. $t_2 = 55.7$ F, $t^\star_2 = 53.3$ F

11.6. 60.8 cu ft

Chapter 12

12.1. 5.36 Btu per (hr)(sq ft)(F)

12.2. $t_2 = 65.7$ F (one row)

$t_2 = 101.3$ F (two rows)

$t_2 = 150.8$ F (four rows)

12.3. (a) 33 in. high, 44 in. long

(b) 1.92 (two rows)

12.4. 37.0 Btu per (hr)(ft)

12.5. (b) 173.5 F

12.9. (a) 42 in. high, 82 in. long

(b) 4.92 (5 rows)

12.10. (a) 32,000 Btu per hr

(b) 41,674 Btu per hr

12.11. (b) $t_2 = 75.1$ F, $t^\star_2 = 61.9$ F (two rows)

$t_2 = 62.1$ F, $t^\star_2 = 56.6$ F (four rows)

$t_2 = 50.6$ F, $t^\star_2 = 49.8$ F (eight rows)

Chapter 13

13.2. (a) 3:56 a.m., 52°03′ east of north

(b) 8:03 a.m., 127°52′ east of north

13.3. 31°27′

13.4. (a) 68.5°

(b) 90.0°

(c) 90.0°

13.9. (a) 45°00′

(b) 8:16 p.m.

13.10. (a) 23°35′

(b) 77°53′

(c) 1°35′

13.11. 9:15 a.m. to 2:45 p.m. (5.50 hr)

13.12. 10,000 F

13.14. 1768 Btu per (day)(sq ft)

13.15. 445 Btu per (day)(sq ft)

13.16. (a) 89 Btu per (hr)(sq ft)

(b) 248 Btu per (hr)(sq ft)

13.17. (a) 295 Btu per (hr)(sq ft)

(b) 290 Btu per (hr)(sq ft)

13.18. $I_D + I_d = 236 + 28 = 264$ Btu per (hr)(sq ft)

Chapter 14

14.1. 0.335 Btu per (hr)(sq ft)(F)

14.2. (a) 0.0939 Btu per (hr)(sq ft)(F)

(b) 0.201 Btu per (hr)(sq ft)(F)

14.3. 0.0479 Btu per (hr)(sq ft)(F)

14.4. 0.446

14.5. Wall A

14.6. 11.94 in.

14.7. $c_1 = 0.275$, $c_2 = 0.725$

14.8. 0.50

14.9. 0.59

14.11. $I_D = 228$ Btu per (hr)(sq ft)

$I_d = 30$ Btu per (hr)(sq ft)

$I_R = 77$ Btu per (hr)(sq ft)

14.12. 17 Btu per (hr)(sq ft)

14.13. 52 Btu per (hr)(sq ft)

14.14. $f = 3.73$ ft, $c = 1.47$ ft

14.17. 101.2 sq ft

14.18. (a) 52 Btu per (hr)(sq ft) gain

(b) 93 Btu per (hr)(sq ft) loss

14.19. 23,400 Btu per hr

14.20. $t_e = 91.1 + 20.6 \cos (15\theta - 203) + 6.7 \cos (30\theta - 15)$

14.21. (a) 1.9 Btu per (hr)(sq ft)

(b) 4.2 Btu per (hr)(sq ft)

14.22. (a) 9.25 Btu per (hr)(sq ft) loss
(b) 0.45 Btu per (hr)(sq ft) gain
(c) 4.93 Btu per (hr)(sq ft) gain
(d) 0.31 Btu per (hr)(sq ft) loss

14.23.   Roof 2:   5.10 Btu per (hr)(sq ft),
5:51 p.m.
Roof 3:   6.60 Btu per (hr)(sq ft),
4:49 p.m.

Chapter 15

15.6.   H.S.I. = 32.4, unfavorable
15.7.   H.S.I. = 24, tolerable
15.8.   107 ft per min

Chapter 16

16.1.   620,324 Btu per hr
16.2.   125,278 Btu per hr
26.99 lb_w per hr
16.4.   123,027 Btu per hr
40.85 lb_w per hr
Reheat not required
16.5. (a) $t_{d,1} = 53.4$ F, $t_1^\star = 55.3$ F
(b) 4512 cu ft per min
16.6. (a) 10,854 cu ft per min
(b) Not acceptable
16.7. (a) $t_1^\star = 74.0$ F, $t_5^\star = 50.6$ F,
$t_7 = 109.0$ F, $t_7^\star = 70.1$ F
(b) 3346 cu ft per min
(c) 41.1 lb_w per hr
16.8. (a) 5528 cu ft per min
(b) 51.5 F
(c) 38.3 lb_w per hr
(d) $_5q_6 = 74,640$ Btu per hr
$_7q_8 = 249,630$ Btu per hr

16.9. (a) $t_5 = 79.3$ F, $t_5^\star = 67.6$ F
(b) 12.3 tons
16.10. (a) $t_1 = 70.0$ F, $t_1^\star = 62.6$ F
(b) 3600 cu ft per min
16.11. (a) $t_1 = 62.1$ F, $t_1^\star = 55.9$ F
(b) 242,720 Btu per hr
16.12. (a) 3270 cu ft per min
(b) 20.7 per cent
(c) 74.6 cents per day
16.13. (a) $t_1^\star = 49.3$ F, $t_5 = 76.5$ F,
$t_5^\star = 61.8$ F, $t_6 = 100.3$ F,
$t_6^\star = 61.8$ F, $t_7 = 86.5$ F
(b) 2721 cu ft per min
16.14. (a) $t_1^\star = 66.8$ F, $t_3^\star = 43.8$ F,
$t_5 = 45$ F, $t_5^\star = 35.3$ F, $t_6 = 45.4$
F
(b) 1080 Btu per lb_w

Chapter 17

17.2.   1098 sq ft
17.4. (a) 58°13′
(b) 8.95 ft
(c) 19.9 ft
(d) 335 sq ft
(e) 7360 F
17.5.   0.649 lb_w per lb_D
17.6.   135 ft
17.7. (a) 117 ft
(b) 704,950 Btu per hr
(c) $q_1 = 15,700$ Btu per hr
$q_2 = 18,110$ Btu per hr
$q_3 = 23,800$ Btu per hr

INDEX

485